Handbook of Entrepreneurial Cognition

Edited by

J. Robert Mitchell

Associate Professor, Ivey School of Business, Western University, Canada

Ronald K. Mitchell

Professor, Rawls College of Business, Texas Tech University, USA

Brandon Randolph-Seng

Assistant Professor of Management, College of Business & Entrepreneurship, Texas A&M University – Commerce, USA

Advisory Panel

Robert Baron, *Oklahoma State University*

Melissa Cardon, *Pace University*

Andrew Corbett, *Babson College*

Joep Cornelissen, *VU University Amsterdam*

Denis Grégoire, *Syracuse University*

Jeff McMullen, *Indiana University*

Edward Elgar

Cheltenham, UK • Northampton, MA, USA

Published by
Edward Elgar Publishing Limited
The Lypiatts
15 Lansdown Road
Cheltenham
Glos GL50 2JA
UK

Edward Elgar Publishing, Inc.
William Pratt House
9 Dewey Court
Northampton
Massachusetts 01060
USA

A catalogue record for this book
is available from the British Library

Library of Congress Control Number: 2014941544

This book is available electronically in the Elgaronline.com
Business Subject Collection, E-ISBN 978 1 78100 659 7

MIX
Paper from
responsible sources
FSC® C013056

ISBN 978 1 78100 658 0

Typeset by Servis Filmsetting Ltd, Stockport, Cheshire
Printed and bound in Great Britain by T.J. International Ltd, Padstow

Contents

Figures

Tables

Boxes

Contributors

Robert A. Baron, Oklahoma State University, USA
David A. Baucus, University of Otago, New Zealand
Melissa S. Baucus, University of Otago, New Zealand
Barbara Bird, American University, Washington, DC, USA
Malin Brännback, Åbo Akademi University, Finland
Melissa S. Cardon, Pace University, New York, USA
Alan L. Carsrud, Åbo Akademi University, Finland
Elsa T. Chan, University of Colorado Boulder, USA
Jean S. Clarke, University of Leeds, UK
Andrew C. Corbett, Babson College, MA, USA
Joep P. Cornelissen, VU University Amsterdam, the Netherlands
Mateja Drnovšek, University of Ljubljana, Slovenia
Maw-Der Foo, University of Colorado Boulder, USA
Daniel P. Forbes, University of Minnesota, USA
Denis A. Grégoire, HEC Montréal, Canada
Mario Hayek, Texas A&M University-Commerce, USA
Jeffery S. McMullen, Indiana University, USA
J. Robert Mitchell, University of Western Ontario, Canada
Ronald K. Mitchell, Texas Tech University, USA
Charles Y. Murnieks, Oregon State University, USA
Leslie E. Palich, Baylor University, TX, USA
Brandon Randolph-Seng, Texas A&M University-Commerce, USA
Michael R. Ryan, Texas Tech University, USA
Saras D. Sarasvathy, University of Virginia, USA
Alenka Slavec, University of Ljubljana, Slovenia
Wallace A. Williams, Jr, Texas A&M University-Commerce, USA
Matthew S. Wood, Baylor University, TX, USA
Miles A. Zachary, Texas Tech University, USA

Foreword

Entrepreneurial cognition has developed into a key stream of research in the field of entrepreneurship and beyond. Initial efforts at distinguishing the cognitions of entrepreneurs from non-entrepreneurs have given way to explaining differences in cognition within the category of entrepreneur. What is particularly interesting is the movement towards explaining variations in cognitions within the individual entrepreneur across time and contexts. These contextual differences include the nature of the task and/or the social situation surrounding it.

This book provides an important historical context for the state-of-the-art in entrepreneurial cognition (Randolph-Seng, Mitchell and Mitchell) and offers a fascinating reflection by (or description of the metamorphoses of) one of the leading experts on entrepreneurial cognition (Baron). The book links cognitions to important outcomes, including behavior (Bird; Carsrud and Brännback), learning (Forbes), and growth (Corbett). We also gain a deeper understanding of different levels of consciousness of cognition and how they can be self-regulated (Randolph-Seng, Williams and Hayek) and how entrepreneurial cognitions can be influenced by affect (Foo, Murnieks and Chan; Grégoire; Drnovsek, Slavec and Cardon), exchange (Mitchell, Mitchell, Zachary and Ryan) and language (Clarke and Cornelissen). Finally, chapters of the book provide some insights into the mechanisms of entrepreneurial cognition, including discussions of neuroscience (McMullen, Wood and Palich; Baucus, Baucus and Mitchell) and the heuristics of effectuation (Sarasvathy). This is a timely book written by the leading scholars in the field.

Dean A. Shepherd, Bloomington, Indiana, USA

Preface

In October of 2010, the *Academy of Management Review* published an article by Joep Cornelissen and Jean Clarke. In this article, Joep and Jean suggested a sensemaking-based process whereby 'entrepreneurs rely on inductive (analogical or metaphorical) reasoning to create and justify a rationale for a novel venture' (2010: 551). In doing so, however, they also highlighted the possibility that entrepreneurship research might overemphasize 'either the individual and his or her present cognitive state or the configuration of the social context and institutional outcomes, at the expense of a more integrative understanding that embeds individual entrepreneurs within their social contexts' (2010: 539).

It was this statement from Joep and Jean that led us to begin envisioning the possibilities of seeing past, present and future entrepreneurial cognition research through a lens of socially situated cognition. This approach encompasses four broad themes, described in more depth in our introductory chapter, that explain 'how social objects not only constitute the content of thought but also shape the process underlying thought and behavior' (Mitchell, Randolph-Seng and Mitchell, 2011: 774). Building on Cornelissen and Clarke (2010), we suggested that, 'when interpreted through the SSC [socially situated cognition] lens,' existing entrepreneurial cognition research may thus serve as a scaffold on which future research in this stream can be comprehensibly and comprehensively built (Mitchell, Randolph-Seng and Mitchell, 2011: 775). This *Handbook* represents an early effort to begin building on this more comprehensive, socially situated cognition scaffold.

I am grateful to Joep and Jean for planting this seed and for the many others who have helped to make this *Handbook* possible. This list includes the Advisory Panel for the *Handbook*, my two co-editors, Dean Shepherd for his encouragement, those individuals who reviewed the manuscripts, Alan Sturmer of Edward Elgar Publishing, Robert Baron and Lowell Busenitz, who suggested to Alan that we produce a *Handbook* in the first place, Chris Ainge, Shawna Chen and Hamid Vahidnia for their work in support of this effort and, of course, the contributors.

Those who know me well know that I love music. I always have. Even as a child I was a frequent visitor to the symphony orchestra and there was much I loved about it. To an eight-year-old, the formality of it all (the dress of the performers and the audience, the magnificence of the performance hall, the hush and respect of the crowd) was exciting. But what I loved most was the wonderful music. I thought it was especially exciting at the beginning of the symphony when each of the different players was warming up and tuning with a cacophony of instruments and sounds until the concert master signaled the principal oboist to play the note for tuning (an A above middle C). I always liked what followed. From the chaos, each musician then aligned his or her instrument with those of others until finally the conductor came to begin a rich and clear performance. So many sounds (individuals) coming together! Each musician came with their own reasons, but as an orchestra they would do something as a community.

Like members of a symphony orchestra, each of the contributors to this *Handbook* has their own reason for contributing. I again thank them for this. There are many purposes of this *Handbook*. Some are expressed in the introductory chapter, others in each of the subsequent chapters, and yet others in the Afterword. For me, though, one of the primary purposes of this *Handbook* is to enable the continued development of a rich research community in entrepreneurial cognition (cf. Grégoire, Corbett and McMullen, 2011). As I look at the list of top scholars who have contributed to this *Handbook*, I see important members of this research community. Their continued contribution is essential to making the entrepreneurial cognition research community accessible to others who wish to be part of it.

I recognize that there are many other members of this community who would have also contributed to this *Handbook* had circumstance, time and space allowed. To them, and to others with interest in this research community, I extend an invitation to join us in further integrating, embedding and situating the cognitions of entrepreneurs within their broader physical and social context (cf. Cornelissen and Clarke, 2010: 551; Mitchell et al., 2011: 774). With this *Handbook*, we have a vision of doing just this, and we invite you to join with us to engage in creating the cacophony of ideas that, when tuned, can result in rich and clear orchestrations (new directions, creative ideas, expansion of perspectives, cutting-edge methods etc.) for research and practice. Welcome to this *Handbook* and to the entrepreneurial cognition research community.

Rob Mitchell, London, Ontario, Canada

REFERENCES

Cornelissen, J.P. and Clarke, J.S. 2010. Imaging and rationalizing opportunities: Inductive reasoning and the creation and justification of new ventures. *Academy of Management Review*, **35**(4): 539–57.

Grégoire, D.A., Corbett, A. and McMullen, J. 2011. The cognitive perspective in entrepreneurship: An agenda for future research. *Journal of Management Studies*, **48**(6): 1443–77.

Mitchell, R.K., Randolph-Seng, B. and Mitchell, J.R. 2011. Socially situated cognition: Imagining new opportunities for entrepreneurship research (Dialogue). *Academy of Management Review*, **36**(4): 774–8.

Introduction: historical context, present trends and future directions in entrepreneurial cognition research

Brandon Randolph-Seng, J. Robert Mitchell and Ronald K. Mitchell

ABSTRACT

In this introduction to the *Handbook of Entrepreneurial Cognition*, we briefly review recent developments in social cognition and entrepreneurial cognition and the context in which these developments have arisen. First, we present a high-level historical review of research in social cognition more generally. Second, we review the roots of entrepreneurial cognition and the current state of the field. Lastly, we provide an overview of the chapters in the *Handbook*.[1]

INTRODUCTION

Entrepreneurial cognition has emerged as an important perspective in entrepreneurship (cf. Ireland and Webb, 2007; Mitchell et al., 2002, 2004, 2007) and has been defined as 'the knowledge structures that people use to make assessments, judgments or decisions involving opportunity evaluation and venture creation and growth' (Mitchell et al., 2002: 97). As research in the area of entrepreneurial cognition has developed, a central research question has emerged: 'how do entrepreneurs think' (Mitchell et al., 2007: 3). While the earlier approaches to entrepreneurial cognition were focused on the psychological processes that underlie behavior (Shaver and Scott, 1991), the area has broadened to focus on heuristic-based logic (e.g. Simon, Houghton and Aquino, 2000), perceptual processes (e.g. Gaglio and Katz, 2001), expertise (e.g. Mitchell, Smith, Seawright and Morse, 2000), and effectuation (e.g. Sarasvathy, 2001), among others. As Grégoire, Corbett and McMullen (2011) highlighted, however, despite the advances made, opportunities remain for developing the area.

In the spirit of their call for further research, a complementary perspective in entrepreneurial cognition is emerging that is based in the idea of socially situated cognition (Smith and Semin, 2004). At the center of the socially situated cognition approach is the idea that cognition is not static, but situated within specific individuals and environments, and the idea that cognition has functional purposes. In the next section, we provide a high-level historical review of research in social cognition more generally. Our reason for doing so is to assist us in setting a foundation that highlights how recent developments in cognition can advance research in entrepreneurial cognition and to illustrate the context in which these recent developments arose. We then provide a brief précis of entrepreneurial cognition specifically. Finally, we provide an overview of the chapters in the *Handbook*, especially as they relate to social cognition research generally and entrepreneurial cognition research specifically.

HISTORICAL CONTEXT

This historical sketch will trace the social-cognitive approach in conjunction with the study of mental processes in psychology. This sketch is admittedly a selective one, and is not meant to present the full spectrum of intellectual thought and research for each time period covered (see Costall, 2006). The purpose of this sketch is to simply place the current review into its historical context.

The Early Years of Mental Process Research

Over 300 years ago, Leibniz concluded in his *New Essays on Human Understanding* that human choice followed by motivated movement toward that choice was largely the result of 'an infinity of perceptions, unaccompanied by awareness or reflection' (1704/1981: 53). The idea that mental processes exist and possibly influence human behavior was not new even in Leibniz's day (see Whyte, 1960); but Leibniz seems to be one of the first to suggest that mental processes preconsciously influence the choices and behaviors we pursue in response to stimuli from the environment (Merikle and Reingold, 1992).

Moving forward 100 years, demonstrations of hypnosis and the advent of evolutionary theory provided additional evidence of mental processes influencing human thought and behavior (Bargh and Morsella, 2008). Toward the end of the 1800s, one of the first reported experimental demonstrations of nonconscious mental processes influencing choices (and the first American psychological experiment; see Kihlstrom, Barnhardt and Tataryn, 1992) was provided by Peirce and Jastrow (1884). Peirce and Jastrow found that, when subjects were forced to choose which of two very similar pressures was heavier, complete guesses were correct more often than chance would allow. A few years later Sidis (1898) reported similar findings using visual stimuli (which were replicated by many others; see Adams, 1957). Helmholtz's (1867/1968) perceptual theory also supported these findings.

However, even in these early years, the idea of unseen mental processes influencing human choice did not go unchallenged, as evidenced by James's (1890/1950) discussion of mental structures in chapter VI of Volume I of his book *Principles of Psychology*. Nevertheless, even James supported the view that one's selective interests or habits can shape one's perception of the environment, which in turn can automatically influence behavior (Gollwitzer and Moskowitz, 1996). In fact, a separate line of research began investigating the formation of mental processes as the result of skill development and habit (see Bryan and Harter, 1899; Solomons and Stein, 1896). Jastrow (1906), for example, proposed that developing skills and habits could lead to thinking withdrawing from conscious awareness.

Around the same time that Sidis and others were conducting their experiments, Breuer and Freud (1893–95/1955) published their book, *Studies in Hysteria*. Freud is often credited as the founder of nonconscious mental process research, but as history reveals, ideas about such processes considerably pre-date Freud. In fact, Freud repeatedly struggled with the problem of explaining mental processes (see Masling, 1992), but as time passed he came to distinguish between primary process thinking (unconscious) and secondary process thinking (preconscious and conscious). Primary processes were thought to be irrational and motivated by pleasure, while secondary processes attempted to achieve rationality and to operate according to reality (Overskeid, 2007). Therefore Freud's contribution was not in his delineation of mental processes, but in his view that perception can be constructed both by emotional (primary process) and intellective (secondary process) considerations (Erdelyi, 1992).

In sum, early work in examining mental processes, including such diverse areas as hypnosis, evolutionary theory, perceptional theory, skills and habits, and unconscious thinking, collectively came to the conclusion that mental processes influence human choice and behavior in conjunction with influences from the environment. Particularly influential in the early years was Freud's view of mental processes; nevertheless, given the difficulty in applying Freud's conceptions to everyday mental functioning, a short time later ideas and research in line with Sidis (1898) and Helmholtz's (1867/1968) work emerged, as reviewed next.

Constructivism to Behaviorism

Consistent with Sidis and Helmholtz, researchers began to view mental processes according to a constructivist perspective. Constructivist views were defined by the assumption that human thinking itself provides meaning to reality. Promoting this line of thought were the Gestalt psychologists, Bartlett, Mead and others. Although the Gestalt approach did not specifically refer to mental processes when explaining perception, the processes by which perceivers were thought to organize the environment were asserted to be automatic and nonconscious (Wittenbrink, 2007). In fact, an early empirical study by Zeigarnik (1927) demonstrated how personal motivations could influence mental activity nonconsciously by influencing what participants could remember. Bartlett (1932), although usually associated with his work on memory, provided research evidence showing that personal motivations and dispositions determined the content of perception (see Erdelyi, 1992). Mead's (1934/1967) social behaviorism attempted to apply empirical methods to the understanding of the constructive nature of social perception (see Bless and Forgas, 2000). Other early social psychological theorists such as Lewin, Sherif, Heider and, later, Festinger all proposed that perceptions of self and others are actively constructed (usually outside of conscious awareness) based on personal motivations (Moskowitz, Skurnik and Galinsky, 1999).

While these important developments were taking place in what would become known as social psychology, behaviorism was increasingly becoming the mainstream approach in psychology (beginning to take root with Watson, 1913). Skinner's (1974) behaviorism, although rejecting a constructivist view of perception, distinguished between rule-governed (conscious) and contingency-shaped (nonconscious) influences on behavior. According to the behaviorist view, people are nonconsciously influenced by their environment without the ability to regulate how they are influenced (Overskeid, 2007). Any mediating internal variables (e.g. interpretations, construals and motivations) between a stimulus and response had no explanatory standing (Bargh and Ferguson, 2000).

In sum, moving forward from Freud's conception of mental processes, researchers focused more on everyday types of mental functioning, such as perception, organization, memory and motivation. Nevertheless, the establishment of behaviorism, which completely rejected references to most mental functions, overshadowed this constructivist view on mental processes. Despite the firm establishment of behaviorism, psychologists were increasingly becoming dissatisfied with the behaviorist explanation of complex human behavior such as speech production (Chomsky, 1959; Hebb, 1949; Lashley, 1951). The problem that the behaviorists constantly faced was that the needed intra-individual psychological mechanisms could not be used as explanatory mediators between the environment and behavior (see Bargh and Ferguson, 2000). The lack of utility in the behaviorist theory for explaining higher-order human thinking and behavior contributed, in part, to the 'New Look' on perception movement of the late 1940s and early 1950s that will be reviewed next (Kihlstrom et al., 1992).

A 'New Look' into Mental Process Research

Starting around the late 1940s, a series of new research findings suggested that even basic human sensory detection and perceptual identification processes could be influenced by higher-order mental processes (e.g. attention, expectation, emotion, motivation; Bruner, 1957). For example, McGinnies (1949) measured participants' galvanic skin responses and found elevated levels (compared to control participants) when exposed to unidentified taboo words. These series of research findings became known as the 'New Look' on perception, since they were in direct contrast to popular behavioral theories of the time.

The 'New Look' was in many ways a 'New Look Back' to the non-behaviorist research and theorizing about the active construction processes humans engage in when viewing their world (e.g., Bartlett, 1932). Proponents of the New Look argued that the reason we actively construct our world is because it allows us, first, to reduce an extremely complex environment in a way that is not perceived as being taxing, and, second, to provide meaning and predictability based on the categorization processes (temporary or chronic; see Kelly, 1955) that are produced (Bruner, 1957). Thus our mental system is powered by the adaptive need to gain perceived 'control over a dynamic social world' (Gollwitzer and Moskowitz, 1996: 377). Importantly, our subjective view of the world appears to exist without mediation by the needs, motivations, desires and so on that actually produced that view (Bless and Forgas, 2000; Moskowitz et al., 1999).

The New Look perspective was met with both conceptual and methodological criticisms (see Kihlstrom et al., 1992). However, as will be discussed further, the general constructivist tenets of mental functioning and the importance of expectations and context on perception have withstood such criticisms (Balcetis and Dunning, 2006). The idea of nonconscious motivation has been less well received (Greenwald, 1992). In fact, the assertion that motivational states can nonconsciously influence perception (based on the research evidence provided by the New Look researchers) was a particularly contentious topic (e.g., Adams, 1957; Eriksen, 1958; Goldiamond, 1958; Ruiz and Krauss, 1968; Zajonc, 1962; for a review see

Dixon, 1981). In retrospect, many of the strong criticisms against nonconscious motivation seem to have been driven by a general discomfort with the idea that human behavior and functioning could be directed by nonconscious motivation and thus consistent with general Freudian conceptions (Bruner, 1992; Dijksterhuis, Aarts and Smith, 2005; Dixon, 1981).

In sum, based in part on the lack of utility in the behaviorist theory for explaining higher-order human thinking and behavior, researchers began to examine the previously forbidden mediators between the environment and behavior. Examination of mediators led to the 'New Look' perspective on perception, which found that such things as attention, expectations and motivations mediate the influence of the environment on human behavior. Nevertheless, based on contested conceptual and methodological issues, researchers began to move beyond New Look research in search of more structured conceptions of mental processing, which, in conjunction with the computer revolution, lead to what we now call cognition.

Cognition as the Mental Process

By 1956 (the year generally accepted as the beginning of the cognitive revolution), research in psychology was increasingly focused on the previously 'off-limits' mediators between stimulus and response. The information-processing approach, which became mainstream by the early 1970s, was the dominant approach adopted in the study of these stimulus–response mediators (Bruner, 1992). The tenets of the New Look perspective concerning the adaptive functioning of selective attention thrived under this information-processing conception of human thinking (see Kahneman and Treisman, 1984). This conception allowed cognitive researchers to attempt to determine experimentally the extent of human preconscious analysis (Bargh, 1992). The cognitive information-processing approach also allowed for experimental demonstrations of early conceptions of skill development and habit (Wegner and Bargh, 1998).

Generally considered to be the manifesto of the cognitive revolution, Neisser's (1967) work claimed that, in order to explain higher-order thinking processes in humans, a person's goals and motivations that are directed toward the current environment must be taken into account. Nevertheless, explanations that focused on motivational and/or personality influences were rare, as they were being replaced with information-processing terminology. Therefore the general constructivist approach survived, albeit in cognitive process form. This new approach was important because it gave researchers a conceptual background free of concepts that had complicated research such as the New Look perspective. In particular, different forms of nonconscious and conscious mental processes were delineated and investigated in ways never conceived of previously (e.g., Logan and Cowan, 1984; Regan, 1981). Priming manipulations (subliminal and supraliminal) became a prominent way to tap into these human mental systems (for other ways see, e.g., Debner and Jacoby, 1994). For example, Marcel (1983) found that subliminal primes associated with target words facilitated lexical responses to targets.

Compared to the New Look movement, mental processes were viewed in a more straightforward and simplistic way (see Bargh and Morsella, 2008; Dijksterhuis et al., 2005). Individuals were thought to nonconsciously construct their world, not according to motivational considerations, but simply based on routine processing of information (see Bargh, 2004; Pyszczynski, Greenberg and Koole, 2004). In many ways, however, cognitive theorists borrowed ideas from a behaviorist backdrop, leading to the above view. One of the main difficulties then came in distinguishing mental processes from application of those processes (e.g. Monteith, Sherman and Devine, 1998). Underlying this problem was a conceptual disconnection between thoughts and actions based on the way mind and behavior were understood by cognitive researchers at the time (Costall, 2006).

In sum, continuing with the focus on the mediators between the environment and human behavior used in the New Look movement, researchers refined conceptions of perception, motivation and individual differences into an information-processing approach. Such an approach allowed researchers to apply computer terminology to the process of human cognition and to more clearly catalog the routine processing of information by humans. Although such an approach provided clear theoretical boundaries to understanding cognition, the application of those cognitive processes to the world was missing. Work by social psychologists helped to remedy this problem with a focus on interpersonal aspects of cognition, as reviewed next.

Social Cognition as the Mental Process

Starting in the early to mid-1970s, social psychologists began to incorporate cognitive methodology and thinking into the study of how individuals construe their social world. Much of this research concluded that the mental workings of the brain are distinct from the subjective experiences of those workings (Nosek, 2007). For example, Bargh and Pietromonaco (1982) showed that subliminally activated trait concepts (e.g. hostility) influenced the impression participants made of a person without the participants being aware of the influence of the prime on their impressions (see also Devine, 1989). The nonconscious activation of social knowledge according to principles of accessibility, applicability and salience were all deemed important (Higgins, 1996). The interaction between the perceiver's mental representations and the environment were thought to determine perception (Bargh, Lombardi and Higgins, 1988). Social-cognitive research also suggested that individuals nonconsciously evaluate all that crosses their paths (Ferguson, 2007). Social-environmental cues were thought to directly guide behavior, with mediation by internal representations being the exception rather than the rule (Dijksterhuis and Bargh, 2001). This early social-cognitive research generally held to the cognitive view of noncomplex mental processes influencing people's construal of their world in a nonmotivational way (Bargh, 2008b). But unlike the cognitive view, there was a focus on the internal interpretation of a situation similar to the New Look perspective (Bargh and Ferguson, 2000). Nevertheless, a major problem that social-cognitive research began to experience was that nonconscious aspects of mental thought were not always simplistic, and motivational considerations (e.g., goals, needs, wants) seemed to be much more important in determining a person's social perception than current theorizing allowed (e.g., Thompson, Roman, Moskowitz, Chaiken and Bargh, 1994). And so, starting in the early 1990s, motivational considerations came into the mainstream of research.

Transitioning from cognitive to motivational considerations still allowed cognitive models such as the parallel-distribution processing model to be predictive, given that many representations (e.g. context, goals, motives) contribute to one psychological state (see Ferguson and Bargh, 2003). But beyond cognitive models, motivation was thought to be more central, determining the kind of processing that occurred. For example, motivation in the form of goals was thought to specify action, persistence and commitment to one's construal of the world (Gollwitzer and Moskowitz, 1996; Kruglanski et al., 2000). As a result, people were predicted to process identical information differently, depending on their currently relevant motivations (Bargh, 2008a; e.g., Tetlock, 2002).

Bargh (1992) posited that goal-directed thoughts and behaviors themselves could be activated and strategically pursued outside of conscious awareness. Many factors may influence whether a nonconscious goal is pursued. As examples, nonconscious attitudes toward a goal (Ferguson, 2007), previous associations with controlling feedback (e.g., Ratelle, Baldwin and Vallerand, 2005), subtle priming of goals (e.g., Rasinski, Visser, Zagatsky and Rickett, 2005), goals of the perceiver (e.g., Jonas and Sassenberg, 2006), and exposure to objects associated with certain goals (e.g., Kay, Wheeler, Bargh and Ross, 2004) can influence whether individuals pursue nonconscious goals. In fact, subliminal persuasion techniques, adapted from the cognitive and social-cognitive literatures, have been found to depend on personal motivational factors. Use of subliminal priming with respect to the brand name of a drink, for example, influences participants' choice and intention to drink that brand only if participants were already thirsty (Karremans, Stroebe and Claus, 2006; see also Strahan, Spencer and Zanna, 2002).

In sum, when social psychologists incorporated cognitive methodology and theory into the study of how individuals construe their social world, the application of cognitive processing theory to the world was better delineated. With a focus on the social-cognitive processing of information came a realization of the central role motivation played in determining the kind of cognitive processing that occurred. Such a focus on motivation brought research in the area of mental processing full circle (reviewed next).

Return to the 'New Look' on Mental Processes

Recent motivational research has supported many of the earlier constructivist views reviewed above (e.g. Bartlett, 1932; Bruner, 1957; Helmholtz, 1867/1968; Mead, 1934/1967; Sidis, 1898). Even the more controversial aspects of the New Look perspective are now finding research support, such that motivations, goals and needs are thought to preconsciously influence perception, thoughts and behaviors. For instance, in research that measures visual gaze, participants have been shown to selectively avoid goal-irrelevant stimuli and tune to goal-relevant stimuli outside of awareness (e.g. Chiao, Heck, Nakayama and Ambady, 2006; Isaacowitz, 2006; Koivisto and Revonsuo, 2007). Individual desires and preferences have also been found to nonconsciously influence interpretation of visual information toward the desired preference (Balcetis and Dunning, 2006). Such findings make sense when considering the numerous and complex stimuli that are present in any one situation at the same time (Gollwitzer and Moskowitz, 1996). Perception and the activation of mental representations must directly interact with individual representations (e.g. current goals and motives) and contextual cues from the environment (Andersen, Moskowitz, Blair and Nosek, 2007; Kunda and Spencer, 2003). In other words, what someone wants (even when not conscious) influences what is mentally accessible, which in turn can lead perceptions of any relevant and ambiguous stimuli to shift based on that accessible knowledge (Fishbach and Ferguson, 2007).

Even in the case of more chronic tendencies like habits, behaviors become linked to the person's desired self-regulatory outcomes (e.g., Neal, Wood and Quinn, 2006). Of importance, these chronic motives may make it less likely that a person will respond to the various influences that come from the environment, allowing for more adaptive and appropriate responses from the person without the need for conscious involvement (Bargh, 2006). In sum, as advances have been made in social-cognitive research, motivations, goals and needs have been shown to preconsciously influence perception, thoughts and behaviors, consistent with a long history of constructivist views of mental functioning, setting the stage for a more integrated approach to understanding human cognition, namely the socially situated cognition approach.

The Socially Situated Cognition Approach

As we have described previously, in this *Handbook* we adopt the lens of socially situated cognition as a generally applicable approach to the examination of the many facets of entrepreneurial cognition research. We have done so, based in part on our historical sketch presented above; specifically, that such an approach brings the evolution of the study of mental processes into a more integrated whole. Four broad themes constitute a socially situated approach to entrepreneurial cognition (Mitchell et al., 2011: 774–5; Smith and Semin, 2004):

(1) Entrepreneurial cognition is *action-oriented* (i.e. captures the positive or negative evaluations of, or motivations toward an object or concept)
(2) Entrepreneurial cognition is *embodied* (i.e. captures the interrelationship between the physical brain and body to capture how the body shapes the mind)
(3) Entrepreneurial cognition is *situated* (i.e. captures the communicative context, relational context, and group context in which cognition and action occur)
(4) Entrepreneurial cognition is *distributed* (i.e. captures the variety that occurs in the distribution of cognition across social agents and the environment).

The socially situated cognition approach (Smith and Semin, 2004, 2007) thus represents an attempt to integrate current insights from social psychology and cognition research, such as those findings reviewed above. In other words, the socially situated cognition approach argues that cognition regularly occurs in a social environment and is: action-oriented, situated, embodied and distributed (for a recent review see Semin and Smith, 2013). According to this view, if cognition is for action and in our body, then motivation and emotion regulate it. Similarly, if cognition is context-specific, then moment-to-moment interactions with the social environment profoundly impact it. And if cognition is distributed, then social objects enable and support it. Social objects therefore not only constitute the content of thought, but also shape the process underlying thoughts and behavior. Such a statement not only provides an accurate representation of the current state of the field of social cognition, but also, as we now describe, is consistent with current trends and directions in entrepreneurial cognition research.

ENTREPRENEURIAL COGNITION

Despite the newness of the area, entrepreneurial cognition research has progressed in unique ways compared to the broader social cognition domain reviewed above. We begin by briefly examining the roots of this area of study and end with a brief review of the current state of the field. Accordingly, in Table I.1 we provide a summary of early work in entrepreneurial cognition up to 1990, and in Table I.2 we provide a summary of work in the field from the 1990s to the present. In doing so, we seek to provide a brief précis of entrepreneurial cognition research, while linking it to work in social psychology more broadly. We do so as a way to link the historical discussion of social cognition research generally and entrepreneurial cognition research specifically to the chapters in this *Handbook*.

Roots of Entrepreneurial Cognition Research

Entrepreneurial cognition research can be traced back to early economists. In particular, Cantillon (1755/1931) suggested that entrepreneurial-like individuals exercise judgment under uncertainty. Likewise, Baudeau (1767/1910) suggested that such exercise of judgment demands specialized knowledge. In his *Treatise on Political Economy* (1821), Jean-Baptiste Say refers to entrepreneurs as individuals who think and act (through experimentation) such that profit may result. We find these links between what exists in entrepreneurs' minds and their resulting action (i.e. exercising judgments in the face of uncertainty through the translation of knowledge into action and experimentation) to be prescient. The idea that entrepreneurs exist and act in uncertain environments captures both the *situated* and *action-oriented* nature of socially situated entrepreneurial cognition (Mitchell, Randolph-Seng and Mitchell, 2011). Entrepreneurship research continues to develop these notions (cf. Alvarez and Busenitz, 2001; Smith, Mitchell and Mitchell, 2009).

Moving forward, early theorists in the area continued to emphasize the importance of cognitive elements of entrepreneurship such as

Table I.1 Earlier work on entrepreneurial cognition (up to 1990)

Year	Reference	Excerpt or summary	Narrative
1755	Cantillon (1755/1931; cited in Hébert and Link, 1989: 42)	. . . someone who engages in exchanges for profit; specifically, he exercises business judgements in the face of uncertainty.	The entrepreneur is someone who exercises judgment under uncertainty
1767	Baudeau (1767/1910: 51)	Nothing is more evident, [than that] we need a numerous race of farmers or chief farmers endowed with the knowledge of their art, moved by a great desire to translate their knowledge into action.	Such an exercise of judgment demands desire and specialized knowledge.
1803	Say (1821/1971: 82–3)	. . . the enormous wealth of Britain is less owing to her own advances in scientific acquirements, high as she ranks in that department, than to the wonderful practical skill of her entrepreneurs[1] in the useful application of knowledge . . . science alone is not sufficient to ensure the progress, without the aid of experiment, which is always attended with more or less of risk, and does not always indemnify the entrepreneur, whose profit, even when successful, is moderated by competition.	The knowledge and skill that the entrepreneur possesses must be applied, especially through the action-oriented process of experimentation.
1920	Dewing (1920: 245–54; cited in Palmer, 1971: 33)	Dewing wrote of the qualities of imagination, initiative, judgment, and restraint.	Other qualities such as imagination, restraint, and taking initiative are also important.

1949	Danhof (1949; cited in Palmer, 1971: 33; emphasis in Palmer)	Danhof divides the function of the entrepreneur into three major roles: *obtaining relevant information, evaluating the information with regard to profit, and setting the operation in motion*. Once the entrepreneur has determined what information should be gathered, the first two functions can be delegated.	Suggests three functions for the entrepreneur. The central role of the entrepreneur is making decisions regarding what types of information are relevant.
1952	Lamb (1952: 91; cited in Palmer, 1971: 33–4)	. . . entrepreneurship is that form of social decision making performed by economic innovators.	Argues that such decision making is social in nature.
1957	Evans (1957: 50; cited in Palmer, 1971: 33)	. . . views the entrepreneur as 'the person or group of persons who has (or assumes) the task of determining the kind of business to be operated.' The decisions germane to this function involve the nature of the goods and services to be offered, the size of the enterprise, and the customers catered to. Once these decisions have been made by the entrepreneur, other decisions, that is decisions to achieve the previous goals set by the entrepreneur, become essentially management's . . . Evans notes that, once these decisions have been made, the role of the entrepreneur does not cease: instead, he must be continually alert and ready to make new decisions in light of changing market conditions and arising opportunities.	Proposes that decision making is directed by being continually alert to changing market conditions and opportunities that arise from those changes.
1964	McGuire (1964: 238; cited in Palmer, 1971: 34)	Over time, and in different societies, there has evidently been a substantial change in entrepreneurial types, and presumably in the entrepreneur function.	The notion that history and culture are also important factors.

Table I.1 (continued)

Year	Reference	Excerpt or summary	Narrative
1971	Palmer (1971: 34)	If the functions of the entrepreneur are to be thoroughly understood, all aspects contributing to his behavioral patterns must be considered social, political, economic, and psychological forces.	Calls for a more holistic approach to the study of the entrepreneur.
1973	Kirzner (1973: 9)	. . . entrepreneurial alertness is crucial to the market process. Disequilibrium represents a situation of widespread market ignorance. This ignorance is responsible for the emergence of profitable opportunities. Entrepreneurial alertness exploits these opportunities when others pass them by.	Further formalizes the idea of alertness by focusing on the role of industry knowledge to exploit profit opportunities.
1975	Schultz (1975: 827)	. . . education and experience influence the efficiency of human beings to perceive, to interpret correctly, and to undertake action that will appropriately reallocate their resources.	The notion that education and experience can improve the efficiency of perceiving and interpreting information.
1988	Smeltzer, Fann and Nikolaisen (1988: 61)	It appears that most small business owner/managers who are responsible for both operational and strategic planning conduct environmental scanning regularly . . . Even though these owners do not have a staff for this function, they do not ignore environmental scanning. These owner/managers believe that personal information is more valuable than impersonal information, and informal personal information is deemed most valuable of all.	Finds that not all types of information are equally valuable for alert entrepreneurs: informal, personal information appears to be the worthiest.

Smith, Gannon, Grimm and Mitchell (1988: 223)	Entrepreneurs from smaller firms are less comprehensive in their decision behavior than professional managers from larger firms, with comprehensiveness defined as the degree to which an individual follows a formal rational decision process . . . As decision comprehensiveness declines, so too does organizational performance, both among entrepreneurs and professional managers.	Provides evidence against formal/ rational decision-making views: entrepreneurs are, in fact, not comprehensive in their decision making.
Cooper, Woo and Dunkelberg (1988: 97)	Although previous evidence on business survival led to the hypothesis that the entrepreneurs would only be cautiously optimistic, this was not the case. They perceived their prospects as very favorable, with 81% seeing odds of 7 out of 10 or better and a remarkable 33% seeing odds of success of 10 out of 10. In considering the prospects for other businesses like their own, they perceived odds which were significantly lower, but still moderately favorable . . . Those who were poorly prepared were just as optimistic as those who were well prepared.	Shows that entrepreneurs are generally optimistic, i.e., perceive low levels of risk.
Corman, Perles and Vancini (1988: 38–9)	Two-thirds of the high-technology entrepreneurs interviewed did not perceive exceptionally high levels of risk when making their decision to venture . . . High-tech entrepreneurs, however, tend to be stable, successful, and highly-educated individuals who are not greatly concerned about their ability to secure and maintain employment. Their skills are currently in high demand, and provide alternatives should the current venture fail. Similarly, their education and training have usually allowed them to demand high salaries, thus further reducing perceived financial risk.	Shows that perceiving low levels of risk can facilitate decision making.

Table I.1 (continued)

Year	Reference	Excerpt or summary	Narrative
1988	Bird (1988: 442)	Entrepreneurial intentions, entrepreneurs' states of mind that direct attention, experience, and action toward a business concept, set the form and direction of organizations at their inception. Subsequent organizational outcomes such as survival, development (including written plans), growth, and change are based on these intentions.	Introduces the concept of entrepreneurial intentions as a necessary condition for any entrepreneurial behavior.
1989	Hébert and Link (1989: 47)	The entrepreneur is someone who specializes in taking responsibility for and making judgemental decisions that affect the location, form, and the use of goods, resources, or institutions.	The idea that entrepreneurs are those who become specialized in judgmental decision making.
1990	Hisrich and Jankowicz (1990: 49)	Principal component analyses . . . reveal relatively low cognitive complexity: essentially, just one or two major areas of emphasis predominate in each venture capitalist's thinking.	Extends the idea of non-comprehensive decision making to the venture capital context. Shows that low cognitive complexity, rather than specialization, governs VCs' decisions.

Note: [1] Prior to the adoption of the French term 'entrepreneur' in English, the word was translated as 'adventurer.' 'Entrepreneur' is used here.

judgment, imagination, decision making and alertness (Danhoff, 1949; Dewing, 1920; Evans, 1957; Lamb, 1952). This early research recognized that entrepreneurship processes interact with the particular time and place in which they occur (McGuire, 1964), and that the social, political, economic and psychological factors that influence entrepreneurial behavior need to be included (Palmer, 1971). Despite all of this, the focus continued to be on the characteristics of the entrepreneur (cf. Brockhaus and Horwitz, 1986). This focus on using personality characteristics/traits to distinguish between entrepreneurs and non-entrepreneurs overlapped with the aforementioned behaviorist movement in psychology in both time and purpose. That is, like the 'behaviorist' view, which asserted that individuals had no control over the influence of the environment on their behaviors, the 'traits' view of entrepreneurship asserted that fairly immutable traits governed the behaviors of entrepreneurs (e.g. McClelland, 1965).

But glimmers of a cognitive view of entrepreneurship were on the horizon. Specialized knowledge, decision making, perceptions and individual differences, for example, were shown to categorize entrepreneurial cognition-like processes (Cooper, Woo and Dunkelberg, 1988; Corman, Perles and Vancini, 1988; Hébert and Link, 1989; Hisrich and Jankowicz, 1990; Kirzner, 1973; Schultz, 1975; Smeltzer, Fann and Nikolaisen, 1988; Smith, Gannon, Grimm and Mitchell, 1988). Toward the end of this period, Bird (1988) introduced her concept of entrepreneurial intentions and the idea of entrepreneurs' cognition (Bird, 1992). Bird's application of the concept of entrepreneurial intentions to cognition was an important theoretical advancement because, to this point in time, the focus had been on behavior as being based on individual differences, rather than on cognitive processes per se. When entrepreneurial intentions were specified as a necessary condition of entrepreneurial behavior, entrepreneurship as a unique form of information processing could be realized, which in turn set the stage for the research in the area starting in the 1990s (see Table I.2).

Current State of the Field of Entrepreneurial Cognition Research

As noted previously, the information-processing view of cognition had long been considered mainstream. With this as a foundation, Shaver and Scott (1991) called for a comprehensive psychological approach to the study of new venture creation and did so by focusing on the choice-based cognitive processes at the entrepreneur level of analysis. Consistent with Shaver and Scott's (1991) call, other researchers were also increasingly focusing on an information-processing approach to understanding entrepreneurial attitudes, decision making and thought processes (Bird, 1992;

Table I.2 Current work on entrepreneurial cognition (1990–present)

Year	Reference	Excerpt	Narrative
1991	Shaver and Scott (1991: 24, 27–8)	As a discipline, psychology is rarely defined as the sum of the activities of its practitioners. Rather, it is distinguished from other social or behavioral sciences, such as sociology, anthropology, and economics, by its emphasis on the individual person as the level of analysis. Within this concentration on the individual (*person*), two of psychology's core theoretical concerns have been the contents of mind (the *process* intervening between external world and observable behavior), and the exercise of free *choice*. (Emphasis in original) . . . A comprehensive psychological portrait of new venture creation will ultimately have to show how the individual's cognitive representations of the world get translated into action. To accomplish this purpose, it will ultimately be necessary to consider general orienting dispositions (such as attitudes), motivational principles (such as subjective expected utility), and personal motives (such as achievement motivation). It should be noted that choice for a psychologist is not always the same as the rational decision making inherent in economic theory . . . A thorough psychological approach to the study of choices involved in new venture creation must incorporate both the rational and the irrational features of decision making.	Calls for a comprehensive psychological approach to the study of new venture creation. Suggests that such an approach will focus mainly on three core issues: (1) the individual entrepreneur, as the level of analysis, (2) the 'processes' (in the mind of the individual) through which the external world is translated into action, and (3) the exercise of 'choice' (both rational and non-rational decision making).
	Robinson, Stimpson, Huefner and Hunt (1991: 13)	The attitude model of entrepreneurship, as it is empirically and conceptually presented here, has ramifications for entrepreneurial education and change programs. Because attitudes are open to change, entrepreneurial attitudes may be influenced by educators and practitioners. The tripartite attitude model suggests ways of initiating	The notion that humans 'process' information is pursued by showing that such processes can be influenced by changing entrepreneurs'

		change by influencing thoughts, feelings, and behavioral intentions (Rosenberg, 1960) with regard to entrepreneurship and related attitudes such as innovation, achievement, self-esteem, and personal control.	attitudes, as attitudes are susceptible to external impacts.
	Kaish and Gilad (1991: 59)	Entrepreneurs do seem to expose themselves to more information and their alertness takes them to the less obvious places . . . Our research suggests that the physical volume of search is one distinguishing characteristic of entrepreneurial behavior.	Shows empirically that entrepreneurs process more information in their decision making.
1992	Bird (1992: 18)	Temporal brackets, pacing, and market events flow from individual differences, environmental rhythms, and entrepreneurs' cognitions. Thus, an entrepreneur, based on his or her personality, motivations, etc. tunes in to changes in the environment and forms thoughts, feelings, and words that describe the venture that is to be created. Some of these thoughts and behaviors act to bracket the time involved in organization creation and to set a pace for that creation.	Refers to the thought processes of the entrepreneurs as 'entrepreneurs' cognition.' Defines the concept in relation to the time-based aspects.
	Manimala (1992: 477–8)	Entrepreneurial heuristics were defined as the thumb-rules guiding the management decisions involved in the start-up and management of a new venture . . . A regression analysis showed that entrepreneurial orientations . . . could explain as much as 50% of the variance in innovativeness.	Expands the research on non-rational choice/decision making by coining the concept of 'entrepreneurial heuristics.' Shows that such heuristics can actually lead to innovativeness.
	Busenitz (1992: iii–iv)	The results of a study of 115 entrepreneurs and 95 managers in large organizations indicate that entrepreneurs are more overconfident in their decision making and that they use the representative heuristic more extensively than managers in large organizations. The results also indicate that the traditional trait approach and situational factors provide only limited help in predicting entrepreneurial activity versus managerial involvement in a large organization.	Provides further evidence for the use of heuristics, such as representative heuristics, in the decisions made by entrepreneurs.

Table I.2 (continued)

Year	Reference	Excerpt	Narrative
1992	Katz (1992: 29–30)	This paper proposes a psychosocial cognitive model (PCM) of employment status choice. The model is psychosocial insofar as it utilizes an individual's psychology in the form of values and decision-making processes, and social insofar as it depends on personal history and social context as factors contributing to the decision process. It is cognitive insofar as the decision processes utilize the cognitive heuristics of availability, representativeness, and in a few cases adjustment from an anchor, to describe the process and decision likelihoods of the individual.	Incorporates heuristic-based decision processes with social and psychological factors to propose a psychosocial cognitive model (PCM) of employment status choice.
	McGrath and MacMillan (1992: 419)	. . . there is a basic set of beliefs that entrepreneurs hold about themselves and about others in their society that, from the perspective of the entrepreneur, differentiate the two. This set of beliefs transcends cultures . . . that even among culturally very different societies there is a core set of perceptions, common across countries, that entrepreneurs hold about others in their countries.	Finds that entrepreneurs believe they are different from non-entrepreneurs and that such beliefs transcend cultures.
	Chandler and Jansen (1992: 223)	. . . although self-assessed proficiency in the entrepreneurial function does not appear to be contingent on the length of previous experience as a founder; it does appear to be a necessary component in the development of high-performance companies.	Shows that successful entrepreneurs think they have high levels of competencies in creation of novel ventures.
	Hansen and Allen (1992: 63)	For intending entrepreneurs, managing the emergence of their pre-organizations may be an important pre-step to the creation of their new organization. Entrepreneurs would appear to improve their ability to create new organizations when they establish pre-organizational information-accessing and processing capabilities that are appropriate to their respective levels of environmental load.	The idea is developed that entrepreneurs' thought processes, i.e., their information-processing capabilities and thinking skills, can be enhanced, even before individuals engage in organization creation activities.

1993	Bull and Willard (1993: 183)	We offer the following tentative entrepreneurship theory, extracted from anecdotal observations and extant literature, in the hope that it will better explain and begin to predict the phenomenon of entrepreneurship: A person will carry out a new combination, causing discontinuity, under conditions of: 1. Task-related motivation, 2. Expertise, 3. Expectation of personal gain, and 4. A supportive environment.	Proposes that expertise is one of the necessary elements for entrepreneurs to succeed.
	McCarthy, Schoorman, and Cooper (1993: 9)	Ongoing research in decision-making suggests that psychological processes may play a role in influencing . . . [reinvestment] decisions. Under certain conditions entrepreneurs may be influenced by a phenomenon termed 'escalation of commitment.' This may lead entrepreneurs to decide to expand the asset bases of their firms, regardless of feedback from the marketplace.	Under certain conditions, entrepreneurs may not process negative feedback/information, due to the phenomenon of escalation of commitment.
	Krueger and Carsrud (1993: 315)	Planned behaviours such as starting a business are intentional and thus are best predicted by intentions toward the behaviour, not by attitudes, beliefs, personality or demographics . . . Intentions fully mediate the relationship between attitudes and the target behaviour, even where attitudes may appear to explain behaviour. Intentions entail an enactive cognitive process which serves to channel beliefs, perceptions and other exogenous factors into the intent to act, then to the action itself.	Further expands intention-based approaches by arguing that intentions channel beliefs and perceptions into the intent to act and, thereby, to the actions.
	Krueger (1993: 17)	This exploratory study found significant support for Shapero's propositions that entrepreneurial intentions derive largely from (1) perceptions of feasibility, (2) perceptions of desirability, and (3) a propensity to act which derives from control beliefs. Path analysis	Finds empirical supports for the proposition that intentions, in fact, channel beliefs and perceptions into actions.

Table I.2 (continued)

Year	Reference	Excerpt	Narrative
1993	Krueger (1993: 17)	demonstrated that the impact of prior entrepreneurial exposure on intentions is indirect, operating through perceived feasibility. The positiveness of those experiences also indirectly influences intentions through perceived desirability.	
1994	Boyd and Vozikis (1994: 63)	This paper further develops Bird's model of entrepreneurial intentionality by suggesting that individual self-efficacy, which has been defined as a person's belief in his or her capability to perform a task, influences the development of both entrepreneurial intentions and actions or behaviors.	Introduces the concept of self-efficacy and argues that it plays in the thought processes of entrepreneurs, as it precedes intentions.
	Krueger and Dickson (1994: 396)	We found support for our hypotheses that an increase in self-efficacy increases perceptions of opportunity and decreases perceptions of threat and that changing opportunity and threat perceptions changes risk taking.	Finds that increased self-efficacy increases perceptions of opportunity and risk-taking behavior, e.g., intentions to act.
	Krueger and Brazeal (1994: 96, 102)	Perceived feasibility in SEE [Shapero's Model of the Entrepreneurial 'Event'] corresponds to perceived behavioral control in TPB [Theory of Planned Behavior] (both correspond to perceived self-efficacy); TPB's other two attitude measures are subsumed by SEE's perceived desirability. Our most important conclusion, though, remains the primacy of perceived feasibility. Given that conclusion, we need to research what factors contribute the most to perceptions of feasibility.	Based on psychological factors such as self-efficacy and attitude, offers an overarching model of entrepreneurial potential.
	Mitchell (1994: 163)	The research conducted in this dissertation (1) investigates three key literature streams in entrepreneurship research and specific	Further formalized the notion of expertise by introducing

		theories within those streams, (2) suggests the in-depth exploration of expert information processing theory (EIPT), and (3) encourages the integration of these two fields to propose a theory of new venture formation expertise.	the term 'entrepreneurial expertise.' Suggests a cognitive-based model by which entrepreneurs can be distinguished from non-entrepreneurs.
	Reuber and Fischer (1994: 372–3)	The findings reported here indicate that 1) owners' expertise is more strongly correlated with firm performance than is owners' experience; 2) different types of expertise are associated with different types of experience; and 3) there is some direct association (net of expertise) of experience on firm performance.	Argues and empirically demonstrates that experience measures are inadequate surrogates for expertise.
1995	Cooper, Folta and Woo (1995: 107)	It was found that those who had no entrepreneurial experience, on the average, sought more, not less, information. However, as expected, those who ventured into fields which were different and those who had higher levels of initial confidence sought less information . . . [N]ovice entrepreneurs searched less extensively in unfamiliar domains, a behavior consistent with bounded rationality. By contrast, experienced entrepreneurs did not vary their search pattern. It was also found that entrepreneurs having high levels of confidence sought less information, as expected.	Shows that novice and experienced entrepreneurs have different search strategies.
	Mitchell and Chesteen (1995: 301, 302)	. . . expertise can be acquired through an individual's participation in specific processes such as significant study, experience, and the exposure to schemata through contact with experts. Whereas the general design of the educational courses described in this study optimizes a student's capability to apply the principles and practices of entrepreneurship in a business setting, the activities of the script-based experiential instructional strategy were tailored specifically to boost the student's readiness to venture by enhancing entrepreneurial expertise.	Demonstrates that entrepreneurial expertise can be enhanced more by experiential entrepreneurial instruction pedagogy than by traditional business plan approach.

Table I.2 (continued)

Year	Reference	Excerpt	Narrative
1995	Mitchell and Chesteen (1995: 301, 302)	. . . the results suggest that venture expertise can be stimulated effectively *within* the instructional setting by the planned series of experiential activities involving contact with experts. (Emphasis in original)	
	Gatewood, Shaver and Gartner (1995: 371)	The purpose of this study was to explore whether certain cognitive factors of potential entrepreneurs (as measured by a personal efficacy scale and the kinds of reasons people offer for their decision to undertake efforts to start a business) can be used to predict their subsequent persistence in business start-up activities and in new venture creation success . . . An analysis of the results found that internal/stable attributions (e.g., 'I have always wanted to be my own boss') were supported for female potential entrepreneurs, whereas external/stable attributions (e.g., 'I had identified a market need') were significant for male potential entrepreneurs.	Shows that self-efficacy is important for subsequent persistence in start-up activities.
	Palich and Bagby (1995: 426)	. . . entrepreneurs categorized equivocal business scenarios significantly more positively than did other subjects, and . . . these perceptual differences were consistent and significant (i.e., entrepreneurs perceived more strengths versus weaknesses, opportunities versus threats, and potential for performance improvement versus deterioration).	Finds that entrepreneurs generally perceive more strengths and opportunities than weaknesses and threats when working on business scenarios.
1996	Busenitz and Lau (1996: 33)	. . . a person's cognition significantly affects start-up intentions. The cross-cultural cognitive model developed in this paper also acknowledges that entrepreneurial cognition is affected by cultural values, social context, and some personal variables. The schema (knowledge structure) of a founder contains the information	Develops a cross-cultural model of venture creation decision. Argues that an individual's cognition significantly affects his/her start-up intentions.

		needed to arrive at the start-up decisions and the cognitive process determines how information is utilized.	
	Kolvereid (1996: 23)	This research developed a classification scheme of reasons given for preferring self-employment versus organizational employment . . . The classification scheme that emerged questions the relevance of earlier models, which have been used to explain and predict occupational status choice.	Individuals provided different reasons for their career choice intentions. A taxonomy is developed based on these reasons.
	Busenitz (1996: 42–3)	. . . Kaish and Gilad (1991) study recently tested Kirzner's (1973) theory of alertness which asserts that entrepreneurs are more alert to new opportunities and use information differently. Because of the lack of generalizable samples and the exploratory nature of the Kaish and Gilad study, this research replicated and further developed some of the scales originally developed by them. The results indicated that little empirical support exists for this theoretical framework, but the measures of entrepreneurial alertness need further development.	Provides empirical evidence that challenges the theory of alertness. Finds that entrepreneurs are not more alert than non-entrepreneurs.
1997	Mitchell (1997: 136)	To insiders, entrepreneurship is not for the few. Entrepreneurship is for the many – albeit the many who have sufficient discipline to learn and abide by the tried-and-true norms of the venturing expert script.	Demonstrates that by learning expert scripts, novice entrepreneurs can become experts.
	Jenkins and Johnson (1997: 895)	This paper uses a causal map methodology to consider the contrasts between entrepreneurial intentions and outcomes. In evaluating a series of propositions drawn from the extant literature the study finds that the elicited causal maps are consistent with contrasts in entrepreneurial intentions, but not outcomes. This suggests that the existing emphasis on entrepreneurial strategies being deliberate, conscious processes may be misplaced: non-deliberate, emergent strategies may be just as influential in producing entrepreneurial outcomes.	Expanding the research on heuristics, demonstrates that emergent, non-deliberate strategies may be just as effective as deliberate ones in producing positive outcomes.

Table I.2 (continued)

Year	Reference	Excerpt	Narrative
1997	Busenitz and Barney (1997: 9)	Under conditions of environmental uncertainty and complexity, biases and heuristics can be an effective and efficient guide to decision-making. In such settings, more comprehensive and cautious decision-making is not possible, and biases and heuristics may provide an effective way to approximate the appropriate decisions.	Further develops the notion of entrepreneurial heuristics. Shows that entrepreneurs are more susceptible to the use of heuristics due to uncertainty and complexity in their environment.
1998	Baron (1998: 288)	The major themes of this article can be summarized as follows: (1) entrepreneurs' thinking may differ, in important ways, from that of other persons; specifically, they may be more susceptible to various kinds of cognitive errors or bias than other persons, and (2) such differences in cognition do not stem primarily from differences between entrepreneurs and other people with respect to personal traits (although such differences may well exist), but rather from the fact that entrepreneurs operate in situations and under conditions that would be expected to maximize such errors or biases.	Argues that entrepreneurs are susceptible to cognitive biases, because of the demand of the situation, i.e., because their environment overloads their information-processing capacity.
	Chen, Greene and Crick (1998: 295)	[Self-efficacy] refers to the strength of a person's belief that he or she is capable of successfully performing the various roles and tasks of entrepreneurship. It consists of five factors: marketing, innovation, management, risk-taking, and financial control.	Further develops and finds empirical support for the construct of entrepreneurial self-efficacy.
	Pineda, Lerner, Miller and Phillips (1998: 60)	The results suggest that the more important the decision and the more the manager perceives himself or herself to be effective in making a particular type of decision, the greater the intensity of the information search, and the greater the use of external information sources during decision-making.	Shows that high self-efficacy in decision making increases the level of information search.

1999	Busenitz (1999: 325)	. . . entrepreneurial risk may be explained by recognizing that entrepreneurs use biases and heuristics more, which is likely to lead them to perceive less risk in a given decision situation.	Demonstrates that a higher level of reliance on heuristics can explain the risk-taking behavior.
	McGrath (1999: 16)	Changing one's perception of failure can require adjusting fundamental assumptions regarding performance. As March and Shapira (1987) observe, failure as manifested in risk taking that goes badly is considered undesirable. Therefore, people seek success and avoid failure, and those efforts can introduce errors in learning and interpretation processes. Paradoxically, such errors often make failure more likely or more expensive than it need have been (see Levinthal & March, 1993). Errors fall into three broad categories: (1) errors caused by extrapolating to the future from past success, (2) errors owing to cognitive bias, and (3) errors introduced through interventions to avoid the occurrence or appearance of failure.	Argues for the importance of failure in the development of the interpretation processes of individuals, as failure provides learning opportunities for entrepreneurs.
2000	Mitchell, Smith, Seawright and Morse (2000: 986)	[T]he study was successful in demonstrating that cognitive scripts explain a significant amount of variance in venture creation decisions . . . [T]he results are consistent with theory that suggests that entrepreneurs in different cultures look first to arrangements scripts to evaluate potential entry into the venture creation decision process, and only then utilize doing-related scripts.	Finds support for the proposition that across cultures, cognitive scripts can explain a significant amount of variance in decisions to create ventures.
	Krueger (2000: 5)	Before we can act on opportunities we must first identify those opportunities. Understanding what promotes or inhibits entrepreneurial activity thus requires understanding how we construct perceived opportunities . . . Based on well-developed theory and robust empirical evidence, we propose an intentions-based model of the cognitive infrastructure that supports or inhibits how we perceive opportunities.	A conceptual model of how entrepreneurs form intentions and construct opportunities based on their perceptions of exogenous factors is developed.

Table I.2 (continued)

Year	Reference	Excerpt	Narrative
2000	Simon, Houghton and Aquino (2000: 113–114)	The study's findings suggest that risk perceptions may differ because certain types of cognitive biases lead individuals to perceive less risk . . . individuals start ventures because they do not perceive the risks involved, and not because they knowingly accept high levels of risks. The belief in the law of small numbers lowered an individual's perceptions of a venture's riskiness . . .	Provides support for the notion that individuals decide to start ventures partly due to the fact that they do not perceive the whole array of risks that are involved.
	Wright, Hoskisson, Busenitz and Dial (2000: 592, 597–8)	We build a new model to explain incentive differences from agency theory (short term versus long term) and describe how fundamental differences in individual cognitive orientation (managerial versus entrepreneurial) can be combined to explain different strategic buyout attributes.	Combines individual cognitive orientation with agency theory to explain different strategic buyout attributes.
		Heuristic-based logic in decision making is very economical and usually provides valuable estimations of the entrepreneurial decisions (Tversky & Kahneman, 1974), especially because of the scarcity of quantifiable information available for a more rational decision . . . [However,] in the face of uncertainty and ambiguity, a heuristic-based logic may lead to bad decisions . . . severe errors (Tversky & Kahneman, 1974) . . . [and] firm failure (. . . Wright et al., 1991).	
	Baron (2000: 80)	Empirical results indicated that entrepreneurs were significantly less likely to engage in counterfactual thinking, experienced significantly less regret over past events, and found it significantly easier to admit past mistakes both to themselves and to others.	Finds that entrepreneurs often take a strong future-oriented perspective, which may reduce their propensity to reflect on past events.

2001	Alvarez and Busenitz (2001: 755)	We extend the boundaries of resource-based theory to include the cognitive ability of individual entrepreneurs. Entrepreneurs have individual-specific resources that facilitate the recognition of new opportunities and the assembling of resources for the venture.	The idea is developed that cognitive abilities of entrepreneurs are one valuable, rare resource for venture creation.
	Gaglio and Katz (2001: 95–6)	Entrepreneurial alertness, a distinctive set of perceptual and information-processing skills, has been advanced as the cognitive engine driving the opportunity identification process . . . To date, empirical support for the construct has been equivocal, leading at least one scholar (Busenitz, 1996) to question its value. However . . . this may follow in part from an unduly narrow approach to the operationalization of theory as well as a potential problem in the match of the psychometric method to the type of phenomenon being studied.	Advances the concept of entrepreneurial alertness as the cognitive engine that drives the process of opportunity identification.
	Sarasvathy (2001: 262)	The essential agent of entrepreneurship, as I argue here, however, is an effectuator: an imaginative actor who seizes contingent opportunities and exploits any and all means at hand to fulfill a plurality of current and future aspirations, many of which are shaped and created through the very process of economic decision making and are not given a priori.	Introduces the logic of effectuation and proposes that expert entrepreneurs follow such logic.
2002	Mitchell, Busenitz, Lant, McDougall, Morse and Smith (2002)	We define entrepreneurial cognitions as follows: *entrepreneurial cognitions are the knowledge structures that people use to make assessments, judgments, or decisions involving opportunity evaluation, venture creation, and growth*. In other words, research in entrepreneurial cognition is about understanding how entrepreneurs use simplifying mental models to piece together previously unconnected information that helps them to identify and invent new products or services, and to assemble the necessary resources to start and grow businesses. (Emphasis in original)	A formal definition for entrepreneurial cognitions is proposed to help direct the future research.

Table I.2 (continued)

Year	Reference	Excerpt	Narrative
2002	Mitchell, Smith, Morse, Seawright, Peredo and McKenzie (2002: 9)	We find . . . that individuals who possess 'professional entrepreneurial cognitions' do indeed have cognitions that are distinct from business non-entrepreneurs . . . . . . we report further confirmation of a universal culture of entrepreneurship . . . . . . we find (a) observed differences on eight of the ten proposed cognition constructs, and (b) that the pattern of country representation within an empirically developed set of entrepreneurial archetypes does indeed differ among countries.	Shows that there exists a cognitive explanation for entrepreneurial phenomena that is universal and cross-cultural.
	Fiet (2002: 221)	I began this book by asserting that the discovery process was at the heart of entrepreneurship . . . If skeptics were saying that acquiring entrepreneurial competence is a passive process that occurs solely as a result of being lucky, this book has taken aim at this argument and hopefully debunked it to your satisfaction. Acquiring competence may occur accidentally to a large extent. However, I have tried to show that it is possible to train entrepreneurs to make discoveries, which increases their competence.	Further develops the idea that the discovery process is crucial to the success of a new venture creation and that such competency can be enhanced using training.
	Simon and Houghton (2002: 105)	. . . entrepreneurs in smaller, younger, firms, who are considering pioneering, are more likely to exhibit illusion of control, law of small numbers, and reasoning by analogy. These biases contribute to underestimating competition, overestimating demand, and overlooking requisite assets.	Suggests that entrepreneurs' decision environments vary greatly and that such variation influences the types of biases that arise.

2003	Markman and Baron (2003: 281)	. . . to the extent entrepreneurs are high on a number of distinct individual-difference dimensions (e.g., self-efficacy, ability to recognize opportunities, personal perseverance, human and social capital, superior social skills) the closer will be the person-entrepreneurship fit and, consequently, the greater the likelihood or magnitude of their success.	Argues that individuals with higher levels of self-efficacy, perseverance, and social skills are more likely to have fit with entrepreneurship phenomena.
2004	Gaglio (2004: 533, 547)	The cognitive processes of mental simulation and counterfactual thinking are proposed as mechanisms by which entrepreneurs identify and develop innovative opportunities. With the right methodology, in which entrepreneurs do the thinking rather than recall previous experience (Ericsson & Simon, 1994; Gaglio & Katz, 2001), it becomes possible to punch a hole in the black box regarding the cognitive work associated with the opportunity identification process and to test the assertions made by the theory of entrepreneurial alertness. The area of opportunity identification can move beyond the descriptive phase and begin to consider questions about dynamics and contingencies.	Suggests that two cognitive heuristics of mental simulations and counterfactual thinking may guide entrepreneurial reasoning and enhance the opportunity identification process.
2005	Baker and Nelson (2005: 329, 356–7)	. . . found that Lévi-Strauss's concept of bricolage – making do with what is at hand – explained many of the behaviors observed in small firms in resource-poor environments that were able to create something from nothing by exploiting physical, social, or institutional inputs that other firms rejected or ignored. The refusal to enact environmental limitations helps firms use bricolage to create something from nothing . . . however, when this is taken to an extreme, identities and communities of practice that are constructed create a new set of limitations that suppress growth. In contrast, when firms use bricolage more narrowly or temporarily . . . they appear to be more likely to grow.	Introduces the concept of entrepreneurial bricolage. Shows that entrepreneurs sometimes refuse to accept environmental limitations, which allows them to meet environmental challenges.

Table I.2 (continued)

Year	Reference	Excerpt	Narrative
2005	Cardon, Zietsma, Saparito, Matherne and Davis (2005: 25)	We suggest that relational metaphors can provide new insight into our understanding of entrepreneurial activities and shed light on aspects of entrepreneurship that seem illogical from a rational perspective, that is, cognitive biases that reduce the perception of risks (love may be blind), entrepreneurial persistence despite poor results (some parents never give up, even when perhaps they should), often extreme devotion to the business entailing self-sacrifice and delayed gratification, and the separation problems that sometimes accompany founder succession. A parenting metaphor highlights the importance of passion (i.e., strong emotions and enthusiasm) and identification (i.e., close association and connection) between an entrepreneur and a venture.	Argues that using the lens of affect can inform prior research on cognition through use of a metaphor that illustrates the limits of a solely rational perspective of cognition.
	Corbett (2005: 473, 484, 485)	This article makes connections between knowledge, cognition, and creativity to develop the concept of learning asymmetries. Individuals with different learning preferences may identify and exploit opportunities differently. Individuals with a convergent learning preference will be more likely to develop an initial solution or idea. Individuals with assimilative learning preference will be more likely to develop more options or opportunities for products from a platform of initial ideas . . . Individuals with divergent learning preference will be more likely to develop a workable business prototype from a number of different options; individuals with accommodative learning preference will be more likely to successfully exploit working prototypes.	A conceptual model of entrepreneurial learning is proposed and the concept of learning asymmetries is introduced.

Forbes (2005: 623)	Results show that individual age, firm decision comprehensiveness and external equity funding affect the degree to which entrepreneurs are overconfident. In addition, founder-managers are shown to be more overconfident than are new-venture managers who did not found their firms. The results suggest that entrepreneurs' cognitive biases are a function of both individual and contextual factors.	Demonstrates that entrepreneurs tend to display overconfident bias due to both differences in individual characteristics and the organizational contexts in which they operate.
Mitchell (2005: 187)	I . . . demonstrate that as a global society we have, in certain ways, been wrong in our approach to entrepreneurship education (both informal and formal), and that a course correction (pun intended) is needed . . . I outline . . . the relationship between education and high-performance to support the argument that entrepreneurs are special, but are not created in the way that is commonly believed: that there is, in actuality, a general process for creating them. I present and discuss the international implications of the emerging 'practice school' of entrepreneurship education for reforming the creation of global entrepreneurs.	Suggests that there is a common process known for creating entrepreneurial expertise: while entrepreneurs are special, creating them is general.
Mitchell, Friga and Mitchell (2005: 653)	Entrepreneurial intuition is poorly defined in the literature that the intuitive is confused with the innate, what is systematic is overlooked, and unexplained variance in entrepreneurial behavior remains high . . . We (1) bound and define the construct of entrepreneurial intuition within the distinctive domain of entrepreneurship research; (2) apply a levels-of-consciousness logic and process dynamism approach to; (3) organize definitions, antecedents, and consequences; and (4) produce propositions that lead to a working definition of entrepreneurial intuition.	Points to the importance of intuition in the thought processes of entrepreneurs. Renders entrepreneurial intuition more usable for future studies.

Table I.2 (continued)

Year	Reference	Excerpt	Narrative
2005	Shepherd and DeTienne (2005: 91)	Results suggest that while prior knowledge of customer problems leads to the identification of more opportunities and opportunities that are more innovative, it also moderates the relationship between potential financial reward and opportunity identification . . . the less knowledgeable an individual was about customer problems, the more positive the effect that potential financial reward had on the number of opportunities identified and the innovativeness of those opportunities.	Finds supports for the argument that prior knowledge of a field allow individuals to identify more 'valuable' opportunities.
2006	Baron (2006: 104)	. . . entrepreneurs identify opportunities for new business ventures . . . by using cognitive frameworks they have acquired through experience to perceive connections between seemingly unrelated events or trends in the external world . . . This pattern recognition perspective on opportunity identification . . . helps integrate into one basic framework three factors that have been found to play an important role in opportunity recognition: engaging in an active search for opportunities; alertness to them; and prior knowledge of an industry or market . . . helps explain why some persons, but not others, identify specific opportunities . . . [It] suggests specific ways in which current or would-be entrepreneurs can be trained to be better at recognizing opportunities.	Proposes that entrepreneurs identify opportunities by using cognitive frameworks that allow them to identify patterns among seemingly unrelated events, i.e., to connect the dots.
	Baron and Ensley (2006: 1331)	New business opportunities are identified when entrepreneurs, using relevant cognitive frameworks, 'connect the dots' between seemingly unrelated events or trends and then detect patterns in these connections suggestive of new products or services . . . The prototypes of experienced entrepreneurs were more clearly defined,	The pattern recognition framework proposed above is supported empirically.

		richer in content, and more concerned with factors and conditions related to actually starting and running a new venture (e.g., generation of positive cash flow) than the prototypes of novice entrepreneurs.	
	Pech and Cameron (2006: 61)	This [information-processing] framework demonstrates how various entrepreneurial needs and attitudes, as well as entrepreneurial motivators, impact on the diagnosis and assessment of informational cues. It describes how opportunity-related information is processed by entrepreneurs in order to reach a decision of acceptance or rejection of potential business opportunities . . . Entrepreneurs have a heightened ability and awareness for recognizing and audaciously exploiting business opportunities. They persistently and continually seek opportunity-laden information in order to satisfy internal motivators such as need for achievement and the fulfillment of competitive urges.	Constructs an information-processing framework of entrepreneurial opportunity recognition.
2007	Bingham, Eisenhardt and Furr (2007: 40)	Our core contribution is the insight that heuristics are at the heart of high performing organizational processes, and so are central to firm capabilities. Specifically, we find that high performing organizational processes consist of heuristics – i.e., informal rules-of-thumb that center on the capture of opportunities within flows of process-specific opportunities (e.g., new countries, acquisition targets, or product development projects). We also find that more heuristics relate to higher process performance. Moreover, high performing organizational processes consist of particular types of heuristics.	Applies the concept of heuristics in the context of firm capability and finds significant relationships between the two.

Table I.2 (continued)

Year	Reference	Excerpt	Narrative
2007	Brigham, De Castro and Shepherd (2007: 29)	Regression analyses indicated higher satisfaction and lower intentions to exit for owner-managers whose dominant decision-making style complemented the levels of formalization and structure in their firms. In addition . . . both satisfaction and intentions to exit were significantly associated with actual turnover.	Introduces the concept of cognitive styles to the literature and finds that entrepreneurs' cognitive styles affect their decisions.
	Busenitz and Arthurs (2007: 147–8)	One of the benefits of looking at both entrepreneurial and dynamic capabilities in the context of entrepreneurial ventures is that we are able to better articulate the nature of dynamic capabilities . . . Although making firm-level adjustments to fit with the environment has been an anchor of the dynamic capabilities literature, what parts of those adjustments are specific to dynamic capabilities and which ones are a regular part of organizational life has been less clear.	Draws a firmer distinction between the entrepreneurial and dynamic capabilities.
	Corbett and Hmieleski (2007: 103)	. . . we examine the interplay and divergence between the role schema of individuals in corporations and the expert event schemas necessary to launch a new venture . . . We then construct a theoretical framework for explaining why this tension results in corporate entrepreneurs emphasizing certain event schemas in a manner that is distinct from independent entrepreneurs' role schemas.	Proposes a framework about how the corporate context can create tension between corporate entrepreneurs' role schemas and the event schemas necessary for new venture emergence.
	Corbett (2007: 97)	Discovering entrepreneurial opportunities requires that individuals not only possess some form of prior knowledge, but that they also have the cognitive abilities that allow them to value and exploit that knowledge . . . After analyzing the empirical data, the article develops the concept of learning asymmetries and explains how the manner in which people learn may affect their ability to identify entrepreneurial opportunities.	Further develops the concept of learning asymmetries and finds a significant relationship between opportunity identification and learning.

Corbett, Neck and DeTienne (2007: 829)	. . . we advance the literature on entrepreneurial human capital by linking cognitive scripts used by corporate entrepreneurs in project termination decisions to corresponding levels of learning . . . [L]ongitudinal investigation suggests that corporate entrepreneurs use three types of termination scripts: (1) undisciplined termination, (2) strategic termination, and (3) innovation drift.	Finds that organizational learning is dependent upon the type of termination script employed.
Dimov (2007a: 713)	This article helps develop the creativity perspective within entrepreneurship in two ways. First, it elaborates on the nature of opportunity as a creative product. Rather than viewing opportunities as single insights, it suggests that they are emerging through the continuous shaping and development of (raw) ideas that are acted upon. Second, rather than attributing them to a particular individual, it highlights the contextual and social influences that affect the generation and shaping of ideas. This helps move entrepreneurship research beyond the single-person, single-insight attribution that currently permeates it.	Coins the concept of opportunity development – as a socially constructed process of venture creation in which initial ideas are elaborated, refined, changed, or discarded as a result of learning that arises from continually shaping, discussing and interpreting ideas.
Dimov (2007b: 579)	This paper reinforces the importance of the intersection between learning and entrepreneurship by showing that the individuality of learning and its situated nature play key enabling roles in the formation of opportunity intentions. Even if equally motivated toward entrepreneurial achievements, individuals differ not only in the nature of ideas they generate in a particular situation, but also in the perceived feasibility of these ideas. In addition, it also bridges the different ontological traditions in which the study of opportunities is currently based . . . Finally, this paper shows that integrating individual and situational factors in the study of entrepreneurial learning adds an important dimension to our understanding of opportunity development as a learning process.	Demonstrates how the subjective nature of learning and its situated character play important and enabling roles in the formation of intentions.

Table I.2 (continued)

Year	Reference	Excerpt	Narrative
2008	Baron (2008: 329)	First, the environments in which entrepreneurs function are often highly unpredictable and filled with rapid change . . . Research on the influence of affect suggests that it is most likely to exert powerful effects on cognition and behavior in precisely this type of situation. In contexts involving high uncertainty and unpredictability, affect can readily tip the balance toward specific actions or decisions – effects it might not produce in environments that are more certain and predictable . . . A second reason why affect may often exert strong effects in the domain of entrepreneurship relates to the specific tasks entrepreneurs perform in starting new ventures. These tasks are highly varied in nature and change significantly as the process unfolds . . . [and] are ones that have previously been shown to be strongly influenced by affect . . . affect has been shown to exert strong effects on creativity . . . on persuasion . . . on decision making and judgments . . . and on the formation of productive working relationships with others . . .	Suggests that affect is important to entrepreneurial cognition and behavior and that its importance is shaped by both the environment and the task.
	Cardon (2008: 78)	Drawing from the psychological literature on emotions and feelings . . . and emotional contagion . . . as well as literature on transformational leadership . . . we build a model of emotional contagion within the entrepreneurial context, from entrepreneurs to employees . . . we discuss entrepreneurial passion in general . . . how it leads to the emotional displays of entrepreneurs . . . build a model of emotional contagion, drawing a distinction between contagion through primitive emotional mimicry and through social comparison processes. We . . . ultimately suggest that for contagion of passion	Proposes a model of how affect (entrepreneurial passion in particular) is transferred to employees through physical mimicry and social comparison.

		from entrepreneur to employee to occur, employees must experience both positive intense feelings for their activities and a sense of meaningfulness or identity connection to those activities within the entrepreneurial firm.	
	Hmieleski and Corbett (2008: 45)	. . . entrepreneurial intentions are found to be significantly associated with measures of personality, motivation, cognitive style, social models, and improvisation . . . The results of hierarchical regression show that improvisation accounts for a significant amount of variance in entrepreneurial intention above and beyond what is accounted for by the other variables.	Demonstrates a strong relationship between entrepreneurial intentions and improvisation.
2009	Bingham (2009: 321)	Firms with more successful foreign market entries decrease improvisation in opportunity selection but increase improvisation in opportunity execution . . . Intriguingly, data suggest that increased improvisation in opportunity execution may be influenced by decreased improvisation in opportunity selection, whereas decreased improvisation in opportunity execution may be influenced by increased improvisation in opportunity selection.	Finds a dynamic relationship between the level of improvisation in opportunity selection and in opportunity execution.
	Cardon et al. (2009: 511–12)	We first address what passion is by proposing a definition of the entrepreneurial passion concept based on psychological research on emotions . . . and on identity . . . as well as grounded work in entrepreneurship. Second, we address what passion does by proposing a conceptual framework to theorize the mechanisms that coordinate the influence of role-identity-specific passion on entrepreneurs' cognitions and behaviors in the pursuit of entrepreneurial effectiveness . . . We use self-regulation as an overall theoretical framework to extract empirically testable propositions.	Applies self-regulation theory to develop a theory of the cognitive and behavioral consequences of entrepreneurial passion.

Table I.2 (continued)

Year	Reference	Excerpt	Narrative
2009	Dew, Read, Sarasvathy and Wiltbank (2009: 287, 296)	. . . entrepreneurial experts frame decisions using an 'effectual' logic (identify more potential markets, focus more on building the venture as a whole, pay less attention to predictive information, worry more about making do with resources on hand to invest only what they could afford to lose, and emphasize stitching together networks of partnerships); while novices use a 'predictive frame' and tend to 'go by the textbook.' Overall results show significant evidence of expert–novice differences as well as differences in logical framing.	Finds that expert entrepreneurs are more likely to use effectual logic than novice entrepreneurs.
	Foo, Uy and Baron (2009: 1086–1087)	. . . we use the affect-as-information perspective as a theoretical foundation for understanding how affect influences entrepreneurial effort . . . [and] clarify the nature and direction of the affect–effort relationship . . . We find that negative affect predicts new venture effort . . . we also find that positive affect predicts new venture effort. [and suggest] . . . that although positive affect signals that all is going well, it does not necessarily reduce effort. Instead, we clarify the mechanism behind the affect–effort relationship by showing that positive affect is linked to increased effort through a future temporal focus . . . this is one of the first studies to show that affect matters in the domain of entrepreneurship.	Empirically demonstrates the importance of both positive and negative affect to entrepreneurial action.
	Read, Dew, Sarasvathy, Song and Wiltbank (2009: 1)	The results show significant differences in heuristics used by expert entrepreneurs and by managers to approach marketing in the face of uncertainty. While managers rely primarily on predictive techniques to make marketing decisions, expert entrepreneurs tend to . . . use an	Provides further support for effectual logic. Shows that expert entrepreneurs are more likely to use effectual logic,

	effectual or non-predictive logic to tackle uncertain market elements and to construct novel markets with committed stakeholders.	e.g., focus on intangible resources, co-creation of value, and stakeholder relationships.
Read, Song and Smit (2009)	In this study, we conduct a meta-analysis of the articles published in the *Journal of Business Venturing*, summarizing data on 9897 new ventures to connect three of the principles of effectuation positively with new venture performance. In so doing, we offer both specific insight into precisely measuring effectuation and a general method for extracting variables from prior work to measure new constructs.	Develops a way to measure effectuation logic. Finds that most of the principles of effectual logic are positively related to new venture performance.
Mitchell, Mitchell and Mitchell (2009: 131)	. . . successful new venture formation is associated with individual knowledge-based scripts . . . As the previous 15 years have demonstrated, the link between expertise and new venture formation is very useful in helping entrepreneurship researchers illuminate the underlying dynamics of new venture formation.	Describes a prototypical approach for identifying the script-based components of new venture formation expertise and for distinguishing entrepreneurial expertise in individuals.
Shepherd and Cardon (2009: 923)	We develop an emotion framework of project failure that relies on self-determination to explain variance in the intensity of the negative emotions triggered by project failure and self-compassion to explain variance in learning from project failure.	Suggests a framework to explain how self-compassion can bridge the gap between negative emotions from failure and the resultant potential learning.
Ucbasaran, Westhead and Wright (2009: 102, 111)	An inverse U-shaped relationship was detected between the proportion of failed businesses relative to the number of businesses owned and the number of opportunities identified in a given period. Business failure experience was not associated with the innovativeness of exploited opportunities. . . . beyond a certain level, the benefits associated with prior business ownership experience may be outweighed by the biases that can stem from experience.	Finds an inverse U-shaped relationship between experience and failure and cautions about the over-reliance on prior experience in opportunity identification.

Table I.2 (continued)

Year	Reference	Excerpt	Narrative
2010	Grégoire, Barr and Shepherd (2010: 413, 424–5)	In contrast to prior research [e.g., Baron and Ensley, 2006], the qualitative and quantitative data do not provide evidence that individuals use prototypes to recognize opportunities. Instead, we find that different kinds of mental connections – structural alignment – play different roles in the process of recognizing opportunities, with different consequences. . . . the significance of our findings lies not in observing that executive entrepreneurs find opportunities by matching technology with market, but rather that their matching of technology and market involves their aligning the superficial features and structural relationships of technology and market to one another.	Challenges the notion that entrepreneurs employ prototypes to recognize opportunities. Demonstrates, instead, that recognizing opportunities involves cognitive processes of structural alignment.
	Haynie, Shepherd, Mosakowski and Earley (2010: 217–18)	We develop a framework of entrepreneurial metacognition to investigate the foundations of an 'entrepreneurial mindset' . . . [This model] offers important insights into entrepreneurial thinking and behaviors; the dynamism and uncertainty inherent in an entrepreneurial context requires entrepreneurs to continuously 'rethink current strategic actions, organization structure, communications systems, corporate culture, asset deployment, investment strategies, in short every aspect of a firm's operation and long-term health (Hitt et al., 1998: 26)' in the context of a contemporary business environment.	Develop a conceptual framework of entrepreneurial metacognition. Suggests that entrepreneurial mindset is metacognitive in nature; thus it enables entrepreneurs to formulate 'higher-order' cognitive strategies and to promote adaptable cognitions.

2011	Mitchell, Randolph-Seng and Mitchell (2011: 774)	To date, entrepreneurial cognition has been explained largely in terms of what social cognition researchers commonly term boxologies: seemingly static representations of abstract, disembodied cognitive structures (e.g., biases, heuristics, scripts, etc., as described in Mitchell et al., 2007) . . . We . . . extend Cornelissen and Clarke's [2010] contribution by discussing how their analysis of sensemaking through explicit language can also illustrate the components of a broader explanatory process emerging within entrepreneurship research [that is, the socially situated approach (Smith and Semin, 2004)].	Suggests that an approach based on socially situated cognition supplies a broad and fruitful organizing framework that can be used to integrate a diverse set of research streams in entrepreneurial cognition.
2012	Haynie, Shepherd and Patzelt (2012: 255)	Our findings indicate that individuals inexperienced in the entrepreneurial process who engage metacognitive resources use feedback more effectively than others, suggesting that metacognitive ability represents an important mechanism related to the development of expert performance.	Finds that metacognitive ability is an important factor in the development of expertise.

Kaish and Gilad, 1991; Robinson, Stimpson, Huefner and Hunt, 1991). Further, the focus on an individual-level information-processing approach to understanding entrepreneurial cognition was exemplified in research on entrepreneurial-specific perceptions, heuristics and biases and entrepreneurial expertise. Lowell Busenitz's (1992) dissertation specifically focused on the role that cognitive biases played in the decision making of entrepreneurs as compared to managers, whereas Ron Mitchell's (1994) focused on the composition and classification of entrepreneurial expertise. In both cases, differences were found in the thinking of entrepreneurs versus non-entrepreneurs.

Likewise, other researchers at the time were looking at the impact of perceptions on entrepreneurial intentions (Krueger, 1993; Krueger and Carsrud, 1993; Krueger and Dickson, 1994) and the idea that self-perceptions are important, especially in terms of a belief by entrepreneurs that they think differently than non-entrepreneurs (McGrath and MacMillan, 1992) and believe that they have high levels of competencies in venture creation (Chandler and Jansen, 1992). These concepts, associated specifically with entrepreneurial intentions, continued to develop (Kolvereid, 1996). Krueger and Carsrud (1993), for example, expanded intention-based approaches by arguing that intentions channel beliefs and perception into the intent to act, which was followed by subsequent empirical support for this argument (Krueger, 1993). Self-efficacy was also found to be an important component to consider in understanding entrepreneurial intentions (Boyd and Vozikis, 1994; Chen, Greene and Crick, 1998; Gatewood, Shaver and Gartner, 1995; Krueger and Brazeal, 1994; Krueger and Dickson, 1994; Pineda, Lerner, Miller and Phillips, 1998).

As the field began to mature, there was an increased focus on expanding on earlier work in entrepreneurial cognition. For instance, Mitchell, Friga and Mitchell (2005) suggested how the development of expertise might be responsible for the development of entrepreneurial intuition. Similarly, Mitchell, Mitchell and Mitchell (2009) provided an approach for identifying the expert script-based components of new venture formation expertise. Baron and Markman (2000) likewise suggested that expertise also applied to the area of social skills in their findings that financial success is contingent on an entrepreneur's social competence (Markman and Baron, 2003). Baron and Ensley (2006) further contributed to research on entrepreneurial expertise in highlighting that expert entrepreneurs differed from novice entrepreneurs in how they identified patterns among seemingly unrelated events. And importantly, based on the work of Sarasvathy (2001) that developed the idea of effectuation in entrepreneurship, extensive research has demonstrated how expert entrepreneurs are more likely to use effectual logic than novice entrepreneurs (Dew, Read, Sarasvathy and Wiltbank, 2009; Read, Dew, Sarasvathy, Song and Wiltbank, 2009; Read, Song and Smit, 2009). Given the development of research on entrepreneurial expertise, an ancillary question has also emerged regarding how entrepreneurial expertise can be developed. For example, Corbett (2005, 2007) proposed a conceptual model of entrepreneurial learning and Mitchell (2005) suggested that there is a common process known for creating entrepreneurial expertise.

Likewise, in the area of cognitive biases and perceptions (Baron, 1998; Busenitz, 1992, 1996, 1999; Busenitz and Barney, 1997; Jenkins and Johnson, 1997; Katz, 1992; Manimala, 1992; Palich and Bagby, 1995; McGrath, 1999; McCarthy, Schoorman and Cooper, 1993), researchers found that individuals decide to start ventures in part due to biased processing of risk (Simon, Houghton and Aquino, 2000), and that entrepreneurs' biases vary depending on the decision environments that arise (Simon and Houghton, 2002). Similarly, entrepreneurs are more likely to have a future-oriented perspective (Baron, 2000) and are more likely to rely on mental simulation and counterfactual thinking to guide their entrepreneurial thinking (Gaglio, 2004). Furthermore, this research found that entrepreneurs tend to display overconfidence bias due to both personal and environmental factors (Forbes, 2005).

The focus on broader environmental factors increased as the area of entrepreneurial cognition developed. Although much of the entrepreneurial cognition research published in the 1990s focused on the cognition of the entrepreneur (for an exception see Busenitz and Lau, 1996), the next decade saw an expanded research focus on factors that externally influence entrepreneurial cognition (Mitchell et al., 2002). That is, research began to address the *socially situated* nature of entrepreneurial cognition. This research included investigations into the cognitive scripts that underlie venture creation across cultures (Mitchell, Smith, Morse, Seawright, Peredo and McKenzie, 2002; Mitchell, Smith, Seawright and Morse, 2000), and into the influence of exogenous, situational factors on entrepreneurial intentions (Krueger, 2000). Consistent with a greater focus on external factors influencing individual-level entrepreneurial cognition, research also began to investigate the role of entrepreneurial cognition at higher levels of analysis. For instance, work in this area investigated:

- how socially constructed resource environments influenced entrepreneurial behavior (Baker and Nelson, 2005);
- how an entrepreneurial cognitive orientation influenced strategic buyouts (Wright, Hoskisson, Busenitz and Dial, 2000);
- what a resource-based view of entrepreneurial cognitive abilities might look like (Alvarez and Busenitz, 2001);

- how organizational heuristics can contribute to the development of firm capabilities related to opportunity (Bingham, Eisenhardt and Furr, 2007);
- how firm-level 'corporate' entrepreneurial cognition can influence new venture emergence and termination (Corbett and Hmieleski, 2007; Corbett, Neck and DeTienne, 2007); and
- how improvisation can influence firm opportunity selection and execution (Bingham, 2009).

What we find most interesting about each of these studies is the way in which they highlight the *combined* importance of the *situated* and *distributed* aspects of entrepreneurial cognition. That is, we notice how the area of entrepreneurial cognition is becoming increasingly socially situated.

A more recent trend is growing interest in the area of embodied cognition. That is, research is just beginning to capture how the physical brain and body shape the mind. For instance, in their work on analogical and metaphorical reasoning, Cornelissen and Clarke (2010: 547) suggest that 'the inductive creation of metaphorical meaning is directed and constrained . . . [where] human motor actions involving physical movement or physically holding or manipulating an object,' shape how entrepreneurs think and speak about new ventures. Likewise, work in entrepreneurship has increasingly begun to focus on the topic of entrepreneurial affect (Baron, 2008). Underlying work in this area, especially as it relates to entrepreneurial cognition, is the idea that affect influences entrepreneurial cognition and behavior, especially as a result of the situated nature of affect and cognition. Work in this area largely began with research on entrepreneurial passion, linking passion to entrepreneurial thinking and action (Cardon, Wincent, Singh and Drnovšek, 2009; Cardon, Zietsma, Saparito, Matherne and Davis, 2005). Key to the embodiment theme of socially situated cognition is the suggestion that physical mimicry matters in the contagion of passion (Cardon, 2008). Other research has broadened the focus from passion to both positive and negative affect and their consequences for entrepreneurial action generally (Baron, 2008; Foo, Uy and Baron, 2009). The importance of the situational context to affect and the importance of affect, as an embodied experience, to action illustrate the need to adopt a socially situated approach to entrepreneurial cognition.

In this way, similar to recent areas of research within social cognition more generally, and providing a culmination of work in entrepreneurial cognition to date, we hope that both our discussion and Tables I.1 and I.2 illustrate the parallels between the progression in psychology research and the progression in entrepreneurship research to a more socially situated approach to cognition. Specifically, we hope to highlight the need to consider this full spectrum of cognition within the entrepreneurial context by adopting a socially situated approach (Mitchell, Randolph-Seng and Mitchell, 2011). The socially situated cognition approach to entrepreneurial cognition reflects how, as the entrepreneurial process unfolds, 'social objects not only constitute the content of thought but also shape the process underlying thought and behavior' (Mitchell et al., 2011: 774).

We see three broad approaches to applying socially situated cognition to entrepreneurship. In the first approach, entrepreneurial cognition research addresses and develops one of the specific themes from socially situated cognition (e.g. the embodiment theme) using an entrepreneurship lens. It is likely that research adopting this first approach will either: (a) be grounded in specific areas of past entrepreneurial cognition research, but will address how these research areas might be reconsidered and rethought from the perspective of a socially situated cognition; or (b) import socially situated concepts and ideas from social and cognitive psychology that are then extended to, and blended with, research in entrepreneurship. In the second approach, entrepreneurial cognition research might bridge two or more themes from socially situated cognition research (i.e. extending research in one theme to other themes). This type of research might synthesize different elements of the socially situated approach (e.g. combining embodied activity with mental states, cognitive processes with content, and individual and collective levels of analysis in entrepreneurship research). In the third approach, entrepreneurial cognition research will extend beyond the themes in the socially situated cognition view (either in a broader or narrower context) that can begin to address antecedent, process, levels and interdisciplinary linkage-focused perspectives (cf. Grégoire, Corbett and McMullen, 2011). Research adopting this approach will likely exist as expansive and reflective contributions on the insights and explanations provided by the socially situated approach, and reflect on its position and research agenda in relation to past research on entrepreneurial cognition and social cognition. In the following section, we seek to highlight how the chapters in this *Handbook* begin to accomplish these purposes.

Organization of the *Handbook*

Participating in an academic conference – whether planning, presenting or attending the sessions – provides an apt metaphor for the organization of this *Handbook*. Many conferences have a theme. There is often a keynote presentation. Organizers seek to group papers by common topics or interests. The goal of conference organization is to empower the participant to craft a singular experience from within a multifaceted conceptual environment. In this respect the organization of this *Handbook of Entrepreneurial Cognition* loosely resembles that of a conference.

The theme of this *Handbook* has not been in doubt (Entrepreneurial Cognition Research) – although as editors we have sought to shed light on the transition from static to dynamic cognitive research conceptualizations through some degree of emphasis on socially situated cognition. The 'keynote' paper in this *Handbook*, prepared by Robert Baron, appears in Chapter 1. While Robert does not require the reader to assume a scope of vision for the field that is limited to his view, he offers readers a revealing essay on the centrality of cognition to understanding complex human behavior – and especially to understanding many of the central aspects of entrepreneurship. This 'big tent' approach to envisioning our field provides an overarching umbrella under whose shelter much that appears in this *Handbook* can be found to benefit.

So then, in extending the metaphor; if we were to gather papers for sessions that contribute to the theory and framing of some of the more concrete conceptualizing, we would attend to Chapters 2–4. The framing task is one of constant creativity, revision and update. In Chapter 2 Alan Carsrud and Malin Brännback continue their linkage-focused work (referenced therein) with a contribution that connects cognitive factors such as intentions and motivations to subsequent behaviors such as goal setting. Chapter 3 has been prepared by Barbara Bird, and in it she makes the case for: (1) the important role that entrepreneurial behavior plays as a concrete outcome: one of cognition's 'most observable concrete outcomes'; and (2) a concerted focus on entrepreneurial behavior as under-researched – in need of taxonomic and measurement contributions to 'true situational theories of the entrepreneur's behavior.' And then in Chapter 4, Brandon Randolph-Seng, Wallace Williams and Mario Hayek import the *non*-conscious cognition literature from the broader social cognition domain to begin to integrate insights within the non-conscious cognition research literature with research in entrepreneurial intentions and intuition. This process of integration permits a broader-scope palate of concepts to become available to entrepreneurial cognition researchers, introducing and situating concepts spanning both the conscious and the non-conscious; and it thereby helps to suggest – in part at least – ways to address the relative neglect of non-conscious cognition theory within the entrepreneurial cognition research domain.

Another cluster of concepts to be found in this *Handbook* centers on research motivated by the continuing scholarly (and popular-press) interest in entrepreneurial affect, for example entrepreneurial emotion and passion. This interface of feeling with thinking is important in several respects – and the contributions in Chapters 5–7 examine some of these.

For example, in Chapter 5, Maw-Der Foo, Charles Murnieks and Elsa Chan examine the *role* of affect – consisting of both emotions *and* moods – on entrepreneurial cognition. Dynamic views of cognition are especially concerned with the source of the dynamism; and understanding the emotional environment within which entrepreneurs face moment-by-moment ups and downs is important to theory building at the feeling/thinking interface. Now consider also the affective/cognitive connection in terms of the forces in play. Such forces are not mono-dimensional. Rather, they exist and exert influence across both time and levels of analysis. In Chapter 6 Denis Grégoire unpacks this bundle of forces, and draws scholarly attention to different classes of affective/cognitive forces in entrepreneurship depending on their enduring versus episodic nature and their plane of influence; and then he illustrates the importance of these distinctions for studying opportunity identification. Another respect in which the interface of feeling with thinking is important concerns context. Mateja Drnovšek, Alenka Slavec and Melissa Cardon, in Chapter 7, advance a persuasive argument for a concept (new to entrepreneurial cognition research) of 'situated emotions.' In this chapter these authors propose and test a culturally situated model that relates entrepreneurial emotions/passion and cognitions/self-efficacy to explore how these factors impact venture performance, and what implications might follow for research and practice.

Of course, as many cognition scholars (entrepreneurship and beyond) are realizing, the 'hardware' that complements the 'software' of human thinking/feeling and acting must also figure prominently in dynamic views of cognition – especially with respect to the notion of embodied cognition: where the body influences the mind. The study of entrepreneurial mental 'hardware' produces the entrepreneurial neuroscience research stream. Chapters 8 and 9 offer assistance to *Handbook* readers in beginning to penetrate this physiologically, conceptually and theoretically complex (but very rich) research domain. Chapter 8, by David Baucus, Melissa Baucus and Ronald Mitchell, offers somewhat of a bridge from the affect-centered research discussion to that of entrepreneurial neuroscience. In this chapter the authors trace in explicit detail the seeing-to-feeling pathways that virtually instantaneously transmit visual stimulus: to produce emotion from within the brains of entrepreneurs. In this process, the extensive physiology involved is plotted for the ambitious and patient reader, as the authors demonstrate how entrepreneurs' brains are physiologically the same as most people's but are different in terms of their experiences and knowledge. Then, in Chapter 9, further helping entrepreneurial neuroscience to become approachable for interested readers (beginning with a primer on brain anatomy and neuroscience methods), Jeffery McMullen,

Matthew Wood and Les Palich partition the entrepreneurial cognition literature into four distinct explanations of the formation and successful implementation of opportunity beliefs; and they further identify a fifth explanation that points to the future of entrepreneurial neuroscience research.

With behavioral, emotional and neurological factors having been offered thus far in this *Handbook* as a platform for penetrating some of the key streams in entrepreneurial cognition research, we may then move, in the next chapter-cluster in this *Handbook*, toward better understanding the shaping of entrepreneurial thought. Chapters 10 and 11 provide these insights. In Chapter 10 Daniel Forbes identifies and situates within the entrepreneurial cognition research literature the new and rich set of learning 'resources that help people acquire entrepreneurship-related knowledge.' In particular he proposes a new way of thinking about the advances in large-scale codification processes (media etc.) and in network formation (markets and other social structures). And because, in part at least, such advanced and specialized symbol systems (as well as even the conceptualization of new opportunity) depend upon language as primary to idea transmission and understanding, we also present Chapter 11, by Jean Clarke and Joep Cornelissen. Their contribution focuses on 'the formative role of language in shaping the ideas of entrepreneurs and their attempts to gain a broader understanding and recognition for a new venture from relevant stakeholders and resource providers.' The provocative idea that language creates opportunities has the scaffolding for such a rationale in this chapter.

When taken as a whole, then, we may say that to this point in the *Handbook* most of the chapters have enabled an understanding of entrepreneurial cognition research for the reader who seeks to become conversant in some of the important conversations under way. A few new or expanded conversations are suggested in the last cluster of chapters: Chapters 12–14. Andrew Corbett suggests in Chapter 12 that, whereas the predominant cognitive perspective extant within both the academic and practitioner world of entrepreneurship research concerns start-ups, this mental representation is limiting in certain respects, and it thus inhibits the development of research targeted toward 'the high and rapid growth ventures that policy makers hope [to stimulate] for their programs and regions.' His essay therefore develops the concept of *entrepreneurial growth cognitions*. Then, in Chapter 13, Ronald Mitchell, Rob Mitchell, Miles Zachary and Michael Ryan further expand the dynamic entrepreneurial cognition conversation by using a simulation to 'disembody' and then 're-embody' a very specific set of exchange cognitions to model the socially situated cognition notion, under highly specified conditions: where dynamism in cognition results from the moment-to-moment interaction of an entrepreneur's inner environment and outer environment – using exchange formation as the relevant outcome. Based on prior probabilities from an experiment (inner environment) and on assumptions developed from the literature (outer environment), these authors explore the boundaries of key theoretical assumptions surrounding opportunity formation. Lastly, in our concluding essay, Chapter 14, Saras Sarasvathy presents ideas that open a multiplicity of pathways for future entrepreneurial cognition research carried forward by the exploration of both formal and informal models as tools for entrepreneurial cognition research. The importance of model veracity is also explained. Then, after the foregoing discussion, the theory of effectuation is utilized as a working example of entrepreneurial-cognition-based model building. As Herbert Simon's last PhD student, Saras is in a unique position to explain how model-building concepts from Simon's notion of 'sciences of the artificial' (1981) can be utilized to unpack behavioral assumptions into models that are high-fidelity embodied representations of entrepreneurs as unique people. We believe this perspective to be highly useful for those looking to the future of entrepreneurial cognition research.

So, like the organizing of an academic conference, the goal of this *Handbook* has therefore been to empower the participant/reader to craft a singular experience from within a multifaceted conceptual environment. It is our hope that this Introduction to the *Handbook* is at least a playing-field leveler; a context provider; an experience creator; and, we hope, even an inclusion enabler for each colleague who responds to this collective invitation into the field of entrepreneurial cognition research. We promise a richly rewarding engagement.

NOTE

1. We offer special thanks to Hamid Vahidnia and Shawna Chen for their invaluable input relating to this introductory chapter, especially in the development of Tables I.1 and I.2 and in comments on earlier drafts.

REFERENCES

Adams, J.K. 1957. Laboratory studies of behavior without awareness. *Psychological Bulletin*, **54**(5): 383–405.

Alvarez, S. and Busenitz, L. 2001. The entrepreneurship of resource-based theory. *Journal of Management*, **27**(6): 755–75.

Andersen, S.A., Moskowitz, G.B., Blair, I.V. and Nosek, B.A. 2007. Automatic thought.

In E.T. Higgins and A. Kruglanski (eds), *Social Psychology: Handbook of basic principles* (2nd edn): 138–75. New York: Guilford Press.
Baker, T. and Nelson, R.E. 2005. Creating something from nothing: Resource construction through entrepreneurial bricolage. *Administrative Science Quarterly*, **50**(3): 329–66.
Balcetis, E. and Dunning, D. 2006. See what you want to see: Motivational influences on visual perception. *Journal of Personality and Social Psychology*, **91**(4): 612–25.
Bargh, J.A. 1992. Does subliminality matter to social psychology? Awareness of the stimulus versus awareness of its influence. In R.F. Bornstein and T.S. Pittman (eds), *Perception without Awareness: Cognitive, clinical, and social perspectives*: 238–55. New York: Guilford Press.
Bargh, J.A. 2004. Being here now: Is consciousness necessary for human freedom? In J. Greenberg, S.L. Koole and T. Pyszczynski (eds), *Handbook of Experimental Existential Psychology*: 385–96. New York: Guilford Press.
Bargh, J.A. 2006. What have we been priming all of these years? On the development, mechanisms, and ecology of nonconscious social behavior. *European Journal of Social Psychology*, **36**(2): 147–68.
Bargh, J.A. 2008a. Free will is un-natural. In J. Baer, J. Kaufman and R. Baumeister (eds), *Are we Free? The psychology of free will*: 128–54. New York: Oxford University Press.
Bargh, J.A. 2008b. Social psychological approaches to consciousness. In P.D. Zelazo, M. Moscovitch and E. Thompson (eds), *The Cambridge Handbook of Consciousness*: 553–67. New York: Cambridge University Press.
Bargh, J.A. and Ferguson, M.J. 2000. Beyond behaviorism: On the automaticity of higher mental processes. *Psychological Bulletin*, **126**(6): 925–45.
Bargh, J.A., Lombardi, W.J. and Higgins, T. 1988. Automaticity of chronically accessible constructs in person X situation effects on person perception: It's just a matter of time. *Journal of Personality and Social Psychology*, **55**(4): 599–605.
Bargh, J.A. and Morsella, E. 2008. The unconscious mind. *Perspectives on Psychological Science*, **3**(1): 73–9.
Bargh, J.A. and Pietromonaco, P. 1982. Automatic information processing and social perception: The influence of trait information presented outside of conscious awareness on impression formation. *Journal of Personality and Social Psychology*, **43**(3): 437–49.
Baron, R.A. 1998. Cognitive mechanisms in entrepreneurship: Why and when enterpreneurs think differently than other people. *Journal of Business Venturing*, **13**(4): 275–94.
Baron, R.A. 2000. Counterfactual thinking and venture formation: The potential effects of thinking about 'what might have been'. *Journal of Business Venturing*, **15**(1): 79–91.
Baron, R.A. 2006. Opportunity recognition as pattern recognition: How entrepreneurs 'connect the dots' to identify new business opportunities. *Academy of Management Perspectives*, **20**(1): 104–19.
Baron, R.A. 2008. The role of affect in the entrepreneurial process. *Academy of Management Review*, **33**(2): 328–40.
Baron, R.A. and Ensley, M. D. 2006. Opportunity recognition as the detection of meaningful patterns: Evidence from comparisons of novice and experienced entrepreneurs. *Management Science*, **52**(9): 1331–44.
Baron, R.A. and Markman, G.D. 2000. Beyond social capital: How social skills can enhance entrepreneurs' success. *Academy of Management Perspectives*, **14**(1): 106–16.
Bartlett, F.C. 1932. *Remembering*. Cambridge, UK: Cambridge University Press.
Baudeau, N. 1910 (1767). *Première introduction à la philosophie économique ou Analyse des états policés*. A. Dubois (ed.). Paris: P. Geuthner.
Bingham, C.B. 2009. Oscillating improvisation: How entrepreneurial firms create success in foreign market entries over time. *Strategic Entrepreneurship Journal*, **3**(4): 321–45.
Bingham, C.B., Eisenhardt, K.M. and Furr, N.R. 2007. What makes a process a capability? Heuristics, strategy, and effective capture of opportunities. *Strategic Entrepreneurship Journal*, **1**(1–2): 27–47.
Bird, B. 1988. Implementing entrepreneurial ideas: The case for intention. *Academy of Management Review*, **13**(3): 442–53.

Bird, B.J. 1992. The operation of intentions in time: The emergence of the new venture. *Entrepreneurship Theory and Practice*, **17**: 11–20.
Bless, H. and Forgas, J.P. 2000. The message within: Toward a social psychology of subjective experiences. In H. Bless and J.P. Forgas (eds), *The Message Within: The role of subjective in social cognition and behavior*: 372–92. Philadelphia, PA: Psychology Press.
Boyd, N.G. and Vozikis, G.S. 1994. The influence of self-efficacy on the development of entrepreneurial intentions and actions. *Entrepreneurship Theory and Practice*, **18**(4): 63–77.
Breuer, J. and Freud, S. 1955 (1893–1895). Studies in hysteria. In J. Strachey (ed.), *The Standard Edition of the Complete Psychological Works of Sigmund Freud*, Vol. 2. London: Hogarth Press.
Brigham, K., De Castro, J. and Shepherd, D. 2007. A person–organization fit model of owner-managers' cognitive style and organizational demands. *Entrepreneurship Theory and Practice*, **31**(1): 29–51.
Brockhaus, R.H., Sr and Horwitz, P.S. 1986. The psychology of the entrepreneur. In D. Sexton and R. Smilor (eds), *The Art and Science of Entrepreneurship*: 25–48. Cambridge, MA: Ballinger.
Bruner, J.S. 1957. On perceptual readiness. *Psychological Review*, **64**(2): 123–52.
Bruner, J.S. 1992. Another look at new look 1. *American Psychologist*, **47**(6): 780–83.
Bryan, W.L. and Harter, L. 1899. Studies on the telegraphic language: The acquisition of a hierarchy of habits. *Psychological Review*, **6**(4): 345–78.
Bull, I. and Willard, G.E. 1993. Towards a theory of entrepreneurship. *Journal of Business Venturing*, **8**(3): 183–95.
Busenitz, L.W. 1992. Cognitive biases in strategic decision-making: Heuristics as a differentiator between managers in large organizations and entrepreneurs, Ph.D. Dissertation, Texas A&M University.
Busenitz, L.W. 1996. Research on entrepreneurial alertness. *Journal of Small Business Management*, **34**(4): 35–44.
Busenitz, L.W. 1999. Entrepreneurial risk and strategic decision making: It's a matter of perspective. *The Journal of Applied Behavioral Science*, **35**(3): 325–40.
Busenitz, L.W. and Arthurs, J.D. 2007. Cognition and capabilities in entrepreneurial ventures. In J.R. Baum, M. Frese and R. Baron (eds), *The Psychology of Entrepreneurship*: 309–30. Mahwah, NJ: Lawrence Erlbaum Associates.
Busenitz, L.W. and Barney, J.B. 1997. Differences between entrepreneurs and managers in large organizations: Biases and heuristics in strategic decision-making, *Journal of Business Venturing*, **12**(1): 9–30.
Busenitz, L.W. and Lau, C.M. 1996. A cross-cultural cognitive model of venture creation: Understanding entrepreneurial and immigrant decision making. *Entrepreneurship Theory and Practice*, **20**(4): 25–37.
Cantillon, R. 1931 (1755). *Essai sur la Nature du Commerce en Général*. H. Huggs (ed. and trans.), London: Macmillan.
Cardon, M.S. 2008. Is passion contagious? The transference of entrepreneurial emotion to employees. *Human Resource Management Review*, **18**(2): 77–86.
Cardon, M.S., Wincent, J., Singh, J. and Drnovšek, M. 2009. The nature and experience of entrepreneurial passion. *Academy of Management Review*, **34**(3): 511–32.
Cardon, M.S., Zietsma, C., Saparito, P., Matherne, B.P. and Davis, C. 2005. A tale of passion: New insights into entrepreneurship from a parenthood metaphor. *Journal of Business Venturing*, **20**(1): 23–45.
Chandler, G.N. and Jansen, E. 1992. The founder's self-assessed competence and venture performance. *Journal of Business Venturing*, **7**(3): 223–36.
Chen, C.C., Greene, P.G. and Crick, A. 1998. Does entrepreneurial self-efficacy distinguish entrepreneurs from managers? *Journal of Business Venturing*, **13**(4): 295–316.
Chiao, J.Y., Heck, H.E., Nakayama, K. and Ambady, N. 2006. Priming race in biracial observers affects visual search for black and white faces. *Psychological Science*, **17**(5): 387–92.
Chomsky, N. 1959. A review of BF Skinner's *Verbal Behavior*. *Language*, **35**(1): 26–58.

Cooper, A.C., Folta, T.B. and Woo, C. 1995. Entrepreneurial information search. *Journal of Business Venturing*, **10**(2): 107–20.
Cooper, A., Woo, C. and Dunkelberg, W. 1988. Entrepreneurs' perceived chances for success. *Journal of Business Venturing*, **3**(2): 97–108.
Corbett, A.C. 2005. Experiential learning within the process of opportunity identification and exploitation. *Entrepreneurship Theory and Practice*, **29**(4): 473–91.
Corbett, A.C. 2007. Learning asymmetries and the discovery of entrepreneurial opportunities. *Journal of Business Venturing*, **22**(1): 97–118.
Corbett, A.C. and Hmieleski, K.M. 2007. The conflicting cognitions of corporate entrepreneurs. *Entrepreneurship Theory and Practice*, **31**(1): 105–24.
Corbett, A.C., Neck, H.D. and DeTienne, D.R. 2007. How corporate entrepreneurs learn from fledgling initiatives: Entrepreneurial cognition and the development of a termination script. *Entrepreneurship: Theory and Practice*, **31**(6): 829–52.
Corman, J., Perles, B. and Vancini, P. 1988. Motivational factors influencing high-technology entrepreneurship. *Journal of Small Business Management*, **26**(1): 36–42.
Cornelissen, J.P. and Clarke, J.S. 2010. Imaging and rationalizing opportunities: Inductive reasoning and the creation and justification of new ventures. *Academy of Management Review*, **35**(4): 539–57.
Costall, A. 2006. 'Introspectionism' and the mythical origins of scientific psychology. *Consciousness and Cognition*, **15**(4): 634–54.
Danhof, C.H. 1949. Observations on entrepreneurship in agriculture. In A.H. Cole (ed.), *Change and the Entrepreneur*: 20–24. Cambridge, MA: Harvard University Press.
Debner, J.A. and Jacoby, L.L. 1994. Unconscious perception: Attention, awareness, and control. *Journal of Experimental Psychology: Learning, Memory, and Cognition*, **20**(2): 304–17.
Devine, P.G. 1989. Stereotype and prejudice: Their automatic and controlled components. *Journal of Personality and Social Psychology*, **56**(1): 5–18.
Dew, N., Read, S., Sarasvathy, S.D. and Wiltbank, R. 2009. Effectual versus predictive logics in entrepreneurial decision-making: Differences between experts and novices. *Journal of Business Venturing*, **24**(4): 287–309.
Dewing, A.S. 1920. *The Financial Policy of Corporations* (Vol. 1). New York: Ronald Press.
Dijksterhuis, A., Aarts, H. and Smith, P.K. 2005. The power of the subliminal: On subliminal persuasion and other potential applications. In R. Hassin, J. Uleman and J.A. Bargh (eds), *The New Unconscious*: 77–106. New York: Oxford University Press.
Dijksterhuis, A. and Bargh, J.A. 2001. The perception–behavior expressway: Automatic effects of social perception and social behavior. In M. Zanna (ed.), *Advances in Experimental Social Psychology*, vol. 30: 1–40. New York: Academic Press.
Dimov, D. 2007a. Beyond the single-person, single-insight attribution in understanding entrepreneurial opportunities. *Entrepreneurship Theory and Practice*, **31**(5): 713–31.
Dimov, D. 2007b. From opportunity insight to opportunity intention: The importance of person–situation learning match. *Entrepreneurship Theory and Practice*, **31**(4): 561–83.
Dixon, N.F. 1981. *Preconscious Processing*. New York: Wiley.
Erdelyi, M.H. 1992. Psychodynamics and the unconscious. *American Psychologist*, **47**(6): 784–87.
Eriksen, C.W. 1958. Unconscious processes. In M.R. Jones (ed.), *Nebraska Symposium on Motivation*: 169–227. Lincoln, NE: University of Nebraska Press.
Evans Jr, G.H. 1957. A century of entrepreneurship in the United States with emphasis upon large manufacturing concerns c. 1850–1957. *Explorations in Entrepreneurial History*, **10**: 90–103.
Ferguson, M.J. 2007. On the automatic evaluation of end-states. *Journal of Personality and Social Psychology*, **92**(4): 596–611.
Ferguson, M.J. and Bargh, J.A. 2003. The constructive nature of automatic evaluation. In J. Musch and K.C. Klauer (eds), *The Psychology of Evaluation: Affective processes in cognition and emotion*: 169–88. Mahwah, NJ: Erlbaum.

Fiet, J.O. 2002. *The Systematic Search for Entrepreneurial Discoveries*. Westport, CT: Quorum Books.

Fishbach, A. and Ferguson, M.J. 2007. The goal construct in social psychology. In E.T. Higgins and A. Kruglanski (eds), *Social Psychology: Handbook of basic principles* (2nd edn): 490–515. New York: Guilford Press.

Foo, M.-D., Uy, M.A. and Baron, R.A. 2009. How do feelings influence effort? An empirical study of entrepreneurs' affect and venture effort. *Journal of Applied Psychology*, **94**(4): 1086–94.

Forbes, D.P. 2005. Are some entrepreneurs more overconfident than others? *Journal of Business Venturing*, **20**(5): 623–40.

Gaglio, C.M. 2004. The role of mental simulations and counterfactual thinking in the opportunity identification process. *Entrepreneurship Theory and Practice*, **28**(6): 533–52.

Gaglio, C.M. and Katz, J. 2001. The psychological basis of opportunity identification: Entrepreneurial alertness. *Small Business Economics*, **16**(2): 95–111.

Gatewood, E.J., Shaver, K.G. and Gartner, W.B. 1995. A longitudinal study of cognitive factors influencing start-up behaviors and success at venture creation. *Journal of Business Venturing*, **10**(5): 371–91.

Goldiamond, I. 1958. Indicators of perception: I. Subliminal perception, subception, unconscious perception: An analysis in terms of psychophysical indicator methodology. *Psychological Bulletin*, **55**(6): 373–411.

Gollwitzer, P.M. and Moskowitz, G.B. 1996. Goal effects on action and cognition. In E.T. Higgins and A.W. Kruglanski (eds), *Social Psychology: Handbook of basic principles*: 361–99. New York: Guilford Press.

Greenwald, A.G. 1992. New look 3: Unconscious cognition reclaimed. *American Psychologist*, **47**(6): 766–79.

Grégoire, D.A., Barr P.S. and Shepherd, D.A. 2010. Cognitive processes of opportunity recognition: The role of structural alignment. *Organization Science*, **21**(2): 413–31.

Grégoire, D.A., Corbett, A. and McMullen, J. 2011. The cognitive perspective in entrepreneurship: An agenda for future research. *Journal of Management Studies*, **48**(6): 1443–77.

Hansen, E.L. and Allen, K.R. 1992. The creation corridor: Environmental load and pre-organization information-processing ability. *Entrepreneurship Theory and Practice*, **17**: 57–65.

Haynie, J.M., Shepherd, D., Mosakowski, E. and Early, P.C. 2010. A situated metacognitive model of the entrepreneurial mindset. *Journal of Business Venturing*, **25**(2): 217–29.

Haynie, J.M., Shepherd, D.A. and Patzelt, H. 2012. Cognitive adaptability and an entrepreneurial task: The role of metacognitive ability and feedback. *Entrepreneurship Theory and Practice*, **36**(2): 237–65.

Hebb, D. 1949. *The Organization of Behavior: A neuropsychological theory*. New York: Wiley.

Hébert, R.F. and Link, A.N. 1989. In search of the meaning of entrepreneurship. *Small Business Economics*, **1**(1): 39–49.

Helmholtz, H. 1968 (1867). Concerning the perceptions in general. In W. Warren and R. Warren (eds), *Helmholtz on Perception: Its physiology and development*: 171–203. New York: Wiley.

Higgins, E.T. 1996. Knowledge activation: Accessibility, applicability, and salience. In E.T. Higgins and A.W. Kruglanski (eds), *Social Psychology: Handbook of basic principles*: 133–68. New York: Guilford Press.

Hisrich, R.D. and Jankowicz, A.D. 1990. Intuition in venture capital decisions: An exploratory study using a new technique. *Journal of Business Venturing*, **5**(1): 49–62.

Hmieleski, K.M. and Corbett, A.C. 2008. The contrasting interaction effects of improvisational behavior with entrepreneurial self-efficacy on new venture performance and entrepreneur work satisfaction. *Journal of Business Venturing*, **23**(4): 482–96.

Ireland, R.D. and Webb, J.W. 2007. A cross-disciplinary exploration of entrepreneurship research. *Journal of Management*, **33**(6): 891–927.

Isaacowitz, D.M. 2006. Motivated gaze: The view from the gazer. *Current Directions in Psychological Science*, **15**(2): 68–72.
James, W. 1950 (1890). *The Principles of Psychology (Vol. 1)*. New York: Dover.
Jastrow, J. 1906. *The Subconscious*. Boston, MA: Houghton-Mifflin.
Jenkins, M. and Johnson, G. 1997. Entrepreneurial intentions and outcomes: A comparative causal mapping study. *Journal of Management Studies*, **34**(6): 895–920.
Jonas, K.J., and Sassenberg, K. 2006. Knowing how to react: Automatic response priming from social categories. *Journal of Personality and Social Psychology*, **90**(5): 709–21.
Kahneman, D. and Treisman, A. 1984. Changing views of attention and automaticity. In R. Parasuraman and R. Davies (eds), *Varieties of Attention*: 29–61. New York: Academic Press.
Kaish, S. and Gilad, B. 1991. Characteristics of opportunities search of entrepreneurs versus executives: Sources, interests, general alertness. *Journal of Business Venturing*, **6**(1): 45–61.
Karremans, J.C., Stroebe, W. and Claus, J. 2006. Beyond Vicary's fantasies: The impact of subliminal priming and brand choice. *Journal of Experimental Social Psychology*, **42**(6): 792–8.
Katz, J.A. 1992. A psychosocial cognitive model of employment status choice. *Entrepreneurship Theory and Practice*, **17**: 29–37.
Kay, A.C., Wheeler, C., Bargh, J.A. and Ross, L. 2004. Material priming: The influence of mundane physical objects on situational construal and competitive behavioral choice. *Organizational Behavior and Human Decision Processes*, **95**(1): 83–96.
Kelly, G.A. 1955. *The Psychology of Personal Constructs*. New York: Norton.
Kihlstrom, J.F., Barnhardt, T.M. and Tataryn, D.J. 1992. Implicit perception. In R.F. Bornstein and T.S. Pittman (eds), *Perception without Awareness: Cognitive, clinical, and social perspectives*: 17–54. New York: Guilford Press.
Kirzner, I.M. 1973. *Competition and Entrepreneurship*. Chicago, IL: University of Chicago Press.
Koivisto, M. and Revonsuo, A. 2007. How meaning shapes seeing. *Psychological Science*, **18**(10): 845–9.
Kolvereid, L. 1996. Prediction of employment status choice intentions. Working paper, Henley Management College.
Krueger, N.F. 1993. The impact of prior entrepreneurial exposure on perceptions of new venture feasibility and desirability. *Entrepreneurship Theory and Practice*, **18**(1): 5–21.
Krueger, N.F. 2000. The cognitive infrastructure of opportunity emergence. *Entrepreneurship Theory and Practice*, **24**(3): 5–23.
Krueger, N.F. and Brazeal, D. 1994. Entrepreneurial potential and potential entrepreneurs. *Entrepreneurship Theory and Practice*, **18**(1): 91–104.
Krueger, N.F. and Carsrud, A.L. 1993. Entrepreneurial intentions: Applying the theory of planned behavior. *Entrepreneurship and Regional Development*, **5**(4): 315–30.
Krueger, N.F., and Dickson, P.R. 1994. How believing in ourselves increases risk taking: Perceived self-efficacy and opportunity recognition. *Decision Sciences*, **25**(3): 385–400.
Kruglanski, A.W., Thompson, E.P., Higgins, E.T., Atash, M.N., Pierro, A., Shah, J.Y. and Spiegel, S. 2000. To 'Do the right thing' or to 'Just do it': Locomotion and assessment as distinct self-regulatory imperatives. *Journal of Personality and Social Psychology*, **79**(5): 793–815.
Kunda, Z. and Spencer, S.J. 2003. When do stereotypes come to mind and when do they color judgment? A goal-based theoretical framework for stereotype activation and application. *Psychological Bulletin*, **129**(4): 522–44.
Lamb, R.K. 1952. The entrepreneur and the community. *Men in Business*: 91–119.
Lashley, K.S. 1951. The problem of serial order in behavior. In L.A. Jeffress (ed.), *Hixon Symposium on Cerebral Mechanisms in Behavior* (112–46). New York: John Wiley & Sons.
Leibniz, G.W. 1981 (1704). *New Essays on Human Understanding* (P. Remnant and J. Benett, trans. and eds). Cambridge, UK: Cambridge University Press.
Logan, G.D. and Cowan, W.B. 1984. On the ability to inhibit thought and action: A theory of an act of control. *Psychological Review*, **91**(3): 295–327.

Manimala, M.J. 1992. Entrepreneurial heuristics: A comparison between high PI (pioneering–innovative) and low PI ventures. *Journal of Business Venturing*, **7**(6): 477–504.
Marcel, A. 1983. Conscious and unconscious perception: Experiments on visual masking and word recognition. *Cognitive Psychology*, **15**(2): 197–237.
Markman, G.D. and Baron, R.A. 2003. Person–entrepreneurship fit: Why some people are more successful as entrepreneurs than others. *Human Resource Management Review*, **13**(2): 281–301.
Masling, J.M. 1992. What does it mean? In R.F. Bornstein and T.S. Pittman (eds), *Perception without Awareness: Cognitive, clinical, and social perspectives*: 261–75. New York: Guilford Press.
McCarthy, A.M., Schoorman, F.D. and Cooper, A.C. 1993. Reinvestment decisions by entrepreneurs: Rational decision-making or escalation of commitment? *Journal of Business Venturing*, **8**(1): 9–24.
McClelland, D.C. 1965. Need achievement and entrepreneurship: A longitudinal study. *Journal of Personality and Social Psychology*, **1**(4): 389–92.
McGinnies, E. 1949. Emotionality and perceptual defense. *Psychological Review*, **56**(5): 244–51.
McGrath, R.G. 1999. Falling forward: Real options reasoning and entrepreneurial failure. *Academy of Management Review*, **24**(1): 13–30.
McGrath, R.G. and MacMillan, I.C. 1992. More like each other than anyone else? A cross-cultural study of entrepreneurial perceptions. *Journal of Business Venturing*, **7**(5): 419–29.
McGuire, J.W. 1964. *Theories of Business Behavior*. Englewood Cliffs, NJ: Prentice-Hall.
Mead, G.H. 1967 (1934). *Mind, Self, and Society: From the standpoint of a social behaviorist*. Chicago, IL: University of Chicago Press.
Merikle, P.M. and Reingold, E.M. 1992. Measuring unconscious perceptual processes. In R.F. Bornstein and T.S. Pittman (eds), *Perception without Awareness: Cognitive, clinical, and social perspectives*: 55–80. New York: Guilford Press.
Mitchell, J.R., Friga, P.N. and Mitchell, R.K. 2005. Untangling the intuition mess: Intuition as a construct in entrepreneurship research. *Entrepreneurship Theory and Practice*, **29**(6): 653–79.
Mitchell, R.K. 1994. The composition, classification, and creation of new venture formation expertise. Unpublished doctoral dissertation, University of Utah.
Mitchell, R.K. 1997. Oral history and expert scripts: Demystifying the entrepreneurial experience. *International Journal of Entrepreneurial Behaviour & Research*, **3**(2): 122–39.
Mitchell, R.K. 2005. Tuning up the global value creation engine: The road to excellence in international entrepreneurship education. In J.A. Katz and D. Shepherd (eds), *Cognitive Approaches to Entrepreneurship Research: Advances in entrepreneurship, firm emergence and growth*, vol. 8: 185–248. Greenwich, CT: JAI Press.
Mitchell, R.K., Busenitz, L.W., Bird, B., Gaglio, C.M., McMullen, J.S., Morse, E.A. and Smith, J.B. 2007. The central question in entrepreneurial cognition research 2007. *Entrepreneurship Theory and Practice*, **31**(1): 1–27.
Mitchell, R.K., Busenitz, L.W., Lant, T., McDougall, P.P., Morse, E.A. and Smith, J.B. 2002. Toward a theory of entrepreneurial cognition: Rethinking the people side of entrepreneurship research. *Entrepreneurship Theory and Practice*, **27**(2): 93–104.
Mitchell, R.K., Busenitz, L.W., Lant, T., McDougall, P.P., Morse, E.A. and Smith, J.B. 2004. The distinctive and inclusive domain of entrepreneurial cognition research. *Entrepreneurship Theory and Practice*, **28**(6): 505–18.
Mitchell, R.K. and Chesteen, S. 1995. Enhancing entrepreneurial expertise: Experiential pedagogy and the new venture expert script. *Simulation and Gaming*, **26**(3): 288–306.
Mitchell, R.K., Mitchell, B.T. and Mitchell, J.R. 2009. Entrepreneurial scripts and entrepreneurial expertise: The information processing perspective. In A. Carsrud and M. Brännback (eds), *Understanding the Entrepreneurial Mind: Opening the black box*: 97–140. Heidelberg: Springer.
Mitchell, R.K., Randolph-Seng, B. and Mitchell, J.R. 2011. Socially situated cognition:

Imagining new opportunities for entrepreneurship research (Dialogue). *Academy of Management Review*, **36**(4): 774–8.
Mitchell, R.K., Smith, J.B., Morse, E.A., Seawright, K.W., Peredo, A.M. and McKenzie, B. 2002. Are entrepreneurial cognitions universal? Assessing entrepreneurial cognitions across cultures. *Entrepreneurship Theory and Practice*, **26**(4): 9–32.
Mitchell, R.K., Smith, J.B., Seawright, K.W. and Morse, E.A. 2000. Cross-cultural cognitions and the venture creation decision. *Academy of Management Journal*, **43**(5): 974–93.
Monteith, M.J., Sherman, J.W. and Devine, P.G. 1998. Suppression as a stereotype control strategy. *Personality and Social Psychology Review*, **2**(1): 63–82.
Moskowitz, G.B., Skurnik, I. and Galinsky, A.D. 1999. The history of dual-process notions, and the future of preconscious control. In S. Chaiken and Y. Trope (eds), *Dual-process Theories in Social Psychology*: 12–36. New York: Guilford Press.
Neal, D.T., Wood, W. and Quinn, J.M. 2006. Habits – A repeat performance. *Current Directions in Psychological Science*, **15**(4): 198–202.
Neisser, U. 1967. *Cognitive Psychology*. East Norwalk, CT: Appelton-Century-Crofts.
Nosek, B.A. 2007. Implicit–explicit relations. *Current Directions in Psychological Science*, **16**(2): 65–9.
Overskeid, G. 2007. Looking for Skinner and finding Freud. *American Psychologist*, **62**(6): 590–95.
Palich, L.E. and Bagby, D.R. 1995. Using cognitive theory to explain entrepreneurial risk-taking: Challenging conventional wisdom, *Journal of Business Venturing*, **10**(6): 425–38.
Palmer, M. 1971. The application of psychological testing to entrepreneurial potential. *California Management Review*, **13**(3): 32–8.
Pech, R.J. and Cameron, A. 2006. An entrepreneurial decision process model describing opportunity recognition. *European Journal of Innovation Management*, **9**(1): 61–78.
Peirce, C.S. and Jastrow, J. 1884. On small differences in sensations. *Memoirs of the National Academy of Science*, **3**: 75–83.
Pineda, R.C., Lerner, L.D., Miller, M.C. and Phillips, S.J. 1998. An investigation of factors affecting the information-search activities of small business managers. *Journal of Small Business Management*, **36**(1): 60–71.
Pyszczynski, T., Greenberg, J. and Koole, S. L. 2004. Experimental existential psychology: Exploring the human confrontation with reality. In J. Greenberg, S. L. Koole and T. Pyszczynski (eds), *Handbook of Experimental Existential Psychology*: 3–9. New York: Guilford Press.
Rasinski, K.A., Visser, P.S., Zagatsky, M. and Rickett, E.M. 2005. Using implicit goal priming to improve the quality of self-report data. *Journal of Experimental Social Psychology*, **41**(3): 321–7.
Ratelle, C.F., Baldwin, M.W. and Vallerand, R.J. 2005. On the cued activation of situational motivation. *Journal of Experimental Social Psychology*, **41**(5): 482–7.
Read, S., Dew, N., Sarasvathy, S.D., Song, M. and Wiltbank, R. 2009. Marketing under uncertainty: The logic of an effectual approach. *Journal of Marketing*, **73**(3): 1–18.
Read, S., Song, M. and Smit, W. 2009. A meta-analytic review of effectuation and venture performance. *Journal of Business Venturing*, **24**(6): 573–87.
Regan, J.E. 1981. Automaticity and learning: Effects of familiarity on name letters. *Journal of Experimental Psychology: Human Perception and Performance*, **7**(1): 180–95.
Reuber, A.R. and Fischer, E.M. 1994. Entrepreneurs' experience, expertise, and the performance of technology-based firms. *Engineering Management, IEEE Transactions on*, **41**(4): 365–74.
Robinson, P.B., Stimpson, D.V., Huefner, J.C. and Hunt, H.K. 1991. An attitude approach to the prediction of entrepreneurship. *Entrepreneurship Theory and Practice*, **15**(4): 13–31.
Ruiz, R.A. and Krauss, H.H. 1968. Perceptual defense versus response suppression. *Journal of Psychology*, **69**(1): 33–7.
Sarasvathy, S.D. 2001. Causation and effectuation: Toward a theoretical shift from economic inevitability to entrepreneurial contingency. *Academy of Management Review*, **26**(2): 243–63.

Say, J.-B. 1971 (1821). *A Treatise on Political Economy*. New York: Augustus M. Kelley.
Schultz, T.W. 1975. The value of the ability to deal with disequilibria. *Journal of Economic Literature*, **13**(3): 827–46.
Semin, G.R. and Smith, E.R. 2013. Socially situated cognition in perspective. *Social Cognition*, **31**(2): 125–46.
Shaver, K. and Scott, L. 1991. Person, process, choice: The psychology of new venture creation. *Entrepreneurship Theory and Practice*, Winter: 23–45.
Shepherd, D.A. and Cardon, M.S. 2009. Negative emotional reactions to project failure and the self-compassion to learn from the experience. *Journal of Management Studies*, **46**(6): 923–49.
Shepherd, D.A., and DeTienne, D.R. 2005. Prior knowledge, potential financial reward, and opportunity identification. *Entrepreneurship Theory and Practice*, **29**(1): 91–112.
Sidis, B. 1898. *The Psychology of Suggestion*. New York: D. Appleton and Company.
Simon, H.A. 1981. *The Sciences of the Artificial*. Cambridge, MA: The MIT Press. Further editions in 1982 and 1996.
Simon, M. and Houghton, S. 2002. The relationship among biases, misperceptions, and the introduction of pioneering products: Examining differences in venture decision contexts. *Entrepreneurship Theory and Practice*, **27**(2): 105–24.
Simon, M., Houghton, S.M. and Aquino, K. 2000. Cognitive biases, risk perception, and venture formation: How individuals decide to start companies. *Journal of Business Venturing*, **15**(2): 113–34.
Skinner, B.F. 1974. *About Behaviorism*. New York: Random House.
Smeltzer, L.R., Fann, G.L. and Nikolaisen, V.N. 1988. Environmental scanning practices in small business. *Journal of Small Business Management*, **26**(3): 55–62.
Smith, E.R. and Semin, G.R. 2004. Socially situated cognition: Cognition in its social context. *Advances in Experimental Social Psychology*, **36**: 53–117.
Smith, E.R. and Semin, G.R. 2007. Situated social cognition. *Current Directions in Psychological Science*, **16**(3): 132–5.
Smith, J.B., Mitchell, J.R. and Mitchell, R.K. 2009. Entrepreneurial scripts and the new transaction commitment mindset: Extending the expert information processing theory approach to entrepreneurial cognition research. *Entrepreneurship Theory and Practice*, **33**(4): 815–44.
Smith, K.G., Gannon, M.J., Grimm, C. and Mitchell, T.R. 1988. Decision making behavior in smaller entrepreneurial and larger professionally managed firms. *Journal of Business Venturing*, **3**(3): 223–32.
Solomons, L. and Stein, G. 1896. Normal motor automatism. *Psychological Review*, **3**(5): 495–512.
Strahan, E.J., Spencer, S.J. and Zanna, M.P. 2002. Subliminal priming and persuasion: Striking while the iron is hot. *Journal of Experimental Social Psychology*, **38**(6): 556–68.
Tetlock, P.E. 2002. Social functionalist framework for judgment and choice: Intuitive politicians, theologians, and prosecutors. *Psychological Review*, **109**(3): 451–71.
Thompson, E.P., Roman, R.J., Moskowitz, G.B., Chaiken, S. and Bargh, J.A. 1994. Accuracy motivation attenuates covert priming: The systematic reprocessing of social information. *Journal of Personality and Social Psychology*, **66**(3): 479–89.
Ucbasaran, D., Westhead, P. and Wright, M. 2009. The extent and nature of opportunity identification by experienced entrepreneurs. *Journal of Business Venturing*, **24**(2): 99–115.
Watson, J.B. 1913. Psychology as the behaviorist views it. *Psychological Review*, **20**(2): 158–77.
Wegner, D.M. and Bargh, J.A. 1998. Control and automaticity in social life. In D.T. Gilbert, S.T. Fiske and G. Lindzey (eds), *Handbook of Social Psychology* (4th edn): 446–96. Boston: McGraw-Hill.
Whyte, L.L. 1960. *The Unconscious before Freud*. New York: Basic Books.
Wittenbrink, B. 2007. Measuring attitudes through priming. In B. Wittenbrink and N. Schwarz (eds), *Implicit Measures of Attitudes*: 17–58. New York: Guilford Press.
Wright, M., Hoskisson, R.E., Busenitz, L.W. and Dial, J. 2000. Entrepreneurial growth

through privatization: The upside of management buyouts. *Academy of Management Review*, **25**(3): 591–601.

Zajonc, R.B. 1962. Response suppression in perceptual defense. *Journal of Experimental Psychology*, **64**(3): 206–14.

Zeigarnik, B. 1927. Das Behalten erledigter und unerledigter Handlungen [The retention of completed and uncompleted actions]. *Psychologische Forschung*, **9**(1): 1–85.

1. Thinking about cognition and its central role in entrepreneurship: confessions of a 'reformed' behaviorist

Robert A. Baron

ABSTRACT

Cognition is now widely recognized as an important key to understanding central aspects of entrepreneurship – everything from opportunity recognition and alertness, on the one hand, through planning, improvisation, strategy formulation, and even the decision to exit from a new venture, on the other. However, this was not always the case; the present author started his own career as a 'hard-nosed behaviorist,' believing that cognition was unobservable and therefore not an appropriate topic for research. Only gradually did he come to recognize its central importance in virtually every aspect of complex human behavior, including entrepreneurship. In order to investigate its role in entrepreneurship, we can draw on several different sources of knowledge (e.g., input from successful entrepreneurs). Among these, however, systematic research offers the most effective approach. Future studies can draw on current knowledge of cognition to investigate a myriad of intriguing issues, including the role of self-regulation in entrepreneurs' success, and the role of several cognitive factors in entrepreneurs' high levels of philanthropy – their widely recognized tendency to 'give back' by sharing their wealth with others (e.g., universities, charitable organizations).

> There's nothing better [than] when something comes and hits you and you think 'YES!'
>
> J.K. Rowling

> I never made one of my discoveries through the process of rational thinking.
>
> Albert Einstein

INTRODUCTION – AND MY FIRST EPIPHANY

Cognition – the dictionary defines it as 'the mental faculty or process of acquiring knowledge by the use of reasoning, intuition, or perception,' and in fact, that's close to what we usually mean when we use this term in entrepreneurship: basically, it refers to a very broad range of processes involved in perceiving, noticing, storing (retaining), retrieving, processing and using information – input from the world around us or, in many instances, our own internal systems and states. In this introductory chapter, I will try to explain why cognition underlies – or at least plays a key role in – virtually every aspect of entrepreneurship we wish to study and to understand. So the words that follow are indeed those of a 'true believer' – someone who does often 'think about cognition' and how it can help us gain new insights into the focus of our field, which, at its core, seeks to understand how individuals (or teams of individuals) convert their ideas and vision into reality – how they perform the complex array of tasks involved in 'making the *possible* real' (Baron, 2013).

Before proceeding, though, I want to note that, in one sense, it is somewhat ironic that I have been invited by my colleagues to prepare this opening chapter. Why? Because I was not always convinced that cognition was an important key to understanding entrepreneurship – or any other complex form of human behavior. On the contrary, it was only through an epiphany, or series of personal epiphanies, that I came to hold this view. So, like J.K. Rowling and Einstein (!), I was enlightened along the way!

I now comment briefly on this personal journey and how it unfolded. The story begins many years ago, when I was a graduate student in psychology. My interests had always been in the 'hard, scientific' side of that field – I had no desire to become a therapist. Rather, I was a true behaviorist, which meant, at the time, that I accepted the view that psychology should focus only on overt behavior – actions that can be directly observed and measured. Consistent with this view, I applied only to graduate programs that were known for this focus and was very pleased when I was accepted into the program in experimental psychology at Iowa – and, moreover, was offered an assistantship with Kenneth W. Spence, who, at the time, was second only to B.F. Skinner in terms of prominence among behaviorists. I knew Spence's work and theories, and dimly understood that there was little room for cognition in them; in fact, aside from some work on the nature of memory (which, somehow, *was* acceptable to behaviorists at the time), the very idea of trying to study human thought was viewed as a total 'dead end,' one that would lead the science of psychology badly astray.

So, filled with visions of working with one of the true 'greats' in the field,

I set off for Iowa City. Alas! Life is filled with surprises, and a very big one awaited me. I arrived on campus only to learn that Professor Spence had left suddenly that summer, following a set of unhappy events best left unspecified here. Whatever the precise reasons, though, there I was, 'high and dry,' with my plans dashed on the rocks of academic politics. What should I do? Head for another university? That would mean being out of school for at least a semester, perhaps a year; or stay, and try to make the best of a bad situation? This, I guess, is where fate – or my true, underlying interests – intervened to start me on the road toward personal enlightenment, and appreciation of the central importance of cognition.

I decided to shift into the Social Psychology program because – somewhat unpredictably – I was quite interested in the 'new' theory at the time – cognitive dissonance. Expressing that interest almost ended my graduate career, however, because it was simply too cognitive for the faculty of that behaviorist-oriented department! As things turned out, I never did do research on cognitive dissonance. Instead, inspired by the work of Albert Bandura on observational learning and modeling, I developed interests in a variable well known to entrepreneurship researchers today – locus of control. Placing this in a social-learning context, I developed several ideas (very naïve at the time) to the effect that people who believe that they can control their own outcomes will learn more readily from others (i.e., social learning) than those who do not. That was the topic of my dissertation, and it yielded weak and inconsistent findings – so inconclusive that I never returned to this line of research (although, many years later, this basic idea was much more fully developed by Martin Seligman, under the heading of 'learned helplessness'). I then shifted my focus entirely, and embarked on the study of human aggression – a topic I pursued for more than 20 years. But looking back, it seems clear that locus of control was basically a cognitive variable, focusing on what people believed they could, and could not, control or influence. So even in my dissertation research, I was well on the way toward my first epiphany: the idea that cognition does indeed matter, and must, in fact, be a central focus of efforts to understand human behavior (see also Bird, Chapter 3, this volume).

After graduation, and as I launched my research on aggression, I gradually began to see that cognition played a key role in such behavior. For instance, I soon learned that how people respond to a provocative action by another depends strongly on the perceived causes of the provocation. Was it intentional? Accidental? Not surprisingly, I repeatedly found that the more such actions were seen as intentional, the stronger the anger – and aggression – they provoked. In short, it was not only the provocative action (e.g., insult) from another person that was important; the recipient's interpretation of its causes mattered, too. As Kelly Shaver and others noted many years later, attributions – our explanations for others' behavior – also play a key role in certain aspects of entrepreneurship, but at the time I was unaware of the field of entrepreneurship (which had not yet emerged), and was incapable of making such connections. Other studies I performed also pointed to the importance of cognitive processes, but in the interests of brevity, I shall not describe them here. I think it is sufficient to say that, gradually, I began to recognize just how central cognition is to almost everything we do, think or feel, and that we *cannot* understand any form of complex human behavior without understanding its cognitive foundations.

MY SECOND EPIPHANY: THE 'COGNITIVE REVOLUTION' IN SOCIAL PSYCHOLOGY AND WHERE IT LED ME

All lives are marked by 'turning points' – events that, looking back, had a strong impact on what followed. One of these happened for me in the late 1970s and early 1980s. At that time, the field of social psychology (my 'home base' then) was shaken by what later became known as the 'cognitive revolution.' The major theme of this 'revolution' – which was a kind of paradigm shift in the field – was simply this: 'We cannot understand human social behavior, and all its complexities, unless we include how people think – what and how they think about others, social situations, and themselves. Lacking such information, we will never gain the insights we seek.'

As this view took hold, the major journals in the field shifted their priorities, so that, soon, only papers that incorporated this new perspective had a chance of gaining acceptance. At first, I struggled to understand this shift and how it related to my own work, but I had one very strong incentive to do this quickly: I was then (and still am) the co-author of a textbook on social psychology (now in its 13th edition), and my co-author, Donn Byrne, being my senior, kindly delegated the task of writing about this new cognitive perspective to me! I proceeded by contacting the leading scholars in social cognition (e.g., Susan Taylor), and, with their expert help, managed to gain a basic understanding of what she and others were attempting to accomplish. The result? I soon found that I agreed strongly with the suggestion that behavior was, and almost always is, intimately linked with cognition. In other words, how and what people do depends, to a major extent, on what and how they think.

Up until that time, I had pursued a career-long interest in affect – the role of feelings, moods and emotions in many forms of behavior. But now I came to realize that these reactions, too, were strongly and continuously related to cognition (see also Foo, Murnieks and Chan, Chapter 5, this volume; Grégoire, Chapter 6, this volume). As I put it in my social psychology text, 'Feelings influence thought and thought shapes feelings.'

Perhaps the theoretical framework that most helped me make this transition was one I still draw on today, Bandura's social cognitive theory. I had been strongly influenced by Bandura's research on observational learning and modeling (many of my earliest papers were on those topics), but now, when I read his theory (e.g., Bandura, 1986, 1997), I found I resonated strongly with the perspective he proposed. Briefly, Bandura suggested that the relationships between cognitive processes, environmental conditions and complex human behavior are reciprocal in nature. Cognitive processes shape overt actions, and are, in turn, strongly influenced by these actions. Further, these complex reciprocal relationships occur against a backdrop of moderating and mediating environmental variables (Hmieleski and Baron, 2009). As Bandura (1997, p. 25) puts it: 'What people think, believe, and feel affects how they behave. The . . . effects of their actions, in turn, partly determine their thoughts and affective reactions.' Further, he adds (p. 129): 'People regulate their level and distribution of effort in accordance with the effects they *expect* their actions to have. As a result, their behavior is better predicted by their beliefs than from the actual consequences of their actions.'

Looking back, I cannot imagine a theoretical framework more helpful in prying me loose from my behaviorist origins, and broadening my personal perspective. And, reading this theory, I also came to realize that, as researchers, we should feel free to borrow ideas, theories and research tools from other fields; doing so is certainly preferable to reinventing the wheel – working hard to generate new theoretical frameworks and principles that are, in fact, already available (see, e.g., Shane, 2012; Venkataraman, Sarasvathy, Dew and Forster, 2012, for additional comments on these issues).

From that point on, I began to build cognition into all of my research and theorizing. For instance, when I was shifting from Social Psychology into Industrial/Organizational Psychology, one of my interests was negative feedback, and how to use it – and receive it – effectively (e.g., Baron, 1988, 1990). The results helped me to realize that when people receive criticism from others, their interpretations of the motives behind such feedback are crucial in determining how they respond. Moreover, it quickly became apparent that people give negative criticism to others for many reasons – to assert dominance over them, to shatter their egos, or to enhance their own. In fact, though, the only fully rational reason for giving negative feedback to others is to help them improve, and to the extent that these other motives intrude, those on the receiving end of criticism will quickly recognize this and respond with anger rather than sincere efforts to improve. In short, consistent with Bandura's (1997) suggestions, people's interpretations of (cognitions about) the motives behind others' actions is a powerful determinant of their subsequent reactions to these actions.

THE NEXT STEP: RECOGNITION OF THE CENTRAL ROLE OF COGNITION IN ENTREPRENEURSHIP

To cut this personal retrospective short, I shall simply add that in the late 1990s, after running my own company for several years, and under the expert instruction of my good friend and colleague Venkataraman, my interests shifted to the field of entrepreneurship. In fact, in our personal conversations Venkataraman often remarked: 'Robert, you seem to know quite a bit about cognition, and that is a topic that is highly relevant to entrepreneurship as it is currently developing.' He was being far too kind in his assessment of my knowledge, but since his advice was given in the most constructive terms (!), coupled with my immense respect for him, I listened carefully, and moved in the direction he suggested. Since that time, cognition has remained a major theme in my work – and in the field as a whole. Indeed, as this entire volume suggests, cognition and the processes it encompasses are now an integral part of the field of entrepreneurship, and the terms 'cognition' or 'cognitive' appear in a substantial proportion of articles published in our leading journals.

Having explained how I moved from the artificial restraints of extreme behaviorism to a much more eclectic and, I hope, inclusive approach, I now turn to very different topics – ones, that, perhaps, will be of more interest to my colleagues than these personal reflections. First, I shall comment briefly on a question that I think is central to our field – and to how it can best acquire the knowledge it seeks. Stated directly, this question is: 'To what extent can we base our knowledge primarily on what entrepreneurs – even highly successful and experienced ones – tell us? Or should we, instead, rely primarily on data?' I think even at present this remains a somewhat unresolved issue in the field, but one deserving of careful examination.

Second, rather than try to review the excellent and diverse research that will be reported in this book, I would like to suggest a few additional questions that, I believe, can be addressed from a cognitive perspective but have not, to date, been included in ongoing research. I have chosen these to be quite diverse in nature in order to illustrate the tremendously broad range of issues and topics that can be effectively examined from a cognitive perspective. One topic relates to the role of self-regulation in entrepreneurship – and why the cognitive processes individuals use to guide, evaluate and adjust their own behavior may be especially relevant to our field (see also Randolph-Seng, Williams and Hayek, Chapter 4, this volume). The other relates to two questions that I find intriguing: (1) are entrepreneurs, as is widely believed, especially benevolent – more likely to make large donations or engage in other prosocial behavior (actions designed to help various recipients without obvious benefit to themselves) than persons who acquired their wealth in other ways (e.g., inheritance, investments); and (2) if they are, why is this the case – what motives underlie such actions? But first, I shall address the question noted above: what should be the source of our knowledge about entrepreneurship?

ACQUIRING KNOWLEDGE: DIFFERENT METHODS, DIFFERENT TOOLS

In one sense, the ultimate goal of our field is, and certainly should be, to help entrepreneurs. To do that, we need reliable, valid knowledge on which to base our recommendations and guidance. But how can we best obtain such knowledge? Basically, three different methods have been and are currently being used: (1) asking successful entrepreneurs to tell us; (2) conducting direct, systematic research on these issues; and (3) importing knowledge from other fields. Although these methods differ greatly, they are, I believe, complementary in nature. In other words, each provides a unique kind of information, so that, together, they all serve to enrich our understanding of the diverse array of factors that combine to generate entrepreneurial success. The fact that they are complementary, however, does not imply that we must give them equal weight. I shall suggest here that we should probably assign most weight to (2) and perhaps (3). I am confident that readers of this book will generally agree, so here I shall try to indicate as clearly as I can why I believe that this is the most appropriate course of action.

Asking Entrepreneurs: Can They Really Tell Us the 'Secrets of Their Success?'

At first glance, nothing seems more reasonable than asking highly successful entrepreneurs to share with us how, in their view, they achieved the success they enjoy. In fact, they are often asked to do so: for example,

'Tell us, Bill [Gates], Mark [Zuckerberg], Elon [Musk]: why have you been so successful? What did you do so *right* that others did so *wrong*?' When asked this question, few famous entrepreneurs refuse to answer. And why should they? They are highly successful, so is it not appropriate for them to comment on how they attained their fame – and wealth? Of course it is; but a more important question is: do they really *know* why they succeeded? Unfortunately, the answer is just as likely to be 'no' as 'yes,' and for a simple reason: a vast body of research findings indicates that, as human beings, we are not nearly as good as we generally believe at understanding the motives behind our own behavior, or the factors that influenced both it and the outcomes we experienced. In fact, we are often wrong – dead wrong! – when we make such assessments. Memory, for instance, is fallible and easily distorted. Further, research in cognitive science indicates that each time we recall a past event and then re-enter it into memory, it is changed – often in ways that make it more understandable and consistent, but not necessarily more accurate (Koutstaal and Cavendish, 2006). Perhaps even more disturbing, our understanding of why we performed specific actions in the past is uncertain and prone to error. In other words, we are not nearly as expert in understanding our own motives as we often believe. Further, we are not very good at distinguishing between the factors and conditions that actually *did* influence our thoughts and actions, and ones that did not. Moreover, we often believe that we know more about various topics than we really do – a tendency described by the phrase 'the illusion of knowledge' (Chabris and Simons, 2009). These are very general statements, so perhaps a concrete example will help.

In a classic experiment, one that vividly illustrates the limits of our own understanding of the factors that shape our behavior or thoughts, Nisbett and Wilson (1977) asked college students to memorize a list of word-pairs. Some of these (e.g., ocean–moon) were designed to generate mental associations that would lead to certain outcomes – words related closely to these word-pairs. For instance, after memorizing the lists of words, participants were asked to name a detergent. A very large proportion came up with 'Tide,' which is clearly related to word-pairs such as ocean–moon. When asked if the word-pairs had influenced their answer, however, participants vehemently denied this; instead, they insisted that they had come up with 'Tide' because they like it best, use it all the time, or because it is the most famous brand on the market. Was this true? Evidence indicates it was not, and that the word-pairs had indeed affected their behavior; but – and this is crucial – participants in the research did not recognize this fact. In a follow-up investigation, the same researchers asked participants to choose the best product from an array of products (e.g., the best radio, pens etc.). Results indicated that participants had a strong 'right-side preference' – they chose the items on the right-hand side of the display much more often than ones on the left. When asked if position had any influence on their behavior, however, they reacted with indignation: 'Of course not!'

If we are not very good at even simple judgments such as these, can we be accurate in reporting the factors responsible for our success – especially since they may have existed or operated years in the past? Readers will, of course, draw their own conclusions, but the weight of scientific evidence is clear: 'No, we are not very good at these tasks, and often make important errors.' So, can we believe entrepreneurs when they tell us why and how they succeeded? Perhaps, but a degree of caution seems essential. This does not mean that we should discount such information; far from it. It is often helpful and insightful, and may suggest valuable ideas for research. But these statements should always be evaluated in terms of the evidence concerning their accuracy – evidence gathered in the ways outlined next.

Conducting Systematic Research

If we cannot rely on comments by entrepreneurs to solve the mystery of entrepreneurial success, how can we obtain such knowledge? Hundreds of years of scientific progress offer the following answer: 'Through systematic research that collects actual data on the questions we want to address.' Rather than simply asking famous entrepreneurs how and why they succeeded, we can delve into these issues by studying large numbers of entrepreneurs and new ventures in order to identify the factors, actions or strategies that are actually related to positive outcomes (financial success or other beneficial results). Conducting such research is far from easy; we must obtain the cooperation of large numbers of entrepreneurs – individuals who are extremely busy running their companies. As Hambrick (2007) notes, this reluctance (or inability) to participate in our research is one reason why it is so difficult to identify the processes through which top executives (including entrepneurs) influence the fortunes of the organizations they head (upper echelons theory). My own view is that Hambrick is, perhaps, too pessimistic in this respect. By choosing our measures and methods carefully, we can win the cooperation of these individuals. I agree with his view that the practical constraints on conducting such research are indeed often formidable, but also believe that the benefits of conducting systematic research, no matter how challenging this may be, are clear: the data obtained in this manner are generally far less subject to the kind of cognitive biases mentioned above. Perhaps another example will help to clarify this point.

It has often been suggested that entrepreneurs experience very high levels of stress; after all, they face incredibly high workloads, high levels of uncertainty, limited resources and substantial risks. Exposure to these conditions would be expected to generate intense stress, and stress, as a huge volume of research indicates, is often harmful to both performance and personal health. This suggests that perhaps one important ingredient in entrepreneurial excellence is the ability to resist the adverse effects of stress – and, perhaps, even to thrive in its presence. How can we find out if this is true? One way is to ask successful entrepreneurs about their own ability to manage stress; but, as noted above, the information they provide may be highly subjective and not very accurate. Another approach is to conduct research in which a large number of entrepreneurs participate by providing information on the levels of stress they experience and the extent to which they possess skills or characteristics that, we have reason to believe, might help to mitigate the adverse effects of stress.

One such factor is known as 'psychological capital,' and involves high levels of self-efficacy (confidence in one's own ability to successfully complete various tasks), optimism, hope and resilience – the ability to 'snap back' after setbacks (Peterson, Luthans, Avolio, Walumbwa and Zhang, 2011). Previous research on psychological capital indicates that it is negatively related to stress: the higher individuals are in this cluster of characteristics, the less stress they experience (Avey, Reichard, Luthans and Mhatre, 2011). To find out if this applies to entrepreneurs, we might ask them to complete measures of psychological capital and stress, and also obtain information on the success of their new ventures – or perhaps on their health and well-being. If psychological capital does indeed provide a 'buffer' against the adverse effects of stress, this would indicate that it is one important ingredient in entrepreneurial excellence: entrepreneurs high in this characteristic would be better able to withstand high levels of stress that might ordinarily interfere with their performance or impair their health and therefore, ultimately, the success of their companies. In fact, such research on this possibility has recently been conducted, and offers support for these conclusions (Baron, Franklin and Hmieleski, in press). Findings indicated that the higher entrepreneurs' psychological capital (as assessed by a brief but reliable questionnaire), the lower the levels of stress they reported. In short, carefully conducted research is one way of gaining knowledge about the essential components of entrepreneurial excellence. Of course, in such research we often collect data reflecting entrepreneurs' replies to our questions, or responses to carefully designed surveys. But this is still very different from relying on the informal comments of a few highly successful entrepreneurs as the foundation for our conclusions.

Importing Knowledge from Other Fields

One more means of obtaining information about the nature of entrepreneurial success involves importing such knowledge from other fields and branches of management. It seems reasonable to propose that the factors that facilitate personal success in contexts outside entrepreneurship might have similar effects in the entrepreneurial domain. In fact, such 'importation' has been very useful. For instance, there is a huge body of evidence suggesting that social and political skills – a wide range of skills that help individuals to get along well with others – have a wide range of beneficial effects. People high in such skills attain greater success in their careers, have more friends, higher-quality social networks, and experience fewer interpersonal problems at work and in their personal lives than persons lower in such skills (e.g., Ferris, Treadway, Perrewé, Brouer, Douglas and Lux, 2007). This knowledge has been 'imported' by the field of entrepreneurship, and a growing body of evidence indicates that it is directly applicable to it: the higher entrepreneurs are in social and political skills, the greater their personal success and that of their companies (e.g., Baron and Tang, 2009). Many other examples of such 'importation' of knowledge exist, but I hope that this example will suffice to illustrate the value of this approach.

In sum, there are several ways of acquiring useful knowledge about entrepreneurship and the factors that influence it, and all should be viewed as potentially useful. Overall, though, it seems clear that placing greater weight on the results of systematic research (either in entrepreneurship or in related fields) will put us on safer ground than relying on the comments and informal reflections of entrepreneurs, no matter how eloquent and insightful these appear to be. There are simply too many potential sources of error in such information to make it the basic foundation of our field.

LOOKING AHEAD: FINDING (OR CREATING) THE NEXT WAVE OF OPPORTUNITIES FOR COGNITIVE RESEARCH IN ENTREPRENEURSHIP

Predicting the future – even the immediate future – is a very tricky task, as I have often learned to my dismay. My record at predicting the economy, stock market, sports events, trends in fashion or music, the Academy Awards, or anything else, is mediocre at best. In fact, it sometimes seems that only the local weather forecasters consistently do worse than I do! But recognizing the risks of offering predictions about our field, I will suggest a few ways in which it can use cognition and a cognitive perspective to address intriguing questions that have, to date, received relatively little attention. I have specifically chosen these topics to be very diverse, as a means of illustrating how broad and useful cognition can be as a focal point for our research.

From Desire to Achievement: The Role of Self-regulation in Entrepreneurship

> You have to learn the rules of the game. And then you have to play better than anyone else.
>
> Albert Einstein

> Desire is the key to motivation, but it's the determination and commitment to an unrelenting pursuit of your goal – a commitment to excellence – that will enable you to attain the success you seek.
>
> Mario Andretti

In essence, both of these quotations refer to 'self-regulation' – cognitive processes that individuals use to select, monitor, direct, guide and evaluate their own behavior in order to attain progress toward key goals. Self-regulation, in turn, involves several different components, including the following:

(1) self-control: performing actions (even if they are not enjoyable) that facilitate progress toward important goals while refraining from actions that are enjoyable but impede progress;
(2) demonstrating a combination of focus and persistence (staying focused on key goals and working persistently toward them);
(3) managing emotions and impulses – for instance, avoiding rash behavior or decisions, and delaying gratification so as to maximize its magnitude;
(4) developing accurate metacognition – acquiring insightful self-knowledge about our own cognition, including, for instance, recognizing what we know and do not know.

Clearly, these processes are important for everyone, but there are several reasons why they may be especially valuable for entrepreneurs. First, in their efforts to create something new, entrepreneurs are generally very much on their own. They do not have supervisors, mentors or coaches, and often find themselves in situations with few if any external rules to guide their actions. This suggests that they must rely heavily – perhaps more heavily than many other people – on self-regulation.

Second, and especially for first-time entrepreneurs, they often find themselves performing tasks and filling roles they have not previously performed or occupied. As a result, it is especially crucial for them to acknowledge what they know and do not know, so that they will seek help when it is truly needed. Third, entrepreneurs, as a group, are very high in optimism and in positive affect (i.e., positive moods and emotions). Although these tendencies are often beneficial, they can also generate harmful effects, e.g., a tendency to ignore relevant negative information; an inclination to engage in quick-and-dirty heuristic thinking when more systematic, careful analysis is essential (Baron, Hmieleski and Henry, 2012); a tendency to act *now* when, in fact, delay and reflection would be more effective; and a tendency to persist with failing courses of action or strategies – to get trapped in 'sunk costs' (e.g,. Staw, 1981). For these and other reasons, entrepreneurs must often rely on self-regulation on the way to the success they seek.

How important are these components of self-regulation? Experts on this topic (Baumeister, Vohs and Tice, 2007) have summarized existing data by noting that, after decades of research on the factors that influence success in many different endeavors, only two have been found to play a very general role: intelligence and self-regulation. After more than a century of careful study, quite a lot is known about intelligence, and one basic fact about it is somewhat discouraging: much as we might wish it were otherwise, intelligence is partly shaped by genetic factors and early childhood experiences, both of which cannot be readily changed. In contrast, self-regulation involves a number of specific skills that can be acquired or strengthened. This suggests that efforts to enhance self-regulation can contribute substantially to personal success and achievement, and a growing body of evidence offers strong support for this view: people who develop and practice effective self-regulation often succeed. Indeed, effective self-regulation may be more important in this regard than genetically influenced talent (Baumeister and Tierney, 2011), and that is an optimistic conclusion indeed!

Having commented on the importance of self-regulation generally, I will now focus more specifically on the implications of two key aspects of it for entrepreneurship: self-control and metacognition.

Self-control: doing what we should do, and refraining from doing what we should not

Almost everyone finds certain activities much more enjoyable than others; and sadly, it is often the ones that are bad for us – in the sense of interfering with achieving our key goals – that are the most attractive. A key task individuals face, then, is somehow to regulate their behavior so that they do the things they should do (exercise regularly, eat moderately, keep careful financial records for their company) and avoid the things they realize they should not do, even if they are enjoyable (e.g., eat high-calorie snacks or desserts, spending on impulse, draw a large salary rather than reinvest profits in a new venture). The everyday term for this capacity is 'willpower,' but another – the one generally used in research on this aspect of self-regulation – is 'self-control' (de Ridder, Lensvelt-Mulders, Finkenauer, Stok and Baumeister, 2012). Research on self-control has continued for several decades, but has recently become much more sophisticated and informative. One clear finding of such research is that people differ greatly in their capacity to exert self-control (Tangney et al., 2004). Some are able to resist even powerful temptations to engage in actions, have thoughts, or experience emotions that interfere with their plans and goals, while others are much lower in this capacity (Baumeister and Alquist, 2009).

Another finding is that self-control appears to be a depletable cognitive resource: after exerting it in one situation or with respect to one activity (or temptation), individuals' capacity to exert it again declines. Interestingly, experiencing positive feelings can restore self-control that has been depleted (e.g., Tice, Baumeister, Shmueli and Muraven, 2007). Recent evidence indicates that exerting self-control involves not merely a draining of cognitive resources but also, perhaps, a shifting of attention between different goals (Fujita, 2011) – this issue is still being debated.

Is self-control concerned primarily with doing what we should do (but perhaps do not want to do) or refraining from doing what we should not do (but perhaps want to do)? A review of existing evidence indicates that it operates in both ways (de Ridder et al., 2012). So it both helps individuals to perform actions that assist them in reaching their important goals, and assists them in refraining from actions that block or interfere with progress toward these goals. Further, it appears that people who are high in dispositional self-control (the tendency to exert self-control in many different situations) are especially effective at acquiring 'good habits.' For such people, as for everyone else, engaging in actions that are not enjoyable but facilitate goal achievement initially requires the exercise of self-control. However, these actions quickly become 'automatic,' and can be performed without further depleting cognitive resources.

Is self-control especially helpful to entrepreneurs? There are several reasons for suggesting that it is. First, high levels of self-control may assist entrepreneurs to select appropriate goals, to direct their actions onto paths that facilitate progress toward key goals, and also to remain focused on these goals over extended periods of time. Further, as suggested by Fujita (2011), individuals with higher levels of self-control are better able to shift their focus among various performance goals as current needs and situations shift, and, hence, are better able to manage and attain progress toward multiple goals (Schmidt and Dolis, 2009).

Second, high levels of self-control may assist entrepreneurs to make effective decisions regarding the selection and development of potentially valuable opportunities. Drawing on the structural alignment concept, recent studies (Grégoire et al., 2010; Grégoire and Shepherd, 2012) have conceptualized opportunity recognition as the cognitive process of finding similarities between the superficial features and structural relationships of a new stimulus (opportunity) and those of a relevant source (e.g., information previously stored in memory). Specifically, this research suggests that comparison of superficial features (e.g., characteristics of a new technology; size of a new market space; number of production processes involved etc.) requires less cognitive effort than the comparison of structural relationships, which involve higher-order processing (e.g., potential impact of a new material on a production process or on product function; the potential rate of growth of a new market etc.). High levels of self-control may facilitate efforts to align structural rather than superficial features, and so may result in the identification of higher-quality opportunities (i.e., those with potentially greater benefits). On the basis of this reasoning and previous research, I posit the existence of a positive relationship between entrepreneurs' self-control and the performance of their new ventures.

Third, a key task faced by founding entrepreneurs is making effective decisions – often under high time pressure. This requires that they focus intently on available information, and extract, from what is often a very diverse range of inputs, the information that is especially pertinent to the decisions at hand. A high level of self-control can facilitate these activities, since it allows entrepreneurs to focus on what is essential while minimizing or ignoring potential distractions. As a result, the quality and speed of their decisions may be enhanced, outcomes that can contribute significantly to new-venture performance. Data relevant to these suggestions have recently been reported by Nambisan and Baron (2013). They found that, among founding entrepreneurs, self-control was significantly related to firm performance, although such effects were not direct – rather, they were mediated by two factors described in upper echelons theory: managerial discretion and job demands. Taking past research and theory into account, I suggest the following hypothesis as one potentially worthy of careful investigation:

Hypothesis 1. Entrepreneurs' level of self-control is positively related to the performance of their new ventures.

Metacognition: understanding and regulating our own thoughts

A second aspect of self-regulation that may be especially relevant to entrepreneurship is 'metacognition,' usually defined as cognition about cognition (Efklides, 2008; Flavell, 1979). Basically, metacognition refers to individuals' understanding and knowledge of their own cognition. Metacognition, in turn, is often divided into several broad categories: metacognitive knowledge – knowledge about ourselves, other persons, tasks, and various strategies for performing them; metacognitive experience – individuals' awareness of what they experience when performing a task and processing information related to it; and metacognitive skills – deliberate use of specific tactics to control cognition (e.g., Haynie et al., 2010). Perhaps the most relevant of these for the present discussion is 'metacognitive knowledge,' which includes individuals' understanding of what they, personally, know and do not know, as well as their understanding of their own capacity to perform various tasks (Efklides, 2008).

High levels of metacognitive knowledge may be especially useful to entrepreneurs because such knowledge, by helping them recognize their own strengths and weaknesses, may also assist them to focus their efforts on tasks they can actually accomplish and opportunities they can personally develop (McMullen and Shepherd, 2006). Previous research suggests that metacognitive knowledge underlies individuals' recognition of when to persist in performing a given task through a particular strategy and when, instead, to withdraw because the probability of success by using that strategy is low (McFarlin, Baumeister and Blascovich, 1984). Finally, and just as important, it may help them to stop investing time, effort and resources in tasks or goals unlikely to yield positive outcomes (Haynie et al., 2010) – that is, it may help them to avoid or escape from the trap of 'sunk costs' (e.g., Staw, 1981), as mentioned above. Sticking with ineffective courses of action or strategies can be disastrous for new ventures, which usually have limited resources. Because entrepreneurs are high in both optimism and positive affect, they may be more likely to succumb to the lure of sunk costs than other people, since they are confident (perhaps unreasonably so) that 'all will turn out well in the end.' Avoiding such potentially costly errors, in turn, may have beneficial effects on new-venture performance. In addition, metacognitive knowledge may assist entrepreneurs in responding to rapid changes in technologies and markets by helping them to grasp the limits of their own knowledge, to appreciate the resources needed to cope with changing conditions, and to formulate new strategies for meeting them effectively (Nambisan and Baron, 2013). On the basis of this reasoning, I suggest the following hypothesis for future research:

Hypothesis 2. Entrepreneurs' metacognitive knowledge (i.e., their understanding of what they know and do not know) is positively related to the performance of their firms.

While self-control and metacognitive knowledge are important aspects of self-regulation, they are far from the only aspects of this general process. Additional aspects include the capacity to focus persistently on, and exert effort toward, important goals (e.g., Duckworth, Kirby, Tsukayama, Berstein and Ericsson, 2011), delaying gratification to times when its magnitude will be maximized (e.g., Mischel and Ayduk, 2004) and avoiding rash, impulsive actions (e.g., Kruglanski, Thompson, Higgins, Atash, Pierro, Shah and Spiegel, 2000). Overall, these cognitive processes may play an important role in entrepreneurship and deserve investigation in future research.

Cognitive Factors in Entrepreneurs' Generosity: Is 'Giving Back' All There Is?

At the present time, entrepreneurship – and entrepreneurs – are very much in the 'the public eye.' In fact, the label 'entrepreneur' has an almost romantic tinge, similar to the term 'leader,' and triggers, in the minds of many persons, very positive images. This favorable aura derives from many characteristics (real or merely assumed), including high levels of creativity, intense energy and, perhaps most important, the capacity to generate wealth both for themselves and for their societies. However, one other factor also seems worthy of mention: entrepreneurs' well-known and often highly publicized philanthropy. Entrepreneurs are widely perceived not merely as individuals who have earned a large amount of money, but also as individuals who are very generous with their wealth.

Is this perception accurate? There is some evidence that it is. Data on this question have been gathered in several large-scale, longitudinal studies focused on what is generally termed 'giving by high net worth individuals' (e.g., Bank of America, 2012; Chronicle of Philanthropy, 2010; Center on Philanthropy, Indiana University, 2008). One aspect of these studies involves comparison of the amounts given by people whose wealth derives from various sources (e.g., inheritance, entrepreneurship, real estate, investments, earned income). For instance, a study of gifts of more than $1 000 000 during the years 2000–2007 indicates that a higher proportion of these gifts were provided by entrepreneurs (33 percent) than by persons whose wealth derived from inheritance (10 percent), investments (6 percent), real estate (4 percent), or partnerships (9 percent; see research from the Center on Philanthropy, Indiana University, 2008).

Results of a more recent investigation (Bank of America, 2012) indicate that in each of three recent years (2007, 2009, 2011), entrepreneurs made significantly larger donations, on average, than any other group. Another study (Chronicle of Philanthropy, 2010) also reported that companies founded by, and still run by, entrepreneurs donated more than twice the percentage of profits to charity or other causes than large, established organizations: 3 percent versus 1.2 percent.

It is important to note that the data described above are far from conclusive – important questions concerning methodology, selection of samples and dependent measures remain. For instance, it is not clear that the personal wealth of participants was truly equivalent across all groups, or that these groups were matched on other dimensions (age, education, experience). However, taking these limitations into account, this research seems to offer some support for the propsition that entrepreneurs do indeed demonstrate higher levels of philanthropy than other successful people.

But, assuming that this conclusion is valid, the question of why such differences exist remains. Why are entrepreneurs especially likely to perform generous actions, making large donations to universities, museums, research laboratories and many other bodies? In commenting on such actions, many entrepreneurs suggest a strong desire to 'give back' to their communities. Further, they often add that, having attained huge financial success, they now feel a powerful obligation to share these benefits with others. Is this really the primary source of their notable generosity? Having commented above on the potential dangers of relying on such informal comments, it seems clear that they probably cannot answer this complex question. Further, existing evidence gathered in several different fields indicates that, like many complex human activities, prosocial behavior (generally defined as actions that benefit others without providing any obvious or immediate benefit to the donor), derives from numerous factors – emotional, cognitive, personal and environmental (e.g., Dovidio, Piliavin, Schroeder and Penner, 2006; see also Grant and Dutton, 2012; Grant and Gino, 2010). Consistent with the scope of this discussion, several cognitive factors that may play a key role in entrepreneurs' philanthropy (an important aspect of prosocial behavior) will now be suggested.

Entrepreneurs' philanthropy and the optimistic bias

An important aspect of human cognition involves what is known as the 'optimistic bias' – a strong tendency to anticipate positive outcomes or results in many situations, including ones in which such outcomes are in fact unlikely. Entrepreneurs have been shown to be high in this aspect of cognition – higher, in fact, than many other tested groups (see, e.g., Hmieleski and Baron, 2009). This strong tendency to be optimistic may play an important role in entrepreneurs' high levels of prosocial behavior. Being optimistic, they believe that they will continue to enjoy financial success in the future, and so can readily afford to engage in prosocial behavior – for instance, making large donations to universities or other non-profit organizations. Their optimism leads them to expect that their personal wealth will continue to grow, so they will soon recover the donated resources. On the basis of this reasoning, we propose the following hypothesis:

Hypothesis 3. There is a positive relationship between entrepreneurs' optimism and their tendency to engage in philanthropy (i.e., one important form of prosocial behavior).

Entrepreneurs' philanthropy and positive inequity

Successful entrepreneurs often accumulate huge personal wealth. Contemplating these assets, some, at least, may recognize that other entrepreneurs who have worked just as hard as they have, who are just as talented, and as strongly committed to their new ventures, have experienced far less success, and may have experienced total business failure (Ucbasaran, Shepherd, Lockett and Lyon, 2013). The result may be strong feelings among successful entrepreneurs of being over-rewarded – of having received more, relative to others, than they actually deserve. While no direct empirical evidence concerning this reasoning exists, it is consistent with findings indicating that in organizations that have undergone 'downsizing,' the survivors often feel over-rewarded in that they have retained their jobs, while others, who have worked just as hard, have not. This results in feelings of positive inequity, leading to efforts to offset such reactions by working harder than ever. It is also consistent with entrepreneurs' statements concerning their strong desire to 'give back' to their communities. Doing so could, in some instances, be a means of reducing feelings of positive inequity. On the basis of this reasoning, I offer the following hypothesis:

Hypothesis 4. There is a positive relationship between feelings of positive inequity on the part of entrepreneurs and the frequency and/or magnitude of their philanthropic (i.e., prosocial) actions.

Prosocial behavior as a means of reducing guilt: making reparation for prior harmful actions

When individuals perform actions of which they do not approve, they often experience guilt – negative feelings of self-censure. Research findings indicate that these feelings are most likely to occur when their actions have produced harm, in some way, to other persons – that is, guilt arises primarily in situations involving interpersonal behavior. Additional evidence indicates that one result of such feelings of guilt is a strong desire to make reparation – to somehow 'make up' for these past transgressions (see, e.g., Baumeister, Stuillwell and Heatherton, 1994). I suggest that such feelings may also be an important source of prosocial actions among entrepreneurs.

Many entrepreneurs realize that, on the way to success, they have engaged in actions of which they do not personally approve. For example, they have 'cut corners' in terms of failing to follow ethical standards they accept (e.g., never lie; treat others fairly; do not engage in actions that will harm potential customers; do not falsify financial records), and have, perhaps, engaged in actions bordering on unfair business practices (e.g., obtaining confidential information about competitors' products or strategies by offering payments to people who can provide them, etc.). Faced with such guilt (negative affective states stemming from cognitive evaluation of one's own behavior), such entrepreneurs, may – as previous research suggests – seek to make reparation in order to reduce or eliminate these feelings. The result? They do indeed 'give back' through generous donations and similar actions. Future research could be designed to test this idea by gathering data relevant to the following hypothesis:

Hypothesis 5. There is a positive relationship between the level of guilt experienced by entrepreneurs concerning their past business actions, and the frequency or magnitude of their philanthropic (i.e., prosocial) actions.

Reciprocity: the cognitive basis for 'giving back'

In describing their reasons for engaging in philanthropic actions (e.g., making large donations to various organizations), entrepreneurs often make statements suggesting that they engage in these actions and donations because they want to 'give back.' But what precisely do they mean by the phrase 'giving back'? One possibility is suggested by the fact that reciprocity – the tendency to treat others as they have treated us – is a powerful guiding principle of human behavior. In general, individuals attempt to 'balance the scales,' so to speak, by matching their own behavior with that of others toward them. If they have been treated courteously, they reciprocate such actions; and if they have been treated discourteously, they tend to return such rudeness in kind. Similarly, if they have been helped by others, they tend to reciprocate for such actions by engaging in similar actions themselves. Entrepreneurs' stated desire to 'give back' may reflect this powerful principle: having been helped by others in the past, they feel a strong obligation to reciprocate, either to these persons or to organizations. Thus the entrepreneurs who make large donations to universities are often alumni of these schools, and through their generous actions seek to reciprocate for the training and other benefits they have received – benefits they identify, cognitively, as having contributed to their success. In short, 'giving back' may relate to a powerful underlying cognitive process in which individuals examine the benefits they have previously received and attempt to repay these by engaging in philanthropy or other forms of prosocial behavior. On the basis of these suggestions, we offer the following hypothesis:

Hypothesis 6. There is a positive relationship between entrepreneurs' motivation to reciprocate for past benefits and their current philanthropic (i.e., prosocial) actions.

Other cognitive factors that might play a role in prosocial behavior include the desire to leave a lasting positive impact (Coombs, Shipp and Christensen, 2008), or the desire to surpass others – what is known in the relevant literature as 'competitive altruism.' Given entrepreneurs' high level of competitiveness and strong desire to 'win,' this latter possibility seems quite feasible.

In sum, a cognitive perspective – and, in fact, a perspective that takes full account of the complex interactions between affective and cognitive processes – could well provide a valuable means for investigating the foundations of entrepreneurs' almost legendary generosity – a topic that has not, to date, received systematic attention.

CONCLUDING THOUGHTS

If there is a central theme to this admittedly somewhat rambling discussion, it is this: understanding cognition – how, why and what people think, believe, reason and decide – is central to understanding their behavior and, in fact, everything they do, say or feel, and is, therefore, also central to the field of entrepreneurship. Having made such a sweeping statement, let me close with an illustration of this approach in action.

Many readers are probably familiar with Hambrick's upper echelons theory – a view mentioned briefly earlier that suggests that CEOs (and, by extension, founding entrepreneurs) exert strong effects on the success and performance of the organizations they head (e.g., Hambrick and Mason 1984; Hambrick 2007). A large body of evidence supports this view, and points to the conclusion that the top people in an organization do indeed matter in terms of firm-level performance. But as Hambrick (2007) notes, most research on this perspective has focused on demographic variables – such factors as CEOs' age, experience, field of training and so on. Results indicate that these variables are indeed related to firm performance, but, as Hambrick states, we do not know why this is so. What happens in between CEO characteristics and firm-level performance takes place, in his own words, in 'a mysterious black box.' What is needed, therefore, is careful probing of this 'black box' to uncover the cognitive and social processes occurring within it – the processes that connect a CEO's perceptions, decisions and characteristics to firm performance. In recent research, my colleague Satish Nambisan and I (Baron and Nambisan, 2013) have suggested that part of the contents of that 'black box' involve self-regulation – the cognitive processes discussed at length in an earlier portion of this chapter. Our initial data provide support for this view, but whether our proposals are correct is not the main point. Rather, it is this: a cognitive perspective can help us to gain insights into the 'black box' of human thought and action. We already have part of the knowledge and some of the tools needed to continue this work (e.g., Baron and Ward, 2004; Grégoire and Shepherd, 2012; Mitchell et al., 2007; Tang, Kacmar and Busenitz, 2012); what remains now is the task of applying them effectively. That is the subject of the field of entrepreneurial cognition – and this book. To put it more concretely, to the extent that our field adopts a cognitive perspective as a central one, we will continue to advance toward our major goal: gaining full and accurate understanding of the complex processes through which enterprising individuals convert the *possible* (i.e., their creativity and the ideas it generates) into the *real* (Baron, 2013; Shane, 2012).

REFERENCES

Avey, J.B., Reichard, R.J., Luthans, F. and Mhatre, K.H. 2011. Meta-analysis of the impact of positive psychological capital on employee attitudes, behaviors, and performance. *Human Resource Development Quarterly*, **22**(2): 127–52.

Bandura, A. 1986. *Social Foundations of Thought and Action: A social cognitive theory*. Englewood Cliffs, NJ: Prentice-Hall.

Bandura, A. 1997. *Self-efficacy: The exercise of control*. New York: Freeman.

Bank of America. 2012. Study of High Net Worth Philanthropy. http://www.philanthropy.iupui.edu/research-by-category/the-2012-study-of-high-net-worth-philanthropy.

Baron, R.A. 1988. Negative effects of destructive criticism: Impact on conflict, self-efficacy, and task performance. *Journal of Applied Psychology*, **73**: 199–207.

Baron, R.A. 1990. Countering the effects of destructive criticism: The relative efficacy of four potential interventions. *Journal of Applied Psychology*, **75**: 235–45.

Baron, R.A. 2013. *Enhancing Entrepreneurial Excellence: Tools for making the possible real*. Cheltenham, UK and Northampton, MA, USA: Edward Elgar.

Baron, R.A., Franklin, R. and Hmieleski, K.M. Why entrepreneurs often experience low, not high, levels of stress: The joint effects of selection and psychological capital. *Journal of Management*. In press.

Baron, R.A., Hmieleski, K.M. and Henry, R.A. 2012. Entrepreneurs' dispositional positive affect: The potential benefits and potential costs of being 'up.' *Journal of Business Venturing*, **27**(3): 310–24.

Baron, R.A., and Nambisan, S. 2013. Entrepreneurs' self-regulation and firm performance: Cognitive processes inside the 'Black Box.' Working paper.

Baron, R.A. and Tang, J. 2009. Entrepreneurs' social competence and new venture performance: Evidence on potential mediators and cross-industry generality. *Journal of Management*, **35**(2): 282–306.

Baron, R.A. and Ward, T.B. 2004. Expanding entrepreneurial cognition's toolbox: Potential contributions from the field of cognition science. *Entrepreneurship Theory and Practice*, **28**(6): 553–73.

Baumeister, R.F. and Alquist, J.L. 2009. Self-regulation as a limited resource: Strength model of control and depletion. In J.P. Forgas, R.F. Baumeister and D.M. Tice (eds), *Psychology of Self-regulation: Cognitive, affective, and motivational processes*: 21–35. New York: Taylor & Francis.

Baumeister, R.F., Stuillwell, A.M. and Heatherton, T.F. 1994. Guilt: An interpersonal approach. *Psychological Bulletin*, **115**(2): 132–67.

Baumeister, R.F. and Tierney, J. 2011. *Willpower: Rediscovering the greatest human strength*. New York: Penguin Press.

Baumeister, R.F., Vohs, K.D. and Tice, D.M. 2007. The strength model of self-control. *Current Directions in Psychological Science*, **16**(6): 351–5.

Center on Philanthropy, Indiana University, 2008. An analysis of million dollar gifts, 2000–2007.

Chabris, C. and Simons, D. 2009. *The Invisible Gorilla: How our intuitions deceive us*. New York: Random House.

Chronicle of Philanthropy. 2010. Largest individual donations in the United States. December 31.

Coombs, J.E., Shipp, A.J. and Christensen, L. J. 2008. Entrepreneur as change agent: Antecedents and moderators of individual-level philanthropic behavior. *Frontiers of Entrepreneurship Research*, **28**(21): 1–10.

De Ridder, D., Lensvelt-Mulders, G., Finkenauer, C., Stok, F.M. and Baumeister, R.F. 2012. A meta-analysis of how trait self-control relates to a wide range of behaviors. *Personality and Social Psychology Review*, **16**(1): 76–99.

Dovidio, J.F., Piliavin, J.A., Schroeder, D.A. and Penner, L.A. 2006. *The Social Psychology of Prosocial Behavior*. Mahwah, NJ: Erlbaum.

Duckworth, A.L., Kirby, T.A., Tsukayama, E., Berstein, H. and Ericsson, K.A. 2011. Deliberate practice spells success: Why grittier competitors triumph at the National Spelling Bee. *Social Psychological and Personality Science*, **2**(2): 174–81.

Efklides, A. 2008. Metacognition: Defining its facets and levels of functioning in relation to self-regulation and co-regulation. *European Psychologist*, **13**(4): 277–87.

Ferris, G.R., Treadway, D.C., Perrewé, P.L., Brouer, R.L., Douglas, C. and Lux, S. 2007. Political Skill in Organizations. *Journal of Management*, **33**(3): 290–320.

Flavell, J. 1979. Metacognitive knowledge and cognitive monitoring: A new area of cognitive-developmental inquiry. *American Psychologist*, **34**(10), 906–11.

Forgas, J.P., Baumeister, R.H. and Tice, D.M. 2009. *Psychology of Self-regulation: Cognitive, affective, and motivational processes*. New York: Psychology Press.

Fujita, K. 2011. On conceptualizing self-control as more than the effortful inhibition of impulses. *Personality and Social Psychology Review*, **155**(4): 342–66.

Grant, A. and Dutton, J. 2012. Beneficiary or benefactor: Are people more prosocial when they reflect on receiving or giving? *Psychological Science*, **23**(9): 1033–9.

Grant, A.M. and Gino, C. 2010. A little thanks goes a long way: Explaining why gratitude expressions motivate prosocial behavior. *Journal of Personality and Social Psychology*, **98**(6): 946–55.

Grégoire, D.A., Barr, P.S. and Shepherd, D.A. 2010. Cognitive processes of opportunity recognition: The role of structural alignment. *Organization Science*, **21**(2): 413–31.
Grégoire, D.A. and Shepherd, D.A. 2012. Technology–market combinations and the identification of entrepreneurial opportunities: An investigation of the opportunity–individual nexus. *Academy of Management Journal*, **55**(4): 753–85.
Hambrick, D.C. 2007. Upper echelons theory: An update. *Academy of Management Review*, **32**(2): 334–43.
Hambrick, D.C. and Mason, P.A. 1984. Upper echelons: The organization as a reflection of its top managers. *Academy of Management Review*, **9**(2): 193–206.
Haynie, J.M., Shepherd, D., Mosakowski, E. and Early, P.C. 2010. A situated metacognitive model of the entrepreneurial mindset. *Journal of Business Venturing*, **25**(2): 217–29.
Hmieleski, K.M. and Baron, R.A. 2009. Entrepreneurs' optimism and new venture performance: A social cognitive perspective. *Academy of Management Journal*, **52**(3): 473–88.
Koutstaal, W. and Cavendish, M. 2006. Using what we know: Consequences of intentionally retrieving gist versus item-specific information. *Journal of Experimental Psychology: Learning, Memory, and Cognition*, **32**(4): 778–91.
Kruglanski, A.W., Thompson, E.P., Higgins, E.T., Atash, M.N., Pierro, A., Shah, J.Y. and Spiegel, S. 2000. To 'Do the right thing' or to 'Just do it': Locomotion and assessment as distinct self-regulatory imperatives. *Journal of Personality and Social Psychology*, **79**(5): 793–815.
McFarlin, D.B., Baumeister, R.F. and Blascovich, J. 1984. On knowing when to quit: Task failure, self-esteem, advice, and nonproductive persistence. *Journal of Personality*, **52**(2): 138–55.
McMullen, J.S. and Shepherd, D.A. 2006. Entrepreneurial action and the role of uncertainty in the theory of the entrepreneur. *Academy of Management Review*, **31**(1): 132–52.
Mischel, W. and Ayduk, O. 2004. Willpower in a cognitive–affective processing system: The dynamics of delay of gratification. In R.F. Baumeister and K.D. Vohs (eds), *Handbook of Self-regulation: Research, theory, and applications*: 99–129. New York: Guilford Press.
Mitchell, R.K., Busenitz, L.W., Bird, B., Gaglio, C.M., McMullen, J.S., Morse, E.A. and Smith, J.B. 2007. The central question in entrepreneurial cognition research 2007. *Entrepreneurship Theory and Practice*, **31**(1): 1–27.
Nambisan, S. and Baron, R.A. 2013. Entrepreneurship in innovation ecosystems: Entrepreneurs' self-regulatory processes and their implications for new venture success. *Entrepreneurship Theory and Practice*, **37**(5): 1071–97.
Nisbett, R.E. and Wilson, T.D. 1977. Telling more than we can know: Verbal reports on mental processes. *Psychological Review*, **84**(3): 231–59.
Peterson, S.J., Luthans, F., Avolio, B.J., Walumbwa, F.O. and Zhang, Z. 2011. Psychological capital and employee performance: A latent growth modeling approach. *Personnel Psychology*, **64**(2): 427–50.
Schmidt, A.M. and Dolis, C.M. 2009. Something's got to give: The effects of dual-goal difficulty, goal progress, and expectancies on resource allocation. *Journal of Applied Psychology*, **94**: 678–91.
Shane, S. 2012. Reflections on the 2010 AMR decade award: Delivering on the promise of entrepreneurship as a field of research. *Academy of Management Review*, **37**(1): 10–20.
Staw, B.M. 1981. The escalation of commitment to a course of action. *Academy of Management Review*, **6**(4): 577–87.
Tang, J., Kacmar, K.M. and Busenitz, L. 2012. Entrepreneurial alertness in the pursuit of new opportunities. *Journal of Business Venturing*, **27**(1): 77–94.
Tangney, J.P., Baumeister, R.F and Boone, A.L. 2004. High self-control predicts good adjustment, less pathology, better grades, and interpersonal success. *Journal of Personality and Social Psychology*, **72**: 2–21.
Tice, D.M., Baumeister, R.F., Shmueli, D. and Muraven, M. 2007. Restoring the self: Positive affect helps improve self-regulation following ego depletion. *Journal of Experimental Social Psychology*, **43**(3): 379–84.

Ucbasaran, D., Shepherd, D., Lockett, A. and Lyon, J. 2013. Life after business failure: the process and consequences of business failure for entrepreneurs. *Journal of Management*, **39**: 163–202.

Venkataraman, S., Sarasvathy, S.D., Dew, N. and Forster, W.R. 2012. Reflections of the 2010 AMR decade award: Whither the promise? Moving forward with entrepreneurship as a science of the artificial. *Academy of Management Review*, **37**(1): 21–33.

2. Linking achievement motivation to intentions, goals and entrepreneurial behaviors

Alan L. Carsrud and Malin Brännback

ABSTRACT

This chapter builds on previous work on entrepreneurial intentions and achievement motivation within the context of goal setting to show how these cognitions subsequently impact entrepreneurial behaviors. This chapter, like Carsrud and Brännback (2011) and Carsrud et al. (2009), should not be considered a comprehensive review of the literature on motivation, intentions, goal setting or other entrepreneurial behaviors. It is rather our attempt to provide potential models linking cognitive factors such as intentions and motivations to subsequent behaviors such as goal setting. In addition, we discuss how such behaviors can then have reciprocal impacts on cognitive and motivational processes that in turn affect future behaviors.

WHO IS AN ENTREPRENEUR?

As we have noted (Carsrud and Brännback, 2011), the entrepreneur is the core of the field, even when the individual is not the unit of analysis in a study. The entrepreneur is an 'innovator,' 'creator' (Schumpeter, 1934), 'locator' and 'implementer' of ideas through exercise of leadership (Baumol, 1968). The entrepreneur is the actor in the process-conscious market theory who exhibits deliberate behaviors (Kirzner, 1973, 1979). Others see the entrepreneur as the 'possessor' of idiosyncratic knowledge that impacts opportunity recognition (Shane and Venkataraman, 2000; Gaglio and Katz, 2001; Shane, 2003; Eckhart and Shane, 2003). Thus what drives and directs the entrepreneur internally becomes important in understanding how to encourage more people to become entrepreneurs with the goal of starting a venture.

WHY STUDY THE INDIVIDUAL?

With the publication of *Job Creation in America: How our smallest companies put the most people to work* (Birch, 1987), it became obvious that small entrepreneurial firms were the drivers of economies. As a result, research in small business and entrepreneurship focused on understanding the entrepreneur, including finding ways of discovering potential entrepreneurs and motivating people to be entrepreneurs. This was a focus of researchers and policy makers (Carland, Hoy, Boulton and Carland, 1984) and continues to be so (Brännback and Carsrud, 2008; Carsrud and Brännback, 2011).

Research on the motivation of entrepreneurs can be traced to two volumes that appeared even before Birch's work. These are *The Art and Science of Entrepreneurship* (Sexton and Smilor, 1985) and *Managing Take-off in Fast Growth Firms* (Smilor and Kuhn, 1986), in which studies on the motivation of the entrepreneur appeared, along with early papers in *Entrepreneurship & Regional Development* in the same time period. A few years later the study of entrepreneurial intentions began to appear (Krueger and Carsrud, 1993; Krueger, Reilly and Carsrud, 2000), which also continues (Carsrud, Brännback, Elfving and Brandt, 2009). Despite extensive research on intentions, little work has explicitly tied motivation to intentions (and other cognitive factors). Even less research has tied intentions to various entrepreneurial behaviors, including goal setting and attainment.

SEARCHING FOR THE ELUSIVE ENTREPRENEURIAL TRAIT

As noted in Carsrud and Brännback (2011) and Carsrud, Brännback, Elfving and Brandt (2009), entrepreneurship researchers borrow heavily from the existing theories in the behavioral and social science disciplines in attempts to answer the question of what makes someone an entrepreneur. These included not only theories on achievement motivation (Maslow, 1943; McClelland, 1961, 1985), intentions-related work such as the theory of planned behavior (Ajzen, 1991, and its various modifications) and increasingly work on goal-directed behavior (Bagozzi, Dholakia and Basuroy, 2003). We have yet to see the application of approaches such as force-field analysis (Kruglanski, Bélanger, Chen and Köpetz, 2011), but it may well have utility. This borrowing by entrepreneurship researchers is in fact often done with little real understanding of the theory and prior work using it (McMullen and Kenworthy, 2014).

Most of the early research on the entrepreneurial personality suffered from the flawed assumption that there was some single trait, or set of traits, unique to entrepreneurs. Many assumed it to be an easy task to identify entrepreneurial personality traits and thus to easily identify a potential entrepreneur. When studies showed no such unique traits, this line of research was largely abandoned. What many researchers did not grasp was that cognitive factors such as personality traits, cognitions, motivations and intentions could still be a way to understand what leads to entrepreneurial behavior. As has been shown, what drives success in other professions can and does drive entrepreneurs (Carsrud and Brännback, 2011; Carsrud et al., 2009; Carsrud, Olm and Thomas, 1989).

The failure to find something unique occurred because many entrepreneurship researchers do not have an adequate understanding of the personality or cognitive social psychology literature. Most were looking for simple direct models rather than accepting that there is a high degree of complexity in human cognition and subsequent behaviors that must be understood (see also Sarasvathy, Chapter 14, this volume). For example, it is important to recognize that personality traits are exhibited through both cognitions and behaviors, and thus, in order to determine underlying personality traits, we are often dependent on assessing cognitions or observing behaviors (see also Baron, Chapter 1, this volume; Bird, Chapter 3, this volume).

We also believe that the attempt by researchers to look for cognitions in entrepreneurs that are different from those of managers fails to take into account that entrepreneurs are also managers. Therefore, once again researchers are hunting for the elusive single unique feature rather than accepting that entrepreneurial cognitions and behaviors may be driven by the same factors that drive any other business-related behavior by managers or employees, including goal setting. Finally, it is important to remember that managers and employees can be entrepreneurial in their behaviors (Renko et al., 2014).

The focus of attention then turned to the firm, in particular entrepreneurial processes and activities at the firm level (Gartner, 1988). Yet it is the entrepreneur (and his/her team) who still makes the decisions for the firm and sets strategy. Firm-level 'behaviors' are the direct result of individual-level decisions and behaviors. Entrepreneurial strategy is set by individuals acting on behalf of the firm. With that realization, the entrepreneur has once again become a focal point of research. This time the interest is in motivations, intentions, passions, attributions, attitudes, beliefs, goal setting behaviors and outcomes (Carsrud and Brännback, 2009; see also Drnovšek, Slavec and Cardon, Chapter 7, this volume). An example of this concern with individual behavior being manifest in its impact on others is the work on entrepreneurial leadership, which can occur at all levels of a firm, from the CEO to the line-employee (Renko et al., 2014).

ENTREPRENEURIAL INTENTIONS

Previously a small group of researchers viewed the understanding of intention to be an entrepreneur as an important link between ideas and action (Bird, 1989; Davidsson, 1991; Krueger and Carsrud, 1993; Krueger, Reilly and Carsrud, 2000). Measuring a person's intent in respect of an entrepreneurial activity was the best predictor of that specific future behavior occurring. Models were dominated by variations on Ajzen's (1991) theory of planned behavior (TPB). Intention effects have indeed been widely demonstrated (Davidsson, 1991; Krueger and Carsrud, 1993; Krueger and Brazeal, 1994; Krueger et al., 2000; Elfving et al., 2009). Generally, intentions are perceived to be driven by personal attitudes and personally perceived behavioral control, also known as 'self-efficacy' (Bandura, 1986, 1989). Self-efficacy can vary in intensity and may be both general self-efficacy and specific entrepreneurial self-efficacy, depending on the particular goal (e.g., starting a venture) toward which it might be directed. Yet the role of goal setting in self-efficacy has yet to be fully researched.

In the research on entrepreneurial intentions, the impact of social norms (including family and societal norms, and social capital) on self-efficacy and subsequent intentions has not been consistently shown (Carsrud and Brännback, 2011; Carsrud et al., 2009). In fact, it has recently been shown that the impact of such norms on self-efficacy varies across cultures (Carsrud, Koçak, Lin, Brännback and Mauer, 2014). Additional work needs to be done to discover what drives this factor in some situations and not in others. However, investigating cultural impacts on self-efficacy is not the primary purpose of this chapter. We can conclude, for the purposes of this chapter, that the general model of entrepreneurial intentions is far less clear than many have assumed.

Studies have confirmed that the relationship between attitudes and behavior is explained by the links between attitude and intention and between intention and behavior links (Kim and Hunter, 1993). While some researchers have acknowledged that motivations are linked to entrepreneurial behaviors (Herron and Robinson, 1993), most researchers perceived the motivation linkage as lacking in the literature (Kuratko, Hornsby and Naffziger, 1997). This assertion exists despite strong evidence to the contrary, produced by, for example, Carsrud and Olm (1986) and Carsrud, Olm and Thomas (1989), who clearly showed a decade

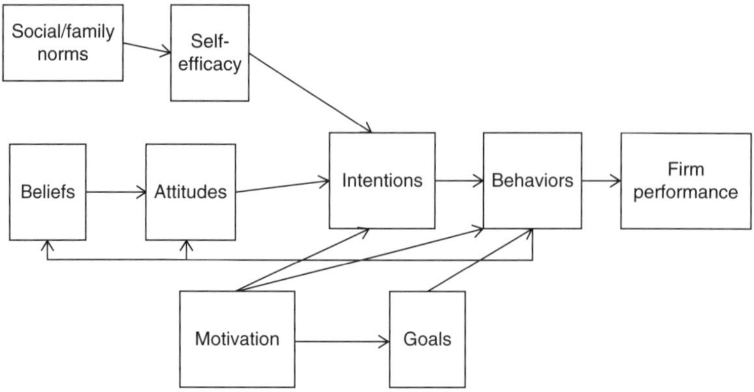

Figure 2.1 Motivations, subsequent behavior and firm performance

earlier the impact of entrepreneurial motivation and subsequent behavior on firm performance. The above discussion is represented by Figure 2.1.

LINKING MOTIVATION, INTENTIONS, GOALS AND BEHAVIORS

Research has also found support for the idea of different stages of intentions (Gollwitzer and Brandstätter, 1997; Gollwitzer and Schaal, 1998), which indicates that the entrepreneurial process may not be linear and suggests the presence of goal-directed behavior with different levels of goals serving as external motivators (Lawson, 1997). We will elaborate on this later in this chapter. For now, we turn to motivation and goals. Bay and Daniel (2003) conceptualized the hierarchy of goals, which implicitly argues for differences in motivational intensity as a requirement for a goal to be acted on. That is, you may intend to start a firm, but if the motivation to do so is weak, the intention is never enacted (see also Bagozzi, 2007). One critique of entrepreneurial intentions studies argues that, while intentions are the best predictors of future action, there is still insufficient understanding of the link between intention and action (Bagozzi and Warshaw, 1990, 1992; Bagozzi, 1992; McBroom and Reed, 1992; Brännback et al., 2007; Bird and Schjoedt, 2009; Edelman et al., 2010). The critical questions are: what triggers subsequent action and under what conditions? In this chapter we argue that it is critical to study the linkage of motivations to intentions.

Simply defined, motivations drive actions. Beliefs are cognitions about what generally exists and these impact attitudes, which are more specific in terms of their reference. Intentions are cognitions about the potential of taking action, which can occur at varying levels. Figure 2.1 tries to show this. We will discuss these concepts in greater detail later in this chapter, as we discuss other issues. However, we need to remind readers that this chapter is not designed to be a thorough review of the above concepts. Nor should it serve as a substitute for an in-depth reading of some 50 years of research in cognitive social psychology.

TWO SCHOOLS OF MOTIVATION THEORIES

There are essentially two schools of motivational theories, one based in economics and the other rooted in psychology (Fisher, 1930), that have been in conflict with each other for decades. Steel and König (2006) and Wilson (1998) called for the use of 'consilience,' which is the linking of facts and fact-based theory across disciplines to create a common framework between these two schools. This approach has brought together various theories of motivation as applied in economics, management and psychology (with a time dimension) into what they call temporal motivational theory (TMT). While this approach may be useful, for this chapter we have chosen to focus on fact-based theories in the psychological tradition.

MOTIVATION RESEARCH IN THE PSYCHOLOGICAL TRADITION

Historically, the psychological tradition in motivation research can be traced to Freud's work on instincts (1900, 1915, 1924), along with the research that followed on needs (Maslow, 1943; Deutsch and Krauss, 1965). Instincts (motives) drive behavior where the goal is to survive, to succeed, and to avoid failure. Traditionally, motivation has been studied in order to answer three main questions: (1) what activates a person; (2) what makes an individual choose one behavior over another; and (3) why do different people respond differently to the same motivational stimuli?

These questions give rise to three important aspects of motivation: activation, selection-direction and preparedness of response (Perwin, 2003). Existing psychologically based motivational theories can be divided roughly into drive theories and incentive theories. Drive theories suggest that there is an internal stimulus, for example hunger or fear, that drives the person, and that the individual seeks a way to reduce the resulting tension. The need for tension reduction thus represents the motivation (Freud, 1924; Murray, 1938; Festinger, 1957). On the other hand, incentive theories emphasize the motivational pull. There is an end point in the form of some kind of goal, which pulls the person towards it. In other words, in drive theories the push factors dominate, while in incentive theories the pull factors dominate. An example of a drive-theory factor would be to discover the cure for cancer, while the incentive-theory view of this example would be the wealth or status that would come from such a discovery.

MOTIVATIONS AND ASPIRATIONS

One such psychological theory is that of achievement motivation, or need for achievement (NAch), which was initiated by Atkinson (1957, 1964). First a unidimensional approach was proposed by McClelland and Winter (1969), and later a multidimensional view by Spence and Helmreich (1978). It was this latter approach that was used to study motivation in entrepreneurs (Carsrud and Olm, 1986; Carsrud, Olm and Thomas, 1989). Interestingly, Atkinson (1957, 1964) builds his model of achievement motivation on the theory of levels of aspiration. Other areas of fact-based motivation research include cognitive dissonance and risk (Cohen and Zimbardo, 1969); and work motivation (Pinder, 1984, 1998), which is the combination of internal and external factors that initiate work-related behaviors and determine its form, direction, intensity and duration (Ambrose and Kulik, 1999). The notion of different levels of aspiration with respect to achievement motivation can also be found in the theory of the hierarchy of goals (Lawson, 1997). Whether a goal leads to action is dependent on its level of abstraction. The more abstract the goal, the less likely it is to be acted on (Bagozzi and Warshaw, 1990, 1992; Bay and Daniel, 2003; Brännback et al., 2007). We will discuss some of the above in more detail later as we relate them to entrepreneurs.

GOALS AND MOTIVATIONS

The importance and impact of goals have gained attention in motivational research (Locke and Latham, 2002, 2004). Goals are mental representations of what the future could be like, encouraging individuals, such as entrepreneurs, not to give up (Perwin, 2003). Goals activate people in ways that often serve as the important link between intention and action (Perwin, 2003; Nuttin, 1984). In fact, being capable of changing goals and

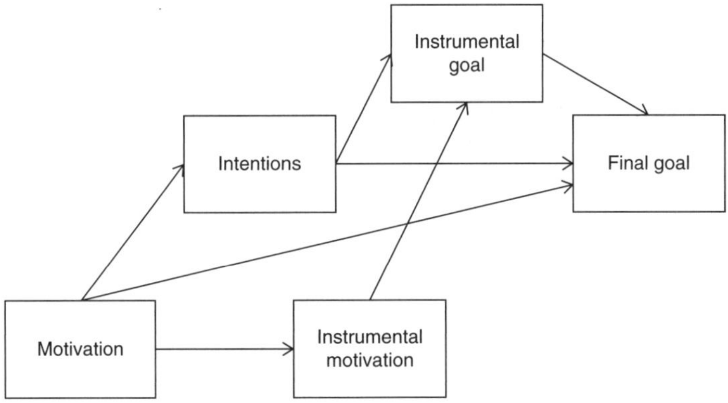

Figure 2.2 The impact of motivations on intentions, and final goals

motives is a way for people to adjust to changing situations or contingencies. This notion is present in the conceptualization of effectuation, which has recently gained attention from entrepreneurship scholars (Sarasvathy, 2001, 2008).

This view also is consistent with Nuttin (1984), who distinguishes between two contextual modes of motivation: final and instrumental motivations. When a person pursues a certain goal, they have a final motivation. When they are doing something that indirectly leads to the final goal, they have an instrumental motivation. Interestingly, entrepreneurship research in this area has been lacking except for work on effectuation in entrepreneurship. Few people have seen various entrepreneurial tasks such as setting up legal status for the firm as an instrumental goal, with the task driven by instrumental motivation. This is represented in Figure 2.2.

MOTIVATIONS AND INTENTIONS

Another view is that of Ryan and Deci (2000), who see motivation as the core of biological, cognitive and social regulation (see also Randolph-Seng, Williams and Hayek, Chapter 4, this volume). They state that motivation involves the energy, direction and persistence of activation, as well as the intention. This indicates that goals and motives play a role in predicting human behavior, and that a link between intentions, motivations and goal-directed behavior indeed exists. This relationship is most likely not linear nor unidirectional. First, there is a time effect. Intentions do not lead to immediate action. This time delay has also been found with respect to NAch (Helmreich, Sawin and Carsrud, 1986). In addition, research has shown that reciprocity exists: attitudes influence behaviors and behaviors influence attitudes (Kelman, 1974), even when looking at entrepreneurial intentions (Brännback et al., 2007). Motivations *may* be the spark that transforms a latent intention into real action, and are therefore the missing link between intentions and action. However, this is a very under-researched area within the existing entrepreneurship literature, although some research is being done in this area (Edelman et al., 2010). The above sections can be summarized in Figure 2.2.

MOTIVATION AND THE ENTREPRENEUR

Traditionally, the reasons for starting a firm (the entrepreneurial goal) have been thought to be economic (Schumpeter, 1934). In the nineteenth century Jean-Bertrand Say defined the entrepreneur as a person who does something for economic gain, and that notion has persisted ever since (Carsrud and Brännback, 2009). Recent research has shown that there are other motives for a person to create a venture, especially in the area of social entrepreneurship. Here the social gains are the primary motivators. It also is acknowledged that lifestyle entrepreneurs are driven by goals and motives that may indeed be economic, but are not necessarily to maximize economic gains. Often their motivation may be far more personal, as we try to show in Figure 2.3.

We know that artists or craftsmen certainly hope to make a living based on their art or crafts; they may not define themselves as entrepreneurs. Rather they define themselves by what motivates them to do what they do (Elfving, 2008). Take, for example, a musician who wants to play

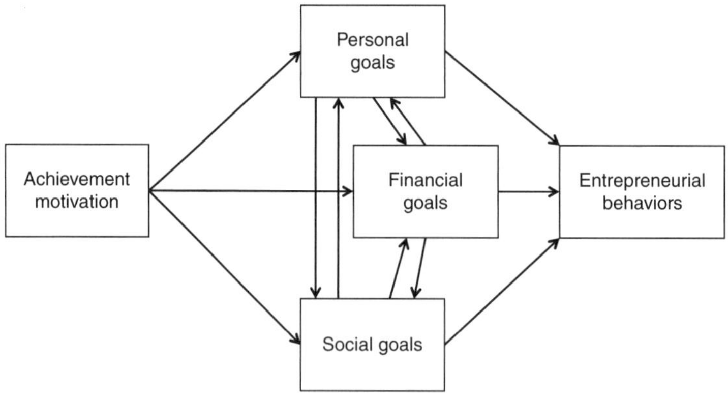

Figure 2.3 Achievement motivations, goals and behavior

music and is prepared to pursue this goal at any cost. As Elfving (2008) has shown, motivations and goals may change over time. The musician who initially played his music at any cost may become motivated to play increasingly for economic reasons over time as a result of initial commercial success and acclaim by audiences. Initial success is the proof that 'I can do it' and provides further encouragement to lift the aspiration level, thus changing goals. See Figure 2.3.

NEED FOR ACHIEVEMENT/ACHIEVEMENT MOTIVATION

As mentioned earlier, success may encourage an individual to lift their aspiration level; this is familiar from the motivational construct of achievement motivation (NAch) (Atkinson, 1957, 1964). While NAch is seen as an important element in entrepreneurial behavior, research results have shown considerable variation (McClelland et al., 1953; McClelland, 1961, 1965; Brockhaus, 1980, 1982; Gasse, 1982; Carland et al., 1984; Carsrud and Olm, 1986; Carsrud et al., 1986; Carsrud et al., 1989). Carland et al. (1984) argued that small-business owners perceived their business as an extension of their personality, while the entrepreneur was characterized by innovative business behavior. This claims that small-business owners attribute their success to their own efforts and entrepreneurs to another part of themselves – their innovative behaviors. Thus this particular distinction is somewhat meaningless, as personality drives behaviors. McClelland and Winter (1969) found that NAch was the differentiating factor between small-business entrepreneurs and other business leaders. Small-business owners were higher in NAch. The issue is that at some point a small-business owner was probably an entrepreneur who started the firm; is McClelland therefore saying that NAch declines once a business is up and running? There is a huge puzzle in this research that few have recognized.

Carsrud, Olm and Eddy (1986) studied multidimensional NAch in samples of male and female entrepreneurs, finding patterns similar to other successful professionals, such as university professors and airline pilots. The role of NAch in entrepreneurial behavior has continued to attract interest among entrepreneurship scholars (Langen-Fox and Roth, 1995; Tuuanaen, 1997; Collins et al., 2004; Lumpkin and Erdogan, 2004; Steward and Roth, 2007; Hart et al., 2007; Carsrud et al., 2009), but more research is needed.

Studies using multidimensional NAch show that the best performance is exhibited by individuals scoring high in mastery needs and work

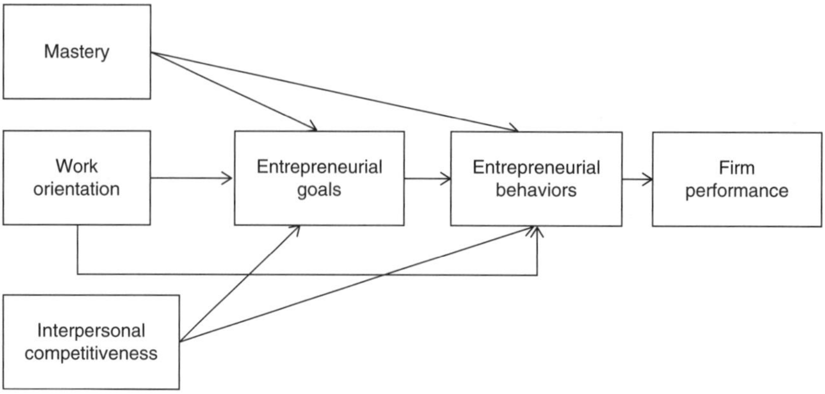

Figure 2.4 Multidimensional NAch and entrepreneurial behavior

orientation, but low in interpersonal competitiveness. This has been found in research scientists, airline pilots and other professionals. Moreover, interpersonal competitiveness, popularly considered a trait of entrepreneurs, is not related to actual entrepreneurial success (Carsrud and Olm, 1986; Carsrud, Olm and Thomas, 1989; Carsrud et al., 2009). The above discussion is summarized in Figure 2.4.

MEASURING MULTIDIMENSIONAL NEED FOR ACHIEVEMENT

At this point we think it necessary to discuss measuring NAch. While Maslow and others used projective techniques, a multidimensional scale of NAch is found in the 'Work and Family Orientation Inventory' (WOFO) (Helmreich and Spence, 1978). The WOFO contains three subscales relevant to the study of entrepreneurs that go beyond the 'lifestyle' concerns of the more unidimensional scales of Mehrabian (1968) and Komives (1972). The WOFO subscales refer to 'mastery needs,' 'work orientation' and 'interpersonal competitiveness.' These dimensions are assessed through questions such as 'I like to work hard' (work orientation), 'I prefer to work in situations that require a high level of skill' (mastery needs) and 'I feel that winning is important in both work and games' (interpersonal competitiveness).

To us, clearly these scales tap into some underlying motivational characteristics of the entrepreneur. The concept of 'mastery' has a great deal in common with the concept of self-efficacy, which is a key antecedent of

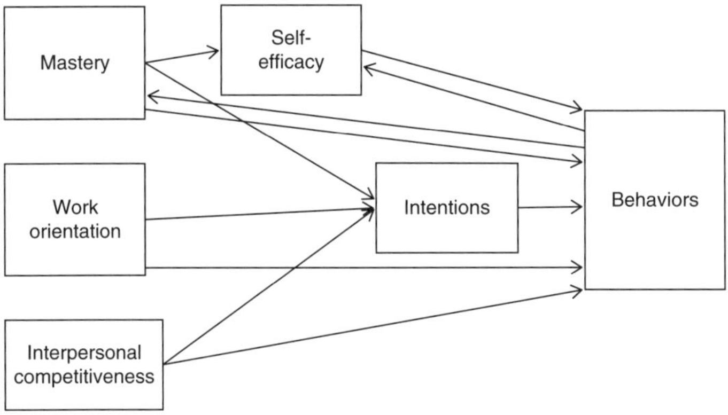

Figure 2.5 WOFO and intentions

entrepreneurial intentions (Krueger, Reilly and Carsrud, 2000; Bandura and Locke, 2003; Zhao, Siebert and Hills, 2005; Wong, Lee and Leung, 2006). Self-efficacy is one example of where a motivationally related concept has been used in entrepreneurial intentions research, but where motivation has not been directly and empirically studied. To show how this now ties motivation to intentions, see Figure 2.5.

MOTIVATION IN NECESSITY VERSUS OPPORTUNISTIC ENTREPRENEURS

A basic assumption, in our view, is that entrepreneurs have the same motivations as anyone for fulfilling their needs and wants in the world. However, they use those motivations in a different manner – they create ventures rather than just work in them. Some become entrepreneurs, even when other even more attractive options for employment exist. They recognize an opportunity and act. The opportunistic entrepreneur (Reynolds et al., 2002) is driven by the achievement of success through exploiting an opportunity for some form of gain, often economic. Commercially oriented entrepreneurs work to earn money, power, prestige and/or status, but these might not be their only motivations. For example, in biotechnology, the search for a cure for a disease may be a far more powerful motivator than personal wealth creation. See again Figure 2.3.

While opportunistic entrepreneurs may be motivated by a need to achieve economically, other entrepreneurs are driven by what could be described as survival-oriented motivations. These are people commonly known as necessity entrepreneurs (Reynolds et al., 2002). Necessity entrepreneurs are more concerned with avoiding failure, which could mean starvation. The central motivation is to earn enough money to be able to support oneself and one's family. When focused on survival, one may ignore opportunities that have a longer payback period. Some necessity entrepreneurs simply cannot afford to wait to achieve a bigger goal as they might starve to death waiting. Thus necessity entrepreneurship could inhibit opportunistic entrepreneurship, rather than foster it. Failing could mean death, a risk not worth taking. Thus understanding risk as a motivator means understanding both success and failure.

COGNITIVE DISSONANCE, RISK, SUCCESS AND FAILURE

Necessity entrepreneurs seek to avoid failure as it is potentially life-threatening for them, their family, and even their communities. It is the very lack of resources that drives these people. This is not what we mean by NAch. It is important to note that, while NAch is high among opportunistic entrepreneurs, they also seek to avoid failure, but in this case this refers to avoiding losing access to future resources or to their personal self-esteem. Frequent failure might well modify an entrepreneur's self-efficacy. The complexity of motivations is exhibited in cognitive dissonance and risk avoidance, both of which are strong motivators for opportunistic entrepreneurs (Monsen and Urbig, 2009).

In addition, research on cognitive dissonance and the need to avoid failure (Cohen and Zimbardo, 1969) could be used to explain why entrepreneurs often do anything to avoid failure in their venture, and why they show higher tenacity (Baum, Locke and Smith, 2001; Baum and Locke, 2004). Moreover, research shows that individuals with high NAch will show greater cognitive dissonance the greater the probability of failure (Cohen and Zimbardo, 1969). This could explain why some individuals commit themselves to a high-risk venture, while others do not. It could also explain the behaviors of investment bankers, venture capitalists and angel investors.

FAILURE AND SUCCESS MOTIVATIONS IN ENTREPRENEURS

Atkinson (1957) showed that failure and success motivations are separate and have different implications for behavior. This distinction appears to have been ignored by entrepreneurship researchers. Risk-taking propensity was treated by entrepreneurship researchers as a personality trait and not as two parts of a motivational paradigm that included dissonance. Even recent commentaries on risk-taking behavior in entrepreneurs (Lumpkin and Erdogan, 2004; Segal et al., 2005) have not applied this broader perspective on success and failure (Carsrud et al., 2009; Carsrud and Brännback, 2011). Building on Atkinson (1957) and Deci (1975), the relationship between success and risk thus includes the motivation of success. Motivation of success is constant in an individual and has an incentive value. The incentive value is higher when a difficult goal is pursued and achieved (such as starting a new firm).

Therefore a person with a strong tendency to create a venture that is considered moderately risky will have a high motive for success (Carsrud et al., 2009). For opportunistic entrepreneurs, the fear of failure is the motive to avoid personal disappointment. There are also expectancies about failure, and an incentive value for failure as well as for success. The motive to avoid failure has been found to be relatively stable (Deci, 1975), and the emotions of shame and embarrassment accompanying failure as an entrepreneur are greater, the easier the task. In other words, the greater the shame, the greater the incentives to avoid failure, such as might occur in starting a new business (Carsrud et al., 2009). There is a need to look at those entrepreneurs who have failed as well as at those that succeed. A more complete discussion on risk taking can be found in Monsen and Urbig (2009). The above sections are represented in Figure 2.6.

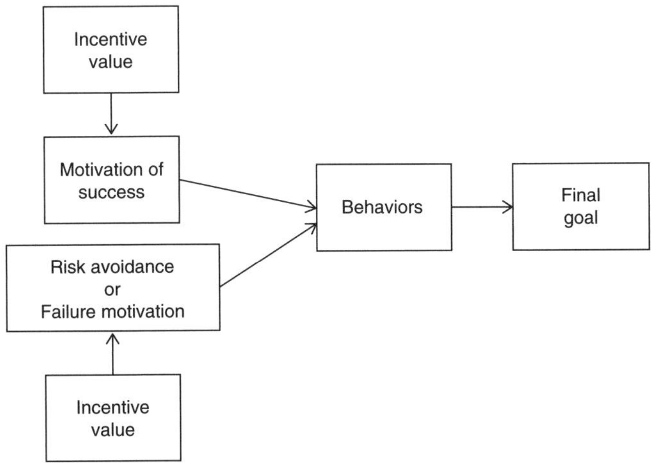

Figure 2.6 How failure and success impact entrepreneurial behavior

INTRINSIC AND EXTRINSIC MOTIVATIONS IN ENTREPRENEURS

Another way to look at motivation is that it can be *intrinsic*, *extrinsic*, or both. Intrinsic motivation refers to a personal interest in the entrepreneurial task, as seen in studies on multidimensional achievement motivation in entrepreneurs (Carsrud and Olm, 1986; Carsrud et al., 1989; Carsrud et al., 2009). Extrinsic motivation refers to an external reward that follows certain behavior. Intrinsic and extrinsic motivations are not mutually exclusive. One can be motivated by both internal and external motives to perform an entrepreneurial act (Elfving, 2008). Internally, entrepreneurs may be motivated to succeed and accomplish a goal, while externally they may be motivated to obtain wealth and status. While most economic-based entrepreneurial research assumes the entrepreneur is motivated by external rewards such as money, power and status, one is left with the reality that some people engage in entrepreneurial activities as an end in themselves. Thus the classic definition of intrinsic motivation could play a role in explaining why social entrepreneurs start social ventures even when there is no apparent reward for doing so other than some internally generated satisfaction. Figure 2.7 represents this discussion.

The idea that an individual engages in entrepreneurial behaviors because of the need for stimulation (a form of intrinsic motivation) is not revolutionary. However, the fact that serial entrepreneurs do this habitually may provide some interesting insights into such behavior. Once an entrepreneur has had the stimulation of starting a firm, he/she frequently returns to that behavior because of intrinsic motivation and the internal and external rewards received from that behavior in the past. Such people

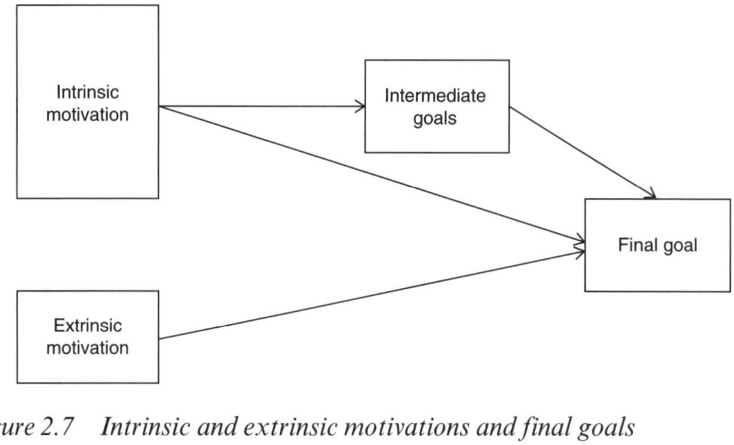

Figure 2.7 Intrinsic and extrinsic motivations and final goals

might even persist in trying for internal reasons, despite never having been rewarded externally through a successful venture. They reduce the cognitive dissonance of perceived possible failure by believing they can be successful this time around. Finally, external motivations include relatively intangible things such as status, power, social acceptance and so on, with the more tangible eternal rewards being money, stock options and other forms of compensation (Carsrud et al., 2009).

ENTREPRENEURIAL WORK MOTIVATION AND VENTURE DESIGN

Of more concern is that entrepreneurship researchers have largely ignored the extensive literature on work motivation (Pinder, 1984, 1998). We also find it interesting that Hackman and Oldham's (1976) research on work design has not been applied to how entrepreneurs design their ventures. From our own experience, some entrepreneurs are motivated by the kinds of firms they could build to create firms in which they would want to work. This might well come from their prior work experiences. Work motivation may offer a viable link to the literature on intentions, goals, goal setting, entrepreneurial leadership, job enrichment and the organizational design of a new venture (Gächter and Falk, 2000; Quigley and Tymon, 2006). One has only to look at the popular media, which is full of stories of the unique organizational structures, perks and incentives that entrepreneurs create for their new ventures to attract and keep employees.

CHANGING ENTREPRENEURIAL GOALS AND MOTIVATIONS

The importance of goals in motivational research is well known (Locke and Latham, 2002; Bagozzi and Warshaw, 1990, 1992; Bay and Daniel, 2003). If people are capable of changing goals, motives and goal-specific intentions, they can adjust to changing situations. We all know that entrepreneurs often have to change direction rapidly and with minimal resources. That is, entrepreneurs' intentions, goals and motives change over time. As Nuttin (1984) points out, motivation is shaped in the individual–environment context. That is, one may intend to start a small business to support the family and end up with the intention to become a global powerhouse in retailing (think Sam Walton here). The changing context over time will impact entrepreneurial motivations and intentions.

This area requires further exploration (Elfving, Brännback and Carsrud, 2009; Carsrud et al., 2009; Edelman et al., 2010).

ENTREPRENEURS' GOALS AND THE THEORY OF PLANNED BEHAVIOR

Interestingly, behavioral goals are neither entirely ignored nor explicitly included in the theory of planned behavior (TPB) of Ajzen and Fishbein (1977) on which most entrepreneurial intentions research is based. Essentially, all entrepreneurial behaviors could be labeled as goals in TPB. Goals can be defined as every positive outcome that one seeks to gain through reasoned behavior (e.g., Ajzen and Fishbein, 1977). But not all goals are the same or hold the same value. For example, if an entrepreneur seeks to raise venture capital funding for his/her firm, the act of approaching the venture capitalist is planned behavior and the goal is the potential economic gain. However, Bagozzi and Warshaw (1990, 1992) would argue that TPB only explains performances, which are solely dependent on an intention and where no impediments prevent the implementation of the intention. Yet we all know that the entrepreneurship literature is full of work on the impediments to raising capital.

For example, if the venture capitalist declines to invest, it does not necessarily mean that the entrepreneur ceases to try to start the firm. He/she may choose to 'boot-strap' the operation. In another example, one may have the intention to start a venture, but the intention may not be acted upon because of any number of reasons, such as lack of social support, insufficient skills, cognitive dissonance and/or insufficient motivation. An entrepreneurial intention does not always lead directly to entrepreneurial behaviors. It is this time lag and how motivations can potentially shorten, or prolong, action that is not yet adequately understood within entrepreneurship research. TBP does not explain why people claim that they intend to start a firm but do not do so. What is missing in TBP is a motivational component.

ENTREPRENEURIAL GOALS AND THE THEORY OF TRYING

To explain the above behaviors, Bagozzi and Warshaw (1990) developed the theory of trying (ToT). While Ajzen and Fishbein (1977) treat action as a single performance, Bagozzi (1992) prefers to view action as an attempt, or a sequence of attempts, to reach final performance. ToT accommodates both intermediate goals and end-state goals (Bagozzi and Warshaw, 1990; Gollwitzer and Brandstätter, 1997; Gollwitzer and Schaal, 1998). With respect to entrepreneurial venture creation, sometimes there is a significant time lag between when the decision is made to start a firm and an opportunity arises, or when motivation is high enough to mandate action at least to try (Bagozzi, Dholakia and Basuroy, 2003; Shane, 2008). Bagozzi and Warshaw (1990) also added the impact of past behavior and additional background factors to their model of ToT. In TPB, intentions and performance are influenced by past behavior only through background factors (Ajzen and Madden, 1986; Ajzen and Fishbein, 2005).

However, Bagozzi and Warshaw (1990) argued that past behavior could make a substantial contribution to understanding future behavior, and could also possibly influence behavior directly without impacting the formation of intentions. This insight may be useful in explaining the behavior and underlying motivations among serial entrepreneurs. The old adage that 'the best predictor of future behavior is past behavior' once again has its place, even in entrepreneurship research. Studies based on ToT have been carried out on fairly low-level goals, such as losing weight or mastering a new piece of software. There is still a need for empirical studies involving higher-level goals such as those involved in new venture creation. It is important to note that various forms of goal-directed behavior can be placed on a continuum and that goals affect behavior differently depending on their position in the hierarchy.

STRENGTH OF ENTREPRENEURIAL MOTIVATIONS AND GOALS

Goals are mental representations of what the future could be like, enabling individuals to persist (Perwin, 2003; Bagozzi and Dholakia, 1999; Bagozzi and Kimmel, 1995). This persistence is a behavior observed in many entrepreneurs. Goals are central in Bandura's (1989) social cognitive theory, where self-efficacy partly determines what people intend to achieve and what kind of goal they set for themselves. As noted earlier, goals activate people, as they are directive and energizing, and they impact persistence. Goals can lead to arousal, discovery and emergence of strategies to achieve those goals (Locke and Lathman, 2002).

However, the strength of the activation is determined by the strength of the motivation. Weak motivation will not transfer into real action, especially if the task is perceived as difficult, not feasible (self-efficacy) or desirable. This may explain some of the findings in the PSED data, where a significant portion of the samples continue to try without either succeeding or failing. Hence we argue that motivation represents an important link between intention and action. This has been noted by Bird and Schjoedt (2009) when discussing the linkage of entrepreneurial cognitions to entrepreneurial behaviors. This indicates that specific goal setting and strength of motivations play a role in predicting entrepreneurial behavior. Once again, there is little direct research on this to date.

RECIPROCAL IMPACT OF FEEDBACK

The existence of feedback is another important factor in goal theory. Entrepreneurs need to be able to check where they stand in relation to their goals so that they can determine whether they need to make adjustments in their behavior in order to attain those goals (Locke, Lathman and Erez, 1988; Locke and Latham, 2002; Lent, Brown and Hackett, 1994; Lent and Brown, 2006). Social cognitive theory also implies that there is a reciprocal relation between self-efficacy, outcome expectations and goal systems (Bandura, 1986). There is some evidence that this is true with respect to entrepreneurial intentions. This reciprocal relationship has yet to be fully explored in the entrepreneurial literature.

INTENTIONS AND GOALS: FOUR PHASES OF INTENTIONS

The work of Gollwitzer and Brandstätter (1997) illustrates the linkage between intentions, motivation and goals by presenting the ideas of 'implementation intentions' and 'goal pursuit.' They describe people's goal pursuits as a continuum including four action phases. The first phase, the pre-decisional phase, is an awakening of desires and wishes (realizing one could be an entrepreneur). In the second phase, the pre-actional phase, goal-directed behavior is initiated (starting to look for opportunities or learning what it takes be an entrepreneur; see also Forbes, Chapter 10, this volume). In the third phase, the actional phase, the goal-directed actions are brought to a successful conclusion (actually starting a firm). Finally, in the fourth phase, the post-actional phase, the outcome is evaluated by comparing what has been achieved with what was originally desired (whether the new venture was a success, or met the expectations of the entrepreneur). It is important to remember that goals serve as motivators, the strength of which will increase as the goals are met or achieved and thus reinforced. This also has a reciprocal impact on self-efficacy and other cognitive factors.

The four action phases are connected through crucial transition points. Gollwitzer and Brandstätter (1997) label the first transition point 'goal intention.' A goal intention, for example, can be 'I intend to become an entrepreneur.' However, an intention is not enough to lead to an action as there might be several impediments along the way. There may also be different ways of achieving the goal that one may have to choose between in order to avoid the risk of failing to seize a specific opportunity. An 'implementation intention' can then function as a mediator and take the goal pursuit one step further. It serves to translate the goal state from a higher level of abstractness to a lower level, and to link a certain goal-directed behavior to a situational context.

An implementation intention could be 'I intend to start my own company when I have finished my studies.' An implementation intention results in a commitment to perform a specified goal-directed behavior once a critical step has been achieved. Furthermore, people who have formed an implementation intention should possess the cognitive structures needed to recognize additional opportunities when they emerge. Gollwitzer and Brandstätter (1997) conclude that a goal is more likely to be achieved if an implementation intention exists. While the idea of being or becoming an entrepreneur is a goal, the absence of an implementation intention will not result in entrepreneurial behavior (Elfving, 2008).

QUESTIONS FOR FUTURE RESEARCH

Building on the above discussion on entrepreneurial motivation, we propose a number of potential research questions that we believe have either been neglected, or for which further research is required. These also have been discussed in Carsrud et al. (2009) and in Carsrud and Brännback (2011). We present some of them again here, not as research hypotheses but as questions for future researchers. As we noted in our earlier work, more questions can certainly be proposed than are suggested below. We have attempted to cluster these by major focus, but several bridge different issues and foci.

General Motivation Questions

The overview of motivational research and motivational research in entrepreneurs leads to these questions:

1. Could aspiration level explain why some people choose to build high-growth firms and others choose lifestyle firms?

2. Could aspiration theory address the tendency of individuals to both achieve success and avoid failure?
3. What motivates different types of entrepreneurs (e.g., social, technology, lifestyle, opportunistic or serial entrepreneurs)?
4. How does motivation impact the decision *not* to create a venture?

Motivation and Opportunity Recognition Questions

The review of the role of motivation in the opportunity recognition process led to these questions:

1. What motivations drive opportunity recognition and how do they vary across different types of entrepreneurs?
2. How do motives, values and skills interact to determine the behaviors of entrepreneurs, especially in opportunity recognition?

Motivational Context Questions

All behaviors, cognitions and motivations exist within a context. Hence the following questions:

1. How does business context impact entrepreneurial motivation?
2. How do self-efficacy, outcome expectation, interest, goals and contextual settings interact to impact entrepreneurial behaviors?

Linking Motivations to Intentions and Behavior Questions

Critical to understanding the importance of motivations is to look at the role motivations may play in linking cognitions, like intentions, to actual behaviors. This leads to the following questions:

1. How does commitment on the part of entrepreneurs depend on the importance of the outcome?
2. How does the likelihood of success impact the entrepreneur's self-efficacy?
3. How does NAch impact self-efficacy in entrepreneurs?
4. How do goals activate entrepreneurs and how are they linked to intentions and actions?

Motivation and Work/Firm Design Questions

The role that an entrepreneur's motivations and cognitions play in how they design and set up their new ventures remains largely unexplored. The review of the work design literature leads to the following questions:

1. Could entrepreneurs who set out with a particular vision of their future success be motivated by the goal of potential future rewards, even though the present work might not be as internally satisfying or externally rewarding?
2. How are intentions, goals, goal setting, leadership and job enrichment linked in the entrepreneurial firm?
3. How do an entrepreneur's motives impact how they design work within their venture?

CONCLUSIONS

In summary, this chapter has explored various aspects of the role of entrepreneurial motivations. Building on our previous work (Carsrud et al., 2009; Carsrud and Brännback, 2011), we have shown that entrepreneurial motivations are important explanatory mechanisms for a variety of entrepreneurial behaviors. We have paid special attention to how such motivations may impact both intentions and subsequent behaviors, including goal setting. We have shown that motivation is implied, or assumed, in articles on entrepreneurial intentions, scripts and cognitive maps of entrepreneurial behaviors (Carsrud and Brännback, 2009; Brännback and Carsrud, 2009; Mitchell, Mitchell and Mitchell, 2009; Krueger, 2009). We see the area still lacking sufficient research despite its critical importance to predicting and explaining entrepreneurial behaviors. Our hope in this chapter is to provide sufficient rationale for entrepreneurship researchers to rediscover the rich complexity of motivations and explore in greater detail the role they play in entrepreneurial behaviors.

REFERENCES

Ajzen, I. 1991. Theory of planned behavior: Some unresolved issues. *Organizational Behavior and Human Decision Processes*, **50**(2): 179–211.

Ajzen, I. and Fishbein, M. 1977. Attitude–behavior relations: A theoretical analysis and review of empirical research. *Psychological Bulletin*, **84**(5): 888–918.

Ajzen, I. and Fishbein, M. 2005. The influence of attitudes on behavior. In D. Albarracin,

B. Johnson and M. Zanna (eds), *The Handbook of Attitudes*: 173–222. Mahwah, NJ: Lawrence Erlbaum Associates.
Ajzen, I. and Madden, T.J. 1986. Prediction of goal-directed behavior: Attitudes, intentions, and perceived behavioral control. *Journal of Experimental Social Psychology*, **22**(5): 453–74.
Ambrose, M.L. and Kulik, C.T. 1999. Old friends, new faces: Motivation research in the 1990s. *Journal of Management*, **25**(3): 231–92.
Atkinson, J.W. 1957. Motivational determinants of risk-taking behavior. *Psychological Review*, **64**(6 pt 1): 359–72.
Atkinson, J.W. 1964. *An Introduction to Motivation*. New York: Van Nostrand.
Bagozzi, R. 1992. The self-regulation of attitudes, intentions, and behavior. *Social Psychology Quarterly*, **55**(2): 178–204.
Bagozzi, R. 2007. The legacy of the technology acceptance model and a proposal for a paradigm shift. *Journal of the Association for Information Systems*, **8**(4): 244–254.
Bagozzi, R. and Dholakia, U. 1999. Goal setting and goal striving in consumer behavior. *Journal of Marketing*, **63**(4): 19–32.
Bagozzi, R., Dholakia, U. and Basuroy, S. 2003. How effortful decisions get enacted: The motivating role of decision processes, desires, and anticipated emotions. *Journal of Behavioral Decision Making*, **16**(4): 273–95.
Bagozzi, R. and Kimmel, P. 1995. A comparison of leading theories for the prediction of goal-directed behaviours. *British Journal of Social Psychology*, **34**(4): 437–61.
Bagozzi, R. and Warshaw, P. 1990. Trying to consume. *Journal of Consumer Research*, **17**(2): 127–40.
Bagozzi, R. and Warshaw, P. 1992. An examination of the etiology of the attitude–behavior relation for goal-directed behaviors. *Multivariate Behavioral Research*, **27**(4): 601–34.
Bandura, A. 1986. *Social Foundations of Thought and Action: A social cognitive theory*. Englewood Cliffs, NJ: Prentice-Hall.
Bandura A. 1989. Human agency in social cognitive theory, *American Psychologist*, **44**(9): 1175–84.
Bandura, A. and Locke, E.A. 2003. Negative self-efficacy and goal effects revisited. *Journal of Applied Psychology*, **88**(1): 87–99.
Baum, J. and Locke, E.A. 2004. The relationship of entrepreneurial traits, skill, and motivation to subsequent venture growth. *Journal of Applied Psychology*, **89**(4): 587–98.
Baum, J., Locke, E.A. and Smith, K. 2001. A multidimensional model of venture growth. *Academy of Management Journal*, **44**(3): 292–303.
Baumol, W.J. 1968. Entrepreneurship in economic theory. *American Economic Review*, **58**(2): 64–71.
Bay, D. and Daniel, H. 2003. The theory of trying and goal-directed behavior: The effect and moving up the hierarchy of goals. *Psychology and Marketing*, **20**(8): 669–84.
Birch, D. 1987. *Job Creation in America: How our smallest companies put the most people to work*. New York: Free Press.
Bird, B. 1989. *Entrepreneurial Behavior*. Glenview, IL: Scott Foresman & Co.
Bird, B. and Schjoedt, L. 2009. Entrepreneurial behavior: Its nature, scope, recent research, and agenda for future research. In A. Carsrud and M. Brännback (eds), *Understanding the Entrepreneurial Mind: Opening the black box*: 327–58. Heidelberg: Springer.
Brännback, M. and Carsrud, A. 2008. Do they see what we see? A critical Nordic tale about perceptions of entrepreneurial opportunities, goals and growth. *Journal of Enterprising Culture*, **16**(1): 55–87.
Brännback, M. and Carsrud, A. 2009. Cognitive maps in entrepreneurship: Researching sense making and action. In A. Carsrud and M. Brännback (eds), *Understanding the Entrepreneurial Mind: Opening the black box*: 75–96. Heidelberg: Springer.
Brännback, M., Carsrud, A., Kickul, J., Krueger, N. and Elfving, J. 2007. Trying to be an entrepreneur? A goal-specific challenge to the intentions model. *Frontiers of Entrepreneurship Research*, **27**(6): Article 8.

Brockhaus, R.H. 1980. Risk taking propensity of entrepreneurs. *Academy of Management Journal*, **23**(3): 509–20.

Brockhaus, R.H. 1982. The psychology of the entrepreneur. In C.A. Kent, D.A. Sexton and K.H. Vesper (eds), *Encyclopedia of Entrepreneurship*: 39–55. Englewood Cliffs, NJ: Prentice-Hall.

Carland, J.W., Hoy, F., Boulton, W.R. and Carland, J.A.C. 1984. Differentiating entrepreneurs from small business owners: A conceptualization. *Academy of Management Review*, **9**(2): 351–9.

Carsrud, A. and Brännback, M. (eds). 2009. *Understanding the Entrepreneurial Mind: Opening the black box*. Heidelberg: Springer.

Carsrud, A. and Brännback, M. (2011), Reflections on twenty years of research on entrepreneurial motivation: Have we learned anything at all? *Journal of Small Business Management*, **49**(1): 9–26.

Carsrud, A., Brännback, M., Elfving, J. and Brandt, K. 2009. Motivations: The entrepreneurial mind and behavior. In A. Carsrud and M. Brännback (eds), *Understanding the Entrepreneurial Mind: Opening the black box*: 141–65. Heidelberg: Springer.

Carsrud, A.L., Koçak, A., Lin, X., Brännback, M. and Mauer, R. 2014. From parental role models to entrepreneurial intentions: Cross-national comparisons on the mediating effects of self-efficacy and social capital. Paper presented at the United States Association for Small Business and Entrepreneurship, Fort Worth, TX.

Carsrud, A.L. and Olm, K.W. 1986. The success of male and female entrepreneurs: A comparative analysis of the effects of multi-dimensional achievement motivation and personality traits. In R. Smilor and R.L. Kuhn (eds), *Managing Take-off in Fast Growth Firms*: 147–63. New York: Praeger.

Carsrud, A.L., Olm, K.W. and Eddy, G.G. 1986. Entrepreneurship: Research in quest of a paradigm. In D. Sexton and R. Smilor (eds), *The Art and Science of Entrepreneurship*: 367–78. Cambridge, MA: Ballinger.

Carsrud, A.L., Olm, K.W. and Thomas, J.B. 1989. Predicting entrepreneurial success: Effects of multi-dimensional achievement motivation, levels of ownership, and cooperative relationships. *Entrepreneurship and Regional Development*, **1**(3): 237–44.

Cohen, A.R. and Zimbardo, P.G. 1969. Dissonance and the need to avoid failure. In P.G. Zimbardo (ed.), *The Cognitive Control of Motivation: The consequences of choice and dissonance*: 23–41. Glenview, IL: Scott Foresman & Company.

Collins, C.J., Hanges, P.J. and Locke, E.A. 2004. The relationship of achievement motivation to entrepreneurial behavior: A meta-analysis. *Human Performance*, **17**(1): 95–117.

Davidsson, P. 1991. Continued entrepreneurship. *Journal of Business Venturing*, **6**(6): 405–29.

Deci, E.L. 1975. *Intrinsic Motivation*. New York: Plenum Press.

Deutsch, M. and Krauss, R.M. 1965. *Theories in Social Psychology*. New York: Basic Books.

Eckhart, J.T. and Shane, S. 2003. Opportunities and entrepreneurship. *Journal of Management*, **29**(3): 333–49.

Edelman, L.F., Brush, C.G., Manolova, T.S. and Greene, P.G. 2010. Start-up motivations and growth intentions of minority nascent entrepreneurs. *Journal of Small Business Management*, **48**(2): 174–96.

Elfving J. 2008. *Contextualizing Entrepreneurial Intentions: A multiple case study on entrepreneurial cognitions and perceptions*. Turku, Finland: Åbo Akademi förlag.

Elfving, J., Brännback, M. and Carsrud, A. 2009. Toward a contextual model of entrepreneurial intentions. In A. Carsrud and M. Brännback (eds), *Understanding the Entrepreneurial Mind: Opening the black box*: 23–33. Heidelberg: Springer.

Festinger, L. 1957. *A Theory of Cognitive Dissonance*. Evanston, IL: Row, Peterson.

Fisher, I. 1930. *The Theory of Interest*. New York: Macmillan.

Freud, S. 1900. *The Interpretation of Dreams* (standard edns). London: Hogarth Press.

Freud, S. 1915. Instincts and their vicissitudes. In S. Freud (ed.), *Collected Papers*, vol. 4: 60–83. New York: Basic Books.

Freud, S. 1924. *A General Introduction to Psychoanalysis*. New York: Permabooks.

Gächter, S. and Falk, A. 2000. Work motivation, institutions, and performance. In R. Zwick and A. Rapoport (eds), *Experimental Business Economics*: 351–72. Dordrecht: Kluwer Academic Publishers.

Gaglio, C.M. and Katz, J. 2001. The psychological basis of opportunity identification: Entrepreneurial alertness. *Small Business Economics*, **16**(2): 95–111.

Gartner, W.B. 1988. 'Who is an entrepreneur?' is the wrong question. *American Journal of Small Business*, **12**(4): 11–32.

Gasse, Y. 1982. Elaborations on the psychology of the entrepreneur. In C.A. Kent, D.A. Sexton and K.H. Vesper (eds), *Encyclopedia of Entrepreneurship*: 57–71. Englewood Cliffs, NJ: Prentice-Hall.

Gollwitzer, P.M. and Brandstätter, V. 1997. Implementation intentions and effective goal pursuit. *Journal of Personality and Social Psychology*, **73**(1): 186–99.

Gollwitzer, P.M., and Schaal, B. 1998. Metacognition in action: The importance of implementation intentions. *Personality and Social Psychology Review*, **2**(2): 124–36.

Hackman, J.R. and Oldham, G.R. 1976. Motivation through the design of work: Test of a theory. *Organizational Behavior and Human Performance*, **16**(2): 250–79.

Hart, J.W., Stasson, M.F. and Mahoney, J.M. 2007. The big five and achievement motivation: Exploring the relationship between personality and a two-factor model of motivation. *Individual Differences Research*, **5**(2): 267–74.

Helmreich, R.L. and Spence, J.T. 1978. The work and family orientation questionnaire: An objective instrument to assess components of achievement motivation and attitudes towards family and career. *JSAS Catalog of Selected Documents in Psychology*, **8**: 35 (Ms. No. 1677).

Helmreich, R.L., Sawin, L.L. and Carsrud, A.L. 1986. The honeymoon effect in job performance: Temporal increases in the predictive power of achievement motivation. *Journal of Applied Psychology*, **71**(2): 185–8.

Herron, L. and Robinson, R.B. 1993. A structural model of the effects of entrepreneurial characteristics on venture performance. *Journal of Business Venturing*, **8**(3): 281–94.

Kelman, H.C. 1974. Attitudes are alive and well and gainfully employed in the sphere of action. *American Psychologist*, **29**(5): 310–24.

Kim, M. and Hunter, J. 1993. Relationships among attitudes, intention and behaviour. *Communications Research*, **20**(3): 331–64.

Kirzner, I.M. 1973. *Competition and Entrepreneurship*. Chicago, IL: University of Chicago Press.

Kirzner, I.M. 1979. *Perception, Opportunity and Profit*. Chicago, IL: University of Chicago Press.

Komives, J.L. 1972. A preliminary study of the personal values of high technology entrepreneurs. In A.C. Cooper and J.L. Komives (eds), *Technical Entrepreneurship: A symposium*: 231–42. Milwaukee, WI: Center for Venture Management.

Krueger, N.F. 2009. Entrepreneurial intentions are dead: Long live entrepreneurial intentions. In A. Carsrud and M. Brännback (eds), *Understanding the Entrepreneurial Mind: Opening the black box*: 51–73. Heidelberg: Springer.

Krueger, N.F. and Brazeal, D. 1994. Entrepreneurial potential and potential entrepreneurs. *Entrepreneurship Theory and Practice*, **18**(1): 91–104.

Krueger, N.F. and Carsrud, A.L. 1993. Entrepreneurial intentions: Applying the theory of planned behavior. *Entrepreneurship and Regional Development*, **5**(4): 315–30.

Krueger, N.F., Reilly, M.D. and Carsrud, A.L. 2000. Competing models of entrepreneurial intentions. *Journal of Business Venturing*, **15**(5/6): 411–32.

Kruglanski, A.W., Bélanger, J. J., Chen, X. and Köpetz, C. 2011. The energetics of motivated cognition: A force-field analysis. *Psychological Review*, **119**(1): 1–20.

Kuratko, D.F., Hornsby, J.S. and Naffziger, D.W. 1997. An examination of owner's goal in sustaining entrepreneurship. *Journal of Small Business Management*, **35**(1): 24–33.

Langen-Fox, J.L. and Roth, S. 1995. Achievement motivation and female entrepreneurs. *Journal of Occupational and Organizational Psychology*, **68**(3): 209–18.

Lawson, R. 1997. Consumer decision making within a goal-driven framework. *Psychology & Marketing*, **14**(5): 427–49.

Lent, R. and Brown, S. 2006. On conceptualizing and assessing social cognitive constructs in career research: A measurement guide. *Journal of Career Assessment*, **14**(1): 12–35.
Lent, R., Brown, S. and Hackett, G. 1994. Toward a unifying theory of career and academic interests, choice, and performance. *Journal of Vocational Behavior*, **45**(1): 79–122.
Locke, E.A. and Latham, G.P. 2002. Building a practically useful theory of goal setting and task motivation. *American Psychologist*, **57**(9): 705–17.
Locke, E.A. and Latham, G.P. 2004. What should we do about motivation theory? Six recommendations for the twenty-first century. *Academy of Management Review*, **29**(3): 388–403.
Locke, E.A., Latham, G.P. and Erez, M. 1988. The determinants of goal commitment. *Academy of Management Review*, **13**(1): 23–39.
Lumpkin, G.T. and Erdogan, B. 2004. If not entrepreneurship, can psychological characteristics predict entrepreneurial orientation? A pilot study. *ICFAI Journal of Entrepreneurship Development*, **1**: 21–33.
Maslow, A.H. 1943. A theory of human motivation. *Psychological Review*, **50**(4): 370–96.
McBroom, W.H. and Reed, F.W. 1992. Toward a reconceptualization of attitude–behavior consistency. *Social Psychology Quarterly*, **55**(2): 205–16.
McMullen, E. and Kenworthy, T. 2014. From philosophy of science to theory testing: Generating practical knowledge in entrepreneurship research. In A. Carsrud and M. Brännback (eds), *Handbook of Research Methods and Application in Entrepreneurship and Small Business*: 20–55. Cheltenham, UK and Northampton, MA, USA: Edward Elgar.
Mehrabian, A. 1968. Male and female scales of the tendency to achieve. *Educational and Psychological Measurement*, **28**(2): 493–502.
McClelland, D.C. 1961. *The Achieving Society*. Princeton, NJ: D. Van Nostrand Company. Reprinted 1967, New York: Free Press.
McClelland, D.C. 1965. Achievement motivation can be developed. *Harvard Business Review*, 43(6): 6–25.
McClelland, D.C. 1985. How motives, skills, and values determine what people do. *American Psychologist*, **40**(7): 812–25.
McClelland, D.C., Atkinson, J.W., Clark, R.A. and Lowell, E.L. 1953. *The Achievement Motive*. New York: Appleton-Century-Crofts.
McClelland, D.C. and Winter, D.G. 1969. *Motivating Economic Achievement*. New York: Free Press.
Mitchell, R.K., Mitchell, B.T. and Mitchell, J.R. 2009. Entrepreneurial scripts and entrepreneurial expertise: The information processing perspective. In A. Carsrud and M. Brännback (eds), *Understanding the Entrepreneurial Mind: Opening the black box*: 97–140. Heidelberg: Springer.
Monsen, E. and Urbig, D. 2009. Perceptions of efficacy, control, and risk: A theory of mixed control. In A. Carsrud and M. Brännback (eds), *Understanding the Entrepreneurial Mind: Opening the black box*: 259–84. Heidelberg: Springer.
Murray, H.A. 1938. *Explorations in Personality*. New York: Oxford University Press.
Nuttin, J. 1984. *Motivation, Planning, and Action*. Leuven: Leuven University Press and Lawrence Erlbaum Associates.
Perwin, L. 2003. *The Science of Personality*. Oxford: Oxford University Press.
Pinder, C.C. 1984. *Work Motivation: Theory, issues, and applications*. Glenview, IL: Scott, Foresman.
Pinder, C.C. 1998. *Work Motivation in Organizational Behavior*. Upper Saddle River, NJ: Prentice Hall.
Quigley, N.R. and Tymon, W.G. Jr. 2006. Toward an integrated model of intrinsic motivation and career self-management. *Career Development International*, **11**(6): 522–43.
Renko, M., El-Tarabishy, A., Carsrud, A. and Brännback, M. 2014. Understanding and measuring entrepreneurial leadership style. *Journal of Small Business Management*. Forthcoming.
Reynolds, P.D., Bygrave, W.D., Autio, E., Cox, L. and Hay, M. 2002. *Global Entrepreneurship Monitor: 2002 executive report*. Kansas City, MO: Kauffman Center for Entrepreneurial Leadership.

Ryan, R.M. and Deci, E.L. 2000. Self-determination theory and the facilitation of intrinsic motivation, social development, and well-being. *American Psychologist*, **55**(1): 68–78.
Sarasvathy, S.D. 2001. Causation and effectuation: Toward a theoretical shift from economic inevitability to entrepreneurial contingency. *Academy of Management Review*, **26**(2): 243–63.
Sarasvathy, S.D. 2008. *Effectuation: Elements of entrepreneurial expertise*. Cheltenham, UK and Northampton, MA, USA: Edward Elgar Publishing.
Schumpeter, J.A. 1934. *The Theory of Economic Development: An inquiry into profits, capital, credit, interest, and the business cycle*. Cambridge, MA: Harvard University Press.
Segal, G., Bogia, D. and Schoenfeld, J. 2005. The motivation to become an entrepreneur. *International Journal of Entrepreneurial Behaviour and Research*, **11**(1): 42–57.
Sexton, D.L. and Smilor, R. (eds). 1985. *The Art and Science of Entrepreneurship*. Cambridge, MA: Ballinger.
Shane, S. 2003. *A General Theory of Entrepreneurship: The individual–opportunity nexus*. Cheltenham, UK and Northampton, MA, USA: Edward Elgar.
Shane, S. 2008. *The Illusions of Entrepreneurship*. New Haven, CT: Yale University Press.
Shane, S. and Venkataraman, S. 2000. The promise of entrepreneurship as a field of research. *Academy of Management Review*, **25**(1): 217–26.
Smilor, R.W. and Kuhn, R.L. (eds) 1986. *Managing take-off in fast growth companies: Innovations in entrepreneurial firms*. New York: Praeger.
Spence, J.T. and Helmreich, R.L. 1978. *Masculinity and Femininity: Their psychological dimensions, correlates, and antecedents*. Austin, TX: The University of Texas Press.
Steel, P. and König C. J. 2006. Integrating theories of motivation. *Academy of Management Review*, **31**(4): 880–913.
Steward, W.J. Jr and Roth, P.L. 2007. A meta-analysis of achievement motivation differences between entrepreneurs and managers. *Journal of Small Business Management*, **45**(3): 401–21.
Tuuanaen, M. 1997. Finnish and US entrepreneurs' need for achievement: A cross-cultural analysis. In A.J. Carland and J.W. Carland (eds), *Proceedings of the Academy of Entrepreneurship*: 8–20. Cullowhee, NC: Academy of Entrepreneurship Inc.
Wilson, E.O. 1998. *Consilience: The unity of knowledge*. New York: Alfred A. Knopf.
Wong, P.K., Lee, L. and Leung, A. 2006. Entrepreneurship by circumstances and abilities: The mediating role of job satisfaction and moderating role of self-efficacy. MPRA Paper No. 596.
Zhao, H., Seibert, S.E. and Hills, G.E. 2005. The mediating role of self-efficacy in the development of entrepreneurial intentions. *Journal of Applied Psychology*, **90**(6): 1265–72.

3. Toward a taxonomy of entrepreneurs' behavior

Barbara Bird

ABSTRACT

This chapter posits that one of the most important outcomes of entrepreneurs' socially situated cognition is concrete behavior to exploit opportunity through creation of a venture. By looking at cognition's most immediate observable concrete outcomes, research advances beyond the 'merely mental.' Such behavior itself becomes a social context for new cognition and new behavior. The chapter addresses two reasons why the behavior of entrepreneurs is under-researched and describes the broad domain of entrepreneurial behavior, which at times muddies the water about behavior. Recommendations are offered to create a taxonomy of 'basic' levels of entrepreneurs' behavior and observational and psychometric measurement methods are addressed. A call is made to build research that uses the same measures across studies and over time to enable true situational theories of entrepreneurs' behavior.[1]

INTRODUCTION

> How do I know what I think until I see what I say?
>
> (Weick, 1979, p. 165)

Cognition and behavior (e.g., thinking and speaking) are tightly interwoven and the causal chain flows both ways: thoughts and cognitive characteristics (perceptions, intentions, knowledge structures, intelligences, metacognitions) precede and influence behavior (Haynie, Shepherd, Mosakowski and Earley, 2010); and behavior influences and changes cognition and cognitive characteristics through the process of learning, role modeling, and metacognitions such as thinking out loud. The domain of entrepreneurship scholarship shows considerable interest in the interrelations among different cognitions, motivations, perceived opportunity, decisions and strategy (Cornelissen and Clarke, 2010; Haynie et al., 2010; Mitchell et al., 2007; Mitchell, Randolph-Seng and Mitchell, 2011; see also Carsrud and Brännback, Chapter 2, this volume; Randolph-Seng, Williams and Hayek, Chapter 4, this volume). However, when the subject matter is economic outcome or the creation of a new organization, nothing of importance occurs until some action is taken.

Research on entrepreneurs and particularly the start-up of a new independent venture (as individuals, not firms) tends to focus a great deal on cognitions and outcomes (e.g., opportunity identified, decision made, venture started, venture grown), and to a considerably lesser extent on behavior. Behavior taken as a result of cognition is cognition's most proximal dependent variable of practical interest, with performance and outcomes more distal. In addition, since cognitions of entrepreneurs are socially situated and to some extent context dependent, those situations and contexts will also invoke, reinforce, chain together and give meaning to specific behaviors of the actors (see also Baron, Chapter 1, this volume).

The importance of behavior to cognition research will depend a great deal on the cognitive domain of interest, since cognition encompasses intelligence, knowledge (explicit and tacit) across industry and venture domains, learning, perception (specifically problem and opportunity recognition), vision, memory, scripts, schema, biases, heuristics, decision making and metacognition (see also Corbett, Chapter 12, this volume). Behavior-centric research questions could be posed in most of these domains. Most generally, research questions take the form of 'What is the relationship (if any) between specific cognitive factors such as those above and the behaviors that result?' For example, does greater 'successful intelligence,' which is domain-specific, result in more experimentation in decision making than less of this attribute? One study (Baum and Bird, 2010) shows that this works in one industrial context, but does it translate to other contexts? Does successful intelligence result in other important behaviors such as better early-stage hires? Better or earlier protection of intellectual property? Better or earlier initial selling of the product or service? And so forth. The same general process applies to other cognitive constructs. For example, do individuals with a stronger or weaker cognitive bias, for example optimism, do more experimentation, file more patents, hire more technical help, do more personal selling than those without that bias? Do individuals with successful intelligence and/or optimism bias experiment, file more patents, hire more or better technical help or do more or better personal selling in different social contexts, say a conflict zone (e.g., Bullough, Renko and Myatt, 2014), within a government economic zone, or in a developing economy?

If behavior mediates the impact of cognition on venture outcomes, general research questions might also include how these behaviorally mediated cognitions impact venture emergence, survival, growth and entrepreneurs' satisfaction. Are some mediated pathways stronger in impacting performance than others? For example, does the successful intelligence–experimentation pathway result in higher venture performance than the optimism–experimentation pathway? Or does the successful intelligence–personal selling pathway show stronger links to venture performance? Does the combined impact of successful intelligence and optimism result in stronger outcomes than either alone? Does the successful intelligence–experimentation pathway work better for entrepreneurs in different social contexts (e.g., high-technology industries, economic recessions, low-conflict regions)?

These research questions are extremely precise. When we are able to pursue these kinds of question, we can better tailor our interventions to the selection, training/education, coaching and mentoring of entrepreneurs. The state of the art on understanding behavior is not quite advanced enough to make these questions researchable right now. Furthermore, one-off studies using unique variables and/or measures does not advance the field; there is little accumulated knowledge of entrepreneurs.

To pursue these kinds of question systematically as a field across accumulating studies, we need a taxonomy of behaviors associated with venture start-ups. The rest of this chapter examines the state of the art of entrepreneurs' behavior research and takes first steps toward a taxonomy. First, I explore reasons why behavior is understudied and propose the development of a behavior taxonomy. Second, I provide a short overview of the broad domain of entrepreneurial behavior, which can often muddy the waters when behavioral is confused with behavior. Then I focus on three ways to move forward: (1) to observe behavior; (2) to measure behavior; and (3) to use common measures. Finally I conclude with some proposals to my fellow entrepreneurship scholars for a 'school of thought' anchored in behavior.

ENTREPRENEURS' BEHAVIOR REMAINS UNDER-RESEARCHED

Why do researchers focus attention on the antecedents and correlates of cognition and neglect behavior? First I must note that we are not alone in our inattention; areas such as health psychology and health behavior suffer this same inattention and need for focus (Johnston and Dixon, 2008). For our domain, I believe there are two reasons – one is the problem of not having a taxonomy of entrepreneur-specific behavior.

Entrepreneur-specific Behaviors

Entrepreneurs' behaviors are 'theoretically observable actions of individuals (as solo entrepreneurs or as part of a team of entrepreneurs) in the start-up or early stages of organization creation' (Bird, Schjoedt and Baum, 2012, p. 890). In its observability, behavior differs fundamentally from cognition. In identifying what entrepreneurs do, their observable actions, researchers face three related problems: (1) lack of agreement on what behaviors are important; for example, there is no generally accepted taxonomy of behaviors of entrepreneurs; (2) failure to understand or utilize the concept of behavioral repertoire; and (3) ignoring the context of behavior.

Taxonomy

A taxonomy is a classification of important elements in a domain such that general categories contain more particular components. Best known to business scholars is the International Standard Industrial Classification wherein industrial segments are parsed into increasingly specific industrial slices. A taxonomy of entrepreneurs' behavior would first identify the broad categories of behaviors involved in starting a new venture. Following that, the taxonomy would move from broad categories to increasingly specific behavior, and research would anchor on the 'basic' level of behavior.

The basic level of a behavior, as defined by perceptual/cognitive psychologists, rests on the notion of events,[2] or behaviors that have meaning in a sequence of time (Newtson and Engquist, 1976). Event psychology and behavior segmentation are subsumed in process approaches to entrepreneurship (e.g., Moroz and Hindle, 2012; Steyaert, 2007; Van de Ven and Engleman, 2004). In this framework, events/behavior can be segmented and taxonomically organized into superordinate (more inclusive, covering more time and being more abstract; e.g., market planning) or subordinate (a part of the larger or abstract event; e.g., testing the product with an initial customer contact). These levels in essence describe a 'chunk' size of behavior. The basic level is that segment or chunk of event/behavior where 'the number of features listed for events [behaviors] increases greatly from the superordinate level to the basic level, but not very much from the basic level to the [next] subordinate level' (Zacks and Tversky, 2001, p. 6). Basic level may be said to have the requisite variety to be both specific and meaningful.

Behaviors conceptualized as superordinate, such as leadership, market research and planning, and sustaining competitive advantage (which are the domain of our theorists and model builders; see also Sarasvathy,

Chapter 14, this volume), are so abstract and general as to apply to many different economic actors. For example, certainly many non-entrepreneur economic actors do market research, including established small–medium size enterprises (which can be decades old and as large as 500 employees in the USA), multinational corporations, not-for-profit organizations and governmental units. Many of these have sophisticated marketing staffs and professionalized processes and tools, and sizable budgets. When entrepreneurs at the start-up stage research the market for their opportunity, it is with fewer resources and most likely to be done by the single actor(s) of the founding team. In this context the specific marketing-focused behaviors of entrepreneurs are likely to play a more crucial role than the behaviors of marketing personnel in larger established businesses.

When the conceptualized behavior drops to the basic, appropriately specific, level, it may be possible to point to important behaviors that are relatively more common among entrepreneurs than among other actors. When the behavior drops below the basic level, the details of behavior may be less meaningful and again may apply across many other economic actors. See Figure 3.1 for a potential general taxonomic hierarchy.

To create a taxonomy requires multiple scholars (probably senior scholars) collaborating to develop and committing to standardize a defining set of general behavior or behavioral domains. Such 'standards' might be culled from extant literature and scholar experience, refined through some community of scholars using Delphi or other methods. For example, the domains might include the behaviors of: (1) opportunity finding, creating and shaping; (2) developing (often innovative) product/service; (3) market research and planning; (4) competitive analysis; (5) selling initial product/service; (6) making personal financial investments; (7) acquiring external financing for different stages of venture development; (8) hiring and socializing early employees and team members; (9) setting up initial operations; (10) prioritizing and rapidly solving problems; (11) formalizing the venture; and (12) sustaining competitive advantage and so on. Domains would be specific enough to allow theorizing. I would not consider decision making as a behavioral domain since it is nearly wholly cognitive (unless done in a group setting or through an observable method). I do include rapid problem solving, which is a narrowly focused domain that is tightly coupled with decision making and may be open to observation. Generally, decision constitutes a cognitive process applied to most if not all domains of entrepreneurs' behavior. Nor would I consider communication to be a domain, since these disparate behaviors (e.g., speaking, listening, writing, reading etc.) can occur in support of most of the suggested domains.

Once the domains are determined, the scholar group would collaborate

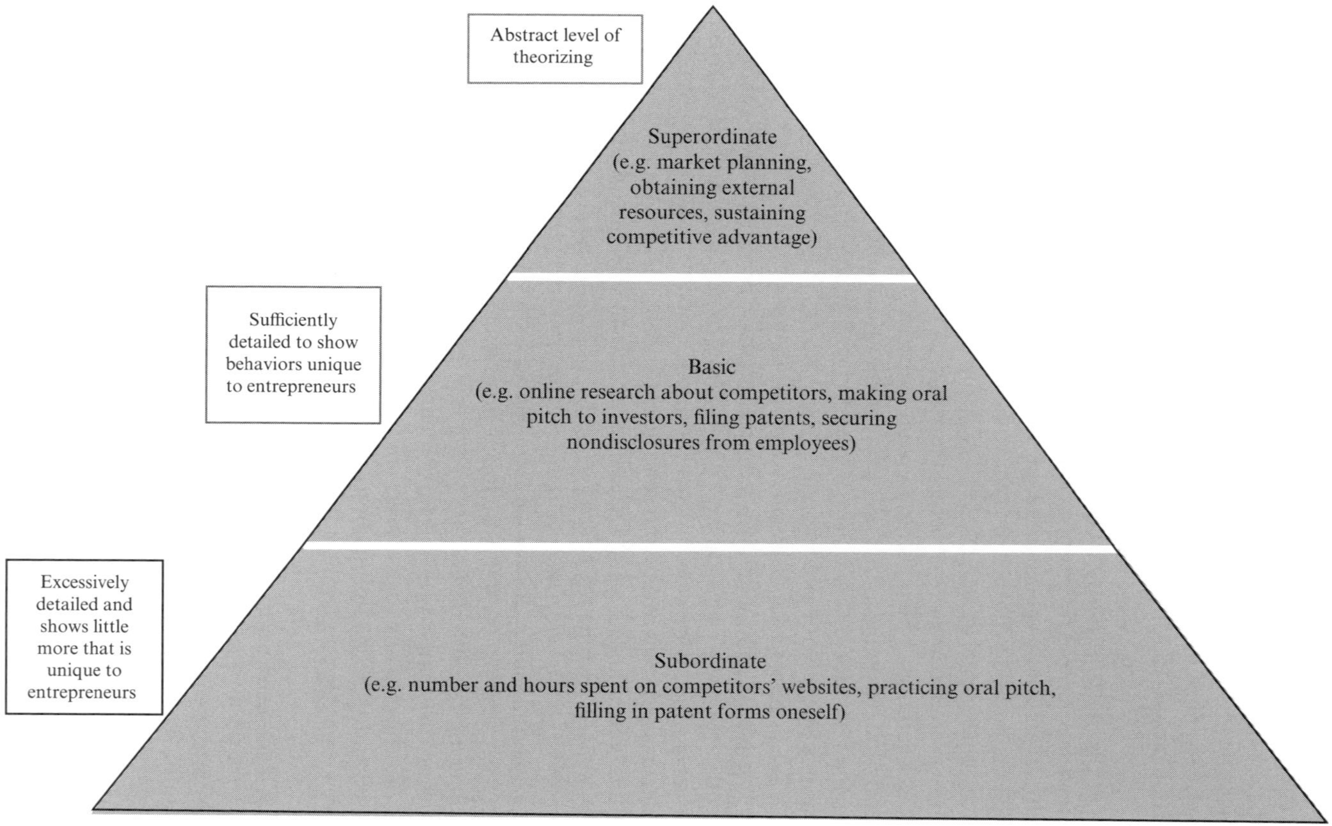

Figure 3.1 Potential taxonomic hierarchy

to define the 'basic' level for each domain, again using collective decision tools such as Delphi. These basic levels of behavior would then become standards for scholars wishing to build a cumulative body of knowledge about what entrepreneurs do. When developed, the commitment of the scholar group would be to use these behaviors for a number of years before revision was addressed. Other specific recommendations to develop cumulative research on entrepreneurs' behavior are offered in the later section on ways forward.

With a taxonomy of basic behaviors (and possibly even agreed-upon measures), we can build research streams where behavior is a key dependent variable for cognition as well as other antecedents such as human and social capital. We can build research where basic behaviors are key antecedents/predictors of team and venture outcomes. We can look at how basic behaviors are learned, chained together and transmitted to others. We can look at how specific types of behaviors influence cognition, and how key stakeholders such as partners, supply chain allies, investors and customers perceive and interpret those behaviors.

Such cumulative research will allow us to become situational in our understanding of entrepreneurs' behavior, just as leadership research developed strong situational theories once the common dimensions of leader behavior such as consideration and initiating structure were identified (Hemphill and Coons, 1957; House, 1996). We could ask and answer questions proposed earlier, and others such as: When are certain behaviors more useful and effective in key outcomes such as gaining funding, gaining sales, hiring talent, survival or growth? How does the effectiveness of behavior vary across these outcomes, over time, and with age and size of the venture? Beside values, what behaviors of founders impact emergence and strength of a new organization's culture? Do certain basic behaviors become 'outdated' over the course of a venture's history or within an industrial or economic 'era'? How might behaviors of entrepreneurs and entrepreneurial teams create a context where different cognitive habits and processes are more or less likely, more or less useful?

Economics of Research

The other impediment to developing a strong behavior-focused research involves the economics of research and researchers' reward systems. Arguably, at least, conceptualizing or theorizing intrinsically motivates some scholars. Conceptual work is by definition abstract (superordinate), and this type of work is considerably harder to publish in scholarly journals. Other scholars pursue theory-derived empirical data, which are often easier to publish. Some of these scholars find data and analytics intrinsically rewarding. Most scholars are extrinsically rewarded (with citations, prestige, publication, tenure and promotion) for placing a 'stake in the ground' by developing a new construct or new measure and hoping that others cite and use that research. Finally, at least some researchers place high value on time-efficiency.

Action research has been abstract

Scholars in the field of entrepreneurship acknowledge the need for attention to behavior. However, the attention has largely been conceptual and framed as 'action research.' The problem with this conceptualizing is the tendency to stay at a superordinate level. This is appropriate for theorizing but does not point to the basic behaviors of entrepreneurs. This can be seen in the summary of three recent manuscripts brought to my attention by one of the early reviewers of this chapter.

First, McMullen and Shepherd (2006) advance a conceptual model about how individual differences in knowledge and motivation in uncertain contexts affect one's perception of an opportunity for someone else and for oneself. The behavior is a 'decision.' I have argued earlier that decision making is a cognition that crosses many domains of behavior. As a cognition, decision making should not be considered a domain of behavior; nor is decision making a distinct domain of entrepreneurs.

Second, Alvarez and Barney (2007) present another conceptual paper on opportunity discovery and creation. The authors point to seven 'actions': (1) leadership; (2) decision making; (3) human resource practices; (4) strategy; (5) finance; (6) marketing; and (7) sustaining competitive advantage. The model mixes actions of individuals and actions of ventures. Leadership (an individual-level construct) is discussed in terms of experience/expertise of/in discovery, and personality/charisma for opportunity creation. Decision making (individual/venture-level construct) is theorized in two forms: behaviors of data collection for opportunity discovery and use of cognitive heuristics and behaviors of iterative experimentation for opportunity creation. Recruitment of employees (individual/venture) is theorized to have different pathways but not necessarily different behaviors. Strategy (venture-level construct) is theorized to be formulated differently in the two contexts but the authors do not offer behaviors associated with each, only that discovery is likely to conform to broad notions of a written business plan. Likewise the authors theorize that the two opportunity modes will lend themselves to different broad sources of venture financing (outside investment and bootstrapping) although the behaviors associated with each are not elaborated. Likewise for marketing (venture-level construct), the authors suggest more precision in marketing plans on the one hand and more experimentation on the other, with behaviors not specified.

In the third article, Mitchell and Shepherd (2010) present an empirical paper that places entrepreneurial decision makers in an experimental situation and the 'action' was their self-reported likelihood of investing in a hypothetical opportunity (measured by a single Likert scale of likelihood). Arguably, this action is really cognition (thoughts about the opportunity), with the behavior choosing a number on a scale.

Behavioral research is hard work

Like entrepreneurs, we researchers seek a rich opportunity that is easy to exploit. Cognition is a rich vein of research and offers ample opportunity to publish. Cognitions are complex, multifaceted and can only (or best) be inferred from self-report measures. The same complexity-nuance exists in 'behavioral' studies of attitudes, values, intentions and so on (more on 'behavioral' below). Complexity allows many researchers to address nuanced relationships among cognitions and other variables (also often self-report) and these many researchers advance the field of study and often their careers with these nuances.

Researchers are lazy, at least some of us are.[3] Self-report measures of behaviors are simply easy (relatively fast and inexpensive) to administer and often achieve acceptable response rates. The response rates drop (and researcher struggle increases) with predictive models and longitudinal studies or with triangulated data (e.g., entrepreneurs and their funders). Other methods of assessing behavior exist (e.g., observation, unobtrusive measures, experimental control), but these methods can be difficult (creating good experimental designs takes several extra steps, may require special training for the researcher, and may be plagued by demand characteristics that impede validity) and the methods are often considerably more time intensive. I discuss observation and measurement in the proposals for ways forward later.

THE BROAD DOMAIN OF ENTREPRENEURIAL BEHAVIOR

While I use the terms 'behavior' and 'action' interchangeably to refer to observable conduct, various 'behavioral' concepts can muddy the focus on what entrepreneurs actually do. As a brief framework, this broader domain of 'behavioral' is now discussed. Then I reconnect this chapter with the larger purpose of this book – the social situation and socially situated cognition.

I define behavior as theoretically observable action of individuals. I suggest the possessive term (entrepreneurs' behavior) for behavior of entrepreneurs and the descriptive term (entrepreneurial behavior) for other related variables, some of which are discussed below, and for the application of entrepreneurs' behaviors to non-entrepreneurs (those who act 'as if,' including organizations). The term behavioral applies to variables associated (theoretically or empirically) with individual behavior and as such is an expansive domain that includes cognitions such as decision making, perception, judgment, intention, attributions and aspects of learning, among others. It also includes human and social capital, affect and emotion, attitudes and values (see also Foo, Murnieks and Chan, Chapter 5, this volume; Grégoire, Chapter 6, this volume). It includes communication, teams, conflict and leadership, among other constructs. Constructs closely related to behavior include competency (enabler of behaviors of different quality), performance (usually a general outcome but, at task level, a function of many behaviors), and tasks (requirements of work that may or may not have observable behaviors in execution). In addition, skills are behaviors from a behavioral repertoire that the person executes at a standard of quality. There is a need for greater clarity of construct use among these myriad terms. Scholars need to understand the differences among them and recognize that not all behavioral constructs are behavior.[4]

Social Situation

The larger domain of entrepreneurs' behavior and entrepreneurial behavior also includes the social situations in which entrepreneurs are located. These contexts include industry, geographic and cultural location (Assudani, 2009; Kwon and Arenius, 2010; Wright, Liu, Buck and Filatotchev, 2008) and temporal location (Bluedorn and Martin, 2008; Hansen and Bird, 1997). Such social contexts are often 'control variables' in studies seeking other modeled effects.[5] These contexts can be causes, correlates and consequences of entrepreneurs' actions and constitute many aspects of 'socially situated cognition.' Situational variables like these can impact the behavioral repertoire of an entrepreneur and his or her situational choice of behavior from that repertoire. For example, entrepreneurs learn behavior (add to their repertoire) and develop skills from socially provided positive and negative reinforcements. Behaviors can also be learned vicariously (e.g., through role modeling and observation). Individuals choose among behaviors based on skills and self-efficacy, habits, metacognitive thoughts and consciously mediated or unconscious social norms.

It is exactly these situational variables that are important to cognition research and that also provide the hope for a 'situational theory' of entrepreneurs' behavior. To get there, I propose we follow the model of leader behavior as mentioned above. Develop and measure entrepreneurs' behavior and then study the situations that either differentially evoke or reward certain behaviors or, perhaps more importantly, look at how the same behavior of an entrepreneur can be more or less effective in different situations and at different times.

Social Desirability and Demand Characteristics

One final aspect of the social situation is the situation of the survey or experiment itself. In survey and interview designs, this is known as the social desirability bias (Crowne and Marlowe, 1960). This refers to respondents' tendency to give socially desirable but probably untrue answers. This bias is particularly relevant when using self-reports to assess thoughts, plans, decisions and actions. In experimental design, the construct is labeled 'demand characteristics' and reflects the subject's awareness of the experiment's purpose and his or her desire to be a good (or bad) experimental subject. Respondents vary in how much they discern the researcher's intention and how much they care about how they appear to others. Such biases can be controlled by design or statistically, but this has generally not been the case in entrepreneurial cognition and behavior studies. Without controls, correlations among even time-lagged variables (cognition and behavior) may exist, in part, because of these response biases.

WAYS FORWARD

I offer three suggestions for ways forward: observation; development of psychometrically sound measurement; and continued use of good measures. The field of entrepreneurship research that deals with the individual entrepreneur (or team) and his or her cognition can improve a great deal on current practice by making basic behaviors key dependent variables.

Observe Behavior

The observation of behavior has a long history in social science, with contributions in psychology, anthropology and sociology. Such studies have value because they do not rely on the subject's self-reports (with the limitations discussed above). Observational methods for entrepreneurship include participant observation (where the researcher participates in the new-venture start-up), real-time longitudinal observation of the start-up (e.g., Bird et al., 2012; Karlsson and Honig, 2009; Lichtenstein, Dooley and Lumpkin, 2006), and in-depth and unstructured interviews (e.g., Brundin, 2007).

Less direct methods include beeper/cellphone logs of activity, diaries and examination of a real email stream (Ziegler, Rosenzweig and Ziegler, 1992). However, some of these time-sampling methods still require self-reports. Other unobtrusive measures could include accessing the web-surfing history of the entrepreneur or nascent entrepreneur. Metadata on the number of visits to a certain site, the links from site to site, the time on a site all are subordinate elements of more abstract concepts. While entrepreneurs may not be willing to submit to such scrutiny, certainly the students and participants in classes and training could participate. Another unobtrusive measure could involve video recording of networking social events, and measures of how many, with whom, and for how long conversations last might be valuable measures.

Many observational studies of entrepreneurs' behavior (as a form of qualitative research) are longitudinal or at least conducted over a period of weeks or months. As mentioned, these studies take considerable time to collect and analyze, and are often difficult to present (and publish). They are often limited in scope to one or a small number of ventures, limiting generalization. However, without these studies the actions and situational demands of entrepreneurs cannot be fully appreciated.

Measure Behavior

If we cannot or will not observe behavior, we need to assess that behavior through some form of self- or other-report using survey-like measurement. These methods are used effectively in paper and web-based surveys and structured interviews. Ideally these measures will be based on understanding what behaviors are basic and will be psychometrically sound, akin to the Leadership Behavior Description Questionnaire that has helped leadership research advance. The measures fall into two broad types – formative and reflective measures.

Whether behavior is best measured by formative (non-correlated indices to define the construct) or reflective (correlated items to reflect the construct) measures is not addressed in the entrepreneurship research (Podsakoff, MacKenzie, Podsakoff and Lee, 2003). However, there are scholarly conversations about these types of measure applied to other entrepreneurship constructs (Covin and Wales, 2012; Salomo, Brinckmann and Talke, 2008).

Formative indicators are used with some regularity. Indicators such as those used in the Panel Survey of Entrepreneurship Dynamics (PSED)[6] allow for fairly quick responses by entrepreneurs (the answer is yes or no), and together actions can be enumerated as an index that some start-up actions have taken place. For the purpose of determining that *some* behavior has taken place, these measures serve research well. These studies have used behavior indices as independent (e.g., Lichtenstein, Carter, Dooley and Gartner, 2007), dependent (e.g., Singh and Lucas, 2005) and selection variables (e.g., Schjoedt and Shaver, 2007), where the greater interest is often the other constructs in the model. What indices lack is the variance and reliability that come with interval-like scales, and thus they need to be evaluated more carefully for reliability and use nonparametric analytic methods (Diamantopoulos and Winklhofer, 2001). In addition, these indices often include behaviors of very different chunk size and concreteness. Even if each indexed behavior is considered basic, as a single item with little variance, there is limited opportunity to explore how that behavior influences and is influenced by other variables, including cognition.

Thus a case can be made for developing psychometrically sound reflective measures. Developing such methods to ensure reliability and validity is also time-intensive. In addition, finding publication outlets for entrepreneurship-centric measures and methods is difficult. At present, the best outlet seems to be a research note in one of the entrepreneurship or small-business journals.

Whichever way behavior is measured, we need to be much more attentive to the possibility of response biases mentioned earlier. Social desirability and demand characteristics are the 'elephants in the room' with self-reports and experimental design respectively. Other biases include leniency (tendency to be generous/critical in evaluations of others) and central tendency (tendency to avoid extremes of a scale).

Use and Reuse Psychometrically Strong Measures of Behavior

While the field of scholarship on entrepreneurs' behavior has not yet formed a consensus on what behavior is basic, some studies have used solid measures of behavior. In preparing for the special issue of *Entrepreneurship Theory and Practice*, my co-author Leon Schjoedt and I reviewed entrepreneurship journals and found a number of measures of start-up behavior in the recent literature (2004–10). Of the journals surveyed and measures used, only a few recommend themselves as psychometrically sound (comprising more than two reflective/interval scaled items), with reliability and at least some validity. Of these, only those in Table 3.1 might conceivably be measuring 'basic' behavior.

This chapter cannot review all of these papers to point to their usefulness. I will, however, briefly note two manuscripts that deal with one possible basic behavior that differentiates entrepreneurs from others (and

Table 3.1 Select entrepreneurs' behaviors[1]

Behavior	Number of items	Citation/measure
Time spent developing ties	2	(Greve and Salaff, 2003)
Encouraging employee participation	4	(Rauch et al., 2005)
Customizing good or service	16	(Cloninger and Oviatt, 2007)
Communicating with customer	6	(De Clercq and Rangarajan, 2008)
Scanning frequency	28	(Stewart et al., 2008)
Decision speed[2]	2	(Talaulicar, Grundei and Werder, 2005)
Debating decision	4	(Talaulicar et al., 2005)
Planning marketing	13	(Gruber, 2007)
Acquiring information[3]	19	(Dyer, Gregersen and Christensen, 2008)

Notes:
1. From Bird et al. (2012).
2. This two-item scale did not reach acceptable reliability but was justified in this case as acceptable for an exploratory study.
3. Subscales were questioning, observing, experimenting, and idea networking.

successful entrepreneurs from less effective ones). One deals with time spent developing social network ties (Greve and Salaff, 2003), which was predicted to vary by the phase of firm development. The authors measured this by hours spent developing and hours spent maintaining these relationships. Hours and frequency are common anchors for behavior measures. The other (Stewart, May and Kalia, 2008) includes, as one dependent variable, information scanning frequency, which was expected to vary by entrepreneur's nation, perceived uncertainty and rate of change. The authors measured scanning frequency by a Likert-like scale across seven environmental sectors.

While the scales and tools provided by papers listed in Table 3.1 are the only interval-like reliable ones we found, and most of these had some modest description of validity, they do not completely reflect the state of the art in understanding entrepreneurs' behavior since they do not reflect several significant qualitative studies (e.g., Elfring and Hulsink, 2007; Hite, 2005; Lichtenstein et al., 2006; Wright et al., 2008), some experimental manipulations (e.g., Burmeister-Lamp, Levesque and Schade, 2012; Corbett, 2007; Grégoire and Shepherd, 2012), and the many studies using indicators or behavior counts (e.g., DeTienne and Chandler, 2007; Grandi and Grimaldi, 2005; Rauch, Frese and Utsch, 2005; Ucbasaran, Westhead and Wright, 2008). Nor did that search for papers look at the much less frequent but important research in mainstream management and psychology journals (e.g., *Academy of Management Journal, Journal of Applied Psychology*). Finally, that review ended in 2010 and other studies have been published since then.

One subsequent set of publications consists of those found in the special issue on entrepreneurs' behavior in *Entrepreneurship Theory and Practice* in 2012. Of particular interest are the behaviors theoretically derived from causation, effectuation and bricolage, which were also empirically sought (and found) in six high-technology cases (Fisher, 2012). The behaviors are described at a basic level for each theoretical framework in the article. Fisher suggests that superordinate behaviors of active problem solving and active engagement with customers/users may be particularly useful for future studies of entrepreneurs, at least in the high-technology sectors. These behaviors could be (and in some cases have been) measured on interval-like scales.

Another study (Mueller, Volery and von Siemens, 2012) looked at six start-up and six growth-stage venture entrepreneurs and used a grounded theoretic approach. Among their conclusions/findings is a taxonomy of entrepreneurs' behavior that explains the 'chunks' of behavior from the lowest level that can be observed (action) to activities that can be seen and understood without interpretation, functions that require interpretation in the organizational context and 'form,' which is the largest chunk (specifically, exploitation and exploration). They found seven activity categories (e.g., exchanges information, consults and sells, and monitors and controls), six functional action categories (e.g., purchasing, marketing, sales and PR, and business and organizational development), and actions associated with exploration and exploitation. They found only five categories of behavior-differentiated start-up and growth-stage entrepreneurs. The taxonomy and behaviors found offer future scholars a significant beginning for other studies of behavior. These behaviors could be measured on interval-like scales.

CONCLUSION

We need to agree that behavior of entrepreneurs deserves much more research attention and that this attention is particularly relevant to cognition scholars. Then we need scholars to undertake the development of a behavior taxonomy for start-up entrepreneurs. That taxonomy needs to have wide support in the scholar community. Then we need to develop or find reliable and valid measures that we agree to use and reuse in our research. With this, we could build a 'school of thought' based on conceptual and empirical research on entrepreneurs' behavior. Such a school of thought is broader than theory. Behavior-centric approaches to entrepreneurship support a myriad of theories, and particularly the theories that guide cognition research and human and social capital research more generally. However, this proposal means doing more than what seems to be both rewarded and normative in our field – building and extending theory – and more than seeking a 'stake in the ground' with a new concept and/or new tool. It means building empirical foundations with common tools. We can use academic conferences as forums to educate scholars in the value of using common tools. We can encourage gatekeepers such as journal editors and reviewers to value the commonalities of studies as much as, if not more than, their novelties. We can discuss the new 'school of thought' with business-school deans at academic and accreditation forums with the intention of shifting the reward system (tenure, merit and promotion) to include incremental contribution to a school of thought.

NOTES

1. I give thanks, praise and recognition to the two anonymous reviewers who helped me so much.
2. Greve (2001) notes that some events fall outside intentional action. However, event psychology provides clarity of behavior in a 'chunk size' that is useful to entrepreneurship scholars.
3. I confess to such laziness.
4. More information on these behavioral concepts can be found in Bird et al. (2012).
5. More information about control variables can be found in Schjoedt and Bird (2014).
6. Similar measures are used in the CAUSEE, the Norwegian parallel to the PSED, and in the GEM surveys.

REFERENCES

Alvarez, S.A. and Barney, J.B. 2007. Discovery and creation: Alternative theories of entrepreneurial action. *Strategic Entrepreneurship Journal*, **1**(1–2): 11–26.

Assudani, R.H. 2009. Ethnic entrepreneurship: The distinct role of ties. *Journal of Small Business and Entrepreneurship*, **22**(2): 197–205.

Baum, J.R. and Bird, B.J. 2010. The successful intelligence of high-growth entrepreneurs: Links to new venture growth. *Organization Science*, **21**(2): 397–412.

Bird, B., Schjoedt, L. and Baum, J.R. 2012. Editors' introduction. Entrepreneurs' behavior: Elucidation and measurement. *Entrepreneurship Theory and Practice*, **36**(5): 889–913.

Bluedorn, A.C. and Martin, G. 2008. The time frames of entrepreneurs. *Journal of Business Venturing*, **23**(1): 1–20.

Brundin, E. 2007. Catching it as it happens. In H. Neergaard and J.P. Ulhoi (eds), *Handbook of Qualitative Research Methods in Entrepreneurship*: 279–307. Cheltenham, UK and Northampton, MA, USA: Edward Elgar.

Bullough, A., Renko, M. and Myatt, T. 2014. Danger zone entrepreneurs: The importance of resilience and self-efficacy for entrepreneurial intentions. *Entrepreneurship Theory and Practice*, **38**(3): 473–99.

Burmeister-Lamp, K., Levesque, M. and Schade, C. 2012. Are entrepreneurs influenced by risk attitude, regulatory focus or both? An experiment on entrepreneurs' time allocation. *Journal of Business Venturing*, **27**(4): 456–76.

Cloninger, P. and Oviatt, B. 2007. Service content and the internationalization of young ventures: An empirical test. *Entrepreneurship Theory and Practice*, **31**(2): 233–56.

Corbett, A.C. 2007. Learning asymmetries and the discovery of entrepreneurial opportunities. *Journal of Business Venturing*, **22**(1): 97–118.

Cornelissen, J.P. and Clarke, J.S. 2010. Imaging and rationalizing opportunities: Inductive reasoning and the creation and justification of new ventures. *Academy of Management Review*, **35**(4): 539–57.

Covin, J.G. and Wales, W.J. 2012. The measurement of entrepreneurial orientation. *Entrepreneurship Theory and Practice*, **36**(4): 677–702.

Crowne, D.P. and Marlowe, D. 1960. A new scale of social desirability independent of psychopathology. *Journal of Consulting Psychology*, **24**(4): 349–54.

De Clercq, D. and Rangarajan, D. 2008. The role of perceived relational support in entrepreneur–customer dyads. *Entrepreneurship Theory and Practice*, **32**(4): 659–83.

DeTienne, D. and Chandler, G. 2007. The role of gender in opportunity identification. *Entrepreneurship Theory and Practice*, **31**(3): 365–86.

Diamantopoulos, A. and Winklhofer, H.M. 2001. Index construction with formative indicators: An alternative to scale development. *Journal of Marketing Research*, **38**(2): 269–77.

Dyer, J.H., Gregersen, H.B. and Christensen, C. 2008. Entrepreneur behaviors, opportunity recognition, and the origins of innovative ventures. *Strategic Entrepreneurship Journal*, **2**(4): 317–38.

Elfring, T. and Hulsink, W. 2007. Networking by entrepreneurs: Patterns of tie-formation in emerging organizations. *Organization Studies*, **28**(12): 1849–72.

Fisher, G. 2012. Effectuation, causation & bricolage: A behavioral approach to comparison of emerging theories in entrepreneurship research. *Entrepreneurship Theory and Practice*, **36**(5): 1019–51.

Grandi, A. and Grimaldi, R. 2005. Academics' organizational characteristics and the generation of successful business ideas. *Journal of Business Venturing*, **20**(6): 821–45.

Grégoire, D.A. and Shepherd, D.A. 2012. Technology-market combinations and the identification of entrepreneurial opportunities: An investigation of the opportunity–individual nexus. *Academy of Management Journal*, **55**(4): 753–85.

Greve, A. and Salaff, J.W. 2003. Social networks and entrepreneurship. *Entrepreneurship Theory and Practice*, **28**(1): 1–22.

Greve, W. 2001. Traps and gaps in action explanation: Theoretical problems of a psychology of human action. *Psychological Review*, **108**(2): 435–51.

Gruber, M. 2007. Uncovering the value of planning in new venture creation: A process and contingency perspective. *Journal of Business Venturing*, **22**(6): 782–807.

Hansen, E. and Bird, B. 1997. The stages model of high-tech venture founding: Tried but true. *Entrepreneurship Theory and Practice*, **22**(2): 111–22.

Haynie, J.M., Shepherd, D., Mosakowski, E. and Early, P.C. 2010. A situated metacognitive model of the entrepreneurial mindset. *Journal of Business Venturing*, **25**(2): 217–29.

Hemphill, J. and Coons, A. 1957. Development of the leader behavior description questionnaire. In R. Stogdill and A. Coons (eds), *Leader Behavior: Its description and measurement*: 6–38. Columbus, OH: Ohio State University.

Hite, J. 2005. Evolutionary processes and paths of relationally embedded network ties in emerging entrepreneurial firms. *Entrepreneurship Theory and Practice*, **29**(1): 113–44.

House, R. 1996. Path–goal theory of leadership: Lessons, legacy, and a reformulated theory. *Leadership Quarterly*, **7**(3): 323–52.

Johnston, M. and Dixon, D. 2008. Current issues and new directions in psychology and health: What happened to behaviour in the decade of behaviour? *Psychology & Health*, **23**(5): 509–13.

Karlsson, T. and Honig, B. 2009. Judging a business by its cover: An institutional perspective on new ventures and the business plan. *Journal of Business Venturing*, **24**(1): 27–45.

Kwon, S.-W. and Arenius, P. 2010. Nations of entrepreneurs: A social capital perspective. *Journal of Business Venturing*, **25**(3): 315–30.

Lichtenstein, B., Carter, N., Dooley, K. and Gartner, W.B. 2007. Complexity dynamics of nascent entrepreneurship. *Journal of Business Venturing*, **22**(2): 236–61.

Lichtenstein, B., Dooley, K. and Lumpkin, G. 2006. Measuring emergence in the dynamics of new venture creation. *Journal of Business Venturing*, **21**(2): 153–75.

McMullen, J.S. and Shepherd, D.A. 2006. Entrepreneurial action and the role of uncertainty in the theory of the entrepreneur. *Academy of Management Review*, **31**(1): 132–52.

Mitchell, J.R. and Shepherd, D.A. 2010. To thine own self be true: Images of self, images of opportunity, and entrepreneurial action. *Journal of Business Venturing*, **25**(1): 138–54.

Mitchell, R.K., Busenitz, L.W., Bird, B., Gaglio, C.M., McMullen, J.S., Morse, E.A. and Smith, J.B. 2007. The central question in entrepreneurial cognition research 2007. *Entrepreneurship Theory and Practice*, **31**(1): 1–27.

Mitchell, R.K., Randolph-Seng, B. and Mitchell, J.R. 2011. Socially situated cognition: Imagining new opportunities for entrepreneurship research (Dialogue). *Academy of Management Review*, **36**(4): 774–8.

Moroz, P.W. and Hindle, K. 2012. Entrepreneurship as a process: Toward harmonizing multiple perspectives. *Entrepreneurship Theory and Practice*, **36**(4): 781–818.

Mueller, S., Volery, T. and von Siemens, B. 2012. Entrepreneurs' behavior: An observational study of early and later-stage entrepreneurs. *Entrepreneurship Theory and Practice*, **36**(5): 995–1017.

Newtson, D. and Engquist, G. 1976. The perceptual organization of ongoing behavior. *Journal of Experimental Social Psychology*, **12**(5): 436–50.

Podsakoff, P.M., MacKenzie, S.B., Podsakoff, N.P., and Lee, J.Y. 2003. The mismeasure of management and its implications for leadership research. *Leadership Quarterly*, **14**(6): 615–56.

Rauch, A., Frese, M. and Utsch, A. 2005. Effects of human capital and long-term human resources development and utilization on employment growth of small-scale businesses: A causal analysis. *Entrepreneurship Theory and Practice*, **29**(6): 681–98.

Salomo, S., Brinckmann, J. and Talke, K. 2008. Functional management competence and growth of young technology-based firms. *Creativity and Innovation Management*, **17**(3): 186–203.

Schjoedt, L. and Bird, B.J. 2014. Control variables in entrepreneurship research: Use, misuse, and recommended use. Paper presented at the Babson Entrepreneurship Research Conference.

Schjoedt, L. and Shaver, K.G. 2007. Deciding on an entrepreneurial career: A test of the pull and push hypotheses using the panel study of entrepreneurial dynamics data. *Entrepreneurship Theory and Practice*, **31**(5): 733–52.

Singh, R. and Lucas, L. 2005. Not just domestic engineers: An exploratory study of homemaker entrepreneurs. *Entrepreneurship Theory and Practice*, **29**(1): 79–90.

Stewart, W.H., May, R. and Kalia, A. 2008. Environmental perceptions and scanning in the United States and India: Convergence in entrepreneurial information seeking? *Entrepreneurship Theory and Practice*, **32**(1): 83–106.

Steyaert, C. 2007. Entrepreneuring as a conceptual attractor? A review of process theories in 20 years of entrepreneurship studies. *Entrepreneurship and Regional Development*, **19**(6): 453–77.

Talaulicar, T., Grundei, J. and Werder, A. 2005. Strategic decision making in start-ups: The effect of top management team organization and processes on speed and comprehensiveness. *Journal of Business Venturing*, **20**(4): 519–41.

Ucbasaran, D., Westhead, P. and Wright, M. 2008. Opportunity identification and pursuit: Does an entrepreneur's human capital matter? *Small Business Economics*, **30**(2): 153–73.
Van de Ven, A.H. and Engleman, R.M. 2004. Event- and outcome-driven explanations of entrepreneurship. *Journal of Business Venturing*, **19**(3): 343–58.
Weick, K. 1979. *The Social Psychology of Organizing*. Reading, MA: Addison-Wesley.
Wright, M., Liu, X., Buck, T. and Filatotchev, I. 2008. Returnee entrepreneurs, science park location choice and performance: An analysis of high-technology SMEs in China. *Entrepreneurship Theory and Practice*, **32**(1): 131–55.
Zacks, J.M. and Tversky, B. 2001. Event structure in perception and conception. *Psychological Bulletin*, **127**(1): 3–21.
Ziegler, M., Rosenzweig, B. and Ziegler, P. 1992. *The Republic of Tea: Letters to a young entrepreneur*. New York: Currency Doubleday.

4. Entrepreneurial self-regulation: consciousness and cognition

Brandon Randolph-Seng, Wallace A. Williams, Jr and Mario Hayek

ABSTRACT

Entrepreneurial intentions involve both conscious and non-conscious cognitive processes. While the understanding of the relationship between these processes has made great strides in the social cognition literature, it has been relatively neglected in the entrepreneurship domain. Therefore, the purpose of this chapter is to introduce entrepreneurial cognition researchers to the area of non-conscious cognition from the broader social cognition literature and integrate this literature with research done in entrepreneurial intentions and intuition. In doing so, we introduce the concept of entrepreneurial self-regulation as the explanatory mechanism between conscious and non-conscious cognitive processes in the entrepreneurial domain. We conclude with a brief review of methods from social cognition research in order to help advance research in the area of non-conscious entrepreneurial cognition.

INTRODUCTION

> It thus remains an open question to which extent unconscious volitional processes are a causal force in human behavior.
>
> (Kuhl and Koole, 2004: 414)

Although researchers have recently taken great strides to understand how entrepreneurial cognitions are formed, little is known about the non-conscious processes underlying such cognitions (see Mitchell, Busenitz, Lant, McDougall, Morse and Smith, 2004). This topic will be examined in the context of Kuhl and Koole's (2004) question above, given that entrepreneurial cognition is a uniquely situated form of cognition. Not only are entrepreneurs required to be involved in many intellectual tasks simultaneously, but such tasks often exist in dynamic contexts imbued with uncertainty, emotion and novelty (Baron, 1998). Entrepreneurial cognition, therefore, is ideal for exploring, 'unconscious volition processes' in that both conscious and non-conscious cognitive processes in and of themselves seem to fail to explain how entrepreneurial cognition is possible, giving credence to entrepreneurial self-regulation as a bridging concept.

First, we discuss socially situated entrepreneurial cognition in order to introduce the idea of non-conscious processes in the context of entrepreneurship. Next, we explore consciousness and cognition, and provide an overview of non-conscious cognitive processes from the broader area of social cognition. We then present a review of both entrepreneurial intentions and intuitions research, followed by the introduction of the entrepreneurial self-regulation concept. We conclude with an outline of methods from social cognition research.

SOCIALLY SITUATED ENTREPRENEURIAL COGNITION

Given the unique environmental conditions faced by entrepreneurs, it may not be surprising to find that entrepreneurs appear to exhibit a general lack of introspection into their own cognitive processes (see Forbes, 2005); however, this lack of insight is not specific to entrepreneurs, but seems to be a consequence of how human beings are cognitively built (e.g., Macrae and Johnston, 1998). In other words, general cognition processes are the same as entrepreneurial cognitions in that both can operate at non-conscious levels, which helps to explain why the processes involved in creating a novel venture are still not clearly understood by researchers or easily explained by entrepreneurs (see Blair, 2010; cf. Cornelissen and Clarke, 2010).

Therefore the idea that entrepreneurial cognition may operate outside of conscious awareness provides an important backdrop for explaining the ways in which entrepreneurs develop entrepreneurial intentions and is consistent with recent work in situated social cognition (Mitchell, Randolph-Seng and Mitchell, 2011). First, central to understanding how entrepreneurs develop intentions is an appreciation of how entrepreneurial cognition is regulated by motivation and emotion within specific environments (see also Carsrud and Brännback, Chapter 2, this volume). Second, given that creating a new venture involves engaging in transactions through interaction and communication with others, the chosen interaction (e.g., with customers, investors, suppliers) will profoundly impact the type of entrepreneurial cognitive processes that will result. Third, the distributed component of entrepreneurial cognition provides a foundation for understanding how entrepreneurs create ventures in industries in which they lack experience. Just as it is not important for one to understand how a shovel is made in order to use it, entrepreneurial cognition can be extended by the social environment. This is important, because the properties of the tools entrepreneurs use to aid their cognition will in turn constrain and influence that cognition.

In order to examine how conscious and non-conscious cognitive processes operate in entrepreneurial cognition, entrepreneurial intentions and intuitions literatures will be examined, followed by an attempt to integrate the two areas by utilizing the concept of entrepreneurial self-regulation. First, however, it is important to understand the broader literature in social cognition by examining the intersection between conscious and non-conscious cognitive processes before investigating the more specific area of entrepreneurship.

CONSCIOUSNESS AND COGNITION

Traditionally, psychologists have assumed that deciding on and proceeding with a specific course of action is the result of conscious higher-order thinking (Carver and Scheier, 1981). More recently, however, psychological research increasingly suggests that many of the choices we make are instead the result of non-conscious thinking instigated by the environment in which we live (Andersen, Moskowitz, Blair and Nosek, 2007). Such non-conscious thinking processes are thought to be highly adaptive because they allow a person to quickly and efficiently respond to and act within a specific environment (Dijksterhuis and Bargh, 2001; cf. Gibson, 1979).

Although laboratory research has consistently provided demonstrations of quick and efficient non-conscious responses to environmental stimuli (e.g., Andersen et al., 2007), actions are finite. How does one non-consciously determine which of the potentially relevant stimuli to respond to? Consider the story of the signmaker's assistant as told by Arnold (1992):

> One day Norman (the signmaker's assistant) looked out of a window over the sign shop and watched the people below. They stopped at the STOP sign. They entered at the ENTER sign. They ate under the GOOD FOOD sign. 'They do whatever the signs say!' said Norman to himself. 'I wonder . . .' He crept into the shop while the signmaker napped. With brush and board he painted a sign of his own. (1992: 4)

It turns out that Norman's big idea was to create signs that either rewarded him personally (e.g., BUY NORMAN A PRESENT posted on the front door of a store) or provided self-amusement (e.g., GARBAGE

DUMP posted over a grocery market sign). Once more people did as the signs said until personal consequences started to accumulate (e.g., loss of money, physical harm). At this point reactance to the signs began in which the townspeople destroyed all of the signs in the town (in addition to Norman's signs). 'Then the real trouble started. Without store signs, shoppers became confused. Without stop signs, drivers didn't know when to stop. Without street signs, Fireman became lost' (1992: 20). Upon Norman's full confession to his part in the creation of the signs the townspeople's anger subsided and a realization of their part in the disaster became apparent: 'We were fools for obeying such signs without thinking' (1992: 26).

The story of the signmaker's assistant points out one of the main adaptive challenges humans face with the ability to non-consciously respond to the 'signs' in their environment: knowing which signs are worthy of a response. Such ability is essential to survival (Baumeister, Schmeichel and Vohs, 2007). If the wrong environmental signs are followed, maladaptive behavior results (e.g., giving Norman all of your money); yet if the signs are ignored, maladaptive behaviors are just as likely (e.g., failing to stop where a STOP sign should be). Generally, humans seem able to successfully navigate their environments, as evidenced by the complete transformation of the planet over the relatively short span of human existence. But on an individual level, how is it that one is able to adaptively sort through the messages coming from without (environment) and balance this with the messages coming from within (goals, motives, desires), and to do so without usurping valuable but limited conscious resources? The emerging answer from research points to the non-conscious regulation of responses to external stimuli by internal factors (Fishbach and Ferguson, 2007; Fitzsimons and Bargh, 2004; Moskowitz, Li and Kirk, 2004; Trope and Fishbach, 2005); specifically, non-conscious self-regulation of responses to the environment based on the goals, motives and desires each individual brings to the situation. In order for this conclusion to become clear, some conceptual clarity to 'non-conscious processes' and other related terms, such as automatic, implicit, subconscious and the like, are needed. Definitional problems in the literature investigating non-conscious processes are not new (see Erdelyi, 1992; Reingold and Merikle, 1993); nevertheless, the literature is consolidating around Moors and De Houwer's (2006) recent analysis, which will be reviewed next.

Definition of Non-conscious Cognitive Processes

Moors and De Houwer (2006; De Houwer and Moors, 2007) have recently suggested that non-conscious processes consist of five key features:

intention, control, autonomy, consciousness and efficiency. Below, we briefly define each feature based on Moors and De Houwer's (2006; De Houwer and Moors, 2007) inductive decompositional analysis.

Intention

An act is considered intentional when it is correctly caused by a mental representation of that act. Intentional acts could also be considered a subclass of controlled actions; however, control refers to more than just a goal to initiate an act.

Control

Control is defined as a goal regarding the act (i.e., avoid, start, stop or change an act), followed by a causal achievement of the goal to act. So, in a controlled process, not only does a person have a goal to act (intention), but they carry out that action. This constitutes an open-loop system of control in which the goal about the act (conscious or not) is necessary and sufficient for the effect to occur (see Wegner and Bargh, 1998).

Autonomy

Autonomous acts are defined as both unintentional and uncontrolled. That is, the occurrence of an act does not include either the goal to engage in the act or the ability to stop or change it. The goal to engage, stop or change an act could be either conscious or not.

Consciousness

Conscious may refer to subjective awareness of an input (e.g., people, places, things) that causes a process (e.g., thoughts, feelings, actions), output of a process, or even the consequences of a process (e.g., how actions specifically influence thoughts; see Gawronski, Hofmann and Wilbur, 2006). For present purposes, however, consciousness will be defined at the functional level (see Natsoulas, 1981). Specifically, a process will be considered conscious if a person is simply able to accurately link a matched output to a specific input (e.g., Johansson, Hall, Sikström, Tärning and Lind, 2006).

Efficiency

A process is considered efficient if it uses a minimal amount of one's attentional capacity or resources, while maintaining fast processing rates. In the case of scarce attentional resources (e.g., a demanding task), these processes should be unaffected (cf. Lien, Ruthruff and Johnston, 2006). Even efficient processes, however, may benefit from additional resources (see Klauer and Teige-Mocigemba, 2007; Schneider, Dumais and Shiffrin, 1984).

With this overview of the broader literature examining non-conscious cognitive process, a review of how conscious and non-conscious cognitive processes have been examined in entrepreneurial cognition will be covered next. First, the entrepreneurial intentions (with a focus on conscious cognitive processes), and second, entrepreneurial intuitions (with a focus on non-conscious cognitive processes) literatures will be examined.

ENTREPRENEURIAL INTENTIONS

Entrepreneurial intention has gained increased attention as the most proximal and strongest predictor of entrepreneurial action (Bird, 1988; Fishbein and Ajzen, 1975; Krueger, Reilly and Carsrud, 2000). Defined as the conscious commitment to pursue a new venture (Shook, Priem and McGee, 2003; Thompson, 2009), entrepreneurial intention models are particularly useful for predicting entrepreneurial behavior when a behavior is 'rare, hard to observe, or involves unpredictable time lags' such as those encountered in the entrepreneurship context (Krueger et al., 2000: 411). Furthermore, intentions model aid in predicting behaviors fraught with uncertainty and ambiguity (Busenitz and Barney, 1997; see also Baron, Chapter 1, this volume; Bird, Chapter 3, this volume). However, as an important first step toward unraveling the mental processes leading to entrepreneurial intentions, a more comprehensive understanding of the conscious and non-conscious elements that may influence this process is warranted.

While consensus has not been reached on a unified theoretical model of entrepreneurial intentions, three theory-based models have dominated the entrepreneurship literature: (1) Ajzen's (1991) Theory of Planned Behavior; (2) Shapero and Sokol's (1982) model of the Entrepreneurial Event (SEE); and (3) Bird's (1988) model of entrepreneurial intentions.

Theory of Planned Behavior (TPB)

Building on the Theory of Reasoned Action (Ajzen and Fishbein, 1980), the TPB posits that intentions are formed by attitudes toward behavior, subjective social norms (i.e., social influences) and perceived behavioral control (i.e., perceptions of control/self-efficacy; Ajzen, 1991). More specifically, attitude toward a behavior 'refers to the degree to which a person has a favorable or unfavorable evaluation or appraisal of the behavior in question' (Ajzen, 1991: 189). Perceived behavioral control refers to the perception of the feasibility of engaging in a particular behavior. Finally, subjective norms refer to social pressures that may influence the decision about engaging in a behavior. The TPB has been the most widely used theory to explain and predict entrepreneurial intentions (Armitage and Conner, 2001).

Shapero and Sokol's Entrepreneurial Event (SEE)

Despite the prevalence of the TPB in the entrepreneurship literature, Shapero and Sokol (1982) were the first to develop one of the dominant models of entrepreneurial intentions. This model suggests that people's lives are driven by inertia until a precipitating (i.e., life-changing) event occurs that requires the individual to engage in a behavior based on the desirability and feasibility of their choice. In addition to assessing the desirability and feasibility of the behavior, Shapero and Sokol's theory suggests that individuals must have a dispositional propensity to act in order to engage in the behavior. Therefore, entrepreneurial intentions are formed by individuals who have a dispositional propensity to act and perceive entrepreneurship as both desirable and feasible.

Shapero and Sokol's (1982) Entrepreneurial Event overlaps considerably with the TPB in that both consider perceived desirability (attitude in the TPB) and feasibility (behavioral control in the TPB; Krueger and Brazeal, 1994). However, in an important difference from other entrepreneurship intention models, Shapiro and Sokol (1982) included the disposition 'propensity to act.' This is interesting because, although entrepreneurship research has shied away from trait-based research (Zhao, Seibert and Lumpkin, 2010), this dominant entrepreneurial intentions model includes a trait. The inclusion of a disposition in this intentions model is particularly relevant to the discussion of conscious and non-conscious decision-making processes because personality traits are considered to include both conscious and non-conscious components (see Rauch and Frese, 2007), while entrepreneurial intention has repeatedly been cited as a conscious decision (Thompson, 2009).

Bird's Model of Entrepreneurial Intentions

While the two dominant intentions models, namely TPB and SEE, have ignored the separation between conscious and non-conscious decision-making influences during the development of intentions, Bird's (1988) model explicitly includes both decision-making processes. She suggests that 'personal and social contexts interact with rational and intuitive thinking during the formulation of entrepreneurial intentions' (Bird, 1988: 443). More specifically, Bird argues that individual and contextual variables influence rational decision making when engaging in tasks such as business plan development, while intuitive thinking shapes/molds a future vision of the venture, thus motivating entrepreneurs to engage and persevere. Furthermore, she posits that both rational and intuitive cognitive processes lead to intentionality of pursuing a new venture, resulting in entrepreneurial behavior.

The three aforementioned theoretical perspectives are largely homologous in that they suggest that entrepreneurial intentions are predominantly driven by attitudes toward pursuing a new venture and a perception of the feasibility of successfully engaging in that behavior. However, while these models are concerned with the proximal antecedents of intentions, they neglect to explain the more distal cognitive mechanisms resulting in these perceptions. Thus it behooves entrepreneurial cognition researchers to understand how conscious and non-conscious cognitive mechanisms influence both attitudes and perceptions. Therefore, while entrepreneurial intentions have been largely defined and measured as a conscious decision, the process leading to this outcome involves both conscious and non-conscious elements (Thompson, 2009).

The ability to articulate the non-conscious processes underlying the formation of entrepreneurial cognition is important for entrepreneurs in order for them to garner support for their ventures with relevant stakeholder groups (Cornelissen and Clarke, 2010; McMullen and Shepherd, 2006). This ability requires entrepreneurs to codify tacit knowledge into a more communicable form of knowledge. Current research has focused on understanding this codification process with hopes of gaining insights into how and under what conditions entrepreneurs translate their knowledge and experience to conscious decision making and, furthermore, to relevant stakeholder groups. For example, in a recent study, Mitchell and Shepherd (2012) found that possessing general versus specific knowledge influences an entrepreneur's ability to translate tacit knowledge into a format that is easier to articulate to stakeholders. Bridging the knowledge gap between conscious and non-conscious decision making can help make important strides in understanding the formation of entrepreneurial intentions. To that end, in the next section we review the entrepreneurial intuition literature, with a focus on non-conscious processes.

ENTREPRENEURIAL INTUITION

In this section we offer a brief review of entrepreneurial intuition research. We begin with an overview of intuition research in management. Then we discuss intuition within the specific context of entrepreneurship.

Intuition in Management

Much of the early work on intuition in management thought can be traced to Chester Barnard (Novicevic, Hench and Wren, 2002), who described intuition as a non-logical mental process that is 'not capable of being expressed in words' (Barnard, 1968: 302). Barnard's conceptualization portrayed intuition as the result of previous experience that has been instilled as 'instinct' stemming from personal habit, societal beliefs and/or cultural norms (Barnard, 1995). Barnard argued that a major part of management decision making relied on habitual, non-conscious behavior (Novicevic, Clayton and Williams, 2011).

Sinclair and Ashkanasy (2005) stated that intuition research falls into two categories. Similar to Barnard's conceptualization of intuition, the first category is based on tacit knowledge that stems from a (non-conscious) recognition of patterns from prior experiences. The second category is derived (at least in part) from feelings, or 'sensory and affective elements' (Sinclair and Ashkanasy, 2005: 356). This second conceptualization views intuition as an emotional response, and is in line with Burke and Miller (1999), who described intuition as a combination of previous experience and emotional inputs (e.g., gut feelings). Similarly, research by Bradley (2010) suggested a psychophysiological approach to intuition, involving both cognitive and emotional (heart-focused) responses to a particular event or decision. Sadler-Smith and Shefy (2004) also acknowledged the notions of both pattern recognition and affective responses, describing the process as interplay between intuition-as-expertise and intuition-as-feeling.

Intuition in Entrepreneurial Research

Specific to the entrepreneurship domain, both practitioners and academicians often discuss the role of intuition in entrepreneurial behavior and decision-making processes (Allinson, Chell and Hayes, 2000; Blume and Covin, 2011; Crossan, Lane and White, 1999). Mitchell, Friga and Mitchell (2005) stated that, while entrepreneurs use intuition to explain their actions (particularly those related to opportunity identification), the term 'entrepreneurial intuition' has myriad definitions that vary widely. Utilizing Gordon's (1992) system of learning and memory, the authors organized the disparate definitions based on their proximity to consciousness, which they describe as deliberate or intentional information retrieval, or 'phenomenal awareness' (Schacter, 1989: 356). Level 1, the furthest from consciousness, consists of non-conscious procedures and systems (e.g., motor skills), and includes Barnard's definition listed above. Level 2 represents a cognitive subsystem based on a response given the situational context (e.g., a response to the current environment). Level 3 contains systems using declarative, explicit (working) memory (i.e., facts and data). Finally, Level 4 entails the deliberate or intentional retrieval of information (e.g., planning and monitoring), and incorporates the executive and control system that 'mediates and organizes information flow to both working memory and the declarative and response systems' (Mitchell et al., 2005: 659) within Levels 2 and 3. Based on this organization (and specific to opportunity identification), Mitchell et al. (2005: 667) defined entrepreneurial intuition as 'the dynamic process by which entrepreneurial alertness cognitions interact with domain competence (e.g., culture, industry, specific circumstances, technology, etc.) to bring to consciousness an opportunity to create new value.'

Related to Mitchell and colleagues' (2005) conception of intuition, non-conscious influences that may affect the formation of attitudes and perceptions leading to entrepreneurial intentions include heuristics and biases. Entrepreneurial environments tend to be characterized as dynamic and uncertain, lending themselves to decision-making processes that use heuristics, biases improvisation, intuition and emotions (Baron and Ward, 2004; Busenitz and Barney, 1997), as opposed to a more planned decision-making schema used by many organizational managers. In general, entrepreneurs have been found to rely heavily on heuristics (Busenitz and Lau, 1996) that are influenced by a number of biases such as overconfidence, representativeness and illusion of control (De Carolis and Saparito, 2006; Keh, Foo and Lim, 2002; Simon, Houghton and Aquino, 2000). These biases have been suggested to influence the intention formation process. For example, Simon et al. (2000: 117) noted that 'a tendency towards overconfidence may have affected entrepreneurs when they were initially evaluating the venture's riskiness.' Similarly, individuals might rely on sparse data and past experiences (i.e., representativeness) to make broad generalizations that do not reflect the objective facts (Kahneman, Slovic and Tversky, 1982). The illusion of control helps reduce anxiety over not being able to control events and may lead individuals to develop a false sense of security, resulting in the intention to pursue a new venture. More specifically, elements such as time and resource constraints, novelty, turbulence (Moorman and Miner, 1998), and organizational dynamism (Brown and Eisenhardt, 1997) influence the extent to which individuals engage in strategic planning, use heuristics and biases, or improvise (Hmieleski and Corbett, 2008). Busenitz and Barney (1997) argued that heuristics and biases are important when justifying deviations from rational decision making, especially for entrepreneurs since there is less accountability for their decisions compared to those of organizational managers (Frydman, Gray, Hessel and Rapaczynski, 1999).

In addition to heuristics and biases, affect may also constitute non-conscious influences on intuition (see also Foo, Murnieks and Chan, Chapter 5, this volume; Grégoire, Chapter 6, this volume). Affect has been defined as feelings and emotions that may be either more stable across situations (dispositional affect) or elicited by environmental/contextual events (state affect; Baron, 2008). These feelings influence cognition by priming 'specific memories and association – ones linked, cognitively, to such feelings' and by 'serving as a heuristic cue – an efficient basis for reactions to a specific person, event, or stimulus' (Baron, 2008: 332). Affective processes are particularly influential in dynamic and uncertain contexts (Quartz, 2009). 'Affect influences perceptions, intentions, and behaviors through unconscious processes, by priming access to and retrieval of memories' (Hayton and Cholakova, 2012: 42). These influences also affect entrepreneurial motivation by regulating the amount of information sought during the decision-making process (Baron, 2008; Cardon, Wincent, Singh and Drnovšek, 2009; Foo, Uy and Baron, 2009). Therefore, in uncertain contexts with strong entrepreneurial motivation, affect may elicit the use of heuristics and biases during the decision-making process (see Cohen, 2005).

More recently, Blume and Covin (2011) extended intuition research by focusing on how and when entrepreneurs actually associate their decisions with an intuitive process. They proposed that entrepreneurs often ascribe intuition to their venture-founding decisions, when other factors may have actually been at play. For example, the entrepreneur may possess a high sense of self-efficacy (which may, in part, be related to their success in previous venture-founding decisions). In other words, other attributes (such as tolerance for ambiguity, or overconfidence bias) may be the 'phenomenon driving the founding decision' (Blume and Covin, 2011: 142), even if the entrepreneur tends to describe their decision as intuitive. This misattribution of intuition helps point out the difficulty of explaining and studying non-conscious processes in entrepreneurial cognition. In order to provide insight into how to move the literature forward, the concept of entrepreneurial self-regulation will be discussed next, followed by methods in social cognition used to measure non-conscious cognitive processes.

ENTREPRENEURIAL SELF-REGULATION

Applying the five features of intention, control, autonomy, consciousness and efficiency to the entrepreneurial intentions and intuition literature reviewed above allows for a clearer understanding of how conscious and non-conscious cognitive processes may interact in forming entrepreneurial intentions and intuitions. First, entrepreneurial self-regulation is intentional: acts are engaged in as a result of specific mental representations of desired outcomes. Second, entrepreneurial self-regulation involves 'preconscious control.' There is a dynamic open-loop relationship between the goal to reach a desired outcome and the achievement of that goal. Third, entrepreneurial self-regulation is not autonomous in that it has both intentional and controlled features. Thus far, entrepreneurial self-regulation does not appear to differ substantially from traditional conceptions of self-regulation (e.g., Carver and Scheier, 1981). However, differences can be found in the last two features: consciousness and efficiency. Entrepreneurial self-regulatory processes are not always conscious. People are often not aware in the moment that their thoughts, feelings and acts strategically reflect some desired business outcome based on their needs, wants and desires. Finally, entrepreneurial self-regulatory processes tend to be efficient. Within a complex real-time environment, effective entrepreneurial self-regulation must operate efficiently.

Entrepreneurial self-regulation provides a middle ground between 'Zombie' (we are usually on autopilot) and 'Commander Data'[1] models (we are seldom on autopilot) of mental functioning (see Cleeremans and Jimènez, 2002; Loftus and Klinger, 1992; Moors and De Houwer, 2006). In other words, entrepreneurial cognition may often be non-conscious, but still under the control of higher-order motivational processes, and thus very complex and flexible. The form of 'strategic automaticity' (Gollwitzer and Schaal, 1998: 124) espoused in this chapter is consistent with current theorizing on non-conscious thought (e.g., Bos, Dijksterhuis and van Baaren, 2007; cf. Schneider et al., 1984), and provides for individual differences in entrepreneurial motivation.

The entrepreneurial self-regulation perspective also provides an example that helps to rectify recent theories of self-regulation (i.e., strength model) with recent theorizing in the automaticity domain (e.g., perception–behavior link). Therefore highly adaptive automatic processes (e.g., Dijksterhuis and Bargh, 2001; Prinz, 2003) and highly functional forms of consciousness (e.g., rational choice; Baumeister, 2008; Bless and Forgas, 2000) may be intimately linked through the preconscious control of the environment not only based on the automatization of formally conscious goals, skills, behaviors and so on (see Bargh, 1992; Moskowitz, Gollwitzer, Wasel and Schaal 1999), but online and in the moment.

Such preconscious regulatory ability may constitute a meta-cognitive process in that what is being consciously perceived is being preconsciously monitored and regulated. Preconscious meta-cognitive experiences, therefore, may be an important link between conscious and non-conscious aspects of information processing (Koriat and Levy-Sadot, 1999), allowing for involvement of conscious self-regulatory processes (see Dijksterhuis, Chartrand and Aarts, 2007). Such a view is similar to Bargh's (2008: 563) proposed purpose of consciousness to create even 'more complex non-conscious processes' to better fit specific environments, needs, desires and so on of the person.

MOVING FORWARD

More research is needed to examine the intersection between conscious and non-conscious cognition in entrepreneurship. First, in almost all cases the research evidence reviewed above does not clearly distinguish conscious from non-conscious cognitive processes. Although it is difficult to explain many of the findings reviewed above outside the context of non-conscious cognitive processes, a more convincing demonstration would involve using procedures that manipulate or measure non-conscious processes. In order to help future researchers accomplish this work, the current state of laboratory procedures used in examining non-conscious processes will be briefly addressed. The discussion will focus first on priming methods (manipulating), followed by implicit measures (measuring).

Priming Methods

Priming is the activation of a person's cognitive representation in one context in order to influence their response in an unrelated context (see Bargh and Chartrand, 2000). Conceptual priming techniques may be either supraliminal (conscious) or subliminal (non-conscious). An important consideration when studying non-conscious processes is that the person being primed is not aware of the influence the prime is having on them, even if they can consciously sense the primed stimuli (Bargh and Morsella, 2008). Phenomenal reports are usually considered the clearest standard for determining the non-consciousness of the prime (Dijksterhuis, Aarts and Smith, 2005; Dixon, 1981), although other less direct methods have been used as well (see Merikle, Joordens and Stolz, 1995; Rydell, McConnell, Mackie and Strain, 2006).

Robust findings show that information (e.g., words, smells) a person is unaware of can influence them in predictable and consistent ways (see Li, Moallem, Paller and Gottfried, 2007; Sparrow and Wegner, 2006); nevertheless, controversy remains (Bargh, 2006; cf. Holender, 1986). For example, questions still arise in the literature about the amount and quality of information that can be processed outside of awareness (Dijksterhuis and Nordgren, 2006; Glaser and Banaji, 1999; Masling, 1992) and applications to real-world settings (e.g., persuasion; Karremans, Stroebe and Claus, 2006). Overall, the real benefit of using priming methods has been in their implications for understanding the non-conscious functional aspects of psychological processing (Bargh, 2006; Dixon, 1981).

Implicit Measures

Although designed to manipulate non-conscious exposure to stimuli, priming techniques led to the use of implicit measures (Nosek, 2007; Wittenbrink, 2007). Implicit measures are thought to tap into people's non-conscious components of their mental structures (Fazio and Olson, 2003). The measure itself is not implicit, but the measurement outcome is thought to be implicit. Depending on the implicit measure used, lack of awareness may come in the form of the stimuli that activate a construct, the construct itself, and if and how the construct influences them (De Houwer and Moors, 2007).

Researchers are increasingly realizing that no implicit measure is 'process pure,' but lies on a continuum from spontaneous to deliberative (Vargas, Sekaquaptewa and von Hippel, 2007: 107). In addition, different types of implicit measures correspond to different types of processes that are thought to be measured (e.g., Gawronski and Bodenhausen, 2005; Ranganath, Smith and Nosek, 2008). Measures of implicit attitudes used in priming studies, for instance, seem to be influenced more by recent rather than early developmental experiences, while the opposite is thought to be true for measures such as the implicit association test (see Rudman, Phelan and Heppen, 2007).

Care must also be taken in matching an explicit measure to one that is thought to be implicit, given that differences can be attributed to a number of different processes besides the implicit or explicit nature of the outcome (Merikle and Reingold, 1998; e.g., Thrash, Elliot and Schultheiss, 2007). One recent example of an attempt to separate out potentially operating processes is the process dissociation approach (Debner and Jacoby, 1994). The process dissociation approach examines the difference between responses when an automatic and controlled process is consistent with an outcome, versus when an automatic but not controlled process is consistent with an outcome (Payne and Stewart, 2007). Nevertheless, even this approach has problems; specifically, in such an approach awareness equals control over thoughts and behaviors (Jacoby, Lindsay and Toth, 1992). Defining awareness in such a way implies that automatic and controlled processes are dichotomous, which leaves little conceptual room for motivational processes such as automatic control (see Glaser and Banaji, 1999). Other approaches, such as the 'quad model of implicit task performance' (Conrey, Sherman, Gawronski, Hugenberg and Groom, 2005) are an improvement on the process dissociation approach; however, where non-conscious motivational processes fit in is still not clear.

Although implicit measures are often used to measure implicit attitudes, researchers are increasingly finding that these implicit attitudes seem to be better conceptualized as 'unintentionally activated evaluations of object-centered contexts' (Ferguson and Bargh, 2007: 217). Instead of implicit attitudes reflecting individuals' unmediated evaluations of their world, such attitudes instead depend on a number of different factors (e.g., memories, instructions, moods, active and chronic goals) that are present at the time of measurement (Andersen et al., 2007; Ajzen and Fishbein, 2000; for recent research examples see Conner, Perugini, O'Gorman, Ayres and Prestwich, 2007; Mitchell, Nosek and Banaji, 2003). Even the subliminal context can influence the outcome of implicit-type measures (Dixon, 1981; Lepore and Brown, 1997). An implicit measure may be better conceptualized, then, as a measure of the relevance of the attitude object as a function of pre-existing personal and external factors without being obscured by introspective thinking (Ferguson and Bargh, 2007; Nosek, 2007). Implicit measures may at times be tapping into self-regulatory representations in that individuals may be automatically sensitive to a specific context in order to better meet their goals (Smith and Conrey, 2007; for recent research evidence see Gregg, Seibt and Banaji, 2006; Legault, Green-Demers, Grant and Chung, 2007; Rudman, Dohn and Fairchild, 2007).

A second issue is to determine at what stage of mental processing entrepreneurial self-regulation occurs. For instance, Isaacowitz (2006) has found that in the case of motivated gazing that occurred outside of awareness, some initial visual scanning occurred first. Therefore it may well be that entrepreneurial self-regulation operates on later stages of cognitive processing, but still before full conscious awareness (see Pylyshyn, 1999). Such avenues of research must rely on accessibility and physiological methodology (see also Baucus, Baucus and Mitchell, Chapter 8, this volume; McMullen, Wood and Palich, Chapter 9, this volume).

Third is the issue of how conscious cognitive processes simultaneously work with non-conscious cognitive processes in forming entrepreneurial self-regulation. Since people are not consciously aware of these processes operating, but are nonetheless aware of the product of such processes, the world they consciously perceive becomes a mere construction of their own entrepreneurial goals, motives and desires that are operating non-consciously in the environment (see Balcetis and Dunning, 2006). Interestingly, one's conscious understanding of the world, based on this non-conscious construction, becomes reality and may in turn be able to feed back into entrepreneurial self-regulation processes (see Pittman,

1992). Future research investigating the mechanisms behind the development of entrepreneurial self-regulation would allow for clearer understanding of the role consciousness plays in the process. Such a line of research would complement advances in psychological theory by providing evidence showing how social cognition can be flexible and situated (Smith and Conrey, 2007) and how 'perception, action, and cognition seem to form a collective community' (Knoblich and Sebanz, 2006: 99). For example, a person's bodily movement may be intimately linked not only to the entrepreneurial cognitive representations of that person's world, but also to their preconscious perception, leading to specific entrepreneurial cognitions, affects and behaviors.

NOTE

1. Referring to the fictional character in the *Star Trek* television series: a machine that is essentially a life form.

REFERENCES

Ajzen, I. 1991. Theory of planned behavior: Some unresolved issues. *Organizational Behavior and Human Decision Processes*, **50**(2): 179–211.

Ajzen, I. and Fishbein, M. 1980. *Understanding Attitudes and Predicting Social Behavior*. Englewood Cliffs, NJ: Prentice-Hall.

Ajzen, I. and Fishbein, M. 2000. Attitudes and the attitude–behavior relation: Reasoned and automatic processes. In W. Stroebe and M. Hewstone (eds), *European Review of Social Psychology*: 1–33. New York: John Wiley & Sons.

Allinson, C., Chell, E. and Hayes, J. 2000. Intuition and entrepreneurial behaviour. *European Journal of Work and Organizational Psychology*, **9**(1): 31–43.

Andersen, S.A., Moskowitz, G.B., Blair, I.V. and Nosek, B.A. 2007. Automatic thought. In E.T. Higgins and A. Kruglanski (eds), *Social Psychology: Handbook of basic principles* (2nd edn): 138–75. New York: Guilford Press.

Armitage, C.J. and Conner, M. 2001. Efficacy of the theory of planned behaviour: A meta-analytic review. *British Journal of Social Psychology*, **40**(4): 471–99.

Arnold, T. 1992. *The Signmaker's Assistant*. New York: Penguin Books.

Balcetis, E. and Dunning, D. 2006. See what you want to see: Motivational influences on visual perception. *Journal of Personality and Social Psychology*, **91**(4): 612–25.

Bargh, J.A. 1992. Does subliminality matter to social psychology? Awareness of the stimulus versus awareness of its influence. In R.F. Bornstein and T.S. Pittman (eds), *Perception without Awareness: Cognitive, clinical, and social perspectives*: 238–55. New York: Guilford Press.

Bargh, J.A. 2006. What have we been priming all of these years? On the development, mechanisms, and ecology of nonconscious social behavior. *European Journal of Social Psychology*, **36**(2): 147–68.

Bargh, J.A. 2008. Social psychological approaches to consciousness. In P.D. Zelazo, M. Moscovitch and E. Thompson (eds), *The Cambridge Handbook of Consciousness*: 553–67. New York: Cambridge University Press.

Bargh, J.A. and Chartrand, T.L. 2000. The mind in the middle: A practical guide to priming and automaticity research. In H.T. Reis and C.M. Judd (eds), *Handbook of Research Methods in Social and Personality Psychology*: 253–85. Cambridge, UK: Cambridge University Press.

Bargh, J.A. and Morsella, E. 2008. The unconscious mind. *Perspectives on Psychological Science*, **3**(1): 73–9.

Barnard, C.I. 1968. *The Functions of the Executive*. Cambridge, MA: Harvard University Press.

Barnard, C.I. 1995. The significance of decisive behavior in social action: Notes on the nature of decision. *Journal of Management History*, **1**(4): 28–87.

Baron, R.A. 1998. Cognitive mechanisms in entrepreneurship: Why and when enterpreneurs think differently than other people. *Journal of Business Venturing*, **13**(4): 275–94.

Baron, R.A. 2008. The role of affect in the entrepreneurial process. *Academy of Management Review*, **33**(2): 328–40.

Baron, R.A. and Ward, T.B. 2004. Expanding entrepreneurial cognition's toolbox: Potential contributions from the field of cognition science. *Entrepreneurship Theory and Practice*, **28**(6): 553–73.

Baumeister, R.F. 2008. Free will in scientific psychology. *Perspectives on Psychological Science*, **3**(1): 14–19.

Baumeister, R.F., Schmeichel, B.J. and Vohs, K.D. 2007. Self-regulation and executive function: The self as controlling agent. In E.T. Higgins and A. Kruglanski (eds), *Social Psychology: Handbook of basic principles* (2nd edn): 138–75. New York: Guilford.

Bird, B. 1988. Implementing entrepreneurial ideas: The case for intention. *Academy of Management Review*, **13**(3): 442–53.

Blair, E.S. 2010. What you think is not what you think: Unconsciousness and entrepreneurial behavior. In A.A. Stanton, M. Day and I.M. Welpe (eds), *Neuroeconomics and the Firm*: 50–65. Cheltenham, UK and Northampton, MA, USA: Edward Elgar Publishing.

Bless, H. and Forgas, J.P. 2000. The message within: Toward a social psychology of subjective experiences. In H. Bless and J.P. Forgas (eds), *The Message Within: The role of subjective in social cognition and behavior*: 372–92. Philadelphia, PA: Psychology Press.

Blume, B.D., and Covin, J.G. 2011. Attributions to intuition in the venture founding process: Do entrepreneurs actually use intuition or just say that they do? *Journal of Business Venturing*, **26**(1): 137–51.

Bos, M.W., Dijksterhuis, A. and van Baaren, R.B. 2007. On the goal dependency of unconscious thought. Poster session presented at the 8th annual conference of the Society for Personality and Social Psychology, Memphis, TN.

Bradley, R.T. 2010. Passionate attention and the psychophysiology of entrepreneurial intuition: A quantum-holographic theory. *International Journal of Entrepreneurship and Small Business*, **9**(3): 324–48.

Brown, S.L. and Eisenhardt, K.M. 1997. The art of continuous change: Linking complexity theory and time-paced evolution in relentlessly shifting organizations. *Administrative Science Quarterly*, **42**(1): 1–34.

Burke, L.A. and Miller, M.K. 1999. Taking the mystery out of intuitive decision making. *Academy of Management Executive*, **13**(4): 91–9.

Busenitz, L.W. and Barney, J.B. 1997. Differences between entrepreneurs and managers in large organizations: Biases and heuristics in strategic decision-making, *Journal of Business Venturing*, **12**(1): 9–30.

Busenitz, L.W. and Lau, C.M. 1996. A cross-cultural cognitive model of venture creation: Understanding entrepreneurial and immigrant decision making. *Entrepreneurship Theory and Practice*, **20**(4): 25–37.

Cardon, M.S., Wincent, J. Singh, J. and Drnovšek, M. 2009. The nature and experience of entrepreneurial passion. *Academy of Management Review*, **34**(3): 511–32.

Carver, C.S. and Scheier, M.F. 1981. *Attention and Self-regulation: A control theory approach to human behaviors*. New York: Springer.

Cleeremans, A. and Jimènez, L. 2002. Implicit learning and consciousness: A graded,

dynamic perspective. In R.M. French and A. Cleeremans (eds), *Implicit Learning and Consciousness*: 1–40. Hove, UK: Psychology Press.
Cohen, J.D. 2005. The vulcanization of the human brain: A neural perspective on interactions between cognition and emotion. *Journal of Economic Perspectives*, **19**(4): 3–24.
Conner, M.T., Perugini, M., O'Gorman, R., Ayres, K. and Prestwich, A. 2007. Relations between implicit and explicit measures of attitudes and measures of behavior: Evidence of moderation by individual difference variables. *Personality and Social Psychology Bulletin*, **33**(12): 1727–40.
Conrey, F.R., Sherman, J.W., Gawronski, B., Hugenberg, K. and Groom, C.J. 2005. Separating multiple processes in implicit social cognition: The quad model of implicit task performance. *Journal of Personality and Social Psychology*, **89**(4): 469–87.
Cornelissen, J.P. and Clarke, J.S. 2010. Imaging and rationalizing opportunities: Inductive reasoning and the creation and justification of new ventures. *Academy of Management Review*, **35**(4): 539–57.
Crossan, M.M., Lane, H.W. and White, R.E. 1999. An organizational learning framework: Learning from intuition to institution. *Academy of Management Review*, **24**(3): 522–37.
De Carolis, D.M. and Saparito, P. 2006. Social capital, cognition, and entrepreneurial opportunities: A theoretical framework. *Entrepreneurship Theory and Practice*, **30**(1): 41–56.
De Houwer, J. and Moors, A. 2007. How to define and examine the implicitness of implicit measures. In B. Wittenbrink and N. Schwarz (eds), *Implicit Measures of Attitudes*: 179–94. New York: Guilford Press.
Debner, J.A. and Jacoby, L.L. 1994. Unconscious perception: Attention, awareness, and control. *Journal of Experimental Psychology: Learning, Memory, and Cognition*, **20**(2): 304–17.
Dijksterhuis, A., Aarts, H. and Smith, P.K. 2005. The power of the subliminal: On subliminal persuasion and other potential applications. In R. Hassin, J. Uleman and J.A. Bargh (eds), *The New Unconscious*: 77–106. New York: Oxford University Press.
Dijksterhuis, A. and Bargh, J.A. 2001. The perception–behavior expressway: Automatic effects of social perception and social behavior. In M. Zanna (ed.), *Advances in Experimental Social Psychology*, vol. 30: 1–40. New York: Academic Press.
Dijksterhuis, A., Chartrand, T.L. and Aarts, H. 2007. Effects of priming and perception on social behavior and goal pursuit. In J.A. Bargh (ed.), *Social Psychology and the Unconscious: The automaticity of higher mental processes*: 51–131. New York: Psychology Press.
Dijksterhuis, A. and Nordgren, L.F. 2006. A theory of unconscious thought. *Perspectives on Psychological Science*, **1**(2): 95–109.
Dixon, N.F. 1981. *Preconscious Processing*. New York: Wiley.
Erdelyi, M.H. 1992. Psychodynamics and the unconscious. *American Psychologist*, **47**(6): 784–87.
Fazio, R.H. and Olson, M.A. 2003. Implicit measures in social cognition research: Their meaning and use. *Annual Review of Psychology*, **54**(1): 297–327.
Ferguson, M.J. and Bargh, J.A. 2007. Beyond the attitude object: Implicit attitudes spring from object-centered contexts. In B. Wittenbrink and N. Schwarz (eds), *Implicit Measures of Attitudes*: 216–46. New York: Guilford Press.
Fishbach, A. and Ferguson, M.J. 2007. The goal construct in social psychology. In E.T. Higgins and A. Kruglanski (eds), *Social Psychology: Handbook of basic principles* (2nd edn): 490–515. New York: Guilford Press.
Fishbein, M. and Ajzen, I. 1975. *Belief, Attitude, Intention, and Behavior: An introduction to theory and research*. New York: Addison-Wesley.
Fitzsimons, G.M. and Bargh, J.A. 2004. Automatic self-regulation. In R.F. Baumeister and K.D. Vohs (eds), *Handbook of Self-regulation: Research, theory, and applications*: 151–70. New York: Guilford Press.
Foo, M.-D., Uy, M.A. and Baron, R.A. 2009. How do feelings influence effort? An empirical

study of entrepreneurs' affect and venture effort. *Journal of Applied Psychology*, **94**(4): 1086–94.

Forbes, D.P. 2005. Are some entrepreneurs more overconfident than others? *Journal of Business Venturing*, **20**(5): 623–40.

Frydman, R., Gray, C., Hessel, M. and Rapaczynski, A. 1999. When does privatization work? The impact of private ownership on corporate performance in the transition economies, *Quarterly Journal of Economics*, **114**(4): 1153–91.

Gawronski, B. and Bodenhausen, G.V. 2005. Accessibility effects on implicit cognition: The role of knowledge activation and retrieval experiences. *Journal of Personality and Social Psychology*, **89**(5): 672–85.

Gawronski, B., Hofmann, W. and Wilbur, C.J. 2006. Are 'implicit' attitudes unconscious? *Consciousness and Cognition*, **15**(3): 485–99.

Gibson, J.J. 1979. *The Ecological Approach to Visual Perception*. Boston, MA: Houghton-Mifflin.

Glaser, J. and Banaji, M.R. 1999. When fair is foul and foul is fair: Reverse priming in automatic evaluation. *Journal of Personality and Social Psychology*, **77**(4): 669–87.

Gollwitzer, P.M., and Schaal, B. 1998. Metacognition in action: The importance of implementation intentions. *Personality and Social Psychology Review*, **2**(2): 124–36.

Gordon, S.E. 1992. Implications of cognitive theory for knowledge acquisition. In R.R. Hoffman (ed.), *The Psychology of Expertise: Cognitive research and empirical AI*: 99–120. New York: Springer.

Gregg, A.P., Seibt, B. and Banaji, M.R. 2006. Easier done than undone: Asymmetry in the malleability of implicit preferences. *Journal of Personality and Social Psychology*, **90**(1): 1–20.

Hayton, J.C. and Cholakova, M. 2012. The role of affect in the creation and intentional pursuit of entrepreneurial ideas. *Entrepreneurship Theory and Practice*, **36**(1): 41–68.

Hmieleski, K.M. and Corbett, A.C. 2008. The contrasting interaction effects of improvisational behavior with entrepreneurial self-efficacy on new venture performance and entrepreneur work satisfaction. *Journal of Business Venturing*, **23**(4): 482–96.

Holender, D. 1986. Semantic activation without conscious identification in dichotic listening, parafoveal vision, and visual masking: A survey and appraisal. *Behavior and Brain Sciences*, **9**(1): 1–66.

Isaacowitz, D.M. 2006. Motivated gaze: The view from the gazer. *Current Directions in Psychological Science*, **15**(2): 68–72.

Jacoby, L.L., Lindsay, D.S. and Toth, J.P. 1992. Unconscious influences revealed: Attention, awareness, and control. *American Psychologist*, **47**(6): 802–9.

Johansson, P., Hall, L., Sikström, S., Tärning, B. and Lind, A. 2006. How something can be said about telling more than we can know: On choice blindness and introspection. *Consciousness and Cognition*, **15**(4): 673–92.

Kahneman, D., Slovic, P. and Tversky, A. 1982. *Judgment under Uncertainty: Heuristics and biases*. Cambridge, UK: Cambridge University Press.

Karremans, J.C., Stroebe, W. and Claus, J. 2006. Beyond Vicary's fantasies: The impact of subliminal priming and brand choice. *Journal of Experimental Social Psychology*, **42**(6): 792–8.

Keh, H.T., Foo, M.-D. and Lim, B.C. 2002. Opportunity evaluation under risky conditions: The cognitive processes of entrepreneurs. *Entrepreneurship Theory and Practice*, **27**(2): 125–48.

Klauer, K.C. and Teige-Mocigemba, S. 2007. Controllability and resource dependence in automatic evaluation. *Journal of Experimental Social Psychology*, **43**(4): 648–55.

Knoblich, G. and Sebanz, N. 2006. The social nature of perception and action. *Current Directions in Psychological Science*, **15**(3): 99–104.

Koriat, A. and Levy-Sadot, R. 1999. Processes underlying metacognitive judgments. Information-based and experienced-based monitoring of one's own knowledge. In S. Chaiken and Y. Trope (eds), *Dual Process Theories in Social Psychology*: 483–502. New York: Guilford Press.

Krueger, N.F. and Brazeal, D. 1994. Entrepreneurial potential and potential entrepreneurs. *Entrepreneurship Theory and Practice*, **18**(1): 91–104.

Krueger, N.F., Reilly, M.D. and Carsrud, A.L. 2000. Competing models of entrepreneurial intentions. *Journal of Business Venturing*, **15**(5/6): 411–32.

Kuhl, J. and Koole, S.L. 2004. Workings of the will: A functional approach. In J. Greenberg, S.L. Koole and T. Pyszczynski (eds), *Handbook of Experimental Existential Psychology*: 411–30. New York: Guilford Press.

Legault, L., Green-Demers, I., Grant, P. and Chung, J. 2007. On the self-regulation of implicit and explicit prejudice: A self-determination theory perspective. *Personality and Social Psychology Bulletin*, **33**(5): 732–49.

Lepore, L. and Brown, R. 1997. Category and stereotype activation: Is prejudice inevitable? *Journal of Personality and Social Psychology*, **72**(2): 275–87.

Li, W., Moallem, I., Paller, K.A. and Gottfried, J.A. 2007. Subliminal smells can guide social preferences. *Psychological Science*, **18**(12): 1044–9.

Lien, M.-C., Ruthruff, E. and Johnston, J.C. 2006. Attentional limitations in doing two tasks at once: The search for exceptions. *Current Directions in Psychological Science*, **15**(2): 89–93.

Loftus, E. and Klinger, M.R. 1992. Is the unconscious smart or dumb? *American Psychologist*, **47**(6): 761–5.

Macrae, C.N. and Johnston, L. 1998. Help, I need somebody: Automatic action and inaction. *Social Cognition*, **16**(4): 400–417.

Masling, J.M. 1992. What does it mean? In R.F. Bornstein and T.S. Pittman (eds), *Perception without Awareness: Cognitive, clinical, and social perspectives*: 261–75. New York: Guilford Press.

McMullen, J.S. and Shepherd, D.A. 2006. Entrepreneurial action and the role of uncertainty in the theory of the entrepreneur. *Academy of Management Review*, **31**(1): 132–52.

Merikle, P.M. and Reingold, E.M. 1998. On demonstrating unconscious perception: Comment on Draine and Greenwald. *Journal of Experimental Psychology: General*, **127**(3): 304–10.

Merikle, P.M., Joordens, S. and Stolz, J.A. 1995. Measuring the relative magnitude of unconscious influence. *Consciousness and Cognition*, **4**(4): 422–39.

Mitchell, J.P., Nosek, B.A. and Banaji, M.R. 2003. Contextual variations in implicit evaluations. *Journal of Experimental Psychology: General*, **132**(3): 455–69.

Mitchell, J.R., Friga, P.N. and Mitchell, R.K. 2005. Untangling the intuition mess: Intuition as a construct in entrepreneurship research. *Entrepreneurship Theory and Practice*, **29**(6): 653–79.

Mitchell, J.R. and Shepherd, D.A. 2012. Capability development and decision incongruence in strategic opportunity pursuit. *Strategic Entrepreneurship Journal*, **6**(4): 355–81.

Mitchell, R.K., Busenitz, L.W., Lant, T., McDougall, P.P., Morse, E.A. and Smith, J.B. 2004. The distinctive and inclusive domain of entrepreneurial cognition research. *Entrepreneurship Theory and Practice*, **28**(6): 505–18.

Mitchell, R.K., Randolph-Seng, B. and Mitchell, J.R. 2011. Socially situated cognition: Imagining new opportunities for entrepreneurship research (Dialogue). *Academy of Management Review*, **36**(4): 774–8.

Moorman, C. and Miner, A.S. 1998. Organizational improvisation and organizational memory. *Academy of Management Review*, **23**(4): 698–723.

Moors, A. and De Houwer, J. 2006. Automaticity: A theoretical and conceptual analysis. *Psychological Bulletin*, **132**(2): 297–326.

Moskowitz, G.B., Gollwitzer, P.M., Wasel, W. and Schaal, B. 1999. Preconscious control of stereotype activation through chronic egalitarian goals. *Journal of Personality and Social Psychology*, **77**(1): 167–84.

Moskowitz, G.B., Li, P. and Kirk, E. 2004. The implicit volition model: On the preconscious regulation of temporarily adopted goals. In M.P. Zanna (ed.), *Advances in Experimental Social Psychology*, vol. 36: 317–413. San Diego, CA: Academic Press.

Natsoulas, T. 1981. Basic problems of consciousness. *Journal of Personality and Social Psychology*, **41**(1): 132–78.

Nosek, B.A. 2007. Implicit–explicit relations. *Current Directions in Psychological Science*, **16**(2): 65–9.

Novicevic, M.M., Clayton, R.W. and Williams, W.A. 2011. Barnard's model of decision making: A historical predecessor of image theory. *Journal of Management History*, **17**(4): 420–34.

Novicevic, M.M., Hench, T.J. and Wren, D.A. 2002. 'Playing by ear' . . . 'in an incessant din of reasons': Chester Barnard and the history of intuition in management thought. *Management Decision*, **40**(10): 992–1002.

Payne, B.K. and Stewart, B.D. 2007. Automatic and controlled components of social cognition: A process dissociation approach. In J.A. Bargh (ed.), *Social Psychology and the Unconscious: The automaticity of higher mental processes*: 293–315. New York: Psychology Press.

Pittman, T.S. 1992. Perception without awareness in the steam of behavior: Processes that produce and limit nonconscious biasing effects. In R.F. Bornstein and T.S. Pittman (eds), *Perception without Awareness: Cognitive, clinical, and social perspectives*: 277–96. New York: Guilford Press.

Prinz, W. 2003. How do we know about our own actions? In S. Maasen, W. Prinz and G. Roth (eds), *Voluntary Action: Brains, minds, and sociality*: 21–33. New York: Oxford University Press.

Pylyshyn, Z.W. 1999. Is vision continuous with cognition? The case for cognitive impenetrability of visual perception. *Behavioral and Brain Sciences*, **22**(3): 341–423.

Quartz, S.R. 2009. Reason, emotion and decision making: risk and reward computation with feeling. *Trends in Cognitive Sciences*, **13**(5): 209–15.

Ranganath, K.A., Smith, C.T. and Nosek, B.A. 2008. Distinguishing automatic and controlled components of attitudes from direct and indirect measurement methods. *Journal of Experimental Social Psychology*, **44**(2): 386–96.

Rauch, A. and Frese, M. 2007. Let's put the person back into entrepreneurship research: A meta-analysis on the relationship between business owner's personality traits, business creation, and success. *European Journal of Work and Organizational Psychology*, **16**(4): 353–85.

Reingold, E.M. and Merikle, P.M. 1993. Theory and measurement in the study of unconscious processes. In M. Davies and G. Humphreys (eds), *Consciousness*: 40–57. Oxford: Basil Blackwell.

Rudman, L.A., Dohn, M.C. and Fairchild, K. 2007. Implicit self-esteem compensation: Automatic threat defense. *Journal of Personality and Social Psychology*, **93**(5): 798–813.

Rudman, L.A., Phelan, J.E. and Heppen, J.B. 2007. Developmental sources of implicit attitudes. *Personality and Social Psychology Bulletin*, **33**(12): 1700–713.

Rydell, R.J., McConnell, A.R., Mackie, D.M. and Strain, L.M. 2006. Of two minds: Forming and changing valence-inconsistent implicit and explicit attitudes. *Psychological Science*, **17**(11): 954–8.

Sadler-Smith, E. and Shefy, E. 2004. The intuitive executive: Understanding and applying 'gut feel' in decision-making. *Academy of Management Executive*, **18**(4): 76–91.

Schacter, D.L. 1989. On the relation between memory and consciousness: Dissociable interactions and conscious experience. In H.L. Roediger and F.I.M. Craik (eds), *Varieties of Memory and Consciousness: Essays in honour of Endel Tulving*: 355–89. Hillsdale, NJ: Lawrence Erlbaum Associates.

Schneider, W., Dumais, S.T. and Shiffrin, R.M. 1984. Automatic and controlled processing and attention. In R. Parasuraman and D.R. Davies (eds), *Varieties of Attention*: 1–27. New York: Academic Press.

Shapero, A. and Sokol, L. 1982. The social dimensions of entrepreneurship. In C. Kent, D. Sexton and K. Vesper (eds), *Encyclopedia of Entrepreneurship*: 72–90. Englewood Cliffs, NJ: Prentice-Hall.

Shook, C.L., Priem, R.L. and McGee, J.E. 2003. Venture creation and the enterprising individual: a review and synthesis. *Journal of Management*, **29**(3): 379–99.

Simon, M., Houghton, S.M. and Aquino, K. 2000. Cognitive biases, risk perception, and venture formation: How individuals decide to start companies. *Journal of Business Venturing*, **15**(2): 113–34.

Sinclair, M. and Ashkanasy, N.M. 2005. Intuition myth or a decision-making tool? *Management Learning*, **36**(3): 353–70.

Smith, E.R. and Conrey, F.R. 2007. Mental representations are states, not things: Implications for implicit and explicit measures. In B. Wittenbrink and N. Schwarz (eds), *Implicit Measures of Attitudes*: 247–64. New York: Guilford Press.

Sparrow, B. and Wegner, D.M. 2006. Unpriming: The deactivation of thoughts though expression. *Journal of Personality and Social Psychology*, **91**(6): 1009–19.

Thompson, E.R. 2009. Individual entrepreneurial intent: Construct clarification and development of an internationally reliable metric. *Entrepreneurship Theory and Practice*, **33**(3): 669–94.

Thrash, T.M., Elliot, A.J. and Schultheiss, O.C. 2007. Methodological and dispositional predictors of congruence between implicit and explicit need for achievement. *Personality and Social Psychology Bulletin*, **33**(7): 961–74.

Trope, Y. and Fishbach, A. 2005. Going beyond the motivation given: Self-control and situational control over behavior. In R.R. Hassin, J. Uleman and J.A. Bargh (eds), *The New Unconscious*: 537–65. New York: Oxford University Press.

Vargas, P.T., Sekaquaptewa, D. and von Hippel, W. 2007. Armed only with paper and pencil: 'Low-Tech' measures of implicit attitudes. In B. Wittenbrink and N. Schwarz (eds), *Implicit Measures of Attitudes*: 103–24. New York: Guilford Press.

Wegner, D.M. and Bargh, J.A. 1998. Control and automaticity in social life. In D.T. Gilbert, S.T. Fiske and G. Lindzey (eds), *Handbook of Social Psychology* (4th edn): 446–96. Boston: McGraw-Hill.

Wittenbrink, B. 2007. Measuring attitudes through priming. In B. Wittenbrink and N. Schwarz (eds), *Implicit Measures of Attitudes*: 17–58. New York: Guilford Press.

Zhao, H., Seibert, S.E. and Lumpkin, G.T. 2010. The relationship of personality to entrepreneurial intentions and performance: A meta-analytic review. *Journal of Management*, **36**(2): 381–404.

5. Feeling and thinking: the role of affect in entrepreneurial cognition

Maw-Der Foo, Charles Y. Murnieks and Elsa T. Chan

ABSTRACT

This chapter examines the role of affect, a general term that comprises emotions and moods, on entrepreneurial cognition. First, we define affect and provide a summary of theories and empirical research on it and its relation to cognition and behaviors. Following that, we review research on affect and entrepreneurial cognition. Affect is particularly important in the study of entrepreneurship because entrepreneurs often face an emotional and dynamic environment in the venture creation process. Previous studies have suggested that affect exerts a significant influence across various entrepreneurial cognitions and behaviors. In addition to reviewing this literature, throughout the chapter we highlight exciting future research opportunities. We also recognize the role of emotion regulation strategies in influencing feelings and consider how these can help entrepreneurs take control of their emotions and subsequently impact their cognitive processes. At the end of the chapter, we recognize the challenges of studying affect and discuss methods that can be employed to overcome these challenges in entrepreneurial research.

INTRODUCTION

> We think too much and feel too little.
>
> Charlie Chaplin

The statement above is not exactly accurate, as, when we feel, we are also thinking. Feeling and thinking, or affect and cognition, are inextricably linked for most people. Beyond the phenomenological experience of joy or pain, affect influences how we think, what our preferences are, what stimuli we pay attention to, what decisions we make, and ultimately how we behave and act. The goal of this chapter is to review and explore the role of affect, which includes emotions and moods (Forgas and George,

2001; Russell and Barrett, 1999), on entrepreneurial cognition (see also Grégoire, Chapter 6, this volume). We begin by defining affect and briefly reviewing its impact on cognition. Following that, we review research on affect in entrepreneurship specifically and highlight areas for future research. Even though affect infuses many aspects of cognition, it is important to remember that we are not slaves to our feelings, as we can use emotion regulation strategies to impact how we feel and react to these feelings (Gross, 1998). In the affective process (Elfenbein, 2007), an individual first encounters a stimulus and carries out emotional registration with the stimulus. Next, the individual feels and experiences the accompanying emotion and expresses it. This automatic process can influence an individual's attitudes, cognitions and behaviors. We close the chapter by reviewing techniques for studying affect in entrepreneurship, acknowledging the strengths and limitations of each technique.

DEFINITION OF AFFECT

Studies examining individual subjective experience related to feelings often use the terms 'affect,' 'emotion' or 'mood.' Affect is typically employed as a broad or family term that includes both moods and emotions (Forgas and George, 2001; Frijda, 1993; Russell and Barrett, 1999; Weiss, 2002). Although the terms 'affect' and 'emotion' have been used interchangeably at times (e.g., Cardon, Foo, Shepherd and Wiklund, 2012), we prefer to separate them. Thus, following the tradition of most studies, including those in entrepreneurship (e.g., Baron and Tang, 2011; Podoynitsyna, Van der Bij and Song, 2012), we reserve the use of affect as the more general term that comprises both moods and emotions. Herein, we use the term 'cognition' broadly to refer to mental operations associated with information processing in general (Johnson, Chang and Lord, 2006; Newell, 1990). These mental operations include processes such as attention, memory, reasoning, learning, problem solving and decision making (Blanchette and Richards, 2010). All of these mental operations are important in the entrepreneurial context, as individual entrepreneurs draw on them to varying extents to identify, evaluate and exploit opportunities (Baron, 1998; Shane and Venkataraman, 2000).

With respect to the intersection between cognition and affect, many theories of the latter include cognitive elements because appraisal of information related to environmental stimuli often plays a central role in the generation and experience of emotions and moods (e.g., Gross, Sheppes and Urry, 2011; Lambie and Marcel, 2002). For example, in Lambie and

Marcel's (2002) model of the microgenesis of emotion states, any event that draws one's attention can either (1) initiate further cognitive evaluation that results in attitude formation accompanied by genesis of affect, or (2) trigger affect and attitudes immediately and automatically, as in the case of certain fight-or-flight reactions. Similarly, in Gross et al.'s (2011) model of emotion generation, emotions emerge from a process initiated when person–situation interactions compel attention. In both models, cognitive processes of attention and evaluation are necessary components in the genesis and experience of affect. However, it is important to note that, despite their intricate interaction within the brain, cognition and affect are still separate constructs. Whereas cognition is characterized by mental operations focused on information processing, affect is defined by the physiological changes and sensations associated with feeling states.

Within the affective realm, moods and emotions have some common features. They are both affective experiences that include physiological responses (Larsen, 2000). Despite these similarities, though, key differences still exist between them. Previous research on mood and emotion (Elfenbein, 2007; Frijda, 1993; Russell and Barrett, 1999; Weiss, 2002) generally makes distinctions between them in three ways. First, the major distinction is that emotions are related to specific targets (objects, Weiss; stimulus, Elfenbein) or have a specific cause, whereas moods are not target-specific (Weiss, 2002). Second, emotions are more intense with respect to the experience of feelings, whereas moods are more diffuse. Third, emotions generally have a short duration, whereas moods last longer.

Affect impacts how people think and subsequently what they do (Baumeister, Vohs, DeWall and Zhang, 2007). Baron (2008) suggests two reasons why affect plays a key role in the entrepreneurial process. The first is that entrepreneurs often face unpredictable environments in which affect can have a strong influence on entrepreneurs' cognitions and behaviors (see also Baron, Chapter 1, this volume; Bird, Chapter 3, this volume). This occurs because entrepreneurs cannot follow well-rehearsed scripts of behavior, but rather must spend significant time searching for cues and deliberating how to proceed. Substantive cognitive processing is required and this affords the opportunity for affect to infuse one's thoughts in a significant way. The other reason is that the range of activities and tasks that entrepreneurs perform has a strong relationship with affect. Entrepreneurs rely heavily on cognitions related to creativity, persuasion and decision making, and a host of research demonstrates that affect influences these various processes in myriad ways. In the next section, we summarize some of these impacts of affect on cognition. Following that, we discuss how affect influences entrepreneurial outcomes.

GENERAL SUMMARY OF EMOTION RESEARCH

Numerous studies have examined the role of affective experiences on cognitive processes and found that affect influences individuals' cognition and decision making and their behaviors in many different ways (Baumeister et al., 2007; Beal, Weiss, Barros and MacDermid, 2005; Bodenhausen, 1993; Bower, 1991; Clore, Schwarz and Conway, 1994; Forgas, 1995; Gendolla, 2000; Schwarz and Clore, 1983, 1996). Below, we discuss some of the more prominent models concerning the interaction between affect and cognition and briefly review some empirical evidence surrounding affect and cognition in general (see also Sarasvathy, Chapter 14, this volume). Several models examine different aspects of the relationship between affective experience and cognition. The affect-as-information model (Schwarz, 1990; Schwarz and Clore, 1988) proposes that individuals use their feelings as a piece of data to make cognitive evaluations. The feelings experienced at the time of evaluation may be incorporated as content into one's cognition, even if the feelings are not necessarily relevant to the item under evaluation. The AIM, or affect infusion model (Forgas, 1995; Forgas and George, 2001), theorizes that affect impacts both the content and process of cognition. When individuals adopt heuristic or substantive cognitive processing strategies, their moods serve as information (similar to the affect-as-information model) and infuse their behaviors. The AIM has been applied in entrepreneurship research numerous times because entrepreneurs are thought to employ heuristic and substantive cognitions often.

Affective events theory (Weiss and Cropanzano, 1996) proposes that events in the workplace create different emotional experiences, and these emotions influence attitudes and behaviors at work. Beal et al. (2005) suggest that affective experiences that vary from time to time are worth studying. They propose an episodic process model that relates individuals' affective states to their performance. Affective experiences that are relevant to the task are related to cognitive processes such as appraisal, and through these cognitive processes performance is affected. The broaden-and-build theory (Fredrickson, 2001; Fredrickson and Branigan, 2005) suggests that positive emotions broaden individuals' cognitions, thoughts and actions, and build durable intangible resources for the future. The emotion-as-a-feedback-system perspective (Baumeister et al., 2007) also demonstrates that emotions influence behavior through cognitive processes such as feedback and appraisals. Hence different theoretical perspectives on affect, including the affect-as-information model, affect infusion model, affective events theory, and the broaden-and-build theory all acknowledge the role that affect plays in cognitive processes.

Research on the influence of affect on cognitive processes and behaviors in organizations has examined a variety of work-related outcomes. First, affect is known to influence job satisfaction. Using an experience-sampling methodology, Judge and Ilies (2004) studied the relationships between trait or dispositional affect, mood and job satisfaction. They found that mood state mediated the relationship between trait affect and job satisfaction. Mood and job satisfaction affected each other, and the influence was stronger for positive mood. Studies in this realm have also found a spill-over effect between the mood experienced at work and the mood experienced at home (Song, Foo and Uy, 2008). Beyond the influence of affect on evaluations of the job situation, studies have found that affect also plays a role in performance. Kaplan, Bradley, Luchman and Haynes (2009) conducted a meta-analysis on the relationship between positive affect and negative affect on job performance. They found that positive affect predicted task performance and organizational citizenship behaviors. In comparison, negative affect predicted lower task performance, fewer organizational citizenship behaviors, and more counterproductive work behaviors, withdrawal behaviors and occupational injury. Staw and Barsade (1993) found that positive emotions led to greater job achievement as measured by higher pay and higher supervisor evaluations. These relationships were mediated by perceptions of fairness, job satisfaction and stress. In summary, affect, through cognition, has a strong and consistent impact on work-related experiences and work outcomes.

Although most research has adopted the valence-based perspective, focusing on positive or negative feelings, it is important to note that many studies have analyzed the effect that specific but prevalent emotions can have on cognition and decision making. This stream of research shows that effects of emotions do not differ only according to their relative degree of positivity and negativity, but that specific discrete emotions also have their unique effects (DeSteno, Petty, Wegener and Rucker, 2000; Keltner, Ellsworth and Edwards, 1993; Lerner and Keltner, 2000, 2001; Tiedens and Linton, 2001). Anger, for instance, has been found to be related to cognitive processes and outcomes, including attributions, attitudes, risk perception, attention and depth of processing (Lerner and Tiedens, 2006). The appraisal tendency framework (Lerner and Keltner, 2000, 2001) proposes that each specific emotion has its appraisal dimensions, which are linked to the outcomes of judgment and decision making. Appraisal dimensions are factors that characterize the core meanings or themes inherent in groups of affect (Lerner and Keltner, 2000). Smith and Ellsworth (1985) identified six separate dimensions of appraisal among emotions: certainty, control, pleasantness, attentional activity, anticipated effort and responsibility. For example, fear and anger are both negative emotions but they differ according to the appraisal themes of control and certainty (Lerner and Keltner, 2000). Whereas fear is associated with appraisal tendencies and cognitions related to uncertainty and lack of control, anger is associated with appraisal tendencies and cognitions related to certainty and control. Studying the effects of fear and anger on risk perception, Lerner and Keltner (2001) found that individuals experiencing fear estimated risk pessimistically and made risk-averse choices, whereas those experiencing anger estimated risk in an optimistic manner and were risk-seeking when making choices. The relationship between emotions experienced and judgment was mediated by appraisal tendencies. Studies such as this provide evidence of the various ways in which different aspects of affect may influence cognition. In the next section, we explore how affect, through cognition, impacts entrepreneurial outcomes.

Studies on Affect and Entrepreneurial Cognition

In entrepreneurship, it has been demonstrated that affect has a definite influence on entrepreneurial cognition and behavior. Although there are few empirical studies on affect in entrepreneurship (Baron and Tang, 2011), the importance of emotions in the entrepreneurial context has been recognized, and research on the topic is increasing (Cardon et al., 2012; Foo, Uy and Baron, 2009). Across a myriad of studies, affect has been found to impact entrepreneurs' cognitions and behaviors among opportunity recognition processes, team processes and interactions with stakeholders. We summarize a number of these studies below.

Among the small but growing group of studies that have attempted to measure affect in entrepreneurship empirically, Foo et al. (2009) drew on the affect-as-information theory to study the relationship between entrepreneurs' affect and venture effort. Employing experience-sampling methodology, they recorded entrepreneurs' reported feelings and effort multiple times daily throughout the study. They found a relationship between positive affect and effort on future-oriented tasks that was beyond what was immediately required for the venture. In fact, future temporal focus fully mediated the relationship between positive affect and venture effort. Moreover, negative affect directly predicted effort on tasks that were immediately required. Baron and Tang (2011) studied the effects of entrepreneurs' positive affect, creativity and environmental dynamism on firm-level innovation. Firm-level innovation was operationalized in terms of the number of innovations as well as their radicalness. The authors found that creativity mediated the relationship between entrepreneurs' positive affect and firm-level innovation. There was also a moderating effect of the environment. In more dynamic environments, positive affect was more strongly related to creativity, and creativity was more strongly related to innovation. Baron, Tang and Hmieleski (2011) examined the effects of dispositional positive affect on firm performance. They found a curvilinear relationship between entrepreneurs' dispositional positive affect and performance outcomes such as product innovation and sales growth rate. Positive affect was initially positively related to firm performance outcomes, but the relationships became negative beyond a certain inflection point. These relationships were stronger when firm size was small. The curvilinear relationship may apply not only to firm performance, but also to other entrepreneurial tasks in opportunity recognition and decision making (Baron, Hmieleski and Henry, 2012). Hayton and Cholakova (2012) also used the affect-as-information perspective to propose that affect influences entrepreneurs' intention to develop and pursue their ideas across a variety of cognitive processes.

Continuing the stream of research surrounding affect-as-information, some researchers have studied the impact of affect from the perspective of specific emotions, and how these emotions influence opportunity recognition, evaluation and implementation. For example, Foo (2011a) used the appraisal tendency framework and examined the effects of four specific emotions on perceived risk. The emotions analyzed – hope, fear, happiness and anger – differed across both valence and appraisal dimensions. He found that individuals perceived lower risk when they experienced happiness or anger, which are emotions that align with the appraisal tendencies of high uncertainty and control. Individuals perceived higher risk when they experienced hope and fear, which are emotions that align with appraisal tendencies related to low uncertainty and control. Entrepreneurs who possessed high trait anger and trait happiness were also more likely to prefer uncertain but high-value outcomes. Extending this line of inquiry, Welpe, Spörrle, Grichnik, Michl and Audretsch (2012) found that, in addition to opportunity evaluation, emotions influence exploitation decisions. In their investigation, fear led to a lower tendency to exploit a business opportunity, while joy and anger increased it. These emotions also affected the relationship between evaluation and exploitation: fear reduced, while joy and anger increased, the positive influence of evaluation on exploitation.

Beyond studying the influence of individual discrete emotions, Podoynitsyna et al. (2012) studied whether mixed emotions, emotions that conflict in terms of valence or cognitive appraisal dimensions, influence entrepreneurial decision making. For example, happiness and hope are two positive emotions with conflicting appraisal dimensions (Podoynitsyna et al., 2012: 116). Happiness is appraised as certain and controllable, while hope is appraised as uncertain and uncontrollable. Podoynitsyna et al. (2012) hypothesized and found that entrepreneurs who experienced mixed emotions perceived more risk. Using affect-as-information theory, they explained that people felt uncomfortable when they experienced mixed emotions, and such discomfort spilled over into their evaluation of the business opportunity. Taken together, the studies reviewed here provide convincing evidence that affect, across many forms, influences the types of ventures entrepreneurs prefer, the amount of risk they are willing to take, and the kinds of ventures they implement. In Table 5.1, we summarize some of the more prominent entrepreneurship studies investigating the relationships between affect and cognition (see also Drnovšek, Slavec and Cardon, Chapter 7, this volume).

Despite the explosion of affect studies in entrepreneurship in the last few years, there is still much to be studied and we expound below on what we view to be the most promising areas for future research. Studies on affect and entrepreneurship tend to be conducted at the individual level, probably since affective experiences originate within individuals. However, many high-growth ventures are started by a team instead of an individual (Foo, Wong and Ong, 2005; Foo, Sin and Yiong, 2006; see also Corbett, Chapter 12, this volume) and firms often reflect the values, preferences and inclinations of the top management team (Hambrick and Mason, 1984; Leung, Foo and Chaturvedi, 2013). Thus there is much to explore concerning affect's influences on the venture team. Through emotion contagion, team members tend to converge on their state affect (Barsade, 2002). Hence, consistent with group polarization (Myers and Lamm, 1975), venture teams can make decisions that are more extreme than the initial inclinations of team members. Teams in which the collective members' dispositional tendency leans towards positive/negative affect would make more/less risky investments or evaluate ventures more favorably/unfavorably than each individual would. Given the importance of entrepreneurial team processes for entrepreneurial outcomes (Foo, 2010, 2011b; Foo et al., 2006), affective influences on the venture should move beyond studying individuals to studying team processes and outcomes.

Another area that is fruitful to explore is the study of affective influences on processes beyond business opportunity identification and evaluation. Most studies on affect and cognition have focused on business opportunity identification and evaluation. Much less is known about how affect impacts implementation of business opportunities, such as getting stakeholders to commit to the venture. One important group of stakeholders is venture employees. It can be difficult to attract qualified people to join a startup as salaries are low, promotion prospects are uncertain, and resources, administrative support, perks and benefits are limited. Moreover, new ventures face uncertainty and a high probability of failure. One technique often used by startups is to hire people with whom they

Table 5.1 Summary of entrepreneurship studies on affect and cognition

Authors	Construct(s)	Type of research	Findings
Baron (2008)	Positive affect and Negative affect	Conceptual	Positive affect (PA) enhances creativity, acquisition of financial/human resources, ability to respond to dynamic environments, expansion of skills and social networks, adoption of efficient decision making (satisficing), toleration of stress, and positively moderates the influence of alertness and active search on opportunity recognition. Negative affect (NA) negatively influences acquisition of financial/human resources, the ability to respond to dynamic environments, and impedes expansion of skills and social networks. NA promotes adoption of slower, thorough decision making (maximizing) and negatively moderates the influence of alertness and active search on opportunity recognition.
Baron, Hmieleski and Henry (2012)	Dispositional positive affect	Conceptual	Dispositional positive affect (DPA) and cognition have a curvilinear (inverted U-shaped) relationship. Up to an inflection point, the effects of affect on processes such as memory, heuristics, attention, perception accuracy, decision making, cognitive flexibility, self-regulation and processing of negative material are enhanced, but afterwards, these processes degrade.
Baron and Tang (2011)	Positive affect	Empirical	PA is significantly related to creativity. This relationship is stronger in dynamic environments.
Baron, Tang and Hmieleski (2011)	Dispositional positive affect	Empirical	DPA is significantly related to product innovation and sales growth rate. These relationships are curvilinear (inverted U-shaped). These effects are stronger in small firms.
Baum and Locke (2004)	Passion	Empirical	Passion is significantly related to setting higher goals for venture growth.

Cardon, Wincent, Singh and Drnovšek (2009)	Entrepreneurial passion	Conceptual	Passion involves intense positive feelings related to salient entrepreneurial role identities. Passion activates and coordinates goal related cognitions (goal challenge, goal commitment, goal striving) and entrepreneurial behaviors (creative problem solving, persistence, absorption).
Foo (2011a)	Emotions of anger, fear, happiness and hope	Empirical	Fear and hope are related to greater risk perceptions than anger and happiness (indicating differences based on appraisal dimensions). Trait anger and trait happiness are related to preferences for higher-value and uncertain (vs lower-value but certain) outcomes.
Foo, Uy and Baron (2009)	Positive affect and negative affect	Empirical	PA is related to future temporal focus, and to increased effort on venture tasks beyond what is immediately required. NA is related to effort on tasks that are immediately required.
Grichnik, Smeja and Welpe (2010)	Emotions of joy and fear	Empirical	Joy leads to more positive evaluation of opportunities, but negatively influences exploitation. Fear leads to more negative evaluations of opportunities, and negatively influences exploitation.
Hayton and Cholakova (2012)	Positive affect and negative affect	Conceptual	PA increases perception of entrepreneurial ideas, working memory, memory recall, as well as creative combination of knowledge in memory. PA increases expectancy beliefs between seeking knowledge and reducing uncertainty. PA increases instrumentality beliefs of pursuit of entrepreneurial ideas and reducing uncertainty. PA increases the range of outcomes considered and their valence. NA decreases perception of entrepreneurial ideas. Overall, the influence of affect is greater for complex and unusual ideas.
Morris, Kuratko, Schindehutte and Spivack (2012)	Positive affect and negative affect	Conceptual	PA drives global and explorative learning. NA drives local and exploitative learning. More positive and intense PA will be associated with higher satisfaction, a greater tendency

Table 5.1 (continued)

Authors	Construct(s)	Type of research	Findings
			to improvise, greater innovation and risk taking, a higher likelihood of developing an entrepreneurial mindset, and a higher likelihood of creating innovative, growth-oriented ventures.
Podoynitsyna, Van der Bij and Song (2012)	Mixed emotions (16 discrete emotions)	Empirical	In terms of appraisal tendency, conflicting emotions (emotions opposite in appraisal tendency and lower in number compared to dominant emotions) have a stronger association to risk perception than dominant emotions (emotions higher in number compared to conflicting emotions). Conflicting emotions based on appraisal tendency have a stronger association to risk perception than conflicting emotions based on valence. Conflicting emotions based in both valence and appraisal tendency play a greater role in risk judgments of serial versus novice entrepreneurs.
Shepherd (2004)	Grief	Conceptual	Grief interferes with learning from business loss. This interference is greater at higher levels of information. Employment of restoration and loss orientations can mitigate these detrimental effects.
Shepherd, Covin and Kuratko (2009)	Grief	Conceptual	Grief interferes with learning from failure. This relationship is moderated by coping self-efficacy, which can increase learning from failure up to a certain inflection point.
Welpe, Sporrle, Grichnik, Michl and Audretsch (2012)	Emotions of fear, joy and anger	Empirical	Fear reduces opportunity exploitation. Anger and joy increase opportunity exploitation. Fear negatively moderates (and anger and joy positively moderate) the direct relationship between opportunity evaluation and exploitation.

have close family or business ties (Leung, Zhang, Wong and Foo, 2006). Using affective appeals can be another way to make it appear exciting for employees to join the fledging company. A classic example is how Steve Jobs persuaded John Sculley to leave PepsiCo and join Apple Inc. Jobs asked Sculley whether he wanted to sell sugared water for the rest of his life, or whether he wanted to change the world. Sculley pointed out that Jobs had a unique ability to appeal to others on a profoundly emotional level, and to convince them of his entrepreneurial vision. At the time of Jobs's pitch to Sculley, Apple was experiencing significant growth, but also significant challenges. Ultimately, Sculley was persuaded by Jobs's emotional pitch.

The example of Steve Jobs highlights how affect can be used deliberately to achieve desired work outcomes. There is a burgeoning field, showing that emotion skills can be used to influence others (Foo, Elfenbein, Tan and Aik, 2004; Elfenbein, Foo, White, Tan and Aik, 2007), and more importantly, these skills can be developed (Wong, Foo, Wang and Wong, 2007). However, only a limited number of studies have attempted to examine this realm surrounding how emotions can be manipulated by entrepreneurs. Baron and Markman (2000) argued that socially skilled entrepreneurs have higher-quality interactions with external stakeholders. Consequently, these stakeholders are more willing to invest resources in the entrepreneurs' ventures. Baron and Markman (2000) contended that social skills could be enhanced through training, and that entrepreneurs should avail themselves of such training opportunities. Other evidence supports their assertion that social skills can be learned, especially with activities that incorporate life experiences (Wong et al., 2007). The use of emotion skills to influence others is further supported by research on emotional labor in the workplace (Hochschild, 1983). Through emotional labor practices, such as expressed emotions or deep acting, service providers can influence customers' service evaluations (Grandey, 2003; Tan, Foo, Chong and Ng, 2003; Tan, Foo and Kwek, 2004). Using these studies as a springboard, scholars in entrepreneurship could look into how entrepreneurs use emotional labor strategies to improve interactions with team members, employees and other stakeholders, and ultimately the willingness of these stakeholders to commit resources.

In daily decisions, entrepreneurs can face many ethical challenges. For example, should an entrepreneur disclose to a customer that the product has not been fully tested, or should s/he disclose this at the risk of losing a sale? In these ethical scenarios, emotions of a moral nature may impact the decisions entrepreneurs make. Moral emotions are related to the interests of the society or other individuals, and include those such as shame, guilt, gratitude, pride and disgust (Haidt, 2003). These emotions drive individuals' moral decision making and moral behaviors (Tangney, Stuewig and Mashek, 2007). The role of moral emotions can be particularly important in entrepreneurial cognitive processes. Entrepreneurs often face resource constraints, high risks and stress (Uy, Foo and Song, 2013). In such situations – without self-regulation and moral awareness to maintain personal integrity (Bryant, 2009) – in the interests of venture survival and to expedite actions, entrepreneurs can be tempted to take short cuts, make ethically questionable decisions, and justify the need to do whatever it takes to make the venture successful. Studying entrepreneurs' moral emotions and cognitions would bring about a better understanding of affect and cognition, and subsequently behaviors, from the moral perspective. For example, we speculate that first-time entrepreneurs make more ethically ambiguous decisions than portfolio entrepreneurs. First-time entrepreneurs may view the venture as their 'baby,' while portfolio entrepreneurs can view each venture as part of a bigger assemblage of investments. Our speculation is supported by neuroscience research showing that people who are emotionally engaged are more likely to make morally questionable decisions (Greene, Sommerville, Nystrom, Darley and Cohen, 2001).

While most studies concerning affect in entrepreneurship explore specific affect–cognition relationships directly, Morris, Kuratko, Schindehutte and Spivack (2012) propose that the entrepreneurship process is an affective experience that involves many unfolding and continuous events. They present an experiential perspective based on affective events theory (Weiss and Cropanzano, 1996). In a dynamic environment, entrepreneurs encounter various unpredictable events that influence their affect and behaviors. To make sense of these events, entrepreneurs often engage both cognitive and affective resources. This leads to affective and learning outcomes, and in turn influences behavior and performance. Future studies can explore these interdependencies among affect, cognition, physiological reactions, sense making and learning (see also Clarke and Cornelissen, Chapter 11, this volume; Forbes, Chapter 10, this volume).

On a larger scale, the entrepreneurial context offers the opportunity to expand our understanding of the interaction between affect and cognition longitudinally in a dynamic environment. Currently, theorizing about affect and cognition in entrepreneurship takes an episodic approach. Extant studies tend to focus on how isolated experiences of either positive or negative affect, or how specific emotions like anger or fear, influence cognitions individually. Scholars examine how one specific instance of affect or emotion reacts to one specific cognitive process such as evaluation or decision making. Although this approach provides valuable insight into how affect and cognition interact, it falls short of mirroring the context in which entrepreneurs work. The process of building a venture is long, spanning years in some cases. As entrepreneurs make this journey, they are likely to experience an entire constellation of moods and emotions, ranging in valence and intensity. Extrapolating current research and theory to the real world requires us to assume that each affective experience is independent, and we know this is not the case. The affect experienced as a result of one event or action can leave a lasting impression on the individual, and influence cognitions or judgments about subsequent events. Social psychological studies documenting anticipated affect (Ajzen and Sheikh, 2013; Greitemeyer, 2009; Richard, Van der Plight and De Vries, 1996) and rumination (Curci, Lanciano, Soleti and Rime, 2013; Martin and Tesser, 1996) provide evidence that humans can and do allow affective experience from one event to bleed over into cognitions about another. Work surrounding the implicit activation of goals and attitudes further indicates that this influence of affect on cognition can occur below the level of conscious deliberation (Custers and Aarts, 2005). The field of entrepreneurship provides a ripe context to draw upon these findings from social psychology to build new theories explaining how repeated affective experiences could accumulate within individuals across time to influence cognition.

We call the field of entrepreneurship a 'ripe' context because it is characterized by extremely intense affective experiences (Baron, 1998). Many entrepreneurs invest significant personal and financial resources in their attempts to start ventures. They bet their own, and often their family's, livelihood on the new business being a success. This makes the entrepreneurial context unique when compared to other occupational environments. Workers and managers in large, stable companies undoubtedly experience cycles of positive and negative affect. Weiss and Cropanzano's (1996) affective events theory is predicated on this premise. That said, the majority of these workers and managers do not experience the range and intensity of the affective events repeatedly experienced by entrepreneurs. Whereas workers may suffer through a downturn in a depressed economy or possibly the dissolution of one firm in which they are employed, entrepreneurs stare bankruptcy and success in the face repeated times on their journeys to build a venture. Workers and managers are surrounded by many individuals in their companies, whereas entrepreneurs work in very small teams with fewer people to insulate them against these forces. Entrepreneurs tend to experience greater stress and fatigue as these pressures build over time. Thus we need more theorizing that accounts for the fact that entrepreneurs repeatedly experience extremely intense affective events coupled with significant stressors that strain them physiologically over extended periods of time. We need answers to questions about how these forces may accumulate longitudinally.

Up to this point, we have taken the perspective that affect is a particular emotion or feeling experienced by a person, and this emotion or feeling influences how the person thinks and acts. But we are mindful of the perspective that people can take an active role in regulating their emotions and feelings. In the next section, we review the literature on emotion regulation and its impact.

Emotion Regulation

Emotion regulation refers to the processes by which individuals influence the emotions they experience, when they experience these emotions, and how they express them (Gross, 1998: 275). Although most emotion regulation focuses on the reduction of one's emotions, emotion regulation can involve increasing, maintaining or reducing the experience of one's emotions (Gross, 1998; Parrott, 1993). Emotion regulation is important because it presents a method by which entrepreneurs can exert greater control over their emotional states. Process models of emotion generation often begin with some type of cue or trigger that elicits an emotional response involving behavioral, physiological and experiential components (Gross, 1998; Lambie and Marcel, 2002). Among these process models, most include an appraisal or modulation component whereby individuals have the opportunity to engage cognitions to regulate their experience and response. This agentic perspective offered by emotion regulation theories allows some degree of control over one's emotions and their effects. Given that emotions can impact cognitions in a potent manner, it stands to reason that entrepreneurs who exert strategic and intentioned control over the emotions they experience possess a powerful tool to augment and aid their cognitive processes. In organizational settings, suppression of unpleasant emotions has been linked to decreased job satisfaction and higher intentions to quit, whereas the amplification of positive emotions through emotion regulation strategies results in increased job satisfaction (Cote and Morgan, 2002). Despite a wealth of studies in psychology, the examination of emotion regulation in the entrepreneurial context has been extremely limited. Huy and Zott (2007) conducted an inductive theory-building study that offered insights into how entrepreneurs use emotion regulation strategies to influence other key stakeholders, but no research that we know of has examined emotion regulation intra-individually, within entrepreneurs themselves.

Emotion regulation strategies can be categorized broadly into two groups: (1) cognitive reappraisal and (2) response modulation or expressive suppression (Cote and Morgan, 2002; Heilman, Crisan, Houser, Miclea and Miu, 2010; Webb, Miles and Sheeran, 2012). Cognitive reappraisal is antecedent-focused, which means it attempts to alter the emotion experienced or the trajectory of the emotional response by reformulating the meaning of the situation (Heilman et al., 2010) before the full emotional episode unfurls. In contrast, response modulation is response-focused (as the name implies), and attempts to alter or repress outward displays of emotion or behaviors that might be associated with the emotion. In everyday life, these strategies manifest themselves in myriad ways (Cisler, Olatunji, Feldner and Forsyth, 2010), such as attentional deployment (directing one's attention with the goal of modifying the emotion experienced), cognitive change (reinterpreting the meaning of the situation to alter emotion) or simple suppression (deliberately hiding physical reactions or displays associated with the emotion being felt at the moment). From a neurological standpoint, research indicates that emotion regulation is centered in the brain regions associated with cognitive control, which highlights why ties between emotion regulation and cognition exist (Gross et al., 2011; see also Baucus, Baucus and Mitchell, Chapter 8, this volume; McMullen, Wood and Palich, Chapter 9, this volume). Being subsumed under the umbrella of emotion regulation, both cognitive reappraisal and response modulation or suppression involve self-regulation mechanisms (Gross, 1998) and also imply some degree of choice by the focal individuals as to which strategy they pursue.

The choice between cognitive reappraisal and response modulation is important because these two processes produce different results. Although both reappraisal and modulation can reduce the expression of all emotions as well as the experience of positively valenced emotions, certain research indicates that only reappraisal reduces the experience of negatively valenced emotions (Heilman et al., 2010). Thus individuals may be good at hiding the outward display of negative emotions, but underneath the calm façade, they still experience those negative feelings. Modulation and suppression require an individual to consciously exert self-control while experiencing the full emotion, whereas cognitive reappraisal aims to reduce the experience of the emotion itself (Yurtserver, 2008). This may lead to modulation and suppression strategies increasing cognitive load and reducing cognitive resources available for other types of information processing.

Studies show that when individuals use expressive suppression in an attempt to hide the display of negative emotions, sympathetic activation or arousal actually increases (Richards and Gross, 1999). Whereas some degree of activation is beneficial because it energizes cognitive processes (De Dreu, Baas and Nijstad, 2008), high levels of activation can narrow one's attention span, cause sensory overload, and interfere with cognitive processes (Clore et al., 2001; Easterbrook, 1959; Janis, 1982; Foo, Uy and Murnieks, in press; Ohman and Mineka, 2001). Research shows that modulation strategies aimed at hiding negative emotions can impair memory (Richards and Gross, 1999), and that if these emotions possess high arousal, they can result in impulsive decision making (Leith and Baumeister, 1996). Additional studies show that attempts to suppress negative emotions under conditions of high cognitive load (which are prevalent in the entrepreneurial context) could increase the level of negative feelings experienced (Wegner, Erber and Zanakos, 1993). Sheldon, Ryan, Rawsthorne and Ilardi (1997) argue that this may occur because the focal individual experiences an intra-individual discrepancy between the emotion they are experiencing (negative) and what they portray to others (positive or neutral). A number of studies seem to point to the advantages of reappraisal and antecedent-focused strategies over modulation and suppressive strategies (Loewenstein, 2007; Webb et al., 2012; Wranik, Barrett and Salovey, 2007). In sum, the suppression of emotions, particularly negative emotions, is generally ineffective.

The fact that modulation or expressive suppression is not effective at reducing the experience of negative emotions is relevant in the entrepreneurial context. The entrepreneurial journey is turbulent, filled with both wins and losses, which correspond to affective highs and lows (Boyd and Gumpert, 1983; Huy and Zott, 2007; Shepherd, 2004). Thus most entrepreneurs should expect to encounter strong positive and strong negative emotions in their endeavors. Entrepreneurs who attempt to suppress negative emotions may find their emotional experience deepened instead. Many scholars have documented the effects that strong emotions can have on cognition and behavior germane to entrepreneurs. For example, Shepherd (2003, 2009) documents the detrimental effects that emotions such as grief can have on the ability of entrepreneurs to learn from setbacks in the entrepreneurial process. Strong negative feelings draw a disproportionate focus toward certain emotional events, and prevent a detailed analysis of the other factors that might have contributed to the failure of a venture (Shepherd, 2003). If grief is suppressed, the impairment in learning is likely to continue and adversely affect the entrepreneur's ability to start a new venture successfully. Lerner, Small and Loewenstein (2004) show how sadness biases decision making. Specifically, they document that sadness lowers the prices at which individuals are willing to sell an item, but raises the prices at which they are willing to choose (or buy) an item. As entrepreneurs are likely to experience some degree of sadness at some point in their journey, these results directly relate to the cognitive processes associated with purchasing raw materials or selling their products. Sadness induced by some adverse event, which is then suppressed rather than reappraised, could result in entrepreneurs overpaying for critical resources or underpricing their product offering.

Effective emotion regulation strategies can help entrepreneurs process information more clearly and quickly, which can facilitate building bridges to key influencers of the venture, such as investors, customers or strategic partners. Research demonstrates that poor emotion regulation strategies coupled with negative emotions can lead to decreased social competence and lower peer acceptance (Eisenberg and Fabes, 1992; Gross, 1998). Studies that analyze rejection shed light on the importance of emotion regulation strategies for entrepreneurs. As entrepreneurs pitch their business to key customers to gain valued accounts, rejection is likely to be encountered at some point. Rejection usually involves negative feelings and a sense of social exclusion by the focal individual. How entrepreneurs regulate these emotions associated with rejection can lead to effective and prosocial responses, or less positive withdrawal and antisocial responses (Smart Richman and Leary, 2009). Suppressing rejection-related emotions is not likely to remove these feelings, and research concerning social exclusion indicates how the experience of these feelings could impair entrepreneurial cognition. Individuals who experience feelings of social exclusion in experiments show evidence of reduced intelligent thoughts, lower performance on tasks requiring effortful logic and complex reasoning, as well as reduced cognitive speed and accuracy (Baumeister, Twenge and Nuss, 2002). Experiments in this realm also link feelings of social exclusion to increased probabilities of taking irrational, self-defeating risks as well as longer procrastination (Twenge, Catanese and Baumeister, 2002).

Whereas these effects are unlikely to be alleviated through expressive suppression, cognitive reappraisal can counter the negative effects of rejection (Smart Richman and Leary, 2009). By engaging in self-affirmation, entrepreneurs can use cognitive reappraisal to reduce negative feelings associated with rejection and engage in prosocial responses aimed at persisting in building ties to valuable investors and customers. Yurtserver (2008) shows empirically that, when individuals experience negative emotions in negotiations, those who regulate these emotions using cognitive reappraisal reap significantly higher economic profits than those who use expressive suppression. Taken together, these research results suggest that entrepreneurs would be better served by using cognitive reappraisal to manage the negative emotions associated with the entrepreneurial process. Starting a new venture is difficult, and attracting investors is likely to be fraught with more rejections than acceptances; negative emotions are inevitable in the entrepreneurial process. Most individuals do not enjoy the experience of negative emotions and work to avoid them. If entrepreneurs try to suppress these negative emotions (i.e., attempt to put on a 'brave face') because they do not want co-founders or employees to see the effects of stress and strain on their faces, cognitive processes are likely to suffer. That said, if entrepreneurs engage in cognitive reappraisal, they can address these negative emotions and prosper. Huy and Zott (2007) note that many entrepreneurs attempt to manage the emotions of their employees in a positive manner. For example, they may display cheerful emotions in order to help calm employees during turbulent periods (Huy and Zott, 2007: 12). To display positive emotions in turbulent times may require the entrepreneur to regulate his or her own emotions first.

We started this chapter by explaining how affect can impact cognitive processes and next extended our review to cover the impact of affect on the early stages of a business venture. Beyond identifying and evaluating business opportunities, we highlighted a range of affective influences during the implementation phase, including that of getting important stakeholders, such as investors and potential employees, to commit to the venture. We then explained that entrepreneurs can take control over their emotions and moods, and in so doing impact their motivation, thought processes and ability to influence others (see also Carsrud and Brännback, Chapter 2, this volume). We close this chapter by discussing some ways to study affect, highlighting challenges and possibilities along the way.

METHODS TO STUDY AFFECT IN ENTREPRENEURSHIP

We have reviewed extant literature surrounding how affect and cognition are linked, and throughout this chapter we have suggested ways to extend this literature. We recognize that studying affect is challenging and that doing so in the context of entrepreneurship is even more challenging. Affect can be fleeting, and measuring it requires capturing it when it occurs (Uy, Foo and Aguinis, 2010). Methods that rely on retrospective reports may be contaminated with rationalization and recall issues. Moreover, entrepreneurs are busy; they may not be willing to participate in surveys that they view as intrusive or time-consuming. Despite overwhelming evidence of the key role that affect plays in influencing our thoughts, actions and behaviors, entrepreneurs may be reluctant to acknowledge the importance of this type of research. Below, we discuss some of the ways to study affect, together with the challenges, and our recommendations.

Affect can be an ephemeral experience. Many organizational, team and individual phenomena are persistent or objective in nature, making it possible to capture these phenomena more easily. At the organizational level, financial metrics, strategies and cultures persist over time as well, and can be recorded. At the team level, norms, team structures and member relationships exhibit some level of stability. At the individual level, goals, traits and preferences endure long enough to be measured. But affect, such as a wave of anger that rises and falls rapidly, can be harder to capture and report accurately. Even moods, which are general feelings of positive or negative valence, usually last for comparatively short periods of time, such as a day or two. Given the ephemeral nature of affect, capturing it as it occurs is difficult. That said, researchers have found several ways to conduct studies that address its fleeting nature.

Laboratory studies have been used to manipulate affect so that its observation can be recorded more reliably. Even small manipulations, such as offering a bowl of candy while a participant completes a task, have been shown to influence how people think, how they make decisions, and how they act (Isen, 2001). Other ways of manipulating affect include giving 'fake' feedback to elicit positive or negative emotions. For example, positive/negative emotions can be elicited by telling participants that they did/did not do well on tasks that are central to their core identities (Kleinke, Peterson and Rutledge, 1998). Such tasks are easily manipulated in a laboratory setting but care must be taken during debriefing so that the participants are fully aware that the task outcomes were randomly assigned and did not reflect their actual performance. One fascinating way of inducing emotions is to get participants to activate facial muscles (Ekman, Roper and Hager, 1980). Participants who are asked to place a pencil between their teeth start to feel happy as the upward lift of muscles around the mouth mimics what happens when people smile. In comparison, asking participants to pull their eyebrows together leads to negative emotions, as this action mimics facial muscles that are triggered when people experience anger or disgust.

One prevalent concern with laboratory studies revolves around external validity. However, such limitations are counterbalanced by higher internal validity. Field studies of entrepreneurs, while providing more external – or ecological – validity, frequently suffer from lack of internal validity. More recently, it is increasingly common to see papers in top journals use multiple studies. In these designs, one study, often a laboratory study, provides internal validity, and the other, such as a field study, complements it by providing external validity. If we view the study of entrepreneurship as an ongoing conversation, different researchers should be encouraged to tackle different parts of the affect–cognition research question. As a group, we can develop much richer and stronger theories of the affect–cognition interface than any one study can provide.

More recent and sophisticated methods of studying affect use imaging technologies such as fMRI (functional Magnetic Resonance Imaging) to accurately pinpoint what happens in the brain when people are exposed to different stimuli or when they make decisions or regulate their emotions (Grecucci, Giorgetta, Van't Wout, Bonini and Sanfey, 2013). While such methods hold much promise, there are limitations concerning which researchers have access to these sophisticated types of equipment. Although it is possible to pinpoint what happens in the brain when people make decisions and thus analyze links between the affective and cognitive neural regions precisely, the small sample sizes of such studies usually preclude more sophisticated multi-level analyses. For example, research questions related to the interaction of environmental factors with individual characteristics to predict how people make decisions are difficult to operationalize with fMRI studies because only a small number of individuals, and consequently a small number of environments, are available to test such hypotheses. Less equipment-intensive methods are available, such as using saliva to test for cortisol, a marker of stress (Kirschbaum and Hellhammer, 1999). Stress, a concept that is closely linked to affect, is beginning to make inroads into entrepreneurship research. Instead of asking participants to report their stress levels or how they cope with a stressful situation (e.g., Uy, Foo and Song, 2013), cortisol levels can be used as a stress measure. Taking cardiovascular or heart-rate readings from a blood pressure monitoring device is another promising way in which entrepreneurship scholars can obtain affect-related measurements when they occur. Each experience of fear, anger, joy and sadness leads to unique patterns of changes in diastolic and systolic blood pressure (Sinha, Lovallo and Parsons, 1992), which are easily measured and obtained from a blood pressure monitor. Unfortunately, though, the use of physiological devices necessary to gather these types of data can be viewed by some participants as cumbersome or intrusive.

One concern commonly raised in affect studies is whether or not people can accurately report their emotions. Fortunately, this issue is not as troubling as one might presume. Most operationalizations of affect allow for individuals to be aware (at least minimally) of the phenomenological feelings involved. We may not be aware of the cause or target of our feelings, but we are aware that we feel a certain way (or that we feel ambivalent) and can report those sensations. Another concern revolves around construct validity. Namely, some people question whether existing research scales measure affect accurately. Fortunately again, at least where positive affect and negative affect are concerned, established and validated scales have existed for decades (e.g., Watson, Clark and Tellegen, 1988). The greater concern, from our viewpoint, especially concerning positive affect and negative affect, is whether existing scales capture the full range of affective influences. Affect has two dimensions: a valence component and an activation component. One critique of Watson et al.'s (1988) PANAS scale is that it captures high activation for both positive and negative affect, but neglects affect of low activation. Valence refers to whether the feeling is pleasurable or not, while activation refers to the level, or intensity, of the feeling (De Dreu et al., 2008). Activation is as important as valence, and in fact, without a certain level of activation, people are not motivated to take any action. Fortunately, more recent measures allow for the assessment of affect valence and activation (e.g., Diener et al., 2010) and we encourage greater employment of such measures.

We close our methods discussion by highlighting the use of experience sampling methodology (ESM). This is a group of techniques whereby participants complete surveys, often multiple times, in a field setting in an attempt to gauge a given phenomenon more accurately. This technique tracks intra-individual affective experiences as they occur in real time, thereby enabling us to have greater confidence in modeling causal relationships between affect, behavior, thoughts and actions. Some entrepreneurship researchers have begun to use ESM for studying affective processes, such as Foo et al. (2009), who showed that affect influences the amount of effort entrepreneurs put into their ventures and the focus of these efforts (either toward tasks of an immediate nature or toward future-oriented tasks). They used cellphones programmed with surveys, assessed repeatedly throughout the day. A drawback of their technique is that only short surveys, often a few sentences or short phrases, can be used since the results are sent via SMS (short message service). Another limitation of this particular method is that Java programming is needed to customize the survey to each phone manufacturer, and often to specific models within the same manufacturer. Fortunately, today, smartphones are widely owned. These phones easily connect to the Internet and allow researchers to administer longer surveys. Hence, with the adoption of smartphone technologies, more complex ESM issues can be studied. The use of smartphone applications (apps) also enables researchers to develop surveys with fewer compatibility issues. We refer readers to the paper by Uy et al. (2010), where the authors elaborated the use of ESM in entrepreneurship research, together with a detailed example of how one ESM study was implemented.

In common parlance, we often ask how a person feels about something when what we actually want to know is what they think about it. As it turns out, this difference between thinking and feeling is not as great as one might expect. Our chapter summarizes the ways affect influences how entrepreneurs think, behave and act. Most of the studies of affect in entrepreneurship have focused on the early stage of a business venture, when entrepreneurs have to identify business ventures, evaluate them, and decide which venture to exploit. We believe it is time to target more research on affective influences at the later stages of a business, such as how entrepreneurs attract key stakeholders to make long-term commitments to the venture, and also how affect plays a role in the day-to-day running of the business venture. Though affect plays important roles in our thought processes, through self-regulation, we have a say in how affect impacts us. This is a promising avenue for future research. We also recognize the challenges of studying affect in entrepreneurship. However, numerous techniques exist to help us overcome such challenges. In fact, we are already witnessing a burgeoning of affect studies in entrepreneurship, and our hope in this chapter is to encourage more such studies. The affect–cognition frontier represents an exciting new realm in entrepreneurship research. We have only just begun to explore this territory and it is our firm belief that more effort is needed in this area.

REFERENCES

Ajzen, I. and Sheikh, S. 2013. Action versus inaction: Anticipated affect in the theory of planned behavior. *Journal of Applied Social Psychology*, **43**(1): 155–62.

Baron, R.A. 1998. Cognitive mechanisms in entrepreneurship: Why and when enterpreneurs think differently than other people. *Journal of Business Venturing*, **13**(4): 275–94.

Baron, R.A. 2008. The role of affect in the entrepreneurial process. *Academy of Management Review*, **33**(2): 328–40.

Baron, R.A., Hmieleski, K.M. and Henry, R.A. 2012. Entrepreneurs' dispositional positive affect: The potential benefits and potential costs of being 'up.' *Journal of Business Venturing*, **27**(3): 310–24.

Baron, R.A. and Markman, G.D. 2000. Beyond social capital: How social skills can enhance entrepreneurs' success. *Academy of Management Perspectives*, **14**(1): 106–16.

Baron, R.A. and Tang, J. 2011. The role of entrepreneurs in firm-level innovation: Joint effects of positive affect, creativity, and environmental dynamics. *Journal of Business Venturing*, **26**(1): 49–60.

Baron, R.A., Tang, J. and Hmieleski, K.M. 2011. The downside of being 'up': Entrepreneurs' dispositional positive affect and firm performance. *Strategic Entrepreneurship Journal*, **5**(2): 101–19.

Barsade, S.G. 2002. The ripple effect: Emotion contagion and its influence on group behavior. *Administrative Science Quarterly*, **47**(4): 644–75.

Baumeister, R.F., Twenge, J.M. and Nuss, C.K. 2002. Effects of social exclusion on cognitive processes: Anticipated aloneness reduces intelligent thought. *Journal of Personality and Social Psychology*, **83**(4): 817–27.

Baumeister, R.F., Vohs, K.D., DeWall, C.N. and Zhang, L. 2007. How emotion shapes behavior: Feedback, anticipation, and reflection, rather than direct causation. *Personality and Social Psychology Review*, **11**(2): 167–203.

Beal, D.J., Weiss, H.M., Barros, E. and MacDermid, S.M. 2005. An episodic process model of affective influences on performance. *Journal of Applied Psychology*, **90**(6): 1054–68.

Blanchette, I. and Richards, A. 2010. The influence of affect on higher level cognition: A review of research on interpretation, judgment, decision making and reasoning. *Cognition and Emotion*, **24**(4): 561–95.

Bodenhausen, G.V. 1993. Emotions, arousal, and stereotypic judgments: A heuristic model of affect and stereotyping. In D.M. Mackie and D.L. Hamilton (eds), *Affect, Cognition, and Stereotyping: Interactive processes in group perception*: 13–37. San Diego, CA: Academic Press.

Bower, G.H. 1991. Mood congruity of social judgments. In J.P. Forgas (ed.), *Emotion and Social Judgments*: 31–53. Oxford, UK: Pergamon.
Boyd, D.P. and Gumpert, D.E. 1983. Coping with entrepreneurial stress. *Harvard Business Review*, **61**(2): 46–56.
Bryant, P. 2009. Self-regulation and moral awareness among entrepreneurs. *Journal of Business Venturing*, **24**(5): 505–18.
Cardon, M.S., Foo, M.-D., Shepherd, D.A. and Wiklund, J. 2012. Exploring the heart: Entrepreneurial emotion is a hot topic. *Entrepreneurship Theory and Practice*, **36**(1): 1–10.
Cisler, J.M., Olatunji, B.O., Feldner, M.T. and Forsyth, J.P. 2010. Emotion regulation and the anxiety disorders: An integrative review. *Journal of Psychopathology and Behavioral Assessment*, **32**(1): 68–82.
Clore, G.L., Schwarz, N. and Conway, M. 1994. Affective causes and consequences of social information processing. In R.S. Wyer Jr and T.K. Srull (eds), *Handbook of Social Cognition* (2nd edn), vol. 1: 323–417. Hillsdale, NJ: Erlbaum.
Clore, G.L., Wyer, R.S., Dienes, B., Gasper, K., Gohm, C. and Isbell, L. 2001. Affective feelings as feedback: Some cognitive consequences. In L.L. Martin and G.L. Clore (eds), *Theories of Mood and Cognition: A user's guidebook*: 27–62. Mahwah, NJ: Erlbaum.
Cote, S. and Morgan, L.M. 2002. A longitudinal analysis of the association between emotion regulation, job satisfaction and intentions to quit. *Journal of Organizational Behavior*, **23**(8): 947–62.
Curci, A., Lanciano, T., Soleti, E. and Rime, B. 2013. Negative emotional experiences arouse rumination and affect working memory capacity. *Emotion*, **13**(5): 867–80.
Custers, R. and Aarts, H. 2005. Positive affect as implicit motivator: On the nonconscious operation of behavioral goals. *Journal of Personality and Social Psychology*, **89**(2): 129–42.
De Dreu, C.K.W., Baas, M. and Nijstad, B.A. 2008. Hedonic tone and activation level in the mood–creativity link: Toward a dual pathway to creativity model. *Journal of Personality and Social Psychology*, **94**(5): 739–56.
DeSteno, D., Petty, R.E., Wegener, D.T. and Rucker, D.D. 2000. Beyond valence in the perception of likelihood: The role of emotion specificity. *Journal of Personality and Social Psychology*, **78**(3): 397–416.
Diener, E., Wirtz, D., Tov, W., Kim-Prieto, C., Choi, D.-W., Oishi, S. and Biswas-Diener, R. 2010. New well-being measures: Short scales to assess flourishing and positive and negative feelings. *Social Indicators Research*, **97**(2): 143–56.
Easterbrook, J.A. 1959. The effect of emotion on cue utilization and the organization of behavior. *Psychological Review*, **66**(3): 183–201.
Eisenberg, N. and Fabes, R.A. 1992. Emotion, regulation, and the development of social competence. In M.S. Clark (ed.), *Review of Personality and Social Psychology*: 119–50. Beverly Hills, CA: Sage.
Ekman, P., Roper, G. and Hager, J.C. 1980. Deliberate facial movement. *Child Development*, **51**(3): 886–91.
Elfenbein, H.A. 2007. Emotion in organizations. *Academy of Management Annals*, **1**(1): 315–86.
Elfenbein, H.A., Foo, M.-D., White, J., Tan, H.H. and Aik, V.C. 2007. Reading your counterpart: The benefit of emotion recognition accuracy for effectiveness in negotiation. *Journal of Nonverbal Behavior*, **31**(4): 205–23.
Foo, M.-D. 2010. Member experience, use of external assistance and evaluation of business ideas. *Journal of Small Business Management*, **48**(1): 32–43.
Foo, M.-D. 2011a. Emotions and entrepreneurial opportunity evaluation. *Entrepreneurship Theory and Practice*, **35**(2): 375–93.
Foo, M.-D. 2011b. Teams developing business ideas: How member characteristics and conflict affect member-rated team effectiveness. *Small Business Economics*, **36**(1): 33–46.
Foo, M.-D., Elfenbein, H., Tan, H. and Aik, V. 2004. Emotional intelligence and negotiation: The tension between creating and claiming value. *International Journal of Conflict Management*, **15**(4): 411–29.

Foo, M.-D., Sin, H.P. and Yiong, L. 2006. Effects of team inputs and intrateam processes on new venture team effectiveness. *Strategic Management Journal*, **27**(4): 389–99.
Foo, M.-D., Uy, M.A. and Baron, R.A. 2009. How do feelings influence effort? An empirical study of entrepreneurs' affect and venture effort. *Journal of Applied Psychology*, **94**(4): 1086–94.
Foo, M.-D., Uy, M.A. and Murnieks, C. Beyond affective valence: Untangling valence and activation influences on opportunity identification. *Entrepreneurship Theory and Practice*. In press.
Foo, M.-D., Wong, P.K. and Ong, A. 2005. Do others think you have a viable business idea? Team conflict and judges' evaluation of ideas in a business plan competition. *Journal of Business Venturing*, **20**(3): 385–402.
Forgas, J.P. 1995. Mood and judgment: The affect infusion model (AIM). *Psychological Bulletin*, **117**(1): 39–66.
Forgas, J.P. and George, J.M. 2001. Affective influences on judgments and behavior in organizations: An information processing perspective. *Organizational Behavior and Human Decision Processes*, **86**(1): 3–34.
Fredrickson, B.L. 2001. The role of positive emotions in positive psychology: The broaden-and-build theory of positive emotions. *American Psychologist*, **56**(3): 218–26.
Fredrickson, B.L. and Branigan, C. 2005. Positive emotions broaden the scope of attention and thought–action repertoires. *Cognition and Emotion*, **19**(3): 313–32.
Frijda, N.H. 1993. Moods, emotion episodes, and emotions. In M. Lewis and J.M. Haviland (eds), *Handbook of Emotions*: 381–403. New York: Guilford Press.
Gendolla, G.H.E. 2000. On the impact of mood on behavior: An integrative theory and a review. *Review of General Psychology*, **4**(4): 378–408.
Grandey, A.A. 2003. When 'the show must go on': Surface acting and deep acting as determinants of emotional exhaustion and peer-rated service delivery. *Academy of Management Journal*, **46**(1): 89–96.
Grecucci, A., Giorgetta, C., Van't Wout, M., Bonini, N. and Sanfey, A.G. 2013. Reappraising the ultimatum: An fMRI study of emotion regulation and decision making. *Cerebral Cortex*, **23**(2): 399–410.
Greitemeyer, T. 2009. The effect of anticipated affect on persistence and performance. *Personality and Social Psychology Bulletin*, **35**(2): 172–86.
Greene, J.D., Sommerville, R.B., Nystrom, L.E., Darley, J.M. Cohen, J.D. 2001. An fMRI investigation of emotional engagement in moral judgment. *Science*, **293**(5537): 2105–8.
Gross, J.J. 1998. The emerging field of emotion regulation: An integrative review. *Review of General Psychology*, **2**(3): 271–99.
Gross, J.J., Sheppes, G. and Urry, H.L. 2011. Emotion generation and emotion regulation: A distinction we should make (carefully). *Cognition and Emotion*, **25**: 765–81.
Haidt, J. 2003. The moral emotions. In R.J. Davidson, K.R. Scherer and H.H. Goldsmith (eds), *Handbook of Affective Sciences*: 852–70. Oxford: Oxford University Press.
Hambrick, D.C. and Mason, P.A. 1984. Upper echelons: The organization as a reflection of its top managers. *Academy of Management Review*, **9**(2): 193–206.
Hayton, J.C. and Cholakova, M. 2012. The role of affect in the creation and intentional pursuit of entrepreneurial ideas. *Entrepreneurship Theory and Practice*, **36**(1): 41–68.
Heilman, R.M., Crisan, L.G., Houser, D., Miclea, M. and Miu, A. C. 2010. Emotion regulation and decision making under risk and uncertainty. *Emotion*, **10**(2): 257–65.
Hochschild, A.R. 1983. *The Managed Heart: Commercialization of human feeling*. Berkeley, CA: University of California Press.
Huy, Q. and Zott, C. 2007. How entrepreneurs regulate stakeholders' emotions to build new organizations. Faculty and Research Working Paper, INSEAD.
Isen, A.M. 2001. An influence of positive affect on decision making in complex situations: Theoretical issues with practical implications. *Journal of Consumer Psychology*, **11**(2): 75–85.
Janis, I.L. 1982. Decision-making under stress. In L. Goldberger and S. Breznitz (eds), *Handbook of Stress: Theoretical and clinical aspects*: 69–80. New York: Free Press.

Johnson, R.E., Chang, C.H. and Lord, R.G. 2006. Moving from cognition to behavior: What the research says. *Psychological Bulletin*, **132**(3): 381–415.

Judge, T.A. and Ilies, R. 2004. Affect and job satisfaction: A study of their relationship at work and at home. *Journal of Applied Psychology*, **89**(4): 661–73.

Kaplan, S., Bradley, J.C., Luchman, J.N. and Haynes, D. 2009. On the role of positive and negative affectivity in job performance: A meta-analytic investigation. *Journal of Applied Psychology*, **94**(1): 162–76.

Keltner, D., Ellsworth, P.C. and Edwards, K. 1993. Beyond simple pessimism: Effects of sadness and anger on social perception. *Journal of Personality and Social Psychology*, **64**(5): 740–52.

Kirschbaum, C. and Hellhammer, D.H. 1999. Noise and stress – salivary cortisol as a non-invasive measure of allostatic load. *Noise & Health*, **1**(4): 57–65.

Kleinke, C.L., Peterson, T.R. and Rutledge, T.R. 1998. Effects of self-generated facial expressions on mood. *Journal of Personality and Social Psychology*, **74**(1): 272–9.

Lambie, J.A. and Marcel, A.J. 2002. Consciousness and the varieties of emotion experience: A theoretical framework. *Psychological Review*, **109**(2): 219–59.

Larsen, R.J. 2000. Toward a science of mood regulation. *Psychological Inquiry*, **11**(3): 129–41.

Leith, K.P. and Baumeister, R.F. 1996. Why do bad moods increase self-defeating behavior? Emotion, risk taking and self-regulation. *Journal of Personality and Social Psychology*, **71**(6): 1250–67.

Lerner, J.S. and Keltner, D. 2000. Beyond valence: Toward a model of emotion-specific influences on judgement and choice. *Cognition & Emotion*, **14**(4): 473–93.

Lerner, J.S. and Keltner, D. 2001. Fear, anger, and risk. *Journal of Personality and Social Psychology*, **81**(1): 146–59.

Lerner, J.S., Small, D.A. and Loewenstein, G. 2004. Heart strings and purse strings: Carryover effects of emotions on economic decisions. *Psychological Science*, **15**(5): 337–41.

Lerner, J.S. and Tiedens, L.Z. 2006. Portrait of the angry decision maker: How appraisal tendencies shape anger's influence on cognition. *Journal of Behavioral Decision Making*, **19**(2): 115–37.

Leung, A., Foo, M.-D. and Chaturvedi, S. 2013. Imprinting effects of founding core teams on HR values in new ventures. *Entrepreneurship Theory and Practice*, **37**(1): 87–106.

Leung, A., Zhang, J., Wong, P. and Foo, M.-D. 2006. The use of networks in human resource acquisition for entrepreneurial firms: Multiple 'fit' considerations. *Journal of Business Venturing*, **21**(5): 664–86.

Loewenstein, G.F. 2007. Affect regulation and affective forecasting. In J.J. Gross (ed.), *Handbook of Emotion Regulation*: 180–203. New York: Guilford Press.

Martin, L.L. and Tesser, A. 1996. Some ruminative thoughts. In R.S. Wyer (ed.), *Advances in Social Cognition*, vol. 9: 1–47. Mahwah, NJ: Erlbaum.

Morris, M.H., Kuratko, D.F., Schindehutte, M. and Spivack, A.J. 2012. Framing the entrepreneurial experience. *Entrepreneurship Theory and Practice*, **36**(1): 11–40.

Myers, D.G. and Lamm, H. 1975. The polarizing effect of group discussion. *American Scientist*, **63**(3): 297–303.

Newell, A. 1990. *Unified Theories of Cognition*. Cambridge, MA: Harvard University Press.

Ohman, A. and Mineka, S. 2001. Fears, phobias, and preparedness: Toward an evolved module of fear and fear learning. *Psychological Review*, **108**(3): 483–522.

Parrott, W.G. 1993. Beyond hedonism: Motives for inhibiting good moods and for maintaining bad moods. In D.M. Wegner and J.W. Pennebaker (eds), *Handbook of Mental Control*: 278–308. Englewood Cliffs, NJ: Prentice Hall.

Podoynitsyna, K., Van der Bij, H. and Song, M. 2012. The role of mixed emotions in the risk perception of novice and serial entrepreneurs. *Entrepreneurship Theory and Practice*, **36**(1): 115–40.

Richard, R., Van der Plight, J. and DeVries, N. 1996. Anticipated affect and behavioral choice. *Basic and Applied Social Psychology*, **18**(2): 111–29.

Richards, J.M. and Gross, J.J. 1999. Composure at any cost? The cognitive consequences of emotion suppression. *Personality and Social Psychology Bulletin*, **25**(8): 1033–44.

Russell, J.A. and Barrett, L.F. 1999. Core affect, prototypical emotional episodes, and other things called emotion: Dissecting the elephant. *Journal of Personality and Social Psychology*, **76**(5): 805–19.

Schwarz, N. 1990. Feelings as information: Informational and motivational functions of affective states. In E.T. Higgins and R. Sorrentino (eds), *Handbook of Motivation and Cognition: Foundations of social behavior*, vol. 2: 527–61. New York: Guilford Press.

Schwarz, N. and Clore, G.L. 1983. Mood, misattribution, and judgments of well-being: Informative and directive functions of affective states. *Journal of Personality and Social Psychology*, **45**(3): 513–23.

Schwarz, N. and Clore, G.L. 1988. How do I feel about it? Informative functions of affective states. In K. Fiedler and J. Forgas (eds), *Affect, Cognition, and Social Behavior*: 44–62. Toronto: Hogrefe International.

Schwarz, N. and Clore, G.L. 1996. Feelings and phenomenal experiences. In E.T. Higgins and A.W. Kruglanski (eds), *Social Psychology: Handbook of basic principles*: 433–65. New York: Guilford Press.

Shane, S. and Venkataraman, S. 2000. The promise of entrepreneurship as a field of research. *Academy of Management Review*, **25**(1): 217–26.

Sheldon, K.M., Ryan, R.M., Rawsthorne, L. and Ilardi, B. 1997. 'True' self and 'trait' self: Cross-role variation in the Big Five traits and its relations with authenticity and well-being. *Journal of Personality and Social Psychology*, **73**(6): 1380–93.

Shepherd, D.A. 2003. Learning from business failure: Propositions of grief recovery for the self-employed. *Academy of Management Review*, **28**(2): 318–28.

Shepherd, D.A. 2004. Educating entrepreneurship students about emotion and learning from failure. *Academy of Management Learning and Education*, **3**(3): 274–87.

Shepherd, D.A. 2009. Grief recovery from the loss of a family business: A multi- and meso-level theory. *Journal of Business Venturing*, **24**(1): 81–97.

Sinha, R., Lovallo, W.R. and Parsons, O.A. 1992. Cardiovascular differentiation of emotions. *Psychosomatic Medicine*, **54**(4): 422–35.

Smart Richman, L. and Leary, M.R. 2009. Reactions to discrimination, stigmatization, ostracism, and other forms of interpersonal rejections: A multimotive model. *Psychological Review*, **116**(2): 365–83.

Smith, C.A. and Ellsworth, P.C. 1985. Patterns of cognitive appraisal in emotion. *Journal of Personality and Social Psychology*, **48**(4): 813–38.

Song, Z.L., Foo, M.-D. and Uy, M.A. 2008. Mood spillover and crossover among dual-earner couples: A cell phone event sampling study. *Journal of Applied Psychology*, **93**(2): 443–52.

Staw, B.M. and Barsade, S.G. 1993. Affect and managerial performance: A test of the sadder-but-wiser vs. happier-and-smarter hypotheses. *Administrative Science Quarterly*, **38**(2): 304–31.

Tan, H.H., Foo, M.-D., Chong, C.L., and Ng, R. 2003. Situational and dispositional predictors of displays of positive emotions. *Journal of Organizational Behavior*, **24**(8): 961–78.

Tan, H.H., Foo, M.-D. and Kwek, M. 2004. The role of customer personality traits on the display of positive emotions. *Academy of Management Journal*, **47**(2): 287–96.

Tangney, J.P., Stuewig, J. and Mashek, D.J. 2007. Moral emotions and moral behavior. *Annual Review of Psychology*, **58**(1): 345–72.

Tiedens, L.Z. and Linton, S. 2001. Judgment under emotional certainty and uncertainty: The effects of specific emotions on information processing. *Journal of Personality and Social Psychology*, **81**(6): 973–88.

Twenge, J.M., Catanese, K.R. and Baumeister, R.F. 2002. Social exclusion causes self-defeating behavior. *Journal of Personality and Social Psychology*, **83**(3): 606–15.

Uy, M.A., Foo, M.-D. and Aguinis, H. 2010. Using experience sampling methodology to advance entrepreneurship theory and practice. *Organizational Research Methods*, **13**(1): 13–54.

Uy, M.A., Foo, M.-D. and Song, Z.L. 2013. Joint effects of prior start-up experience and coping strategies on entrepreneurs' psychological well-being. *Journal of Business Venturing*, **28**(5): 583–97.

Watson, D., Clark, L.A. and Tellegen, A. 1988. Development and validation of brief measures of positive and negative affect. The PANAS scales. *Journal of Psychology*, **54**(6): 1063–70.

Webb, T.L., Miles, E. and Sheeran, P. 2012. Dealing with feeling: A meta-analysis of the effectiveness of strategies derived from the process model of emotion regulation. *Psychological Bulletin*, **138**(4): 775–808.

Wegner, D.M., Erber, R. and Zanakos, S. 1993. Ironic processes in the mental control of mood and mood-related thought. *Journal of Personality and Social Psychology*, **65**(6): 1093–104.

Weiss, H.M. 2002. Conceptual and empirical foundations for the study of affect at work. In R.G. Lord, R.J. Klimoski and R. Kanfer (eds), *Emotions in the Workplace: Understanding the structure and role of emotions in organizational behavior*: 20–63. San Francisco: Jossey-Bass.

Weiss, H.M. and Cropanzano, R. 1996. Affective events theory: A theoretical discussion of the structure, causes and consequences of affective experiences at work. *Research in Organizational Behavior: An annual series of analytical essays and critical reviews*, vol. 18: 1–74. New York: Elsevier Science/JAI Press.

Welpe, I.M., Spörrle, M., Grichnik, D., Michl, T. and Audretsch, D.B. 2012. Emotions and opportunities: The interplay of opportunity evaluation, fear, joy, and anger as antecedent of entrepreneurial exploitation. *Entrepreneurship Theory and Practice*, **36**(1): 69–96.

Wong, C.S., Foo, M.-D., Wang, C.W. and Wong, P.M. 2007. The feasibility of training and development of EI: An exploratory study in Singapore, Hong Kong, and Taiwan. *Intelligence*, **35**(1): 141–50.

Wranik, T., Barrett, L.F. and Salovey, P. 2007. Intelligent emotion regulation: Is knowledge power? In J.J. Gross (ed.), *Handbook of Emotion Regulation*: 393–407. New York: Guilford Press.

Yurtserver, G. 2008. Negotiators' profit predicted by cognitive reappraisal, suppression of emotions, misrepresentation of information, and tolerance of ambiguity. *Perceptual and Motor Skills*, **106**(2): 590–608.

6. Exploring the affective and cognitive dynamics of entrepreneurship across time and planes of influence

Denis A. Grégoire

ABSTRACT

In spite of the broad interest in studying affect and cognition in entrepreneurship, extant research has struggled to untangle the complex web of affective and cognitive forces influencing entrepreneurial action, and to unpack these forces' influence across time and levels of analysis. To address these challenges, this chapter draws attention to different classes of affective/cognitive forces – depending on their enduring or episodic nature and their plane of influence. After illustrating the import of these distinctions for studying opportunity imagination/identification, the text discusses their implications for broader research on entrepreneurship's affective and cognitive dimensions.

INTRODUCTION

Whether from an economic or psychological perspective, scholars have long advocated the benefits of studying the drivers of entrepreneurial action from entrepreneurs' own standpoint (Carland, Hoy and Carland, 1988 versus Gartner, 1988; Kirzner, 1997; Neergaard and Ulhøi, 2007). Echoing a series of developments that shook the social sciences from the 1950s onward (Bishop, 2005; Gardner, 1985), entrepreneurship researchers eventually began to consider the influence of entrepreneurs' mental representations, abilities and other affective and cognitive resources. From Comegys' (1976) early focus on cognitive dissonance or Busenitz and Barney's (1997) study of entrepreneurs' decision heuristics to a series of monographs and special issues calling attention to the potential contributions of research on affect and cognition (e.g., Baron, 2004, 2008; Cardon, Foo, Shepherd and Wiklund, 2012; Katz and Shepherd, 2003; Krueger, 2003; Mitchell et al., 2002a, 2004, 2007), an increasing number of studies have appeared on the influence of thoughts, emotions and other affective and cognitive forces on entrepreneurship (cf. Forbes, 1999 versus Grégoire, Corbett and McMullen, 2011).

Given current academic knowledge about human action in general (Morsella, Bargh and Gollwitzer, 2009) and entrepreneurial action in particular (Katz and Corbett, 2012; McMullen and Shepherd, 2006; Mitchell and Shepherd, 2010), however, it is hardly insightful anymore to simply argue – or show – that what entrepreneurs think and feel influences their behavior (see also Baron, Chapter 1, this volume; Bird, Chapter 3, this volume). By the same token, it has become redundant to advance that entrepreneurs have cognitive or affective predispositions toward certain attitudes, biases, strategies, styles or behavioral responses – and, thus, that these particular dispositions necessarily influence their choices and behaviors (Grégoire, Corbett and McMullen, 2011; Mitchell, Randolph-Seng and Mitchell, 2011).

By contrast, substantial opportunities for advancing entrepreneurship research remain in exploring how and why mental dynamics unfold the way they do – and the consequent influence this has for entrepreneurial action. But doing so presents important difficulties. Thoughts, feelings, knowledge, creative abilities and other mental resources, strategies and predispositions are subject to a variety of influences. Some of these are external to individuals, whereas others are internal. Likewise, some influences are completely exogenous to entrepreneurial action: they would occur regardless of whether someone was engaged in entrepreneurship or not. By contrast, other influences are endogenous (or at least concurrent) to this action: the very act of engaging in entrepreneurship (and the consequences that ensue) may not only lead to the emergence of particular affective or cognitive variables/dynamics, but may also influence the effects these have. Furthermore, thoughts, emotions and other mental resources might often exhibit important variations in time (Fisher, 2008; Johnson-Laird, 2009). Yet whether they do so consciously or not, individuals use and leverage the knowledge, information, feelings and other mental resources they have; they integrate these with their understanding of the world, and can further transform and transcend all these mental elements in their efforts to launch, develop and lead their new ventures. Because of this, it becomes critically important to study the unfolding of entrepreneurs' mental representations, states and other affective and cognitive influences across time and levels of analyses (Ashkanasy, 2003; Huy, 2012). In other words, it becomes important to study the unfolding of affective and cognitive dynamics (Grégoire and Lambert, 2014; McMullen and Dimov, 2013; see also Foo, Murnieks and Chan, Chapter 5, this volume).

To help advance research on the unfolding of such affective and cognitive influences on entrepreneurial action, this chapter draws attention to two important distinctions between different classes of affective/cognitive forces:

1. The enduring or episodic nature of affective/cognitive forces, that is, their high (or low) susceptibility to exhibit important changes over short periods of time; and
2. The anchoring of different affective/cognitive forces at different planes of influence – from individual-level mental resources, abilities and predispositions and shared mental representations to relevant formal properties of entrepreneurs' tasks, circumstances and information environments that also influence entrepreneurs' affect and cognition.

After explaining these distinctions, the text illustrates their import for extant research on affective and cognitive forces pertinent to the imagination/identification of entrepreneurial opportunities. The chapter concludes by briefly discussing the broader implications of these distinctions for future entrepreneurship research on affect and cognition. By increasing familiarity with these notions, I hope to help stimulate further developments of the conceptual and methodological articulations of research on the affective and cognitive dynamics of entrepreneurial action.

PRELIMINARY ASSUMPTIONS

Past research has traditionally emphasized strict definitions of affect and cognition – and of the distinctions between the two. By such definitions, affect concerns the subjective experience and influence of feelings, emotions, moods and other affective states or dispositions (see Cardon et al., 2012: 2), whereas, in the strictest sense, cognition refers to the mental processes by which the mind transforms, reduces, elaborates, stores, recovers and/or uses sensory stimuli (see Neisser, 1967; Thagard, 2006). In more figurative terms, then, affect is said to concern matters of the heart, whereas matters of the mind define the domain of cognition.

As such, affect and cognition have long been presented as two different aspects of the human experience, separate from one another. Indeed, debates have raged about the primacy of affect over cognition, and vice versa (cf. Lazarus, 1982, 1984, 1991 versus Zajonc, 1980, 1984). In practice, the research literatures on these topics have long evolved in parallel to one another, with few constructive exchanges between the two.

From a neuro-physiological standpoint, it is generally assumed that affect and cognition tend to involve separate neural structures (cf. Baucus, Baucus and Mitchell, Chapter 8, this volume; Gazzaniga, Ivry and Mangun, 2013; McMullen, Wood and Palich, Chapter 9, this volume). Yet contemporary research has shown that, instead of being completely independent from one another, affective and cognitive systems dynamically work together to regulate human behavior (see Damasio, 1994; Dolan, 2002; Forgas, 2008; Isen, 1984; Pham, 2004; Thagard, 2006). As a result, affect and cognition are no longer seen as completely separate phenomena (see Drnovšek, Slavec and Cardon, Chapter 7, this volume).

In similar fashion, contemporary research on cognition is no longer confined to a strict information-processing paradigm (Broadbent, 1958; Simon and Newell, 1964). Within entrepreneurship research and the larger managerial sciences, for instance, the broad umbrella of cognitive research can encompass perspectives as diverse as computational approaches (e.g., Ethiraj and Levinthal, 2004; Gavetti and Levinthal, 2000), sense making (Maitlis and Sonenshein, 2010; Weick, Sutcliffe and Obstfeld, 2005), social constructionism (Gergen, 1985; Rao, 1994), cognitive neuroscience (Laureiro-Martinez, Brusoni, Canessa and Zollo, in press) or situated and/or embodied cognition (Elsbach, Barr and Hargadon, 2005; Harquail and King, 2010; Mitchell et al., 2011).

Whereas adhering to rigorous definitions and clear boundary distinctions is necessary for making sound theoretical developments and empirical contributions, this chapter's purpose is not to develop (or test) a particular perspective, theory, model or set of propositions specific to a particular affective (or cognitive) driver of entrepreneurship. Consistent with the emergent views that the affective and cognitive domains are theoretically distinct but not completely separable from one another (see Drnovšek et al., Chapter 7, this volume), the text that follows generally discusses affective and cognitive forces alongside one another, using both terms. Conversely, the text uses a single term (affect or cognition) when a particular argument or observation applies specifically to one domain.

As this chapter is a contribution to a volume devoted to entrepreneurial cognition research, however, I begin by summarizing the meaning and import that the cognitive perspective has had across the social sciences in general, and for entrepreneurship research more specifically. Doing so is also consistent with the field's historical evolution (cf. Cardon et al., 2012; Forbes, 1999; Grégoire et al., 2011). It also helps set the stage for introducing relevant distinctions on the enduring/episodic nature of different forces, and their planes of influence.

WHAT CAN THE COGNITIVE PERSPECTIVE IMPLY FOR ENTREPRENEURSHIP RESEARCH?

Although it is often portrayed as having replaced behaviorism in psychology (Robins, Gosling and Craik, 1999), the cognitive perspective finds its roots in a series of convergent developments that, in the late 1950s, took place across a broad swath of disciplines ranging from philosophy, psychology, linguistics and anthropology to the fields of neuroscience and artificial intelligence. At issue were long-standing epistemological questions about the nature of knowledge, its components, sources, development and deployment, and about the implications of these questions for research on the human mind and its interactions with others and the environment (cf. Bishop, 2005; Gardner, 1985). As the cognitive perspective gained momentum, it not only impacted the conceptualization of philosophical problems, but also had profound and frame-breaking implications for research in other domains (Turner, 2001).

In practice, a primary implication of the cognitive revolution has been to advance research on the human mind's functioning. Instead of treating the mind as a 'black box' of unobservable phenomena that could not be observed independently from external stimuli and consequent responses, early cognitive works made it legitimate and worthwhile to develop models of mental representations, knowledge, attitudes, motivations and other mental states and processes as they would exist in the mind – and to empirically test the plausibility and import of such models (cf. Fiske and Taylor, 1991; Smith, 2000; Sternberg and Ben-Zeev, 2001; Thagard, 2005).

The case of social psychology research on attitudes illustrates this. Whereas prior research had treated attitudes as individual tendencies to view particular objects, people or ideas in a more or less positive (or negative) light, socio-cognitive psychologists theorized that attitudes were mental constructs subject to the same cognitive structures, processes and phenomena governing knowledge and other mental representations formed from perception and experience. As a result, social-cognitive psychologists began investigating the dynamics leading to the development of attitudes, their storage in long-term memory, their transformation in light of new stimuli/experience, or their use to make choices and decisions, or to initiate particular actions (cf. Cialdini, Petty and Cacioppo, 1981; Fiske and Taylor, 1991; Smith, 2000).

Taken together, these developments fostered extensive research on thoughts, emotions and other mental states, representations, abilities and resources, on the development of such mental constructs, their transformation and storage for subsequent use, and on the processing of internal and external information for reasoning, decision making, action and so on (Smith and Osherson, 1995; Thagard, 2005). In addition, and although other factors also contributed, the cognitive perspective's focus on the unfolding of mental phenomena paved the way for further studies of affective dynamics – such as the emergence, transformation and influence of moods, feelings, emotions and other affective processes, resources and phenomena (cf. Dolan, 2002; Forgas, 2008; Isen, 1984; Pham, 2004; Thagard, 2006).

Given its particular focus on the human experience, research on affect and cognition has naturally paid a great deal of attention to individual-level phenomena. To illustrate this kind of research articulation, one can think of particular attitudes, emotions and thoughts as distinctly belonging to specific individuals, the same way that some individuals can be uniquely characterized as having a particular level of knowledge in a domain, as demonstrating a certain level of a cognitive skill, or as exhibiting a predisposition to use a particular heuristic or form of reasoning. When articulated in this fashion, these variables form distinct 'properties' of (different) individuals.

In entrepreneurship research, scholars have mobilized this individual-level articulation to investigate a rich set of variables, dynamics and other forces of cognitive interest. Among many other examples, one notes:

1. Studies on the nature and influence of entrepreneurs' mental perceptions/representations – and those of relevant stakeholders (e.g., Acedo and Jones, 2007; Cossette and Audet, 1992; Dodd, 2002; Douglas, 2006; Jenkins and Johnson, 1997; Krackhardt, 1990);
2. Studies of the effects of intelligence, experience, knowledge or expertise as cognitive resources fostering entrepreneurial action (e.g., Baron and Ensley, 2006; Chetty, Erksson and Lindberg, 2006; Frese et al., 2007; Halaby, 2003; Shane, 2000; Westhead, Ucbasaran and Wright, 2005);
3. Studies of the nature and effects of particular cognitive abilities or dispositions, such as the reliance on particular cognitive styles (e.g., Brigham, De Castro and Shepherd, 2007; Buttner and Gryskiewicz, 1993; Corbett, 2007; Dutta and Thornhill, 2008; Hmieleski and Corbett, 2006), decision heuristics (e.g., Busenitz and Barney, 1997; Forbes, 2005; Keh, Foo and Lim, 2002; Simon and Houghton, 2000; Wickman, 2003), prevention/promotion foci (e.g., Bryant, 2007, 2009; Fitzsimmons and Douglas, 2011; McMullen and Shepherd, 2002), or self-efficacy (e.g., Arora, Haynie and Laurence, 2013; Chen, Greene and Crick, 1998; Gatewood, Shaver and Gartner, 1995; Luthans and Ibrayeva 2006; Markman, Balkin and Baron, 2002).

These are but a few examples of the many individual-level forces that have been investigated in entrepreneurship cognition research (see other chapters in this volume).

Yet it is important to note that research on affective and cognitive forces need not be restricted to the individual level of analysis. Even if scholarly interest in affect and cognition remains grounded in how individuals feel and think, important benefits can arise from articulating some variables as properties of 'units' anchored at other 'levels' than the individual. For instance, scholars of managerial and organizational cognition have demonstrated the influence of shared mental representations of strategic groups and other competitive constraints on firm-level actions and strategies (Lant and Shapira, 2001; Porac, Thomas and Baden-Fuller, 1989; Reger and Palmer, 1996). Even if data were collected from individual managers, employees and/or other stakeholders, these studies conceptually articulated shared mental representations as cognitive properties of the firms themselves – and demonstrated the methodological validity of doing so (see Daniels, de Chernatony and Johnson, 1995). In the same vein, scholars building on institutional theory have drawn attention to the cognitive legitimacy of particular firms – or industries – within one or more particular audience (Deeds, Mang and Frandsen, 2004; Kennedy, 2008; Sine, Haveman and Tolbert, 2005). Emphasizing the cognitive dimension(s) of broader social structures and institutions (DiMaggio, 1997), some scholars have also explored national discourses about entrepreneurship (Perren and Jennings, 2005; Radu and Redien-Collot, 2008), while others examined national differences in cognitive institutions for entrepreneurship (Busenitz, Gomez and Spencer, 2000; Spencer and Gomez, 2004), or the cognitive scripts underpinning entrepreneurial efforts in different countries (Mitchell et al., 2000, 2002b).

In short, leveraging a cognitive perspective needs not restrict one to studying individual-level phenomena. While affective and cognitive dynamics necessarily rest on what individuals think and feel, a number of variables, dynamics and other forces can be validly conceived (and studied) at other levels of analysis. Table 6.1 illustrates this articulation of entrepreneurial cognition research through specific examples of cognitively relevant variables anchored at different levels. In principle, the same kind of articulation might be possible for studying the role and influence of affective forces in entrepreneurship.

Table 6.1 Examples of cognitively relevant forces anchored at different 'levels'

Levels (in bold) and examples of cognitively relevant forces	Examples of relevant studies
Cognitive properties of cultures, nations, people, societies etc.	
Cognitive dimension(s) of countries' institutional profiles for entrepreneurship	Busenitz, Gomez and Spencer, 2000 Spencer and Gomez, 2004
Cognitive dimension(s) of psychic distance characterizing an internationalization target (relative of a home country)	Dow and Karunaratna, 2006 Håkanson and Ambos, 2010
Cognitive properties of broad industrial sectors, specific industries, strategic groups	
Cognitive dimension(s) of a sector's legitimacy institutions	Sine, Haveman and Tolbert, 2005
Industries' sociopolitical legitimacy	Deeds, Mang and Frandsen, 2004
Sector-level representations of market issues/ constraints	Chataway and Tait, 1993
Shared perceptions about strategic groups' characteristics	Reger and Huff, 1993
Cognitive properties of firms, organizations etc.	
Cognitive dimension(s) of a firm's social capital	Atuahene-Gima and Murray, 2007 Presutti, Boari and Fratocchi, 2007
Customers' perceptions of a new venture's cognitive legitimacy	Choi and Shepherd, 2005
Firms'/new ventures' cognitive embeddedness	Kennedy, 2008
Firms'/new ventures' cognitive legitimacy	Deeds, Mang and Frandsen, 2004
Cognitive properties of groups, teams, families etc.	
Differentiation and integration of strategic constructs	West, 2007

Table 6.1 (continued)

Levels (in bold) and examples of cognitively relevant forces	Examples of relevant studies
Cognitive properties of groups, teams, families etc.	
Shared cognition (strategic orientation)	Ensley and Hmieleski, 2005 Ensley and Pearce, 2001
Team-level decision comprehensiveness	Chowdhury, 2005
Teams' transactive memory systems	Yoo and Kanawattanachai, 2001
Cognitive properties of individuals	
Cognitive style: analytic vs holistic	Dutta and Thornhill, 2008
Cognitive resources (cognitive ability + human capital)	Frese et al., 2007
Complexity of a firm CEO's cognitive map of competition	Calori, Johnson and Sarnin, 1994
Gender identity: masculinity and femininity	Eddelston and Powell, 2008
Learning style preferences (information acquisition: comprehension vs apprehension; information transformation: extension vs intention)	Corbett, 2007
Metacognitive abilities	Haynie and Shepherd, 2009 Haynie, Shepherd and Pazelt, 2012
Perceptions of important issues post takeover	Douglas, 2006

ADDRESSING THE COMPLEXITY OF ENTREPRENEURSHIP'S AFFECTIVE AND COGNITIVE DYNAMICS

Analyzing articles published between 1976 and 2008, Grégoire, Corbett and McMullen (2011) found that 'cognitive research in entrepreneurship has been primarily focused on studying cognition at single levels of human activity (whether the individual, the firm or society) – as opposed to studying the operation of cognitive forces *across* levels of analysis' (2011: 1455). The trend has largely continued. In much the same vein, few entrepreneurship studies consider the unfolding development(s) of affective or cognitive forces across time – whether prior to or during entrepreneurial action – let alone how contextual dynamics might influence (and be influenced by) such unfolding developments (Grégoire and Lambert, 2014).

Unfortunately, these tendencies lead to an incomplete understanding of the interactive dynamics of mind, environment and entrepreneurial action (Davidsson, 2006; McMullen and Dimov, 2013). For instance, the single-time/single-level approach makes it difficult to determine the extent to which the influence of a particular cognitive skill results from genetic predispositions that an individual was born with, from cognitive development and learning dynamics that preceded the entrepreneurial action proper, or from the resulting influence of engaging in entrepreneurial efforts, the particular circumstances of that effort, and/or the results that ensued from it (see also Forbes, Chapter 10, this volume). In other words, the articulation of past entrepreneurship research makes it difficult to unpack the variety of affective and cognitive forces influencing entrepreneurial action and outcomes.

From a research standpoint, addressing this problem calls for leveraging conceptualizations and methods that allow for unpacking the complex tangle of interactive influences between the external environment (physical and social), the individual (from his/her genetic makeup though his/her resources, skills, abilities and other predispositions, both physical and mental), and any relevant mental representations, structures and mechanisms that individuals can leverage – not to mention the actions they initiate and whatever outcomes may ensue.

This is a not a trivial challenge. Past research on the imagination/identification of entrepreneurial opportunities presents an illustrative example of the difficulties posed by separating the influences of different factors – including some that proceed from the individual, and some that originate from other sources external to individual entrepreneurs. As others have observed before (cf. Davidsson, 2003; Dimov, 2007b; 2011; Short et al., 2010), many opportunity identification studies have leveraged cross-sectional research designs that focus on rare and idiosyncratic confluences of individuals, opportunities and circumstances. In such designs, individual resources, opportunity characteristics and broader contextual factors are necessarily confounded with one another. It thus becomes difficult to discern the respective effects of different categories of affective and/or cognitive influences (Grégoire, Shepherd and Lambert, 2010). Because of this, one cannot determine the extent to which opportunity imagination/identification is most influenced by, among other factors and dynamics, the attention allocated toward environmental signals (Ocasio, 1997; Shepherd, McMullen and Jennings, 2007), the perceived characteristics of these signals (Grégoire and Shepherd,

2012; Jackson and Dutton, 1988; Julian and Ofori-Dankwa, 2008), individual/organizational abilities to process such signals (Kuvaas, 2002; Milliken, 1990), the knowledge at their disposal (Dimov, 2007b; Shane, 2000; Shepherd and DeTienne, 2005), a firm's unique resources, organizational slack, or strategy (Chattopadhyay, Glick and Huber, 2001; Thomas and McDaniel, 1990), and/or other individual or organizational resources and abilities (Barnett, 2008; Cattani and Ferriani, 2008).

Interestingly, the problem is not simply a methodological one. In many studies of opportunity identification, little theoretical attention is devoted to the origin and emergence of affective or cognitive forces prior to the opportunity identification events, let alone during the unfolding of such events. Furthermore, the only affective or cognitive forces considered in many studies of opportunity identification are those conceived as individual-level abilities, dispositions or resources. By contrast, few studies consider the potential effect of relevant affective or cognitive forces that proceed from the characteristics of different opportunity ideas (Davidsson, 2012; Dahlqvist and Wiklund, 2012; Grégoire and Shepherd, 2012), and/or from the broader context and circumstances surrounding different opportunities (Lichenthaler, 2009; Thomas and McDaniel, 1990).

The net result is an incomplete understanding of the individual–opportunity nexus (Shane and Venkataraman, 2000) – and particularly with respect to the affective and cognitive forces that may foster opportunity imagination/identification. In spite of important advances, the invisible college of entrepreneurship scholars remains ill-equipped to determine what specific affective or cognitive forces spur opportunity insights in the first place. In turn, advice for entrepreneurs, educators and consultants often boils down to peddling brainstorming and creativity techniques that are valid and useful, but do not readily channel imagination/identification efforts towards opportunity-specific insights, issues or rationales.

Clearly, opportunity identification is a complex phenomenon, where many affective and cognitive forces can play a determinant role. Extant empirical research has indeed documented the influence of several factors (Shepherd and Grégoire, 2012; Short, Ketchen, Shook and Ireland, 2010) – and many conceptual articles point towards a number of others (Ardichvilli, Cardozo and Ray, 2003; Baron, 2007, 2008; Cornelissen and Clarke, 2010; McMullen and Shepherd, 2006; Wood, Williams and Grégoire, 2012). Given this rich literature, then, any effort to better understand the underlying drivers of opportunity imagination/identification demands that one not only seeks to obtain observations that are as rigorously valid as possible, but also tries to account for the effects of different kinds of influencing forces, often articulated at different levels. In the case of entrepreneurs' effort to imagine/identify opportunities, for instance, one must be able to separate the effects of individual resources in abilities, knowledge or motivation (individual-level forces) from the effects of context, time and circumstances where identification takes place, the effects of relevant firm-level resources and other characteristics (if any), not to mention the effects of relevant characteristics and variables associated with the particular opportunities being thought about.

The same kind of research challenges hinder the broader research on the affective and cognitive underpinnings of entrepreneurship. Given current understanding of the factors and dynamics influencing human action, scholars must be able to distinguish between different forces that originate and operate from different sources, often anchored at different levels. For instance, one must distinguish between forces that proceed from the individual entrepreneur himself or herself (e.g., skills, abilities, individual resources etc.) and forces that proceed from the entrepreneur's context and environment (including the influence of culture, social institutions, other individuals etc.). In similar fashion, scholars must distinguish between the influence of mental resources and/or dispositions that vary in the long run (e.g., passion, knowledge, memories) and that of mental states and/or influences that are considerably more episodic (e.g., short-term emotions, particular information stimuli, current conditions and constraints, etc.). Furthermore, scholars must also consider the extent to which different affective states and cognitive representations may evolve in time, influence each other, and possibly be influenced in turn by retrospective feedback from the outcome of one's actions.

To help address these challenges, I propose that future studies of the affective and cognitive underpinnings of entrepreneurship specifically investigate the unfolding of such forces across time, and across different planes of influence. The next two sections develop each point in turn. In doing this, the text draws attention to the theoretical import of distinguishing between different classes of affective/cognitive forces, and illustrates the kind of research questions that could be pursued on these bases. I submit that fruitful advances in entrepreneurship research lie in integrating such distinctions with the extant conceptualization of affect and cognition's influence, and with the articulation of consequent research designs and methods.

EXPLORING AFFECTIVE AND COGNITIVE DYNAMICS ACROSS TIME

To help distinguish between different forces influencing entrepreneurial action, this chapter encourages future research to pay explicit attention to the susceptibility of different affective/cognitive forces to exhibit more (or less) important changes over relatively short periods of time. In other words, this text advocates that future research accounts for the enduring versus episodic nature of different affective/cognitive forces.

At one extreme, one finds forces that are essentially 'fixed': they are unlikely to change over the course of one's life. At the individual level, this is the case of genetic and neuro-physiological bases of intelligence, personality, mental abilities and other affective or cognitive resources. One could also include here basic instinctive phenomena such as fears, fight-or-flight responses, or inner motivations to satisfy basic human needs from hunger and shelter to sexual pursuits – phenomena that are all associated with evolutionary/ancient brain structures (see Baucus et al., Chapter 8, this volume; McMullen et al., Chapter 9, this volume). By extension, one could also include here affective or cognitive elements that are so deeply ingrained in shared psyches, societal institutions and other cultural norms and practices that they are unlikely to exhibit high variations over the course of individual lives. Examples could include broadly shared ideas about the pursuit of happiness, or social prescriptions against cannibalism, murder, rape, incest and so on.

At the opposite extreme, one finds forces such as stimuli-driven emotions, thoughts and other episodic impressions that will spontaneously and effortlessly vary in nature, intensity and other dimensions – depending on individual whims or the circumstances where they arise and/or exert their influence. These forces are the province of short-term memory and associated processes. From a neuro-physiological standpoint, many such forces proceed from structures associated with the limbic system and the mammalian brain (see McMullen et al., Chapter 9, this volume). Among other examples, cognitive processes of attention are particularly sensitive to episodic stimuli. The human mind does not have the processing capacity to deal with the wealth of information stimuli our senses receive. Thankfully, cognitive processes of attention guide which information we attend to, and which we ignore. A particularly important factor in such processes is the salience of an information signal, that is, the degree to which a signal contrasts with the background of what we would otherwise expect. Because of this, we constantly form transient thoughts and emotions about issues that catch our attention – and these can often have profound impact on our choices, decisions and behaviors. Yet the salience of particular signals can vary substantially with time and/or changes in circumstances: as a result, the content of human attention can exhibit considerable variations over short periods of time.

In between these extremes, one finds a broad class of forces that exhibit some susceptibility to change – though such changes might take longer, be more or less gradual, and/or require more or less effort, energy and so on. Examples include learning- and experience-driven abilities, knowledge, expertise, more or less enduring moods and other affective dispositions, cognitive resources acquired through training and education, and the like. In terms of McMullen, Wood and Palich's chapter (Chapter 9 in this volume), these forces are the province of long-term memory and associated neural structures.

Coupled with the above notion that affective and cognitive forces can be articulated at different levels, discussing the enduring or episodic nature of different forces and their susceptibility to exhibit important changes over short periods of time points to the pertinence of distinguishing between different classes of forces relevant to entrepreneurial affect and cognition. Table 6.2 illustrates such distinctions.

These distinctions matter. Affective and cognitive dynamics are not only about the effects of enduring abilities, predispositions and other resources. Entrepreneurs must often make important decisions on the basis of unforeseen circumstances, new information or other emerging/evolving dynamics that draw their attention and/or trigger particular feelings. Yet limited scholarly knowledge exists about the extent to which temporal changes in such forces influence entrepreneurial action – let alone what may prompt such temporal changes in the first place. Useful research advances on the affective and cognitive dimensions of entrepreneurship might thus be realized by explicitly distinguishing different forces in terms of their high (or low) susceptibility to exhibit important changes over short periods of time – and to integrate such distinctions with future conceptual models and empirical studies (see also Sarasvathy, Chapter 14, this volume). For instance, one could build on these distinctions to examine the relative influence of enduring versus episodic components of a particular cognitive resource, or one's predilection for some affective tendencies.

Recent developments on the construct of entrepreneurial passion provide an illustrative example (see Drnovšek et al., Chapter 7, this volume). In a series of recent papers, Cardon and her colleagues advanced a multidimensional conception of entrepreneurial passion, whereby passion concerns the experience of intense positive feelings for activities that are central to an entrepreneur's self-identity (Cardon, Grégoire, Stevens and Patel, 2013; Cardon, Wincent, Singh and Drnovšek, 2009). According to their model, passion requires both the experience of intense positive feelings and activities that are central to one's self-identity. Yet, in principle, one could imagine that the experience of more or less positive feelings for different activities might exhibit more variations in time than the importance of these activities *vis-à-vis* one's self-identity. To date, however, research has not explored the temporal unfolding of entrepreneurial passion. As a

Table 6.2 Examples of forces exhibiting different susceptibilities to change

Levels	Very low susceptibility ←←	**Susceptibility to exhibit important changes over short periods of time**	→→ Very high susceptibility
	Enduring forces: unlikely to exhibit substantial changes over time	*Forces that can exhibit some changes over time, but this requires substantial effort, particular conditions, etc.*	*Episodic forces: likely to exhibit substantial changes over short periods of time*
Cultures, nations, peoples, societies etc.	Deep-seated cultural myths, social norms and prescriptions etc.	Broadly shared representations about a nation's ethos, its place in the world etc.	Broadly shared 'national mood' *vis-à-vis* ongoing issues etc.
Economic sectors, industries, strategic groups etc.	Deep-seated representations people have of the core characteristics of economic sectors, industries, strategic groups etc.	Broadly shared 'understanding' of tacit rules governing exchanges and behaviors in particular economic sectors, industries, strategic groups etc. – as acquired over time from experience	Broadly shared 'impressions' about current issues specific to particular economic sectors, industries, strategic groups etc.
Firms, organizations	Cognitive dimensions/representations of a firm's deep-seated identity, mission, sense of purpose etc. Broadly shared myths and representations people have of a firm's past history etc.	Cognitive dimensions/representations of a firm's routines, identity, legitimacy, and/or reputation – as acquired over time through interactions/experience with this firm	Interpretations arising from ongoing events (e.g., sudden threats; broadly shared issues regarding a strategic course of action, such as a proposed merger, acquisition, innovation, takeover etc.)
Groups, teams	Cognitive dimensions/representations of a team's deep-seated identity, mission, sense of purpose etc.	Cognitive dimensions/representations of a team's routines, identity, legitimacy, and/or reputation etc. – as developed over time through experience	Team-level interpretations arising from ongoing events and issues etc.
Individuals	Genetic bases of intelligence, of personality, of someone's cognitive abilities, affective predispositions etc.	Learned/acquired dimensions of personality, attitudes, affective predispositions etc. Knowledge acquired from education, learning, experience etc.	Stimuli-driven moods, episodic emotions, passing non-consequential thoughts, transient opinions etc.

result, important questions remain about passion's endurance in time – and about the lastingness of its effects on entrepreneurial efforts. Seen in this light, it would appear useful to examine the extent to which the two different components of entrepreneurial passion (intense positive feelings and activities that are central to one's self-identity) might exhibit different levels of temporal variability – and what might cause such variations.

By extension, important insights could be derived from examining the possible interplay of short-term and long-term dynamics. Focusing on entrepreneurs' mental models, among a multitude of other possibilities, one might investigate the extent to which new information about competitors can modify more enduring mental models entrepreneurs may have for success in particular markets or industries. By studying the respective influence of enduring and episodic forces on the unfolding of entrepreneurial action, one would not only help specify the boundary conditions of different affective or cognitive phenomena, but could also identify useful synergies between different affective/cognitive resources or strategies.

In practice, pursuing such projects would likely call for leveraging longitudinal research designs and data collection methods that allow for tracking the unfolding developments of entrepreneurs' thoughts and emotions across time (cf. Grégoire and Lambert, 2014; Uy, Foo and Aguinis, 2010). Before returning to these broader implications for future research, however, I discuss a second axis of development.

EXPLORING AFFECTIVE AND COGNITIVE FORCES THAT PROCEED FROM DIFFERENT PLANES OF INFLUENCE

In addition to examining the interplay of forces characterized by different susceptibility to exhibit important changes over a short period of time (i.e., forces that are enduring in the long run versus subject to more or less short-term variations), important advances could be realized by explicitly considering the role of affective and cognitive forces that proceed from different 'planes of influence.'

I use the term 'plane of influence' in this chapter to explicitly mark important distinctions between (i) a research's particular level of analysis (i.e., the level of what a study is trying to explain); (ii) the level(s) at which one collects data for a study (and which could be different from the level of analysis); and (iii) the level(s) at which relevant affective and/or cognitive forces may be anchored, and from where they exert their influence. For instance, one may study the consequences of senior managers' mental representations of competition for their firm's strategic action and performance (e.g., Porac, Thomas and Baden-Fuller, 1989; Reger and Palmer, 1996). In such studies, individual managers provide the data, but the study articulates its inferences at the firm's level of analysis. From a plane of influence standpoint, however, mental representations of competition could be influenced by relevant forces proceeding from one's individual background, upbringing, education and experience – but also from formal properties of the information environment(s) these individuals are facing (e.g., complex and dynamic), of the circumstances where this person is operating (e.g., the depth/richness of his/her organization's identify, knowledge resources or routines), and/or of the particular tasks this person must perform (e.g., repetitive versus unusual tasks, simple versus complex tasks etc.). This chapter hopes to draw more attention to the latter kind of distinctions, that is, to affective and cognitive forces that are anchored on – and exert their effect from – planes of influence that can be distinct from a study's level of analysis (or data collection).

In this regard, particularly pertinent avenues for future research lie in examining the effects of affective and cognitive forces stemming from other planes of influence than the traditional emphasis on individual abilities, predispositions and other resources – or that of shared representations at the group, firm or broader levels. More specifically, this chapter proposes that entrepreneurs' tasks, circumstances and information environments can have formal properties that exert a critical influence on entrepreneurship's affective and cognitive dynamics. Because these formal properties are intrinsically distinct from individuals' subjective perceptions, representations, interpretations, and other mental abilities, predispositions and resources, the influence they exert on affect and cognition proceeds from different planes than that of individual-level forces. Conceptually, this kind of articulation explicitly acknowledges that the 'world-out-there' has characteristics and properties that are affectively and cognitively relevant for what goes on 'inside the hearts and minds of entrepreneurs.'

Research developments in the area of problem solving help illustrate this argument. In their efforts to better understand problem-solving strategies and the nature of expertise, cognitive scientists have explored the potential import of distinguishing between well- and ill-structured problems (cf. Reitman, 1964, 1965; versus Simon, 1973). From a formal standpoint, the following features characterize well-structured problems (WSPs):

1) A definite criterion for recognizing solutions and a mechanizable process for applying that criterion;
2) At least one problem space in which the successive problem states may be represented;
3) A structure wherein attainable stage changes (legal moves) and considerable moves can be represented in a problem space as transitions;

> 4) A representation for any knowledge a problem solver can acquire, in one or more problem spaces;
> 5) A reflection in the state changes, of the laws that govern the external world and the effects upon a state of applying any operator; and
> 6) Basic processes that require only practicable amounts of computing and only information that is available with practicable amounts of search. (Fernandes and Simon, 1999: 226)

Examples of WSPs include simple puzzles like the 'Tower of Hanoi' or 'Missionaries and cannibals' problems used in classic psychology studies (e.g., Jeffries, Polson, Razran and Atwood, 1977), following the instructions to put together a Lego™ model, or house tasks such as replacing an oven's heating elements (. . .). By contrast, ill-structured problems (ISPs) consist of problems that lack definition in one or more of the above features. Examples range from raising a teenage daughter to developing arguments to satisfy reviewers of an academic manuscript.

Interestingly, Simon (1973) offered a powerful demonstration that the import of distinctions between WSPs and ISPs is *not* to reveal fundamentally different classes of problems altogether, but rather to draw attention to the 'structuring' achievements that efficient problem solvers are able to realize when facing 'real-world' problems.

> Definiteness of problem structure is largely an illusion that arises when we systematically confound the idealized problem that is presented to an idealized (and unlimitedly powerful) problem solver with the actual problem that is to be attacked by a problem solver with limited (even if large) computational capacities. If formal completeness and decidability are rare properties in the world of complex formal systems, effective definability is equally rare in the real world of large problems.
>
> In general, the problems presented to problem solvers by the world are best regarded as ISPs. They become WSPs only in the process of being prepared for the problem solvers. It is not exaggerating much to say that there are no WSPs, only ISPs that have been formalized for problem solvers (. . .). Nevertheless, there is merit to the claim that much problem solving effort is directed at structuring problems, and only a fraction of it at solving problems once they are structured. (Simon, 1973: 186–7)

These observations proved important in subsequent efforts to examine the nature and import of expertise in problem solving (see Chase and Simon, 1973; Chi, Feltovich and Glaser, 1981; Ericsson, Charness, Feltovich and Hofman, 2006; Gobet and Simon, 1996, 2000). Among the most salient characteristics distinguishing experts from novices, one notes experts' larger scope of knowledge, more comprehensive mental models (within their domains of expertise), more articulate problem-solving schemas, as well as superior abilities to represent problems, to categorize instances of problems, and to abstract useful insights from their representations. In turn, such cognitive resources generally lead to performance advantages in problem-solving abilities, efficiency, accuracy, or the ability to concentrate on problems for longer periods of time.

The effects of expertise are particularly potent when facing real-world ill-structured problems. When encountering a complex or ambiguous situation, a person will perceive what he or she is prepared to perceive. It follows that the more complex or ambiguous a problem is, the more its perception will be determined by what an individual already knows, and the less it will be determined by the stimulus itself (Fernandes and Simon, 1999). In this regard, Voss and Post (1988) showed that experts tend to spend more time than novices in developing their representations of problems. In turn, however, experts propose fewer and more abstract solutions, and typically tend to develop more comprehensive arguments in support of their solutions. By contrast, novices tend to propose more but simpler solutions, with more limited argumentation.

The few findings reported above are but a pale reflection of the vast territory covered by extant research on problem solving. From this chapter's perspective, however, the relevant point is that cognitive scientists did not advance scholarly understanding of expertise by studying solely individual differences between experts and novices. They actively integrated research questions about the import of problems with different structures, features, and other ambiguities and contradictions (Ericsson et al., 2006; Schraw, Dunkle and Bendixen, 1995; Wang and Chiew, 2010). By analogy, this chapter proposes that interesting advances in entrepreneurship research could be realized by exploring the effects of relevant properties of entrepreneurs' tasks and problems, circumstances, and/or information environment(s). The following sections briefly discuss research possibilities for each topic, in turn.

Variations in the Formal Properties of Entrepreneurial Tasks and Problems

Building on the above considerations, this chapter encourages future research on the potential import of formal variations in the tasks and problems faced by entrepreneurs – and that at different moments of the entrepreneurial journey (McMullen and Dimov, 2013). Among candidate variations, one could examine to extent to which different tasks and problems inherent to entrepreneurship might vary in complexity – and the impact this may have on the affective and cognitive dynamics of entrepreneurship. For instance, Funke (1991: 186) advanced that six non-orthogonal criteria characterize complex problems:

1) *Intransparency* – the notion that in a complex problem situation, only some variables (and their effect(s) on relevant outcomes) are observable directly;
2) *Polytely* – the notion that multiple goals may be present and contradict one another;
3) *Situational complexity* – the notion that the more elements a problem counts, the more difficult it becomes to identify such elements, their relationships, and the principles governing their relationships;
4) *Connectivity* – the notion that when multiple elements are interconnected with one another, it becomes more difficult to anticipate the multi-order effects and consequences these elements have on one another;
5) *Dynamic developments* – the notion that complex problem situations might evolve incrementally rapidly, forcing problem solvers to act sooner and under time pressure;
6) *Time-delayed effects* – the notion that not every action within a complex problem space will show immediate consequences.

(see Funke, 1991; see also Funke, 2010)

From this chapter's perspective, these observations' interesting import is to draw attention to a formal property of tasks and problems (complexity) that is completely separate from individual abilities, affect or cognitive resources: even if experts will exhibit performance advantages over novices when facing such problems, different reasons explain why a problem is complex or not. Such formal characteristics might thus have important implications for the affective and cognitive dynamics associated with different problem-solving efforts. Seen in this light, exploring the emergence, nature and consequence of such properties offers fruitful avenues to augment scholarly understanding of the complex dynamics between the hearts and minds of entrepreneurs, their environment and their action. Among other possible questions, for instance, are the following. Might there be some particular entrepreneurial tasks and/or problems that exhibit higher levels of complexity than others? Why? What could this imply for entrepreneurs' ability to solve such tasks? Over and above the critical role of expertise, are there other forces – affective, cognitive or otherwise – that might moderate entrepreneurs' strategies for solving complex problems (see Baer, Dirks and Nickerson, 2013)?

By the same token, one may ask whether the particular tasks of entrepreneurship also have their importance, specifically. From opportunity imagination, identification and evaluation to marshaling resources, convincing suppliers, partners, investors, employees and stakeholders to support the venture or selling to customers, are there entrepreneurial tasks with characteristics that foster (or constrain) the affective and/or cognitive drivers of entrepreneurship? What are these characteristics? Is it only a question of task difficulty or complexity, or are there other task-relevant dimensions that are particularly relevant for affect and cognition – like the numbers of other persons that must be involved, the centrality of such tasks to one's self-identify, the emotional salience of these tasks, or the looming consequences of failing to complete such tasks in a timely fashion or with at least some measure of success? To what extent do such affective or cognitive characteristics foster (or hinder) the onset (or persistence) of entrepreneurial action? And how – through what affective and/or cognitive mechanisms?

Focusing more directly on the formal characteristics of different tasks, scholars building on Herbert Simon's (1996) *Sciences of the Artificial* have recently drawn increased attention to fundamental differences between the kind of rational optimization problems traditionally emphasized in the information-processing and decision-making literatures, and the kind of open-ended design and creativity challenges pursued by a variety of people in many aspects of their life – and particularly in tasks that lead to innovations and prospective actions, as in business, commerce and entrepreneurship (Sarasvathy, 2003, 2008; but see also this volume, Chapter 14). Hatchuel (2001: 264) provides an illustrative example of such distinctions.

> Two groups of friends living in a big town have to organize their next Saturday evening. Group 1 is discussing '(going to see a) good movie' and Group 2, '(throwing) a nice party.' Through intuitive means and simple observations we can gain an initial distinction between problem-solving theory, which is well adapted to the 'movie case' and something we can call 'Design theory,' which better captures the 'nice party case.'
>
> First remark: we *can apply to the 'good movie' problem all the classics of bounded rationality* (emphasis in the original) . . .
>
> Second remark: Exactly the same set of problem-solving procedures will be required in Group 2 for the 'nice party' case. Yet, and this is our crucial point, 'party' is an infinitely expandable concept and *different processes will also appear in Group 2* (emphasis in the original).

Among other salient features of such design problems, Hatchuel draws attention to three important ones. The first concerns the unexpected expansions of the initial concepts. In the above example, questions such as 'what *is* a party?', 'what kind of party do we want to have?' and/or 'where should we have this party, when, with whom, etc.?' each present expanding series of 'design choices' the friends will need to address. Because of this, the imagination of new unforeseen possibilities can occur at multiple points along Group 2's efforts to realize their project (or not at all). The second feature concerns the design of learning devices. For the friends in Group 1 (movie), learning emerges from the exploration of known elements from the relevant problem space (i.e., movies, theaters, comedians, reviews, critics etc.). For the friends in Group 2 (party), by contrast, learning can also emerge from the exploration of unpredictable domains: the friends' learning will be different if they choose to throw an opera-themed party than if they decide to have a disco ball (or simply choose to watch professional sports on TV). In design problems, then, learning is not solely determined by the problem's particular domain, but is also contingent on the design choices one makes. Lastly, the third feature refers to the potential social dimension of design problems, and to the role and influence of social interactions as design resources.

> The decision-makers of Group 1 are also the 'clients' of their own choices. In Case 2, this is no more true: Group 2 has to at least take into account the expected judgments and behavior of the selected guests. This means that the success of the party cannot be completely controlled by the designers. (Hatchuel, 2001: 266–7)

Within the domain of entrepreneurship research, Sarasvathy has built on similar notions to develop her own theory of effectuation (2001, 2008). In addition to stressing the distinctions between causation and effectuation processes – and the implications these have for the creation of economic artifacts such as firms and markets (Sarasvathy, 2001) – she demonstrated the strong parallels between entrepreneurial action and Simon's 'Theory of Design' (Sarasvathy, 2003). More directly, perhaps, she showed the import of a 'design logic' to reinvestigate concrete entrepreneurial acts like deciding a new venture's name, learning from failure, or making the decision to 'plunge' (see Sarasvathy, 2004). To date, many empirical efforts on this front have examined whether the use of effectuation logic distinguishes entrepreneurs from non-entrepreneurs (Sarasvathy, Simon and Lave, 1998), expert from novice entrepreneurs (Dew, Read, Sarasvathy and Wiltbank, 2009) or the entrepreneurial processes taking place in family and non-family firms (Hayton, Chandler and DeTienne, 2011).

Building on the above arguments, this chapter proposes that useful advances could result from turning this kind of question around. Instead of asking only whether different individuals use causation or effectuation logics to 'solve' entrepreneurial problems, this chapter encourages future research to also examine the extent to which the nature of different tasks and problems might 'change' individual strategies for solving them (see Sarasvathy, 2001: 252). Among possible questions building on this logic, one might ask: does the encounter of design or optimization tasks prime (or moderate) the reliance on particular strategies or heuristics? Do such encounters with different kinds of tasks and situations moderate the affective and/or cognitive forces and dynamics at play when facing such challenges? Or what sort of affective and/or cognitive resources might mediate entrepreneurs' adaptive responses to different tasks?

Variations in the Formal Properties of Entrepreneurs' Information Environments

In addition to examining the influence of formal properties of different tasks and problems on the affective and cognitive dynamics fostering entrepreneurship, this chapter encourages future research on the impact of entrepreneurs' information environments – and more specifically, of variations in the formal properties of such environments. Of all the information environments that entrepreneurs face, for instance, are there some that exhibit properties that have important consequences for affect and cognition? Building on a similar logic to the one discussed above, what are the effects of a particular information environment's level of complexity (or simplicity) on entrepreneurs' thinking and feelings with respects to their entrepreneurial efforts and actions? Does this place particular constraints on entrepreneurs' affect and cognition? Or does it foster insightful advances – such as innovative novel ideas, intense positive feelings, or particularly strong motivations? And what do the same distinctions about the complexity of relevant information environments imply for the thoughts and feelings of other relevant actors?

By extension, a number of other relevant properties of information environments could also be worth exploring. For instance, what are the effects of facing information environments that are particularly rich (or poor)? Likewise, and still focusing on the import of these differences for entrepreneurship's affective and cognitive dynamics, what difference does it make if entrepreneurs (or other stakeholders) face information environments that exhibit different levels of dynamism (or stability), different levels of emotional intensity, or different levels of uncertainty? Do such differences influence the way entrepreneurs think about their efforts, the mental representations they form, their motivation and persistence, the resources and strategies they leverage to face these environments, and so on (see also Carsrud and Brännback, Chapter 2, this volume)? In other words, to what extent do properties of the information environments influence entrepreneurship's affective and cognitive dynamics?

Variations in the Formal Properties of Entrepreneurs' Circumstances

Turning to the role of entrepreneurs' circumstances (and that of other stakeholders), does it make a difference whether entrepreneurial efforts stem from organizational, institutional or family circumstances that already exhibit rich (or poor) routines (such as information acquisition routines, scanning, integration, learning etc.) or identities (in the organizational identification sense; see Ashforth, Harrison and Corley, 2008;

Rousseau, 1998)? What are the properties of such routines (or identities) that are of particular import for affect and cognition – and why? When considering the potential influence on relevant affective and cognitive forces and dynamics, does it matter whether entrepreneurial actions take place in circumstances characterized by routines that are particularly rich or complex (versus poor or simple)? Why? Do these properties of routines place particular constraints or demands on entrepreneurs' affect and cognition, or do they encourage insightful advances? To what extent? Likewise, does it matter whether entrepreneurial actions take place in circumstances exhibiting strong or weak social identities, or whether social identities are broad or narrow, exclusive or inclusive, deep or shallow? If so, why? What are the implications of these properties of social identities for relevant affective and cognitive dynamics of entrepreneurial action?

Whether one talks about tasks individual entrepreneurs must perform, firm-level routines and identities or the broader levels of one's social institutions, cultural norms or industry- or market-level environments, useful advances could be realized by acknowledging explicitly that entrepreneurship's tasks, information environments and other circumstances have intrinsic affective and cognitive properties that are wholly distinct from individual-level forces, yet can exert a critical influence on individual thoughts, emotions and actions. Much like cognitive scientists have done when studying the import of well- and ill-structured problems, and, more recently, the import of optimization versus design problems, promising avenues for future research lie in directly exploring the implications that formal properties of entrepreneurship's task, circumstances and information environments might have for the affective and cognitive dimensions of entrepreneurial action. To help illustrate this more concretely, the next few sections discuss the case of extant research on the identification/imagination of entrepreneurial opportunities.

THE CASE OF OPPORTUNITY IDENTIFICATION RESEARCH

There has been considerable research on the imagination, identification, recognition, discovery, creation and/or development of entrepreneurial opportunities (Ireland, Reutzel and Webb, 2005; McMullen, Plummer and Acs, 2007; Shepherd and Grégoire, 2012; Short et al., 2010). By and large, this research focuses on entrepreneurs' efforts to 'find or imagine' promising business ideas for their new ventures (Davidsson, 2003; Dimov, 2011). As such, this area of research is concerned with entrepreneurial action's very early stages (Ardichvilli, Cardozo and Ray, 2003; McMullen and Dimov, 2013; McMullen and Shepherd, 2006).

Past research has seen multiple debates about the ontological nature of opportunities (objective or subjective), their origins (exogenous or endogenous to entrepreneurs' efforts), the consequent dynamics leading to entrepreneurial action (discovery versus creation; recognition versus development) and, ultimately, about the proper terminology for writing about the phenomena (cf. Alvarez and Barney, 2007, 2010; Davidsson, 2003, 2012; Hansen, Shrader and Monllor, 2011). Although these debates are important, they remain peripheral to this chapter's topic and objectives. For the present purposes, research on opportunity imagination/identification concerns the processes and dynamics by which enterprising individuals, teams and/or organizations form initial ideas and beliefs to possibly introduce new product(s)/service(s)/business model(s) in particular market(s) (see also Grégoire and Shepherd, in press; Shepherd and Grégoire, 2012; Shepherd, McMullen and Jennings, 2007).

As noted above, one of the most important challenges facing opportunity identification scholars is to disentangle the complex web of influences affecting individual efforts to imagine/identify alternate solutions to supply–demand circumstances, and to form preliminary beliefs about such opportunity ideas (see also Grégoire, Barr and Shepherd, 2010; Grégoire and Shepherd, 2012; and Grégoire, Shepherd and Lambert, 2010). To face this challenge, the present chapter advances the pertinence of simultaneously considering – that is, within the same empirical studies – the concurrent effects of different classes of relevant affective/cognitive forces. More specifically, this chapter encourages further investigations of the interplay of affective/cognitive forces characterized by different susceptibilities to exhibit important changes over short periods of time and/or that exert their influence from different planes (such as relevant properties of entrepreneurs' tasks, circumstances and information environments).

Interestingly, entrepreneurship research on opportunity identification is increasingly testing models that contrast the respective effects of multiple cognitive forces. Among many other examples, for instance, Tumasjan and Braun (2012) examined the independent and joint effects of individual differences in prevention and promotion foci, as well as of differences in creative and entrepreneurial self-efficacy. By examining these different forces alongside one another, within the same research design, these authors were able to unpack their respective effects on opportunity identification. More specifically, they not only reinforced prior findings for the positive role of promotion focus (cf. Bryant, 2007, 2009; Fitzsimmons and Douglas, 2011; Wu, McMullen, Neubert and Yi, 2008), but also pointed to the compensatory dynamics between regulatory focus and self-efficacy.

According to these authors' findings, individuals with low levels of creative/entrepreneurial self-efficacy will benefit from having high levels of promotion focus, just as individuals with low levels of promotion focus will benefit from having high levels of creative/entrepreneurial self-efficacy. In other words, their findings show that multiple cognitive resources can lead to similarly high 'performance' on opportunity identification tasks.

In similar fashion, Welpe and her colleagues (2012) examined the parallel effects of three emotional states: fear, joy and anger. Although their study focused on exploitation decisions (as opposed to the earlier stage of opportunity identification), it casts light on the interplay of each of these different emotions and individuals' evaluation of opportunities with different characteristics. In doing so, these authors are effectively unpacking relationships between individuals' emotions, individuals' evaluations of different opportunities, and the decisions individuals make with respect to such opportunities (see also Foo, 2011; Foo, Uy and Murnieks, in press; Wood, McKelvie and Haynie, 2014).

These efforts to consider multiple affective and/or cognitive forces within the same studies are extremely encouraging. Yet, for all their import, past studies have tended to limit their focus to forces that share similar susceptibilities to exhibit important changes over short periods of time. For instance, the affective and cognitive forces in Welpe and colleagues' (2012) study were primarily episodic in nature (e.g., short-term emotions and stimuli-driven impressions). By contrast, the forces in Tumasjan and Braun's (2012) study were more enduring in nature (e.g., individual tendencies toward promotion and prevention foci, experience-based dimensions of creative and entrepreneurial self-efficacy). The same limitations characterize some of the articles I co-authored. In Grégoire and Shepherd's (2012) study, for instance, we considered the parallel effects of several cognitive resources and predispositions (e.g., dimensions of human capital, creative self-efficacy, prior knowledge). Yet, by and large, these forces not only develop from an individual's past education and experiences, but are also unlikely to exhibit large variations over short periods of time.

To better understand the temporal dynamics surrounding opportunity identification (cf. McMullen and Dimov, 2013; Tumasjan, Welpe and Spörrle, 2012), this chapter proposes that future studies explicitly acknowledge the distinctions between enduring and episodic forces, and integrate these considerations into their models, research designs and results. For instance, one could study the parallel effects of episodic feelings versus those of enduring moods or emotional predispositions. Doing this within the same study, one could then examine the effects of possible interactions of episodic and enduring dimensions. In the same fashion, one could compare the effects of genetic bases of cognition (such as some dimensions of intelligence), acquired but enduring cognitive resources (such as prior knowledge in a domain or depth/breadth of experience), and more episodic aspects of cognition (such as short-term superficial impressions or perceptions). By examining the independent and joint effects of these forces, one would effectively disentangle the operation of different affective and cognitive dynamics across time. In other words, one would foster a better understanding of the temporal articulation of the affective and cognitive dynamics influencing opportunity identification/ imagination.

In parallel fashion, this chapter encourages the development of research programs that explicitly develop and test models of opportunity identification that contrasts, within the same studies, the role of affective and cognitive forces in exerting their effects from different planes of influence. In more specific terms, this chapter encourages future studies to examine in more detail the potential effects of opportunity identification's task(s), circumstance(s) and/or information environment(s) on the affective and cognitive dynamics that otherwise support entrepreneurs' efforts to imagine/identify new opportunities.

In Grégoire and Shepherd (2012), for instance, we advanced the notion that not all opportunity ideas were the same, and theorized that some of these differences have important implications for the beliefs individuals form about such ideas. More specifically, we examined the effects of variations on the superficial and structural similarities characterizing different opportunity ideas. We focused on these particular dimensions because the perception and interpretation of similarities had been shown to play an influential role in creative endeavors (Dahl and Moreau, 2002; Dunbar, 1993; Ward, 1995), but also (and perhaps more fundamentally) because the question of similarity was intrinsically relevant to a conception of entrepreneurial opportunities emphasizing hitherto untested matches between new means of supply and latent market demand (Casson, 1982; Hill and Birkinshaw, 2010; Venkataraman and Sarasvathy, 2001). Seen in this light, our focus on the similarity characteristics of different opportunity ideas acknowledges that a form of information that is particularly relevant to this task (i.e., information about opportunity ideas) concerns the matching of supply-side elements with demand-side considerations (see Grégoire, Barr and Shepherd, 2010). Consistent with this logic, we used an experimental approach to examine the extent to which such variations in the similarity characteristics of different opportunity ideas could influence the preliminary beliefs that experienced entrepreneurs formed about such ideas. At the same time, our research also examined the effects of possible interactions between these similarity variations and individual-level differences in prior knowledge and new startup intentions. Thus our research began to cast light on the interplay of cognitively relevant forces exerting their effects from different planes of influence – in this case, individual-level resources in knowledge and motivation and cognitively relevant variations between different opportunity ideas.

Several other possibilities might be worth exploring (Davidsson, 2012; Dimov, 2007a, 2011). Extant research has already begun considering the performance outcomes of exploiting opportunities with different characteristics (Åstebro and Dahlin, 2005; Åstebro and Michela, 2005; Dencker, Gruber and Shah, 2009; Samuelsson and Davidsson, 2009). Focusing more directly on opportunity identification, prior studies have considered the effects of variations in opportunities' level of tacitness (Smith, Matthews and Schenkel, 2009) or their novelty (Dahlqvist and Wiklund, 2012). From the standpoint of advancing research on the affective and cognitive dynamics leading to opportunity identification, however, the 'trick' is to focus on properties of entrepreneurial opportunities that are relevant for affect and cognition. In other words, one must leverage theory suggesting that particular properties of opportunity signals or ideas should have particular bearings on entrepreneurs' 'thinking and feelings' when they strive to imagine/identify potential opportunity ideas (cf. Davidsson, 2012; Grégoire, Barr and Shepherd, 2010; Grégoire and Shepherd, 2012; Grégoire and Williams, 2012).

By extension, and still building on the above, promising avenues for future research lie in examining the affective and cognitive consequences of entrepreneurs' circumstances – and particularly, in this case, with respect to their efforts to imagine/identify entrepreneurial opportunities. To the extent that different family circumstances/work environments exhibit particular modes of learning, information acquisition routines or social identities with different formal properties that are important for affective or cognitive dynamics, it becomes pertinent to examine if these properties influence individuals' efforts to think about new entrepreneurial ideas. One could then examine the potential interplay of such external variations, individual resources, abilities and other affective/cognitive predispositions.

BROADER IMPLICATIONS FOR FUTURE RESEARCH ON ENTREPRENEURIAL AFFECT AND COGNITION

In spite of a sustained interest in studying affect and cognition in entrepreneurship, extant research has struggled to untangle the complex web of affective and cognitive forces influencing entrepreneurial action, and to unpack these forces' influence across time and levels of analysis. To help address these challenges, this chapter has drawn attention to two important distinctions between different classes of relevant forces:

1. the enduring or episodic nature of affective/cognitive forces, that is, their high (or low) susceptibility to exhibit important changes over short periods of time; and
2. the anchoring of different affective/cognitive forces at different planes of influence – from individual-level mental resources, abilities and predispositions and shared mental representations to relevant formal properties of entrepreneurs' tasks, circumstances and information environments that also influence entrepreneurs' affect and cognition.

Table 6.3 synthesizes these notions by giving examples of different forces, in different classes.

By increasing familiarity with these notions and drawing attention to the import of examining the concurrent effects of different classes of affective and cognitive forces, I hope to help stimulate further developments of the conceptual and methodological articulations of entrepreneurship research on affect and cognition – and on the unfolding of affective and cognitive dynamics supporting entrepreneurial action.

Whereas the text above has explained the meaning of these distinctions for research on the imagination/identification of entrepreneurial opportunities, the same kind of conceptual articulations and research questions could be profitably leveraged in other areas of research concerned with the affective and cognitive dynamics fostering entrepreneurship. More directly, the ideas developed above encourage the articulation of studies examining the independent and joint effects of multiple dynamics of affective/cognitive interest. More importantly, perhaps, the above points to the pertinence of examining side by side the effects of forces that are enduring or episodic, of forces that exert their influence from different planes (such as the effects of individual-level resources, abilities and other predispositions versus the effects of formal properties of entrepreneurs' tasks, circumstances and information environments), and/or of the interactions of all these classes of forces.

Without necessarily using the same language and/or leveraging this kind of distinctions explicitly, a few studies have already begun investigating the unfolding of affective and/or cognitive forces across time and planes of influence. In some of his dissertation-related work, for instance, Jeffery McMullen had begun disentangling the effects of chronic regulatory focus (promotion and prevention) on entrepreneurial intentions, versus that of

Table 6.3 Examples of forces exhibiting different susceptibilities to change, at different planes of influence

	Very low susceptibility ← ←	**Susceptibility to exhibit important changes over short periods of time**	→ → Very high susceptibility
Planes of influence	*Enduring forces: unlikely to exhibit substantial changes over time*	*Forces that can exhibit some changes over time, but this requires substantial effort, particular conditions, etc.*	*Episodic forces: likely to exhibit substantial changes over short periods of time*
Shared mental representations (nations, societies, organizations, groups, teams)	Deep-seated cultural myths, social norms and prescriptions, etc. Representations of a firm's identity, mission, sense of purpose etc.	Broadly shared representations about a nation's ethos, its place in the world etc. Cognitive dimensions/representations of a firm's routines, identity, legitimacy, and/or reputation, etc.	Broadly shared 'national mood' *vis-à-vis* ongoing issues etc. Interpretations arising from ongoing events etc.
Individuals	Genetic bases of intelligence, of personality, of someone's cognitive abilities, affective predispositions etc.	Learned or acquired dimensions of personality, attitudes, affective predispositions etc. Knowledge acquired from education, learning, experience etc.	Stimuli-driven moods, episodic emotions, passing non-consequential thoughts, transient opinions etc.
Relevant properties of entrepreneurial action(s)' information environment(s)	Enduring forces explaining why an information environment is complex (e.g., number of different government regulations/ agencies)	Number of international competitors entering industry, and their varying rate of new product introduction, new strategies etc.	Episodic forces explaining why an information environment is complex (e.g., sudden crisis that reduces the number of regulations, competitors etc.)

Table 6.3 (continued)

	Very low susceptibility ← ←	**Susceptibility to exhibit important changes over short periods of time**	→ → Very high susceptibility
Planes of influence	*Enduring forces: unlikely to exhibit substantial changes over time*	*Forces that can exhibit some changes over time, but this requires substantial effort, particular conditions, etc.*	*Episodic forces: likely to exhibit substantial changes over short periods of time*
Relevant properties of entrepreneurial action(s)' circumstances	Enduring forces fostering a family firm's strong (weak) identity	Changes in the manner each successive generation identifies with the family firm (with time, but also as number of family members increases)	Episodic forces fostering a family business' reputation
Relevant properties of task at hand, problem, challenge	Formal nature of task or problem (well-/ill-structured; design vs optimization; higher/lower levels of complexity)	Rules and constraints that may change as problem/situation evolves	Salience/importance of particular information
Relevant information for this task, problem etc.	Supply-side elements	Importance of latent demand Level of competition	New information and other stimuli that arise and change rapidly: variations in price, market demand etc.

situation-induced manipulations of the same prevention and promotion foci (see McMullen and Shepherd, 2002). In doing so, these authors were effectively casting light on different drivers of regulatory focus (enduring versus episodic), and of drivers exerting their effects from different planes of influence (individual versus situational). In similar fashion, Daniel Forbes's (2005) study of the individual and organizational drivers of overconfidence offers another example of research designed to disentangle the effects of different classes of cognitive forces. Forbes's concluding observations are particularly telling in this regard:

> The findings that organizational decision practices and organizational ownership influence the degree to which entrepreneurs are overconfident support Baron's (1998) contentions that the phenomenon of entrepreneurial cognition is not entirely a function of individual differences filtered through a self-selection process and that entrepreneurial cognition is, at least partly, induced by the context within which entrepreneurs function. At the same time, however, the finding that founder-managers exhibit more overconfidence than do new-venture managers who did not found their firms implies that new-venture managers' cognition is not entirely a function of their present work context and that, as Busenitz and Barney (1997) argue, at least some of entrepreneurial cognition can be explained with reference to the cognitive profile that individuals bring with them to the task of new-venture creation. The finding that younger managers are more overconfident than are older ones reinforces this point but, at the same time, suggests that individuals' cognitive profiles change over time in ways that may affect the presence of entrepreneurial cognition. (Forbes, 2005: 637)

Integrating different classes of affective/cognitive forces into future research will demand increased attention to the theoretical articulation of different forces. In a recent study of the role of organizational identity in family business, for instance, Zellweger and colleagues (2013) argued for the role of (i) the family's visibility as the controlling coalition, (ii) the family's intentions to transfer the firm to future generations, and (iii) the firm's self-enhancement capability for the family as three important drivers of a family firm's organizational concern for its reputation (identity) – and for the latter's influence on the firm's pursuit of non-financial non-family-centered objectives. By and large, the above three drivers of organizational identity are relatively enduring: they have exerted their influence over many years, and are thus unlikely to exhibit sudden changes. By contrast, however, a family firm's reputation might be subject to a host of short-term episodic forces (from a host of poorly handled customer issues to successful product launches, employment growth or philanthropic engagements). To the extent that enduring forces govern family-business members' shared representations of organizational identity whereas more episodic forces might influence a firm's reputation, it thus becomes interesting to investigate the point at which a family business's pursuit of non-financial goals ceases to be influenced by enduring affective/cognitive forces, and starts being influenced by short-term drivers of its reputation. By considering the interplay of enduring and episodic forces affecting the relationships between affect/cognition and a family business's actions, one can better understand the boundary conditions of different phenomena. Conversely, such studies could also uncover interesting tradeoffs or synergies between different forces.

METHODOLOGICAL IMPLICATIONS

By expanding entrepreneurship research's conceptualization of affective and cognitive forces to examine the import of their enduring or episodic nature, as well as their anchoring on different planes of influence, future studies could make powerful advances in unpacking the complex interplay of affective and cognitive dynamics influencing entrepreneurial action. Importantly, however, realizing such promises not only calls for increased attention to the theoretical articulation of different classes of forces; it also calls for leveraging different methodological approaches. Two overarching strategies appear particularly important in this regard:

- The first strategy is to use data collection techniques that allow for documenting thoughts, emotions and other affective and cognitive dynamics directly, as their emergence and subsequent influence unfolds. (. . .)
- The second strategy is to leverage research designs that explicitly seek to disentangle the effects of different sources of variance – such as separating the effects of enduring individual differences from those of more episodic affective and cognitive variables influenced by contextual forces and/or one's encounter with particular information signals from the environment.
(See Grégoire and Lambert, 2014: 461 for elaborations on these two strategies)

Given this chapter's proposal to contrast the effects of affective/cognitive forces that are relatively stable and enduring with those of episodic forces that can exhibit wider variations over short periods of time, it becomes extremely useful to use data collections and techniques that allow for making multiple observations over time. Rather than simply leveraging longitudinal research designs, the first strategy above is about monitoring the emergence, development and/or transformation of thoughts, emotions and other affective/cognitive factors and dynamics over time – and directly as they exert their influence.

Among the many different data collection techniques conducive to this purpose, ethnographic approaches and verbal protocol techniques offer a number of interesting benefits. Ethnographic approaches have the advantage of leveraging 'field-level' observations from multiple informants to capture the intrinsic meaning that individuals attach to different issues and phenomena – and to do so directly as relevant events unfold. As such, the technique allows for documenting individuals' use of a variety of affective/cognitive resources and strategies, as well as their ongoing efforts to integrate affective/cognitive stimuli from their environment (see Johnstone, 2007; Ram, 1999). By asking participants to verbalize their thoughts, emotions, goals and whatever comes to their mind as they complete a particular task, verbal protocol techniques have the advantage of casting light on what individuals 'do' with the stimuli they encounter, as well as the strategies they use to leverage their affective and cognitive resources to solve the task at hand (see Ericsson and Simon, 1993; Grégoire, Barr and Shepherd, 2010).

In addition to ethnographic methods and verbal protocol techniques, a host of emerging technologies now allow for collecting data unobtrusively, as part of participants' regular day-to-day interactions. For instance, Foo and colleagues have introduced the use of event or experience sampling methodology (ESM) in entrepreneurship research with a series of studies conducted through short SMS surveys participants completed directly on their cellphones (Foo, Uy and Baron, 2009; Uy, Foo and Aguinis, 2010). Other potentially useful techniques include the use of socio-metric sensors or dynamic audience-response technologies (see Grégoire and Lambert, 2014). Along the same lines, useful insights could also be gained by leveraging priming methodologies and other implicit measures (Bargh and Chartrand, 2000; Fazio and Olson, 2003). Some scholars have also advanced the potential benefits of using physiological approaches more common in cognitive neuroscience studies, such as measures of cardiovascular activity, skin conductance, electro-encephalograms or brain-imaging techniques (see Scherbaum and Meade, 2013 for a recent review). From this chapter's perspective, the key advantage of these data collection techniques is that they allow for documenting the unfolding of thoughts, emotions and other affective and cognitive forces over time – as they exert their influence.

In parallel to using data collection techniques that allow for tracking the unfolding of thoughts and emotions across time, efforts to disentangle the complex web of affective and cognitive forces influencing entrepreneurial action could also benefit from leveraging more complex research designs. Within-subject and repeated-measures experiments appear particularly pertinent for such purposes (Grégoire and Lambert, 2014). By allowing the same respondents to offer multiple observations about different cases, scenarios or environmental conditions, such approaches can help isolate the effects of different sources of variance (Kirk, 2012; Shadish, Cook and Campbell, 2002) – such as the effects of enduring individual differences – from those of more episodic affective and cognitive forces prompted by sudden changes in circumstances or new information signals. In similar fashion, such within-subject designs allow for observing the same participants' 'behavior' *vis-à-vis* different stimuli. Thus it becomes easier to disentangle the influence of person-centered forces from that of affective or cognitive forces that exert their effect from different planes of influence – including effects that proceed from formal properties of the tasks, problems, circumstances and information environments that these same participants might encounter.

In practice, within-subject designs are perhaps most commonly associated with experimental methods. In the Grégoire and Shepherd (2012) study cited above, for instance, we employed a two-by-two within-subject design to assess the beliefs that experienced entrepreneurs formed about different opportunity cases. Each case consisted of a particular application of a new technology to a particular market. Consistent with our theoretical developments, we had manipulated the scenarios such that each participant saw four different technology–market combinations corresponding to four different patterns of superficial and structural similarities: low–low, low–high, high–low and high–high. By subjecting each participant to each pattern, we were able to 'separate' the effects of participants' cognitive resources in knowledge and motivation on the beliefs they formed about potential opportunities from the effects of variations in the similarity characteristics of opportunity signals (in our case, differences we had specified in the types and levels of similarity shared (or not) by technologies and markets). Because we had also collected additional data from our participants, we were also able to assess these effects while controlling for between-participant differences in other cognitive resources (such as participants' highest level of education obtained, their length of work experience, the number of new ventures they had started, and their innovation/creative self-efficacy). In this sense, the research design we leveraged allowed us to assess the effect of cognitive forces proceeding from different planes of influence.

Interestingly, this second methodological strategy of leveraging more complex research designs is not restricted to experiments: the same kind of within-subject manipulation underpins research using metric conjoint analysis (see Lohrke, Holloway and Woolley, 2010; Shepherd, 1999) and discrete-choice techniques (see Grégoire, Williams and Zacharakis, 2011; Hensher, Rose and Greene, 2005). More generally, the key methodological principle underpinning this second strategy of leveraging more complex research designs (Grégoire and Lambert, 2014) takes advantage of research situations where the same participants encounter different stimuli, tasks or problems, and/or end up making choices, decisions and actions in different circumstances and environments. To the extent that one can associate these different situations with theoretically valid variations in tasks, problems, circumstances and/or information environments that have some sort of affective or cognitive import, it thus becomes possible to assess the potential effects of such variations, and to separate them from the effects of individual-level affective or cognitive forces.

In order to help untangle the complex web of affective and cognitive dynamics influencing entrepreneurial action, this chapter has drawn attention to the theoretical import of distinguishing between different classes of affective/cognitive forces – depending on their enduring or episodic nature and their plane of influence. By increasing familiarity with these notions, illustrating the kind of research questions that could be pursued on these bases, and suggesting the kind of strategies that could be mobilized for doing so, I hope to have helped stimulate further developments of the conceptual and methodological articulations of entrepreneurship research on affect and cognition – and on the unfolding of affective and cognitive dynamics supporting entrepreneurial action.

REFERENCES

Acedo, F. and Jones, M. 2007. Speed of internationalization and entrepreneurial cognition: Insights and a comparison between international new ventures, exporters and domestic firms. *Journal of World Business*, **42**(3): 236–52.

Alvarez, S.A. and Barney, J.B. 2007. Discovery and creation: Alternative theories of entrepreneurial action. *Strategic Entrepreneurship Journal*, **1**(1–2): 11–26.

Alvarez, S.A. and Barney, J.B. 2010. Entrepreneurship and epistemology: The philosophical underpinnings of the study of entrepreneurial opportunities. *Academy of Management Annals*, **4**(1): 557–83.

Ardichvilli, A., Cardozo, R. and Ray, S. 2003. A theory of entrepreneurial opportunity identification and development. *Journal of Business Venturing*, **18**(1): 105–23.

Arora, P., Haynie, J.M. and Laurence, G.A. 2013. Counterfactual thinking and entrepreneurial self-efficacy: The moderating role of self-esteem and dispositional affect. *Entrepreneurship Theory and Practice*, **37**(2): 359–85.

Ashforth, B., Harrison, S. and Corley, K. 2008. Identification in organizations: An examination of four fundamental questions. *Journal of Management*, **34**(3): 325–74.

Ashkanasy, N.M. 2003. Emotions in organizations: A multilevel perspective. In F. Dansereau and F.J. Yammarino (eds), *Research in Multi-level Issues*, vol. 2: 9–54. Oxford, UK: Elsevier/JAI Press.

Åstebro, T.B. and Dahlin, K.B. 2005. Opportunity knocks. *Research Policy*, **34**(9): 1404–18.

Åstebro, T.B. and Michela, J.L. 2005. Predictors of the survival of innovations. *Journal of Product Innovation Management*, **22**(4): 322–35.

Atuahene-Gima, K. and Murray, J. 2007. Exploratory and exploitative learning in new product development: A social capital perspective on new technology ventures in China. *Journal of International Marketing*, **15**(2): 1–29.

Baer, M., Dirks, K.T. and Nickerson, J.A. 2013. Microfoundations of strategic problem formulation. *Strategic Management Journal*, **34**(2): 197–214.
Barnett, M.L. 2008. An attention-based view of real options reasoning. *Academy of Management Review*, **33**(3): 606–28.
Bargh, J.A. and Chartrand, T.L. 2000. The mind in the middle: A practical guide to priming and automaticity research. In H.T. Reis and C.M. Judd (eds), *Handbook of Research Methods in Social and Personality Psychology*: 253–85. Cambridge, UK: Cambridge University Press.
Baron, R.A. 1998. Cognitive mechanisms in entrepreneurship: Why and when enterpreneurs think differently than other people. *Journal of Business Venturing*, **13**(4): 275–94.
Baron, R.A. 2004. The cognitive perspective: A valuable tool for answering entrepreneurship's basic 'why' questions. *Journal of Business Venturing*, **19**(2): 221–39.
Baron, R.A. 2007. Behavioral and cognitive factors in entrepreneurship: Entrepreneurs as the active element in new venture creation. *Strategic Entrepreneurship Journal*, **1**(1–2): 167–82.
Baron, R.A. 2008. The role of affect in the entrepreneurial process. *Academy of Management Review*, **33**(2): 328–40.
Baron, R.A. and Ensley, M.D. 2006. Opportunity recognition as the detection of meaningful patterns: Evidence from comparisons of novice and experienced entrepreneurs. *Management Science*, **52**(9): 1331–44.
Bishop, R.C. 2005. Cognitive psychology: Hidden assumptions. In B. Slife, J. Reber and F. Richardson (eds), *Critical Thinking about Psychology: Hidden assumptions and plausible alternatives*: 151–70. Washington, DC: American Psychological Association.
Brigham, K., De Castro, J. and Shepherd, D. 2007. A person–organization fit model of owner-managers' cognitive style and organizational demands. *Entrepreneurship Theory and Practice*, **31**(1): 29–51.
Broadbent, D.E. 1958. *Perception and Communication*. New York: Pergamon.
Bryant, P. 2007. Self-regulation and decision heuristics in entrepreneurial opportunity evaluation and exploitation. *Management Decision*, **45**(4): 732–48.
Bryant, P. 2009. Self-regulation and moral awareness among entrepreneurs. *Journal of Business Venturing*, **24**(5): 505–18.
Busenitz, L.W. and Barney, J.B. 1997. Differences between entrepreneurs and managers in large organizations: Biases and heuristics in strategic decision-making, *Journal of Business Venturing*, **12**(1): 9–30.
Busenitz, L.W., Gomez, C. and Spencer, J.W. 2000. Country institutional profiles: Unlocking entrepreneurial phenomena. *Academy of Management Journal*, **43**(5): 994–1003.
Buttner, E.H. and Gryskiewicz, N. 1993. Entrepreneurs' problem-solving styles: An empirical study using the Kirton adaption/innovation theory. *Journal of Small Business Management*, **31**(1): 22–31.
Calori, R., Johnson, G. and Sarnin, P. 1994. CEOs' cognitive maps and the scope of the organization. *Strategic Management Journal*, **15**(6): 437–57.
Cardon, M.S., Foo, M.-D., Shepherd, D.A. and Wiklund, J. 2012. Exploring the heart: Entrepreneurial emotion is a hot topic. *Entrepreneurship Theory and Practice*, **36**(1): 1–10.
Cardon, M.S., Grégoire, D.A., Stevens, C.E. and Patel, P.C. 2013. Measuring entrepreneurial passion: Conceptual foundations and scale validation. *Journal of Business Venturing*, **28**(3): 373–96.
Cardon, M.S., Wincent, J., Singh, J. and Drnovšek, M. 2009. The nature and experience of entrepreneurial passion. *Academy of Management Review*, **34**(3): 511–32.
Carland, J.W., Hoy, F. and Carland, J.A. 1988. Who is an entrepreneur? is a question worth asking. *American Journal of Small Business*, **12**(4): 33–9.
Casson, M. 1982. *The Entrepreneur: An economic theory*. Totowa, NJ: Barnes & Noble.
Cattani, G. and Ferriani, S. 2008. A core/periphery perspective on individual creative performance: Social networks and cinematic achievements in the Hollywood film industry. *Organization Science*, **19**(6): 824–44.
Chase, W.G. and Simon, H.A. 1973. Perception in chess. *Cognitive Psychology*, **4**(1): 55–81.

Chataway, J. and Tait, J. 1993. Management of agriculture-related biotechnology: Constraints on innovation. *Technology Analysis & Strategic Management*, **5**(4): 345–67.

Chattopadhyay, P., Glick, W.H. and Huber, G.P. 2001. Organizational actions in response to threats and opportunities. *Academy of Management Journal*, **44**(5): 937–55.

Chen, C.C., Greene, P.G. and Crick, A. 1998. Does entrepreneurial self-efficacy distinguish entrepreneurs from managers? *Journal of Business Venturing*, **13**(4): 295–316.

Chetty, S., Eriksson, K. and Lindbergh, J. 2006. The effect of specificity of experience on a firm's perceived importance of institutional knowledge in an ongoing business. *Journal of International Business Studies*, **37**(5): 699–712.

Chi, M.T., Feltovich, P.J. and Glaser, R. 1981. Categorization and representation of physics problems by experts and novices. *Cognitive Science*, **5**(2): 121–52.

Choi, Y.R. and Shepherd, D.A. 2005. Stakeholder perceptions of age and other dimensions of newness. *Journal of Management*, **31**(4): 573–96.

Chowdhury, S. 2005. Demographic diversity for building an effective entrepreneurial team: Is it important? *Journal of Business Venturing*, **20**(6): 727.

Cialdini, R.B., Petty, R.E. and Cacioppo, J.T. 1981. Attitudes and attitude change. *Annual Review of Psychology*, **32**: 357–404.

Comegys, C. 1976. Cognitive dissonance and entrepreneurial behavior. *Journal of Small Business Management*, **14**(1): 1–6.

Corbett, A.C. 2007. Learning asymmetries and the discovery of entrepreneurial opportunities. *Journal of Business Venturing*, **22**(1): 97–118.

Cornelissen, J.P. and Clarke, J.S. 2010. Imaging and rationalizing opportunities: Inductive reasoning and the creation and justification of new ventures. *Academy of Management Review*, **35**(4): 539–57.

Cossette, P. and Audet, M. 1992. Mapping of an idiosyncratic schema. *Journal of Management Studies*, **29**(3): 325.

Dahl, D.W. and Moreau, C.P. 2002. The influence and value of analogical thinking during new product ideation. *Journal of Marketing Research*, **39**(1): 47–60.

Dahlqvist, J. and Wiklund, J. 2012. Measuring the market newness of new ventures. *Journal of Business Venturing*, **27**(2): 185–96.

Damasio, A.R. 1994. *Descartes' Error: Emotion, reason and the human brain*. New York: Grossett/Putnam.

Daniels, K., de Chernatony, L. and Johnson, G. 1995. Validating a method for mapping managers' mental models of competitive industry structures. *Human Relations*, **48**(9): 975–91.

Davidsson, P. 2003. The domain of entrepreneurship research: Some suggestions. In J.A. Katz and D.A. Shepherd (eds), *Cognitive Approaches to Entrepreneurship Research: Advances in entrepreneurship, firm emergence and growth*, vol. 6: 351–72. Oxford, UK: Elsevier/JAI Press.

Davidsson, P. 2006. Method challenges and opportunities in the psychological study of entrepreneurship. In J.R. Baum, M. Frese and R.A. Baron (eds), *The Psychology of Entrepreneurship*: 287–323. Mahwah, NJ: Lawrence Erlbaum Associates.

Davidsson, P. 2012. Entrepreneurial opportunity and the entrepreneurship nexus: A reconceptualization. Paper presented at the Academy of Management Conference, Boston, MA.

Deeds, D.L., Mang, P.Y. and Frandsen, M.L. 2004. The influence of firms' and industries' legitimacy on the flow of capital into high-technology ventures. *Strategic Organization*, **2**(1): 9–34.

Dencker, J.C., Gruber, M. and Shah, S.K. 2009. Individual and opportunity factors influencing job creation in new firms. *Academy of Management Journal*, **52**(6): 1125–47.

Dew, N., Read, S., Sarasvathy, S.D. and Wiltbank, R. 2009. Effectual versus predictive logics in entrepreneurial decision-making: Differences between experts and novices. *Journal of Business Venturing*, **24**(4): 287–309.

DiMaggio, P. 1997. Culture and cognition. *Annual Review of Sociology*, **23**(1): 263–87.

Dimov, D. 2007a. Beyond the single-person, single-insight attribution in understanding entrepreneurial opportunities. *Entrepreneurship Theory and Practice*, **31**(5): 713–31.

Dimov, D. 2007b. From opportunity insight to opportunity intention: The importance of person–situation learning match. *Entrepreneurship Theory and Practice*, **31**(4): 561–83.
Dimov, D. 2011. Grappling with the unbearable elusiveness of entrepreneurial opportunities. *Entrepreneurship Theory and Practice*, **35**(1): 57–81.
Dodd, S.D. 2002. Metaphors and meaning: A grounded cultural model of US entrepreneurship. *Journal of Business Venturing*, **17**(5): 519–35.
Dolan, R.J. 2002. Emotion, cognition, and behavior. *Science*, **298**(8): 1191–4.
Douglas, D. 2006. Intransivities of managerial decisions: A grounded theory case. *Management Decision*, **44**(2): 259–75.
Dow, D. and Karunaratna, A. 2006. Developing a multidimensional instrument to measure psychic distance stimuli. *Journal of International Business Studies*, **37**(5): 578–602.
Dunbar, K. 1993. Concept discovery in a scientific domain. *Cognitive Science*, **17**(3): 397–434.
Dutta, D.K. and Thornhill, S. 2008. The evolution of growth intentions: Toward a cognition-based model. *Journal of Business Venturing*, **23**(3): 307–32.
Eddelston, K. and Powell, G. 2008. The role of gender identity in explaining sex differences in business owners' career satisfier preferences. *Journal of Business Venturing*, **23**(2): 244–56.
Elsbach, K.D., Barr, P.S. and Hargadon, A.B. 2005. Identifying situated cognition in organizations. *Organization Science*, **16**(4): 422–33.
Ensley, M.D. and Hmieleski, K.M. 2005. A comparative study of new venture top management team composition, dynamics and performance between university-based and independent start-ups. *Research Policy*, **34**(7): 1091–105.
Ensley, M.D. and Pearce, C.L. 2001. Shared cognition in top management teams: Implications for new venture performance. *Journal of Organizational Behavior*, **22**(2): 145–60.
Ericsson, K.A., Charness, N., Feltovich, P. and Hoffman, R.R. 2006. *The Cambridge Handbook on Expertise and Expert Performance*. Cambridge, UK: Cambridge University Press.
Ericsson, K.A., and Simon, H.A. 1993. *Protocol analysis: Verbal reports as data*. Cambridge, MA: MIT Press.
Ethiraj, S.K. and Levinthal, D. 2004. Bounded rationality and the search for organizational architecture: An evolutionary perspective on the design of organizations and their evolvability. *Administrative Science Quarterly*, **49**(3): 404–37.
Fazio, R.H. and Olson, M.A. 2003. Implicit measures in social cognition research: Their meaning and use. *Annual Review of Psychology*, **54**(1): 297–327.
Fernandes, R. and Simon, H.A. 1999. A study of how individuals solve complex and ill-structured problems. *Policy Sciences*, **32**(3): 225–45.
Fisher, C.D. 2008. What if we took within-person performance variability seriously? *Industrial and Organizational Psychology: Perspectives on Science and Practice*, **1**(2): 185–9.
Fiske, S.T. and Taylor, S.E. 1991. *Social Cognition* (2nd edn). New York: McGraw-Hill.
Fitzsimmons, J.R. and Douglas, E.J. 2011. Interaction between feasibility and desirability in the formation of entrepreneurial intentions. *Journal of Business Venturing*, **26**(4): 431–40.
Foo, M.-D. 2011. Emotions and entrepreneurial opportunity evaluation. *Entrepreneurship Theory and Practice*, **35**(2): 375–93.
Foo, M.-D., Uy, M.A. and Baron, R.A. 2009. How do feelings influence effort? An empirical study of entrepreneurs' affect and venture effort. *Journal of Applied Psychology*, **94**(4): 1086–94.
Foo, M.-D., Uy, M.A. and Murnieks, C. Beyond affective valence: Untangling valence and activation influences on opportunity identification. *Entrepreneurship Theory and Practice*. In press.
Forbes, D.P. 1999. Cognitive approaches to new venture creation. *International Journal of Management Reviews*, **1**(4): 415–39.
Forbes, D.P. 2005. Are some entrepreneurs more overconfident than others? *Journal of Business Venturing*, **20**(5): 623–40.
Forgas, J.P. 2008. Affect and cognition. *Perspectives on Psychological Science*, **3**(2): 94–101.
Frese, M., Krauss, S., Keith, N., Escher, S., Grabarkiewicz, R., Luneng, S., Heers, C.,

Unger, J. and Friedich, C. 2007. Business owners' action planning and its relationship to business success in three African countries. *Journal of Applied Psychology*, **92**(6): 1481–98.

Funke, J. 1991. Solving complex problems: Exploration and control of complex systems. In R.J. Sternberg and P.A. Frensch (eds), *Complex Problem Solving: Principles and mechanisms*: 185–222. New York: Lawrence Erlbaum.

Funke, J. 2010. Complex problem solving: a case for complex cognition? *Cognitive Processing*, **11**(2): 133–42.

Funke, J. and Frensh, P.A. 2007. Complex problem solving: The European perspective – 10 years after. In D.H. Jonassen (ed.), *Learning to Solve Complex Scientific Problems*: 25–47. New York: Lawrence Erlbaum.

Gardner, H. 1985. *The Mind's New Science: A history of the cognitive revolution*. New York: Basic Books.

Gartner, W.B. 1988. 'Who is an entrepreneur?' is the wrong question. *American Journal of Small Business*, **12**(4): 11–32.

Gatewood, E.J., Shaver, K.G. and Gartner, W.B. 1995. A longitudinal study of cognitive factors influencing start-up behaviors and success at venture creation. *Journal of Business Venturing*, **10**(5): 371–91.

Gavetti, G. and Levinthal, D. 2000. Looking forward and looking backward: Cognitive and experiential search. *Administrative Science Quarterly*, **45**(1): 113–37.

Gazzaniga, M.S., Ivry, R.B. and Mangun, G.R. 2013. *Cognitive Neuroscience: The biology of the mind* (4th edn). New York: W.W. Norton & Company.

Gergen, K.J. 1985. The social constructionist movement in modern psychology. *American Psychologist*, **40**(3): 266–75.

Gobet, F. and Simon, H.A. 1996. Templates in chess memory: a mechanism for recalling several boards. *Cognitive Psychology*, **31**(1): 1–40.

Gobet, F. and Simon, H.A. 2000. Five seconds or sixty? Presentation time in expert memory. *Cognitive Science*, **24**(4): 651–82.

Grégoire, D.A., Barr P.S. and Shepherd, D.A. 2010. Cognitive processes of opportunity recognition: The role of structural alignment. *Organization Science*, **21**(2): 413–31.

Grégoire, D.A., Corbett, A. and McMullen, J. 2011. The cognitive perspective in entrepreneurship: An agenda for future research. *Journal of Management Studies*, **48**(6): 1443–77.

Grégoire, D.A. and Lambert, L.S. 2014. Getting inside entrepreneurs' heart and mind: Methods for advancing entrepreneurship research on affect and cognition. In E. Baker and F. Welter (eds), *Routledge Companion to Entrepreneurship*: 450–66. Milton Park, Abingdon: Routledge.

Grégoire, D.A. and Shepherd, D.A. 2012. Technology-market combinations and the identification of entrepreneurial opportunities: An investigation of the opportunity–individual nexus. *Academy of Management Journal*, **55**(4): 753–85.

Grégoire, D.A. and Shepherd, D.A. Opportunity identification. In D. Kuratko and M. Morris (eds), *Encyclopedia of Entrepreneurship*. Hoboken, NJ: Wiley. In press.

Grégoire, D.A., Shepherd, D.A., and Lambert, L.S. 2010. Measuring opportunity-recognition beliefs: Illustrating and validating an experimental approach. *Organizational Research Methods*, **13**(1): 114–45.

Grégoire, D.A. and Williams, D.W. 2012. Opportunity identification and structural alignment. In M. Marvel (ed.), *Encyclopedia of New Venture Management*: 356–61. Thousand Oaks, CA: Sage.

Grégoire, D.A., Williams, D.W. and Zacharakis, A. 2011. Entrepreneurs' decisions to internationalize early: A choice experiment. Paper presented at Babson College Entrepreneurship Research Conference, Syracuse, NY: June 9–11.

Håkanson, L. and Ambos, B. 2010. The antecedents of psychic distance. *Journal of International Management*, **16**(3): 195–210.

Halaby, C.N. 2003. Where job values come from: Family and schooling background, cognitive ability, and gender. *American Sociological Review*, **68**(2): 251–78.

Hansen, D.J., Shrader, R. and Monllor, J. 2011. Defragmenting definitions of entrepreneurial opportunity. *Journal of Small Business Management*, **49**(2): 283–304.

Harquail, C.V. and King, A.W. 2010. Construing organizational identity: The role of embodied cognition. *Organization Studies*, **31**(12): 1619–48.
Hatchuel, A. 2001. Towards design theory and expandable rationality: The unfinished program of Herbert Simon. *Journal of Management and Governance*, **5**(3): 260–73.
Haynie, J.M. and Shepherd, D.A. 2009. A measure of adaptive cognition for entrepreneurship research. *Entrepreneurship Theory and Practice*, **33**(3): 695–714.
Haynie, J.M., Shepherd, D.A. and Patzelt, H. 2012. Cognitive adaptability and an entrepreneurial task: The role of metacognitive ability and feedback. *Entrepreneurship Theory and Practice*, **36**(2): 237–65.
Hayton, J., Chandler, G.N. and DeTienne, D.R. 2011. Entrepreneurial opportunity identification and new firm development processes: a comparison of family and non-family new ventures. *International Journal of Entrepreneurship and Innovation Management*, **13**(1): 12–31.
Hensher, D.A., Rose, J.M. and Greene, W.H. 2005. *Applied Choice Analysis: A primer*. Cambridge, UK: Cambridge University Press.
Hill, S.A. and Birkinshaw, J.M. 2010. Idea sets: Conceptualizing and measuring a new unit of analysis in entrepreneurship research. *Organizational Research Methods*, **13**(1): 85–113.
Hmieleski, K.M. and Corbett, A. 2006. Proclivity for improvisation as a predictor of entrepreneurial intentions. *Journal of Small Business Management*, **44**(1): 45–63.
Huy, Q. 2012. Emotions and strategic organization: Opportunities for impactful research. *Strategic Organization*, **10**(3): 240–47.
Ireland, R.D., Reutzel, C.R. and Webb, J.W. 2005. Entrepreneurship research in AMJ: What has been published and what might the future hold? *Academy of Management Journal*, **48**(4): 556–64.
Isen, A.M. 1984. Toward understanding the role of affect in cognition. In R.S. Wyler and T.K. Srull (eds), *Handbook of Social Cognition*: 179–236. Hillsdale, NJ: Erlbaum.
Jackson, S.E. and Dutton, J.E. 1988. Discerning threats and opportunities. *Administrative Science Quarterly*, **33**(3): 370–87.
Jeffries, R., Polson, P.G., Razran, L. and Atwood, M.E. 1977. A process model for missionaries – cannibals and other river-crossing problems. *Cognitive Psychology*, **9**(4), 412–40.
Jenkins, M. and Johnson, G. 1997. Entrepreneurial intentions and outcomes: A comparative causal mapping study. *Journal of Management Studies*, **34**(6): 895–920.
Johnson-Laird, P. 2009. *How we Reason*. Oxford: Oxford University Press.
Johnstone, B.A. 2007. Ethnographic methods in entrepreneurship research. In H. Neergaard and J.P. Ulhøi (eds), *Handbook of Qualitative Research in Entrepreneurship*: 97–121. Cheltenham, UK and Northampton, MA: Edward Elgar.
Julian, S.D. and Ofori-Dankwa, J.C. 2008. Toward an integrative cartography of two strategic issue diagnosis frameworks. *Strategic Management Journal*, **29**(1): 93–114.
Katz, J.A. and Corbett, A.C. (eds). 2012. *Entrepreneurial Action: Advances in entrepreneurship, firm emergence and growth*, vol. 14. Bingley, UK: Emerald.
Katz, J.A. and Shepherd, D.A. 2003. Cognitive approaches to entrepreneurship research. In J.A. Katz and D.A. Shepherd (eds), *Cognitive Approaches to Entrepreneurship Research: Advances in entrepreneurship, firm emergence and growth*, vol. 6: 1–10. Oxford: Elsevier/JAI.
Keh, H.T., Foo, M.-D. and Lim, B.C. 2002. Opportunity evaluation under risky conditions: The cognitive processes of entrepreneurs. *Entrepreneurship Theory and Practice*, **27**(2): 125–48.
Kennedy, M. 2008. Getting counted: Markets, media, and reality. *American Sociological Review*, **73**(2): 270–95.
Kirk, R.E. 2012. *Experimental Design: Procedures for the behavioral sciences* (4th edn). Thousand Oaks, CA: Sage.
Kirzner, I.M. 1997. Entrepreneurial discovery and the competitive market process: An Austrian approach. *Journal of Economic Literature*, **35**(1): 60–85.
Krackhardt, D. 1990. Assessing the political landscape: Structure, cognition, and power in organizations. *Administrative Science Quarterly*, **35**(2): 342–69.

Krueger, N.F. Jr. 2003. The cognitive psychology of entrepreneurship. In Z.J. Acs and D.B. Audretsch (eds), *Handbook of Entrepreneurship Research: An interdisciplinary survey and introduction*: 105–40. New York: Springer.

Kuvaas, B. 2002. An exploration of two competing perspectives on informational contexts in top management strategic issue interpretation. *Journal of Management Studies*, **39**(7): 977–1001.

Lant, T.K. and Shapira, Z. 2001. Introduction: Foundations of research on cognition in organizations. In T.K. Lant and Z. Shapira (eds), *Organizational Cognition*: 1–14. Mahwah: NJ: Lawrence Erlbaum Associates.

Laureiro-Martinez, D., Brusoni, S., Canessa, N. and Zollo, M. Understanding the exploration–exploitation dilemma: An fMRI study of attention control and decision-making performance. *Strategic Management Journal*, DOI: 10.1002/smj.222. In press.

Lazarus, R.S. 1982. Thoughts on the relation between cognition and emotion. *American Psychologist*, **37**(9): 1019–24.

Lazarus, R.S. 1984. On the primacy of cognition. *American Psychologist*, **39**(2): 124–9.

Lazarus, R.S. 1991. Cognition and motivation in emotion. *American Psychologist*, **46**(4): 352–67.

Lohrke, F.T., Holloway, B.B. and Woolley, T.W. 2010. Conjoint analysis in entrepreneurship research: A review and research agenda. *Organizational Research Methods*, **13**(1): 16–30.

Luthans, F. and Ibrayeva, E. 2006. Entrepreneurial self-efficacy in Central Asian transition economies: Quantitative and qualitative analyses. *Journal of International Business Studies*, **37**(1): 92–110.

Maitlis, S. and Sonenshein, S. 2010. Sensemaking in crisis and change: Inspiration and insights from Weick (1988). *Journal of Management Studies*, **47**(3): 551–80.

Markman, G.D., Balkin, D.B. and Baron, R.A. 2002. Inventors and new venture formation: The effects of general self-efficacy and regretful thinking. *Entrepreneurship Theory and Practice*, **27**(2): 149–65.

McMullen, J.S. and Dimov, D. 2013. Time and the entrepreneurial journey: The problems and promise of studying entrepreneurship as a process. *Journal of Management Studies*, **50**(8): 1481–512.

McMullen, J.S., Plummer, L.A. and Acs, Z.J. 2007. What is an entrepreneurial opportunity? *Small Business Economics*, **28**(4): 273–83.

McMullen, J.S. and Shepherd, D.A. 2002. Regulatory focus and entrepreneurial intention: Action bias in the recognition and evaluation of opportunities. In W.D. Bygrave et al. (eds), *Frontiers of Entrepreneurship Research*: 61–73. Babson Park, MA: Babson College.

McMullen, J.S. and Shepherd, D.A. 2006. Entrepreneurial action and the role of uncertainty in the theory of the entrepreneur. *Academy of Management Review*, **31**(1): 132–52.

Milliken, F.J. 1990. Perceiving and interpreting environmental change: An examination of college administrators' interpretation of changing demographics. *Academy of Management Journal*, **33**(1): 42–63.

Mitchell, J.R. and Shepherd, D.A. 2010. To thine own self be true: Images of self, images of opportunity, and entrepreneurial action. *Journal of Business Venturing*, **25**(1): 138–54.

Mitchell, R.K., Busenitz, L.W., Bird, B., Gaglio, C.M., McMullen, J.S., Morse, E.A. and Smith, J.B. 2007. The central question in entrepreneurial cognition research 2007. *Entrepreneurship Theory and Practice*, **31**(1): 1–27.

Mitchell, R.K., Busenitz, L.W., Lant, T., McDougall, P.P., Morse, E.A. and Smith, J.B. 2002a. Toward a theory of entrepreneurial cognition: Rethinking the people side of entrepreneurship research. *Entrepreneurship Theory and Practice*, **27**(2): 93–104.

Mitchell, R.K., Busenitz, L.W., Lant, T., McDougall, P.P., Morse, E.A. and Smith, J.B. 2004. The distinctive and inclusive domain of entrepreneurial cognition research. *Entrepreneurship Theory and Practice*, **28**(6): 505–18.

Mitchell, R.K., Randolph-Seng, B. and Mitchell, J.R. 2011. Socially situated cognition: Imagining new opportunities for entrepreneurship research (Dialogue). *Academy of Management Review*, **36**(4): 774–8.

Mitchell, R.K., Smith, J.B., Morse, E.A., Seawright, K.W., Peredo, A.M. and McKenzie, B. 2002b. Are entrepreneurial cognitions universal? Assessing entrepreneurial cognitions across cultures. *Entrepreneurship Theory and Practice*, **26**(4): 9–32.

Mitchell, R.K., Smith, J.B., Seawright, K.W. and Morse, E.A. 2000. Cross-cultural cognitions and the venture creation decision. *Academy of Management Journal*, **43**(5): 974–93.

Morsella, E., Bargh, J.A. and Gollwitzer, P.M. (eds). 2009. *Oxford Handbook of Human Action*. Oxford: Oxford University Press.

Neergaard, H. and Ulhøi, J.P. (eds). 2007. *Handbook of Qualitative Research in Entrepreneurship*. Cheltenham, UK and Northampton, MA, USA: Edward Elgar.

Neisser, U. 1967. *Cognitive Psychology*. East Norwalk, CT: Appelton-Century-Crofts.

Ocasio, W. 1997. Towards an attention-based view of the firm. *Strategic Management Journal*, **18**(S1): 187–206.

Perren, L. and Jennings, P.L. 2005. Government discourses on entrepreneurship: Issues of legitimization, subjugation, and power. *Entrepreneurship Theory and Practice*, **29**(2): 173–84.

Pham, M.T. 2004. The logic of feeling. *Journal of Consumer Psychology*, **14**(4): 360–69.

Porac, J.F., Thomas, H. and Baden-Fuller, C. 1989. Competitive groups as cognitive communities: The case of Scottish knitwear manufacturers. *Journal of Management Studies*, **26**(4): 397–416.

Presutti, M., Boari, C. and Fratocchi, L. 2007. Knowledge acquisition and the foreign development of high-tech start-ups: A social capital approach. *International Business Review*, **16**(1): 23–46.

Radu, M. and Redien-Collot, R. 2008. The social representation of entrepreneurs in the French press: Desirable and feasible models? *International Small Business Journal*, **26**(3): 259–98.

Ram, M. 1999. Trading places: The ethnographic process in small firms' research. *Entrepreneurship and Regional Development*, **11**(2): 95–108.

Rao, H. 1994. The social construction of reputation: Certification contests, legitimation, and the survival of organizations in the American automobile industry: 1895–1912. *Strategic Management Journal*, **15**(S1): 29–44.

Reger, R.K. and Huff, A.S. 1993. Strategic groups: A cognitive perspective. *Strategic Management Journal*, **14**(2): 103–23.

Reger, R.K. and Palmer, T.B. 1996. Managerial categorization of competitors: Using old maps to navigate new environments. *Organization Science*, **7**(1): 22–39.

Reitman, W.R. 1964. Heuristic decision procedures, open constraints, and the structure of ill-defined problems. In M.W. Shelley and G.L. Bryan (eds), *Human Judgments and Optimality*: 282–315. New York: Wiley.

Reitman, W.R. 1965. *Cognition and Thought*. New York: Wiley.

Robins, R.W., Gosling, S.D. and Craik, K.H. 1999. An empirical analysis of trends in psychology. *American Psychologist*, **54**(2): 117–28.

Rousseau, D.M. 1998. Why workers still identify with organizations. *Journal of Organizational Behavior*, **19**(3): 217–33.

Samuelsson, M. and Davidsson, P. 2009. Does venture opportunity variation matter? Investigating systematic process differences between innovative and imitative new ventures. *Small Business Economics*, **33**(2): 229–55.

Sarasvathy, D.K., Simon, H.A. and Lave, L. 1998. Perceiving and managing business risks: Differences between entrepreneurs and bankers. *Journal of Economic Behavior & Organization*, **33**(2): 207–25.

Sarasvathy, S.D. 2001. Causation and effectuation: Toward a theoretical shift from economic inevitability to entrepreneurial contingency. *Academy of Management Review*, **26**(2): 243–63.

Sarasvathy, S.D. 2003. Entrepreneurship as a science of the artificial. *Journal of Economic Psychology*, **24**(2): 203–20.

Sarasvathy, S.D. 2004. Making it happen: Beyond theories of the firm to theories of firm design. *Entrepreneurship Theory and Practice*, **28**(6): 519–531.
Sarasvathy, S.D. 2008. *Effectuation: Elements of entrepreneurial expertise*. Cheltenham, UK and Northampton, MA, USA: Edward Elgar Publishing.
Scherbaum, C.A. and Meade, A.W. 2013. New directions for measurement in management research. *International Journal of Management Reviews*, **15**(2): 132–48.
Schraw, G., Dunkle, M.E. and Bendixen, L.D. 1995. Cognitive processes in well-defined and ill-defined problem solving. *Applied Cognitive Psychology*, **9**(6): 523–38.
Shadish, W.R., Cook, T.D. and Campbell, D.T. 2002. *Experimental and Quasi-experimental Designs for Generalized Causal Inference*. Belmont, CA: Wadsworth, Cengage Learning.
Shane, S. 2000. Prior knowledge and the discovery of entrepreneurial opportunities. *Organization Science*, **11**(4): 448–69.
Shane, S. and Venkataraman, S. 2000. The promise of entrepreneurship as a field of research. *Academy of Management Review*, **25**(1): 217–26.
Shepherd, D.A. 1999. Venture capitalists' assessment of new venture survival, *Management Science*, **45**(5): 621–32.
Shepherd, D.A. and DeTienne, D.R. 2005. Prior knowledge, potential financial reward, and opportunity identification. *Entrepreneurship Theory and Practice*, **29**(1): 91–112.
Shepherd, D.A. and Grégoire, D.A. 2012. An introduction to state-of-the-art research on entrepreneurial opportunity. In D.A. Shepherd and D.A. Grégoire (eds), *Entrepreneurial Opportunity*: xv–xxvi. International Library of Entrepreneurship. Cheltenham, UK and Northampton, MA, USA: Edward Elgar.
Shepherd, D.A., McMullen, J.S. and Jennings, P.D. 2007. The formation of opportunity beliefs: Overcoming ignorance and reducing doubt. *Strategic Entrepreneurship Journal*, **1**(1–2): 75–96.
Short, J.C., Ketchen, D.J., Shook, C.L. and Ireland, R.D. 2010. The concept of 'opportunity' in entrepreneurship research: Past accomplishments and future challenges. *Journal of Management*, **36**(1): 40–65.
Simon, H.A. 1973. The structure of ill structured problems. *Artificial Intelligence*, **4**(3): 181–201.
Simon, H.A. 1996. *The Sciences of the Artificial* (3rd edn). Cambridge, MA: MIT Press.
Simon, H.A. and Newell, A. 1964. Information processing in computer and man. *American Scientist*, **52**(3): 281–300.
Simon, M. and Houghton, S.M. 2000. Cognitive biases, risk perception, and venture formation: How individuals decide to start companies. *Journal of Business Venturing*, **15**(2): 113–134.
Sine, W.D., Haveman, H.A. and Tolbert, P.S. 2005. Risky business? Entrepreneurship in the new independent-power sector. *Administrative Science Quarterly*, **50**(2): 200–232.
Smith, B.R., Matthews, C.H. and Schenkel, M.T. 2009. Differences in entrepreneurial opportunities: The role of tacitness and codification in opportunity identification. *Journal of Small Business Management*, **47**(1): 38–57.
Smith, E.E. and Osherson, D.N. (eds). 1995. *Thinking: An invitation to cognitive science*, vol. 3 (2nd edn). Cambridge, MA: MIT Press.
Smith, E.R. 2000. Social cognition. In A.E. Kadzin (ed.), *Encyclopedia of Psychology*: 324–9. Oxford: Oxford University Press.
Spencer, J.W. and Gomez, C. 2004. The relationship among national institutional structures, economic factors, and domestic entrepreneurial activity: A multicountry study. *Journal of Business Research*, **57**(10): 1098–107.
Sternberg, R.J. and Ben-Zeev, T. 2001. *Complex Cognition: The psychology of human thought*. Oxford: Oxford University Press.
Thagard, P. 2005. *Mind: Introduction to cognitive science* (2nd edn). Cambridge, MA: MIT Press.
Thagard, P. 2006. *Hot Thought: Mechanisms and applications of emotional cognition*. Cambridge, MA: MIT Press.

Thomas, J.A. and McDaniel, R.R. 1990. Interpreting strategic issues: Effects of strategy and the information-processing structure of top-management teams. *Academy of Management Journal*, **33**(2): 286–306.

Tumasjan, A. and Braun, R. 2012. In the eye of the beholder: How regulatory focus and self-efficacy interact in influencing opportunity recognition. *Journal of Business Venturing*, **27**(6): 622–36.

Tumasjan, A., Welpe, I. and Spörrle, M. 2013. Easy now, desirable later: The moderating role of temporal distance in opportunity evaluation and exploitation. *Entrepreneurship Theory and Practice*, **37**(4): 859–88.

Turner, M. 2001. *Cognitive Dimensions of Social Science*. Oxford: Oxford University Press.

Uy, M.A., Foo, M.-D. and Aguinis, H. 2010. Using experience sampling methodology to advance entrepreneurship theory and practice. *Organizational Research Methods*, **13**(1): 13–54.

Venkataraman, S. and Sarasvathy, S.D. 2001. Strategy and entrepreneurship: Outlines of an untold story. In M.A. Hitt, E. Freeman and J. Harrison (eds), *The Blackwell Handbook of Strategic Management*: 650–68. Oxford: Blackwell Publishers.

Voss, J.F. and Post, T.A. 1988. On the solving of ill-structured problems. In M.T.H. Chi, R. Glaser and M.J. Farr (eds), *The Nature of Expertise*: 261–85. Hillsdale, NJ: Lawrence Erlbaum Associates.

Wang, Y. and Chiew, V. 2010. On the cognitive process of human problem solving. *Cognitive Systems Research*, **11**(1): 81–92.

Ward, T.B. 1995. What's old about new ideas? In S.M. Smith, T.B. Ward and R.A. Finke (eds), *The Creative Cognition Approach*: 157–78. Cambridge, MA: MIT Press.

Weick, K.E., Sutcliffe, K.M. and Obstfeld, D. 2005. Organizing and the process of sensemaking. *Organization Science*, **16**(4): 409–21.

Welpe, I.M., Spörrle, M., Grichnik, D., Michl, T. and Audretsch, D.B. 2012. Emotions and opportunities: The interplay of opportunity evaluation, fear, joy, and anger as antecedent of entrepreneurial exploitation. *Entrepreneurship Theory and Practice*, **36**(1): 69–96.

West, G.P. III. 2007. Collective cognition: When entrepreneurial teams, not individuals, make decisions. *Entrepreneurship Theory and Practice*, **31**(1): 77–102.

Westhead, P., Ucbasaran, D. and Wright, M. 2005. Experience and cognition. *International Small Business Journal*, **23**(1): 72–98.

Wickman, P.A. 2003. The representativeness heuristic in judgements involving entrepreneurial success and failure. *Management Decision*, **41**(2): 156–67.

Wood, M.S., McKelvie, A. and Haynie, J.M. 2014. Making it personal: Opportunity individuation and the shaping of opportunity beliefs. *Journal of Business Venturing*, **29**(2): 252–72.

Wood, M.S., Williams, D.W. and Grégoire, D.A. 2012. The road to riches? A model of the cognitive processes and inflection points underpinning entrepreneurial action. In J.A. Katz and A.C. Corbett (eds), *Entrepreneurial Action: Advances in entrepreneurship, firm emergence and growth*, vol. 14: 207–52. Bingley, UK: Emerald.

Wu, C., McMullen, J.S., Neubert, M.J. and Yi, X. 2008. The influence of leader regulatory focus on employee creativity. *Journal of Business Venturing*, **23**(5): 587–602.

Yoo, Y. and Kanawattanachai, P. 2001. Developments of transactive memory systems and collective mind in virtual teams. *International Journal of Organizational Analysis*, **9**(2): 187–208.

Zajonc, R.B. 1980. Feelings and thinking: Preferences need no inferences. *American Psychologist*, **35**(2): 151–75.

Zajonc, R.B. 1984. On the primacy of affect. *American Psychologist*, **39**(2): 117–23.

Zellweger, T.M., Nason, R.S., Nordqvist, M. and Brush, C.G. 2013. Why do family firms strive for nonfinancial goals? An organizational identity perspective. *Entrepreneurship Theory and Practice*, **37**(2): 229–48.

7. Cultural context, passion and self-efficacy: do entrepreneurs operate on different 'planets'?

Mateja Drnovšek, Alenka Slavec and Melissa S. Cardon

ABSTRACT

We explore how national culture impacts emotional and cognitive processes of entrepreneurs. In particular, we extend the recent conceptual work on entrepreneurial passion by linking it to the theoretical lens on situated emotions. We also extend current work on entrepreneurial cognitions by suggesting that cognition is distributed among group processes and the environment. We propose a culturally situated model of entrepreneurial emotions (e.g., entrepreneurial passion) and cognitions (e.g., self-efficacy), and explore how these impact venture performance. We test the model in two different cultures – Slovenia and the USA – to show that the strength of feelings of entrepreneurial passion varies among entrepreneurs from different environments. In addition, we find that there are cultural differences in the manifestation of entrepreneurial self-efficacy and different domains of entrepreneurial passion, as well as the relationship between them and firm innovation and market share. We discuss implications for research and practice.

CULTURE, EMOTIONS AND COGNITIONS OF ENTREPRENEURS

Entrepreneurship scholars and practitioners have been recently fascinated by a specific emotion – entrepreneurial passion, the 'most observed phenomenon in the entrepreneurial process' (Smilor, 1997: 342). Entrepreneurial passion (EP) involves consciously accessible, intense positive feelings that result from engagement in activities that have identity meaning to an entrepreneur (Cardon, Wincent, Singh and Drnovšek, 2009). As an intense positive emotion, passion contributes to a better understanding of entrepreneurial behaviors and outcomes (Cardon et al.,

2009; Foo, Uy and Baron, 2009). Emotions such as passion often work alongside cognitions, which have long been acknowledged to provide theoretically rigorous and testable explanations of how entrepreneurs think and why they do some of the things they do (Mitchell et al., 2002), as well as how such behaviors lead to specific outcomes (see also Baron, Chapter 1, this volume; Bird, Chapter 3, this volume).

Among entrepreneurial cognitions, entrepreneurial self-efficacy (ESE) has been seen as one of the most important cognitions impacting entrepreneurial behaviors and outcomes. ESE refers to the extent to which an entrepreneur believes that he/she can complete various roles and tasks of entrepreneurship (Boyd and Vozikis, 1994; Chen, Greene and Crick, 1998). If entrepreneurial emotions, specifically EP, are seen in the heart of the entrepreneurial process (Cardon, Foo, Shepherd and Wiklund, 2012), entrepreneurial cognitions, specifically entrepreneurial self-efficacy, may be seen as a 'willpower' of entrepreneurship since it has been shown to provide entrepreneurs with the strength necessary to persist despite the obstacles they encounter on their entrepreneurial journeys (Chen et al., 1998).

The context within which entrepreneurial cognitions and emotions interact while entrepreneurs pursue their goals is also important (Izard, 1978). Because entrepreneurs act and make decisions in response to opportunity under uncertainty, and evaluate the desirability and feasibility of an opportunity and exploit it (McMullen and Shepherd, 2006; Shane and Venkataraman, 2000) within a specific context, some variations in this dynamic process can be ascribed to differences in national contexts within which the entrepreneurial process emerges (Aldrich et al., 1991). Indeed, an aspect of society that critically shapes the environment for organizations has been identified in the literature – culture, including its norms and values. Because national cultures have different values and norms about entrepreneurship, entrepreneurial endeavors and performance are viewed in different ways (Ekman, 1992). By exploring the role of national culture in the entrepreneurial process, we can better understand how culture influences motives, values and beliefs of entrepreneurs (Frijda, 1987). Although extant culture-related research in entrepreneurship has so far provided tremendous insights (e.g., Aldrich et al., 1991; Hayton, George and Zahra, 2002), there are several theoretical and empirical gaps in our understanding about the impact that national culture might have on the emergence and effects of entrepreneurial cognitions and emotions.

One unanswered question is how culture-related factors influence the emergence and development of emotional and cognitive processes of entrepreneurs. To our knowledge, the entire published set of empirical data on entrepreneurial passion has been collected in a North American setting, with one exception, which was collected in Germany (Breugst, Domurath, Patzelt and Klaukien, 2012). Similarly, despite a rich cumulative body of research on entrepreneurial cognitions, most of the empirical studies involving entrepreneurial cognitions have been conducted in Anglo-Saxon environments. In this chapter we seek to address these limitations and to make the following contributions. First, we explore how entrepreneurial passion develops across cultures and whether entrepreneurs in different environments tend to experience passion differently. Central to the nature of entrepreneurial passion is an entrepreneur's self-identity, namely the role identity of inventor, founder and developer. Given that all identities begin as social roles (Stryker and Burke, 2000), their emergence is socially situated. Furthermore, individuals learn what it means to be an entrepreneur by first viewing how society interprets that particular social role (Murnieks, Mosakowski and Cardon, 2014: 7). By exploring in Hypothesis 1 how the strength and focus of entrepreneurial passion differs among countries, we offer some preliminary explanations for the context-related variations in experiencing different types of entrepreneurial passion. We specifically look at the extent to which culture influences the amount of passion that is experienced, as well as the focus of that passion (on the inventor, founder and/or developer roles).

Our second contribution aims to explore how a specific emotion and a specific cognition relate to venture performance in two different cultural contexts. Recently, several authors (e.g., Mitchell, Randolph-Seng and Mitchell, 2011) suggested the need to explore how cognitive, motivational and emotional processes of entrepreneurs interact within specific social situations and with specific social actors. The emerging theory on situated emotions in psychology (Keltner and Haidt, 1999; Parkinson, Fischer and Manstead, 2005) suggests that environment plays an active role in structuring and enabling emotional engagements, and is in turn influenced by the unfolding of emotions. We integrate this theoretical perspective with existing conceptual work on entrepreneurial passion to develop a culturally situated model of entrepreneurial cognitions (e.g., entrepreneurial self-efficacy) and entrepreneurial emotions (e.g., entrepreneurial passion) in Hypotheses 2 and 3 and explore how those in turn affect venture performance on a large data set of practicing entrepreneurs in two different cultural environments. We also follow the idea of distributed cognition (Smith and Semin, 2004), suggesting that cognition is distributed among group processes and the environment so that 'the social context of speaking and interactions with others affect the construction of meaning about a new venture' (Cornelissen and Clarke, 2010: 542). For example, Bandura (1994) suggests that an important means of strengthening one's self-efficacy is through vicarious learning and social persuasion. We look at samples of entrepreneurs in the countries of Slovenia and the USA, and specifically at firm-level market share and innovation associated with entrepreneurial passion and entrepreneurial self-efficacy.

The remainder of the chapter is organized as follows. We begin with a brief review of theories of situated emotions and cognitions that we use as foundations for our arguments about the role of culture in impacting individuals' emotions and cognitions. Specifically, we first investigate differences in experienced entrepreneurial passion in two different cultural contexts, as previous studies did not investigate in detail such cultural distinctions of emotions. We then focus on the relationship between entrepreneurial passion, entrepreneurial self-efficacy and venture performance, and develop propositions supporting these links. The rest of the chapter summarizes our methods, data from samples in Slovenia and the USA, and results. We conclude with a discussion of the study and its implications for future research.

THEORY AND HYPOTHESES DEVELOPMENT

Emotions and Cognitions: A Situated Perspective

The situated perspective of emotions maintains that individuals' emotions are scaffolded by the environment (Griffiths and Scarantino, 2005; Parkinson et al., 2005): emotions are designed to function in a social context, and the environment in which they unfold influences them and is influenced by them. Although emotions are often privately experienced, they are most often expressed and often observed by others, who in turn respond to them (Van Kleef, 2010). By being (visually) expressed, emotions serve social functions, have interpersonal consequences (Van Kleef, 2010), and are shaped by cultural practices and symbol systems (Hareli and Parkinson, 2008).

A specific social role that emotions play is regulation of social interactions, influencing not only the behavior of the 'emoter' but also that of others in the social environment (Elfenbein and Ambady, 2002; Frijda, 2009; Parkinson et al., 2005) because emotions convey information about the situation of the individual and his/her behavioral intentions (Ekman, Friesen and Ellsworth, 1972; Izard, 1978). In essence, emotions 'mobilize the organism to deal quickly with important interpersonal encounters' (Ekman, 1992: 171). Emotions are experienced in response to significant events (Frijda, 1987) and provide a wealth of information not just to oneself (Frijda, 1987) but also to one's social environment (Keltner and Haidt, 1999). That said, the experience of an emotion depends on the current construal of a social situation (Frijda, 2009), and in turn plays a pivotal role in transforming the very nature of the emerging social situation and instigating the person to engage in certain actions (Markus and Kitayama, 1991). Emotional experience systematically varies with the construal of the self (Markus and Kitayama, 1991), and, based on the cultural impact, two broad concepts of the self emerge – independent and interdependent. The independent view is most clearly exemplified in a sizable segment of American culture as well as in many West European cultures. The independent view believes in the inherent separateness of an individual from others, and the behavior of that individual is meaningful primarily by reference to his/her own internal repertoire of thoughts, feelings and actions. In the independent construal of the self, the person is viewed as 'a bounded, unique, more or less integrated motivational and cognitive universe and set contrastively both against other such wholes and against a social and natural background' (Geertz, 1975: 48). The interdependent self is exemplified in Asian cultures, but can also be found in African, Latin American and Southern European cultures (Markus and Kitayama, 1991), including Slovenia, with a history of a socialist political regime strongly rooted in an interdependent view of society. The interdependent view sees an individual as an element of an encompassing social relationship and recognizes that his/her behavior is contingent on what the actor perceives to be the thoughts, feelings and actions of others in the relationship (DeVos, 1985). Overall, the situated perspective on emotions provides conceptual grounds for examining the cultural embededdness of entrepreneurial passion, which we discuss below.

How is the experience of entrepreneurial passion associated with the prior cognitive processes of an entrepreneur? The general hypothesis in psychology is that affect and cognition are separate and partially independent systems that most of the time function conjointly, although there are instances when affect can be generated without a prior cognitive process (Zajonc, 1984). The debate on interdependence versus independence of both systems is conditional on how emotions and cognitions are defined. For the purposes of our analysis we contend that both systems interact in regulating entrepreneurial efforts (Cardon et al., 2012). Supported by neuroscience advances and findings, the emerging frame views 'hot' passions and 'cold' reason as coherent and connected systems that work together to dynamically regulate human behavior toward desired goals (Damasio, 1994; Dolan, 2002; Pham, 2004; see also Baucus, Baucus and Mitchell, Chapter 8, this volume; McMullen, Wood and Palich, Chapter 9, this volume). The existing conceptual work on entrepreneurial passion suggests that it is a central entrepreneurial emotion because entrepreneurs are not only responsive to 'cold' cognitions, but are actively responsive to 'hot' emotions (e.g., entrepreneurial passion). Finally, even economists are beginning to recognize that the 'picture of human behavior that emerges is one that is more symmetric in its treatment of thoughts and feelings' (Romer, 2000: 439). These developments have shifted the debate from asking how 'hot' passion limits 'cold' cognitions to understanding how the hot and cold systems cooperate in directing entrepreneurial pursuits. We discuss this in terms of entrepreneurial passion and self-efficacy below.

Culture and Entrepreneurial Passion

Cultures are sometimes equated with nations or societies, but more often a culture is restricted to a community of shared meanings, such as for example in D'Andrade's (1984: 116) treatment of culture as 'learned systems of meaning, communicated by means of natural language and other symbol systems, having representational, directive, and affective functions, and capable of creating cultural entities and particular senses of reality' (see also Clarke and Cornelissen, Chapter 11, this volume). Cultural norms and values influence cognition, emotions and motivation (Markus and Kitayama, 1991), while government actions and political events create new institutional structures for entrepreneurial action, encouraging some activities and thwarting others (Ekman et al., 1972). Culture not only creates the social world; it guides people in the affective reactions needed to function in that world.

Culture shapes many aspects of emotion processes, including experience, expression, regulation and recognition (Elfenbein and Ambady, 2002; Markus and Kitayama, 1991). Some authors go so far as to propose that emotions (Lutz, 1988: 5) can be viewed as cultural and interpersonal products of naming, justifying and persuading by people in relationship to each other. Emotional meaning is then a social rather than an individual achievement, an emergent product of social life. There are culture-specific valuations of the experience and expression of emotion, for example in relation to gender, age and social status (Lutz, 1988).

Drawing from Hofstede's cultural dimension taxonomy (Hofstede and Hofstede, 2001), there is a frequent observation in the entrepreneurship literature that the entrepreneurial process is facilitated by cultures that are high in individualism, low in uncertainty avoidance, low in power distance and high in masculinity (e.g., Hayton et al., 2002). For instance, Shane (1992) shows that uncertainty avoidance is negatively associated with innovation in both time periods of his longitudinal study, whereas individualism is positively associated with innovation only in one time period. Thomas and Mueller (2000) find that innovativeness and internal locus of control are prevalent in cultures high in individualism and low in uncertainty avoidance, which leads them to suggest that cultures high in individualism and uncertainty avoidance are in general supportive of entrepreneurship. Baker and colleagues (Baker, Gedajlovic and Lubatkin, 2005) find that in countries where regional agglomerations are better developed, the likelihood that individuals who discover good entrepreneurial opportunities achieve success through *de novo* start-ups increases. Finally, Mitchell and colleagues (2007) find that cognitive scripts (e.g., tolerance for commitment and motivation) associated with venture creation decisions vary across cultures and are associated with cultural values of individualism and power distance.

These literatures mostly emphasize intrapersonal aspects of emotions and largely sidestep the broader contextual meaning of emotional experience. Yet, as we have shown earlier, there is a strong theoretical notion that emotions have an important interpersonal function by structuring social relations (Frijda and Mesquita, 1994; Keltner and Haidt, 1999) by providing information to interaction partners about our feelings, intentions, motives and social goals. Hence, in contrast to prior research on culture and entrepreneurship, the focus of this study is on an interpersonal perspective of the entrepreneurial process. Indeed, entrepreneurial passion has been shown to have contagion effects. For example, when entrepreneurs are passionate about developing their firms, this passion also manifests itself behaviorally so that other stakeholders of the firm can readily see it (Chen, Yao and Kotha, 2009). It is manifested in animated facial expressions, energetic body movements, and elevated pitch and speed of voice (Chen et al., 2009), all of which can be perceived by others, such as potential investors (Mitteness, Sudek and Cardon, 2012). Within an organization, employees who perceive a firm's entrepreneur to be passionate about inventing or developing are more committed to the organization (Breugst, Domurath, Patzelt and Klaukien, 2012). This occurs because an entrepreneur's passion has a contagion effect (Cardon, 2008) whereby employees are infected with the positive feelings of passion so that they also experience positive emotion, which leads them to be more committed (Breugst et al., 2012).

We argue that the link between culture and entrepreneurial passion is maintained through the role-identity component of passion. By viewing entrepreneurial passion as a specific emotional experience linked to an entrepreneur's self-identity, prior work has identified three meaningful entrepreneurial role identities of entrepreneurs (Cardon et al., 2009) – inventor, founder and developer – that are associated with three domains of entrepreneurial passions – passion for inventing, passion for founding and passion for developing. Although entrepreneurial passion is an enduring stable emotion (Cardon, Grégoire, Stevens and Patel, 2013), it can change gradually as the entrepreneur's identity shifts (Cardon et al., 2009). Entrepreneurs can thus experience different types of entrepreneurial passion during a course of events, and can therefore experience different impacts on their success as entrepreneurs. Different types of entrepreneurial passion are associated with three meaningful role identities that entrepreneurs undertake (Gartner, 1985): (1) an inventor identity that involves activities related to seeking out new ideas, tinkering with new product development, and scanning the environment for new opportunities; (2) a founder identity that involves activities of assembling the resources necessary to create a firm, and the actual founding of the firm; and (3) a developer identity that concerns activities related to firm and market development (e.g., finding new customers) and firm growth (Cardon et al., 2009: 517; see also Corbett, Chapter 12, this volume). Since socially situated expectations of what entrepreneurs do may vary because of different self-concepts (e.g., independent, interdependent; Markus and Kitayama, 1991), these three types of entrepreneurial passion may be experienced differently across cultures.

Building on the theories of situated emotions in psychology (Parkinson et al., 2005), prior literature in entrepreneurship exploring the impact of culture (e.g., Hayton et al., 2002), and work on entrepreneurial passion, we argue that the strength and type of experienced entrepreneurial passion differ across cultural contexts. This is supported by the following logic. First, the conceptualization of entrepreneurial passion emphasizes an intrinsic link between the role identity of the entrepreneur and a specific experience of entrepreneurial passion (Cardon et al., 2009; Murnieks et al., 2014). Second, the social psychology literature finds that the emergence of role identities is culturally dependent (e.g., Stryker and Burke, 2000). Integration of these two observations provides a lens through which the experience of entrepreneurial passion may be seen as a dynamic process helping entrepreneurs to maintain their relations with continually changing social environments. Similarly to how firm-level dynamic capability involves the firm's ability to integrate, build and reconfigure internal and external competences to address rapidly changing environments (Vogel and Güttel, 2012), entrepreneurial passion may motivate entrepreneurs to develop competencies necessary for the success of their ventures, or to recruit those competencies as needed, and to otherwise adapt to their environment, such as to the rapidly digitized and global business world, and the change from a socialist to a market economy. In short, passion may lead entrepreneurs to do whatever it takes in order to succeed and to fulfill their passion. Because individuals in different countries have strikingly different construals of the self, of others and of the interdependence of the two, these construals can influence and in many cases determine the very nature of individual experience, including cognition, motivation and emotion (Markus and Kitayama, 1991: 224).

Given the scarcity of prior cross-cultural work on emotions between Slovenia and the USA, we do not have empirical grounds to predict which country will experience greater levels of entrepreneurial passion. Yet, drawing from existing theorizing of independent and interdependent self-construals, we speculate that in cultures with prevailing independent self-construals (e.g., the USA), the intensity of experienced entrepreneurial passion may be higher. In contrast, we speculate that the interdependent self-construals in Slovenia may lead to lower levels of experienced entrepreneurial passion. The interdependent view of Slovenian society can be understood by looking at several norms and values regarding entrepreneurial endeavors. For example, being wealthy and owning a great deal of property is stigmatized as a bad thing because of the still-present bitter aftertaste of nontransparent privatization of social property that Slovenia went through in the 1990s. Moreover, experiencing low entrepreneurial passion may come also from the fact that entrepreneurs are often inappropriately seen as tycoons and therefore are less passionate about doing business. In addition, failing with one's own venture is not as negatively judged in the USA as it is in Slovenia, where creditors, financial institutions and the public view such bankruptcy negatively, leading entrepreneurs to be less willing to start a business again, whereas in the USA failing and starting several businesses is more usual. This leads us to propose the following:

Hypothesis 1: There will be significant differences in the level of experienced entrepreneurial passion by groups of entrepreneurs in different cultural contexts. Specifically, entrepreneurial passion will be lower in Slovenia than in the USA.

Entrepreneurial Passion and Venture Performance

A positive link between entrepreneurial emotions and venture performance is supported in prior literature (e.g., Baron and Tang, 2011; Cardon et al., 2012). First, drawing from the evidence in psychology, the experience of entrepreneurial passion has informational value. According to the 'feeling as information hypothesis,' positive feelings (such as passion) provide individuals with information that tells them that their current situation is safe (Schwarz and Clore, 1996). In so doing, positive affect enhances performance because it broadens the thought–action repertoires of individuals (Fredrickson, 1998; see also Foo, Murnieks and Chan, Chapter 5, this volume; Grégoire, Chapter 6, this volume), which leads them to generate and consider more diverse alternatives (Isen and Baron, 1991), search for novel information, and be more receptive to new ideas (Estrada, Isen and Young, 1997). Second, the experience of entrepreneurial passion has motivational value. Positive feelings of entrepreneurial passion motivate entrepreneurs toward activities that are particularly important and salient to the entrepreneur's self-identity. This link is maintained through a self-consistency mechanism – the wish to maintain one's self-identity and protect self-conceptions against change. Such engagement in activities that validate salient identities arouses positive affect, in accordance with affect control theory (Burke and Reitzes, 1981). Therefore, when entrepreneurs experience strong feelings such as passion for activities associated with role identities that are important to them, their thought–action repertoires are broadened as they work to preserve the salient role identities. In doing so, entrepreneurs often engage in more creative problem solving and persistence toward their goals (Cardon and Kirk, 2015). This behavior leads to better venture performance and growth. Taking these together, we suggest the following:

Hypothesis 2a: Entrepreneurial passion will be positively associated with venture performance in terms of market share growth, regardless of the cultural context; this will be true for passion for inventing, passion for founding, and passion for developing.

Hypothesis 2b: Entrepreneurial passion will be positively associated with innovation performance, regardless of the cultural context; this will be true for passion for inventing, passion for founding, and passion for developing.

Entrepreneurial Self-efficacy and Venture Performance

Over the years of research on determinants of venture performance, self-efficacy theory has probably been the most often used conceptual base for explaining variations in individual entrepreneurs' and venture-related performance. The applicability of the concept of self-efficacy to entrepreneurship is appropriate because of its inherent link to goal-driven behaviors. In his grounding work, Bandura (1994) proposes that individuals with varying levels of self-efficacy beliefs differ in the amount of effort they spend on goal-directed tasks, the magnitude of coping activities they initiate to overcome impediments, and the degree to which they pursue goals despite obstacles (Stajkovic and Luthans, 2001); and that these behaviours have positive effects on performance outcomes. Indeed, when the theory of self-efficacy is applied to entrepreneurial contexts, there is a strong view that self-efficacy is a good thing for entrepreneurs to have (Drnovšek, Wincent and Cardon, 2010). Shane and colleagues (2003: 267) went so far as to argue that an entrepreneur who is high in self-efficacy is likely to 'exert more effort for a greater length of time, persist through setbacks, and develop better plans and strategies for the task.'

The entrepreneurial self-efficacy construct has been conceptually and empirically directly linked to important entrepreneurial outcomes such as start-up intentions (e.g., Krueger, Reilly and Carsrud, 2000) and new-venture growth (e.g., Boyd and Vozikis, 1994), as well as personal success of entrepreneurs (Markman, Balkin and Baron, 2002), while indirect effects of entrepreneurial self-efficacy on entrepreneurial behaviours, such as improvisation (Hmieleski and Baron, 2008), have also been explored. Overall, prior evidence suggests that self-efficacy is an important construct in entrepreneurship research and is related to important entrepreneurial outcomes. This leads us to propose:

Hypothesis 3a: Entrepreneurial self-efficacy will be positively associated with venture performance in terms of market share growth, regardless of the cultural context.

Hypothesis 3b: Entrepreneurial self-efficacy will be positively associated with innovation, regardless of the cultural context.

Entrepreneurial Self-efficacy and Entrepreneurial Passion

As discussed above, research in psychology and entrepreneurship views emotion and cognition as intertwined systems, where it is nearly impossible to completely separate the two (Zajonc, 1980). 'In nearly all cases, however, feeling is not free of thought, nor is thought free of feelings' (Zajonc, 1980: 154). Similarly, prior work in entrepreneurship has found significant correlations between entrepreneurial passion and entrepreneurial self-efficacy. For example, Murnieks et al. (2014) found a correlation of 0.30, Cardon et al. (2013) found correlations ranging from 0.22 to 0.54, and Cardon and Kirk (2015) found correlations of 0.19 to 0.47, all significant, between dimensions or domains of entrepreneurial passion and measures of entrepreneurial self-efficacy. We therefore also examine the relationship between these constructs as part of our model. We do not expect any cultural differences between these relationships, however. Therefore we propose:

Hypothesis 4: Entrepreneurial self-efficacy will be positively associated with entrepreneurial passion, regardless of the cultural context.

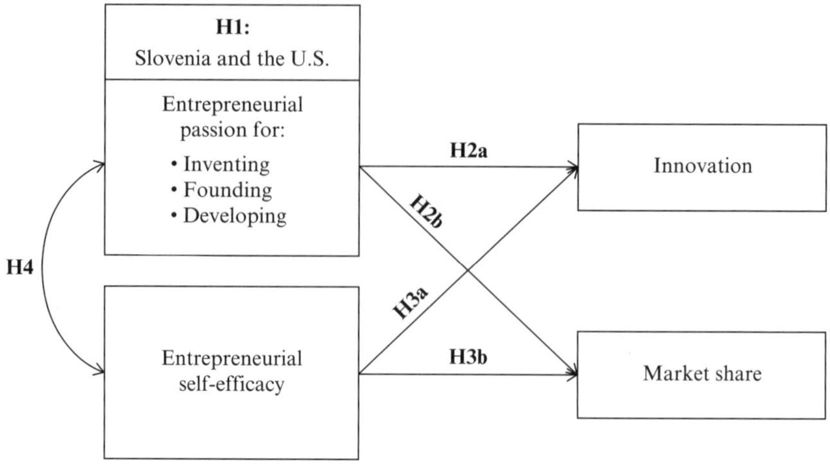

Figure 7.1 Conceptual model

Figure 7.1 summarizes the relationships discussed above. Our model suggests that when effects of entrepreneurial passion and entrepreneurial self-efficacy are explored within the same model, regardless of cultural effect, the experience of entrepreneurial passion and the strength of entrepreneurial self-efficacy beliefs contribute to two important venture-related outcomes – the growth of market share and innovation performance.

METHODS

Data

The sample includes entrepreneurs from Slovenia and the USA. We selected these two countries to examine cultural differences for several reasons. First, we compared two countries that differ substantially in their size (Slovenia with 2.0 million inhabitants and the USA with 317.1 million inhabitants), GDP per capita in 2012 (PPP) (US$29 000 in Slovenia and US$49 000 in the USA) and economic tradition (Slovenia as a recently transitioned country with a post-socialist imprint that can still be felt and the traditional market-economy country of the USA). Second, Slovenia's history and culture are tightly associated with those of Continental European countries, while the USA makes a good representation of the Anglo-Saxon cultural environment. Third, Slovenia is culturally a more homogeneous country, while the USA is a highly multicultural country.

Fourth, the USA is generally considered an extreme example of high individualism, while Slovenia has a less individualistic culture, which derives in part from the socialist regime that was in place in Yugoslavia until 1990.

Respondents for our samples were randomly selected using official databases of incorporated businesses – the PIRS database was used in Slovenia and ReferenceUSA in the USA. In each country a sample of 4000 firms that had 5–249 employees, were privately owned and operated in all sectors except in public administration was initially generated. Three e-mail reminders to participate in our study were sent in the USA, and one reminder was sent in Slovenia. For those respondents who preferred an online version, the reminder contained a link to our web survey. The first reminder was sent two weeks after mailing the paper version questionnaires, the second and third reminders for the US sample were sent four and six weeks after the first mailing. Data for the US sample were collected in the period September–November 2011, whereas data for the Slovenian sample were collected in the period February–April 2012.

A total of 713 useful responses was obtained from the Slovenian sample (response rate of 17.83 percent) and 367 useful responses from the US sample (response rate of 9.18 percent). We analyzed the two data sets in terms of missing data. We determined that data were missing completely at random since no pattern of missing data was found in the two databases.

We tested the data for the presence of common-method variance in several ways. First, we used different sources of information for some key variables (Podsakoff, MacKenzie, Lee and Podsakoff, 2003), as explained below. In addition, common-method variance was reduced *ex ante* by assuring confidentiality of the study and pretesting the questionnaire items for their clearness, unambiguity and familiarity of wording (Chang, van Witteloostuijn and Eden, 2010; Podsakoff et al., 2003). Following additional suggestions by Podsakoff and colleagues (2003), we used different scale endpoints and formats for predictor and criterion measures within the questionnaires. Finally, common-method variance was evaluated *ex post* by conducting Harman's single-factor test in EFA in SPSS and the common factor CFA in AMOS. Both tests provided evidence that common-method variance was not present in the samples.

A detailed description of the samples' characteristics is provided in Table 7.1. Average age of the responding firms in the Slovenian sample was 17 years, while US responding firms were on average five years older. US firms were larger in terms of number of employees, firm owners were older and had more years of experience in general (from their first to their current job), and also more years of experience in the current industry. Approximately the same percentage of women was present in both samples (20 percent).

Table 7.1 Sample characteristics

	Mean		S.D.	
	SLO	USA	SLO	USA
Firm age (years)	17.37	22.29	10.70	18.80
No. of employees	17.43	25.21	25.22	34.73
Overall business tenure (years)	24.97	33.26	10.12	10.63
Business tenure current industry (years)	19.72	24.28	9.43	11.55
Owner age	47.81	54.60	9.60	9.90
Gender (0 = female, 1 = male)	0.80	0.79	0.40	0.41

Measures

Data were collected using self-administered questionnaires, using Dillman and colleagues' (2009) tailored design method. Initially we prepared an English version of the questionnaire that was then translated into the Slovene language and back-translated into English by two translators following recommendations by Brislin (1970). Questionnaires were mailed with a pre-paid return envelope and an accompanying letter of invitation to participate in our study.

Entrepreneurial passion

Entrepreneurial passion was measured with the scale validated by Cardon et al. (2013). The measure captures the three EP domains – inventing, founding and developing. Each domain comprises two dimensions – intense positive feelings (IPF) and identity-centrality (IC). Respondents were asked to assess their extent of agreement with items pertaining to EP dimensions on a five-point Likert scale ranging from strongly disagree to strongly agree. Measures for IPF dimensions consisted of summated average scores for their respective items. By multiplying the respective IPF and IC dimensions we obtained the EP-inv, EP-fnd and EP-dev scores.

Entrepreneurial self-efficacy

Entrepreneurial self-efficacy captures the strength of an individual's capability beliefs about performing the roles and tasks of an entrepreneur (Chen et al., 1998). In our study, we used a shorter version of ESE validated by Prodan and Drnovšek (2010) that refers to ESE for management, ESE for innovation, and ESE for marketing. However, one item pertaining to the dimension of ESE for management ('I am able to establish position in product/service market') loaded on all three dimensions and was excluded from further analyses. We asked respondents to evaluate their degree of certainty in performing tasks on a seven-point Likert scale ranging from completely unsure to completely sure. Reliability was assessed in EQS by the RHO reliability coefficient due to the second-order nature of the measure, resulting in RHO(Slovenian sample) = 0.936, RHO(US sample) = 0.928.

Innovation

Innovation was measured as a second-order construct representing three dimensions – product innovations, process innovations and administrative innovations (Jiménez-Jiménez and Sanz-Valle, 2011). Respondents were asked to indicate how their firms performed compared to competitors in the last three years. They assessed innovation performance on a seven-point Likert scale ranging from much worse than competitors to much better than competitors. Reliability was assessed in EQS by the RHO reliability coefficient due to the second-order nature of the measure, resulting in RHO(Slovenian sample) = 0.945, RHO(US sample) = 0.958.

Market share growth

To assess market share growth, we asked respondents to rate the increase/decrease in market share of their firms in the last three years in comparison to competitors in respect of their firms' age and phase of development. The responses were assessed on a seven-point Likert scale ranging from major decrease through a middle point of remained the same to the highest score of major increase.

Analysis

We began with the single-group analysis to validate EP domains following a confirmatory factor analysis. In the second step, we performed multi-group analysis in order to test for configural and metric invariance of EP domains across different cultures (Byrne, 2010). Configural invariance requires the number of factors and the factor-loading pattern to be the same across the groups of entrepreneurs for each EP domain, while metric invariance requires the equality of factor loadings. We also added constraints on equality of error variances.

After multi-group analyses of EP domains across Slovenian and US entrepreneurs, we turned to the latent-mean analyses of EP domains to test for the equivalence of means related to each underlying EP dimension factor. We introduced the intercepts of observed variables that comprised EP dimensions of IPF and IC as well as the means of IPF and IC EP dimensions. The latent-means analysis is the correct method to test if there are any cultural differences that preclude responding to instruments in similar ways (Vandenberg and Lance, 2000). We used AMOS version 20.0 to compute all statistical tests of invariance in multi-group analyses and in latent-means analyses.

Hypotheses that postulated casual relationships between emotions and cognitions and firm-related outcomes were tested using structural equation models. We used EP-inv, EP-fnd, EP-dev and ESE as independent variables and innovation and market share growth as dependent variables. All models were evaluated in terms of model fit based on the following fit indices: χ^2, NFI, CFI, SRMR and RMSEA.

Results

Results of the single-group CFA analyses of EP domains for Slovenian and US entrepreneurs confirmed the proposed structure with model-fit indices that were above the recommended thresholds (Hair, Anderson, Tatham and Black, 1998). One error covariance was allowed in the dimension EP-fnd and dimension EP-dev. Allowing error covariances among related indicators can reflect the fact that a similar idea underlies the constructs, albeit with different wording.

We used multi-group analysis to test for configural invariance, metric invariance, and invariance of error variances. First, results confirmed configural invariance of all EP domains since data fitted well on the model in which the same number of factors and the same pattern of fixed and free factor loadings were specified in both groups simultaneously, while allowing all parameters to be freely estimated (Kline, 2011).

Second, a stronger form of invariance is represented by metric invariance, which is evidenced when unstandardized factor loadings of each indicator are equal across groups (Kline, 2011: 253). Metric invariance was confirmed for EP-inv and EP-fnd since factor loadings were equivalent across Slovenian and US samples of entrepreneurs and the change in the CFI index between the configural invariance model and the metric invariance model was below the suggested threshold of 0.01. Partial metric invariance was supported for the EP-dev dimension since one item ('Pushing my employees and myself to make our company better motivates me') had non-invariant loadings, suggesting that entrepreneurs in Slovenia and in the USA differ in regard to the level to which they push their employees and themselves to develop their firms. It seems that the performance-oriented culture in the USA drives people to serve the companies' needs better in order to retain their jobs by doing well for the company, while, in Slovenia, employees' rights are extensive and the rigid labor market makes it difficult to lay off workers. It might be that here lies a part of the reason for which entrepreneurs push their employees to a lesser extent in Slovenia than in the USA.

The error variance–invariance testing resulted in full invariance of error variances for EP-inv and EP-dev, whereas for EP-fnd partial invariance of error variances was shown. We checked which is the invariant error variance and found that error variance connected to the indicator 'Owning my own company energizes me' was different in the two groups under study. In interpreting results of invariance tests we are aware that measurement systems are imperfect and that some proportion of variation in targeted indicators may be attributed to influences other than the abstraction of substantive interest, including systematic and nonsystematic measurement error (Vandenberg and Lance, 2000). The results of goodness-of-fit statistics for configural invariance, metric invariance and invariance of error variances are presented in Table 7.2.

Hypothesis 1, which postulated significant differences in the mean levels of experienced entrepreneurial passion between American and Slovenian entrepreneurs was supported. We used latent-means analyses to test it. In performing this analysis we set the Slovenian sample of entrepreneurs as the

Table 7.2 Multi-group analyses of entrepreneurial passion domains

	χ^2	DF	P(ΔP)	NFI	CFI	SRMR	RMSEA (90% confidence interval; pclose)
Entrepreneurial passion for inventing:							
Unconstrained	25.88	10	0.00 (/)	0.98	0.99	0.02	0.04 (0.02–0.06; 0.72)
Model 1a	30.63	13	0.00 (0.19)	0.98	0.99	0.03	0.04 (0.02–0.06; 0.84)
Model 2a	39.78	17	0.00 (0.06)	0.97	0.98	0.03	0.04 (0.02–0.05; 0.88)
Entrepreneurial passion for founding:							
Unconstrained	1.49	2	0.48 (/)	1.00	1.00	0.00	0.00 (0.00–0.06; 0.89)
Model 1b	3.06	4	0.55 (0.46)	1.00	1.00	0.01	0.00 (0.00–0.05; 0.97)
Model 2b	4.36	6	0.63 (0.52)	0.99	1.00	0.01	0.00 (0.00–0.04; 0.99)
Entrepreneurial passion for developing:							
Unconstrained	5.67	2	0.06 (/)	0.99	0.99	0.02	0.05 (0.00–0.09; 0.49)
Model 1c	7.99	3	0.05 (0.13)	0.99	0.99	0.02	0.04 (0.01–0.08; 0.56)
Model 2c	14.87	6	0.02 (0.08)	0.98	0.99	0.02	0.04 (0.01–0.07; 0.69)

Notes:
Model 1a = equality of factor loadings; Model 2a = Model 1a + equality of error variances.
Model 1b = equality of factor loadings; Model 2b = Model 1b + partial equality of error variances.
Model 1c = partial equality of factor loadings; Model 2c = Model 1c + equality of error variances.

Table 7.3 Model-fit statistics for latent-means analyses

	χ^2	DF	P(ΔP)	NFI	CFI	SRMR	RMSEA (90% confidence interval; pclose)
Entrepreneurial passion for inventing:							
Model 1a	49.76	13	0.00	0.96	0.97	0.02	0.06 (0.04–0.07; 0.25)
Model 2a	55.86	16	0.00 (0.11)	0.96	0.97	0.03	0.05 (0.04–0.07; 0.36)
Model 3a	64.81	20	0.00 (0.06)	0.95	0.96	0.03	0.05 (0.04–0.06; 0.48)
Entrepreneurial passion for founding:							
Model 1b	27.65	4	0.00	0.95	0.96	0.00	0.08 (0.05–0.11; 0.03)
Model 2b	29.09	5	0.00 (0.23)	0.95	0.96	0.01	0.07 (0.05–0.10; 0.06)
Model 3b	31.65	7	0.00 (0.28)	0.94	0.96	0.01	0.06 (0.04–0.09; 0.15)
Entrepreneurial passion for developing:							
Model 1c	5.67	2	0.06 (/)	0.99	0.99	0.01	0.05 (0.00–0.09; 0.49)
Model 2c	7.99	3	0.05 (0.13)	0.99	0.99	0.03	0.04 (0.01–0.08; 0.56)
Model 3c	14.87	6	0.02 (0.08)	0.98	0.99	0.02	0.04 (0.02–0.07; 0.69)

Notes:
Model 1a = Equality of intercepts; Model 2a = Model 1a + equality of factor loadings; Model 3a = Model 2a + equality of error variances.
Model 1b = Equality of intercepts; Model 2b = Model 1b + partial equality of factor loadings; Model 3b = Model 2b + partial equality of error variances.
Model 1c = Partial equality of intercepts; Model 2c = Model 1c + partial equality of factor loadings; Model 3c = Model 2c + equality of error variances.

reference group (means were fixed to 0). We introduced intercepts into the analysis and tested for their equality across the two groups (scalar invariance). The results showed invariance of intercepts for EP-inv and EP-fnd, but not for EP-dev. When this constraint was relaxed, we got acceptable results. We also tested the invariance of factor loadings and invariance of indicators' unique variances (error variances) across groups (Vandenberg and Lance, 2000). The results are in line with results from multi-group analyses. We present the results on model-fit statistics in Table 7.3.

Acknowledging (partial) invariance of all tests so far, we turn to the estimation of the latent means. The results presented in Table 7.4 show that there are statistically significant differences in latent means, with means for IPF and IC of EP-inv, EP-fnd and EP-dev being significantly higher for the US sample of entrepreneurs. This allowed us to accept Hypothesis 1 because results show that US entrepreneurs experienced higher degrees of IPF and IC of EP-inv, EP-fnd and EP-dev.

Hypotheses 2a and 2b suggested positive relationships between entrepreneurial passion and performance outcomes. Results of structural equation

Table 7.4 Latent-means statistics for US sample (Slovenian sample as reference group)

	Estimate	Approximate standard error	Critical ratio	P (two-tailed)	Sample means	
					SLO	USA
Entrepreneurial passion for inventing:						
IPF-inv	0.16	0.05	3.62	0.001	4.04	4.19
IC-inv	0.35	0.06	5.83	0.001	3.86	4.21
Entrepreneurial passion for founding:						
IPF-fnd	0.42	0.07	6.23	0.001	3.74	4.13
IC-fnd	0.22	0.07	3.07	0.002	3.92	4.13
Entrepreneurial passion for developing:						
IPF-dev	0.39	0.06	6.73	0.001	3.94	4.23
IC-dev	0.34	0.06	5.70	0.001	3.97	4.31

Note: IPF = intense positive feeling; IC = identity centrality.

analyses show that EP-inv, EP-fnd and EP-dev have slightly different influences on dependent variables in the two samples. While a statistically significant influence was found for EP-inv and EP-dev on innovation and not on market share growth in the Slovenian sample, EP-inv significantly impacts innovation and market share in the US sample. The results show that EP-dev is related only to market share in the US sample. Interestingly, EP-fnd is not significantly related to innovation or market share in either sample. This provides partial support for Hypotheses 2a and 2b.

Hypotheses 3a and 3b suggested positive relationships between entrepreneurial self-efficacy and performance outcomes. ESE had a significant influence on innovation and market share in both samples, which supports both hypotheses. The unstandardized and standardized estimates, along with standard deviations,are presented in Table 7.5. While R^2 statistics for innovation were 0.41 and 0.33 for the Slovenian and US samples, respectively, the R^2 statistics for market share were 0.19 in both samples. The goodness-of-fit statistics for the structural model were the following: (1) for the Slovenian sample $\chi^2 = 295.68$, df = 66, p = 0.00, NFI = 0.95, CFI = 0.96, SMRM = 0.05, RMSEA = 0.07 (90 percent confidence interval from 0.06 to 0.08); and (2) for the US sample $\chi^2 = 198.60$, df = 66, p = 0.00, NFI = 0.94, CFI = 0.96, SRMR = 0.06, RMSEA = 0.07 (90 percent confidence interval from 0.06 to 0.09).

Finally, consistent with H4, we evaluated the interrelatedness of the EP dimensions and ESE to see whether there are some differences in

Table 7.5 AMOS results for structural equation models

Relation	Unstandardized estimate		Standard error		Standardized estimate		P value (two-tailed)	
	SLO	USA	SLO	USA	SLO	USA	SLO	USA
EP-inv→Innovation	0.03	0.09	0.01	0.01	0.20	0.45	***	***
EP-inv→Market share	−0.01	0.06	0.02	0.02	−0.05	0.24	n.s.	**
EP-fnd→Innovation	−0.02	−0.03	0.01	0.01	−0.15	−0.15	n.s.	#
EP-fnd→Market share	−0.01	0.01	0.02	0.02	−0.03	0.07	n.s.	n.s.
EP-dev→Innovation	0.05	0.01	0.02	0.02	0.29	0.04	**	n.s.
EP-dev→Market share	0.05	−0.05	0.03	0.03	0.21	−0.22	#	*
ESE→Innovation	0.35	0.40	0.05	0.07	0.32	0.35	***	***
ESE→Market share	0.54	0.70	0.09	0.10	0.34	0.50	***	***

Note: # $p < 0.10$; * $p < 0.05$; ** $p < 0.01$; *** $p < 0.001$.

Table 7.6 Correlations among variables

	EP-inv		EP-fnd		EP-dev		ESE		Innovation		Market share	
	SLO	USA	SLO	USA	SLO	USA	SLO	USA	SLO	USA	SLO	USA
EP-inv	1	1										
EP-fnd	0.63	0.55	1	1								
EP-dev	0.67	0.62	0.86	0.76	1	1						
ESE	0.51	0.53	0.43	0.36	0.50	0.43	1	1				
Innovation	0.46	0.57	0.36	0.25	0.45	0.36	0.50	0.55	1	1		
Market share	0.24	0.40	0.26	0.22	0.31	0.20	0.40	0.55	0.42	0.56	1	1

Note: All correlations significant at $p < 0.001$, except EP-dev ↔Market share significant at $p < 0.01$.

correlations among these variables between the two samples. The results reported in Table 7.6 show major differences in the strength of the correlations of EP dimensions with ESE in the US sample (correlation coefficients from 0.36 to 0.53), while in the Slovenian sample there were only slight differences in the strength of the correlations of EP dimensions with ESE (correlation coefficients from 0.42 to 0.50). However, in both samples ESE correlates the most with EP-inv, then with EP-dev, and the least with EP-fnd. But still, all the correlations are statistically significant ($p < 0.001$), consistent with H4. Based on the results on these final tests we can assume that EP and ESE correlate with each other in a significant way in both samples, despite different cultural and national backgrounds.

DISCUSSION

In this chapter we extend the recent work on entrepreneurial passion (Breugst et al., 2012; Cardon et al., 2009) by integrating this stream of research with work on situated emotions to empirically test how entrepreneurial passion is experienced across different cultures. We believe that this exploration of entrepreneurial passion in two different cultures – Slovenia and the USA – is important for the following reasons. First, existing research into entrepreneurial passion that has emphasized it as an important motivational resource needs to be culturally embedded. This is because prior literature shows that some variations in the process by which entrepreneurs act and make decisions in response to opportunities at hand can be ascribed to differences in national culture norms and values. Second, given that the concept of entrepreneurial passion is conceptualized in the literature (Mitchell et al., 2007; Mitchell, Smith, Seawright and Morse, 2000), and a measurement instrument has been developed and validated, further studies of this important construct will contribute to a fuller picture of how the construct manifests itself in a nomological network with other important entrepreneurship constructs. We add to this body of work by examining how entrepreneurial passion works alongside entrepreneurial self-efficacy and how they influence venture-related outcomes in terms of innovation performance and market share growth. As we have earlier emphasized, we see emotions (e.g., entrepreneurial passion) and cognitions (e.g., entrepreneurial self-efficacy) as two cooperating systems that dynamically interact in directing behaviors of entrepreneurs and venture-related outcomes in turn. This is consistent with other research that has looked at their interrelationships (Cardon and Kirk, 2015).

Our main theoretical contribution to the entrepreneurship literature stems from exploring how contextual factors influence the emergence and development of emotional and cognitive processes of entrepreneurs. First, we posit reasons why entrepreneurial passion may develop differently across cultural contexts. We empirically test our propositions and provide evidence that the strength of experienced entrepreneurial passion varies among entrepreneurs from different environments. In particular, we found that all three domains of passion were experienced more intensely in the US entrepreneur sample than in the Slovenian sample. This means that, on average, US entrepreneurs self-evaluate their entrepreneurial passion higher than Slovenian entrepreneurs do, with IPF-fnd having the highest difference between the two samples (difference of 0.42) and with IPF-inv having the lowest difference between the two samples (difference of 0.16). In regard to the IC dimension, the highest difference was evidenced with IC-inv (difference of 0.35) and the lowest with IC-fnd (difference of 0.22). But in all cases, entrepreneurs from the USA reported higher IPF and IC values. Second, we add to the literatures that explore how specific emotions and cognitions are associated with venture outcomes. We find an intriguing result here – despite the fact that the average level of reported passion is lower in Slovenia than in the USA, it has more of an effect on innovation in Slovenia than in the USA. On the other hand, the relationship between passion and innovation was significant for all three domains of EP in Slovenia, but only for passion for inventing in the USA. In terms of market share, there was no effect of passion in Slovenia, but a significant effect of passion for inventing and passion for developing on market share growth in the US sample. Further, ESE had a significant effect on market share growth and innovation in both the US and Slovenian samples, and passion and ESE were significantly correlated in both samples.

This raises several important questions and implications for future work. First, why are the reported levels of entrepreneurial passion higher in the USA, on average, than in Slovenia? While the USA has a long entrepreneurial tradition facilitated by the capitalistic system, in Slovenia entrepreneurial ventures sprang up only at the beginning of the 1990s after Slovenia gained its independence and the socialist economic system transitioned into a capitalistic one. It might be that the experience of EP is lower among Slovenian entrepreneurs due to traces of the collectivistic culture that can still be felt and that do not praise entrepreneurs, their effort or their successes in economic development to such an extent as a more capitalistic culture does. Indeed, research suggests (e.g., Ekman, 1984) that national cultures have different values and norms about any phenomenon visibly occurring in the society (e.g., entrepreneurship activity), implying that entrepreneurial endeavors and performance are differently publicly emphasized. Future research should examine in more detail how Slovenians experience passion and whether their lower reports of it are due to perceived cultural expectations or norms for formally reporting emotional experiences or, instead, actual lower levels of feelings experienced, or perhaps less identity-centrality of the roles of inventor, founder or developer in the Slovenian context. The results we report here show clearly that there are significant differences in the levels of EP reported by entrepreneurs from two different nations that also have very different entrepreneurial traditions, with Slovenian entrepreneurs reporting lower levels of experienced feelings and identity-centrality. This adds to the conversation proposed by Cardon et al., in 2013, who suggested further investigation into whether and how the experience of EP varies among entrepreneurs. It is particularly interesting that the relationship between passion and innovation is stronger in Slovenia than in the USA, yet the relationship between passion and market share growth is stronger in the USA than in Slovenia. Further research should expand this study to other countries, specifically other parts of the world that differ substantially from both Slovenia and the USA in terms of general cultural elements as well as the entrepreneurial environment. Future work could also explore cultural norms concerning the two dimensions of passion, positive intense feelings and the identity-centrality of various entrepreneurial roles. In terms of feelings, researchers should examine whether the experience of passion (level) truly differs or just the willingness of people to honestly report their passion. In the USA, passion is seen as a critical factor in entrepreneurship, leading to greater employee engagement and greater potential for investment dollars (e.g., Ekman, 1984). Therefore US entrepreneurs may be quite willing to wear their hearts on their sleeves and openly admit to feelings of passion. Is this also true in other countries, including Slovenia? To answer this type of question, it is important that additional work include not only self-assessments where entrepreneurs report their own experiences of passion, but also independent assessments of others about the level and focus of an entrepreneur's passion. More objective biological measures of passion would also help differentiate between the experience and reporting norms or practices concerning passion.

A second area of future research related to measuring passion concerns whether entrepreneurs are able to determine what they are passionate about. In the USA, the literature suggests three distinct phases of the entrepreneurial process – identifying and/or creating opportunities (inventing), founding firms through a formal legal process, and growing firms (developing). The focus of the literature on these three areas of the process led to the conceptualization and measurement of entrepreneurial passion as being related to these three roles (Cardon et al., 2013). However, it is entirely possible that these three phases are not as clearly delineated or understood as distinct in other cultures, or that other additional role identities or domains of passion are more relevant to entrepreneurs there. Qualitative work in multiple cultural contexts is needed to answer these questions.

Third, we encourage additional work that integrates EP into models with other affective, emotional, cognitive and motivational components (see also Carsrud and Brännback, Chapter 2, this volume). We responded to a call for this type of work by Cardon and colleagues (2009) by investigating the influence of EP and ESE on venture outcomes, but clearly more work is needed in this area. Future research should explore other personality, firm and environmental variables that may interplay with EP and ESE. For example, how does passion interact with the use of biases and heuristics in entrepreneurial decision making? It would also be useful to examine the explicit interaction of passion, ESE, and other emotional and cognitive variables with explicitly measured cultural variables such as individualism/ collectivism, power distance and others. This is especially important as global boundaries continue to recede and individuals often attend school and/or do business in different countries, which can lead to shifts in individual experiences of culture and greater cultural blending.

Finally, future research could explore the social function of entrepreneurial passion at different levels of social interactions: individual, dyadic and group levels. For instance, at the dyadic level (team level) of analysis, researchers could focus on how emotions organize the interactions of individuals in meaningful relationships. The interacting dyad is the system with respect to which the consequences of behaviors are interpreted. Researchers here focus on the communication of emotion in facial, vocal and postural channels. When exploring social functions at a group level of analysis, researchers can focus on how emotions help groups of interacting individuals with common identities during the goal attainment process. Finally, at the cultural level of analysis, researchers can focus on how emotions are shaped by historical and economic factors, on how emotions are embedded in cultural institutions and practices, and on the cultural norms and scripts for the proper expression and experience of emotions.

REFERENCES

Aldrich, H., Birley, S., Dubini, P., Greve, A., Johannisson, B., Reese, P.R. and Sakano, T. 1991. The generic entrepreneur? Insights from a multinational research project. Paper presented at Babson Conference on Entrepreneurship.

Baker, T., Gedajlovic, E. and Lubatkin, M. 2005. A framework for comparing entrepreneurship processes across nations. *Journal of International Business Studies*, **36**(5): 492–504.

Bandura, A. 1994. Self-efficacy. In V.S. Ramachaudran (ed.), *Encyclopedia of Human Behavior*, vol. 4: 71–81. New York: Academic Press.

Baron, R.A. and Tang, J. 2011. The role of entrepreneurs in firm-level innovation: Joint effects of positive affect, creativity, and environmental dynamics. *Journal of Business Venturing*, **26**(1): 49–60.

Boyd, N.G. and Vozikis, G.S. 1994. The influence of self-efficacy on the development of entrepreneurial intentions and actions. *Entrepreneurship Theory and Practice*, **18**(4): 63–77.

Breugst, N., Domurath, A., Patzelt, H. and Klaukien, A. 2012. Perceptions of entrepreneurial passion and employees' commitment to entrepreneurial ventures. *Entrepreneurial Theory and Practice*, **36**(1): 171–92.

Brislin, R.W. 1970. Back-translation for cross-cultural research. *Journal of Cross-Cultural Psychology*, **1**(3): 185–216.

Burke, P.J. and Reitzes, D.C. 1981. The link between identity and role performance. *Social Psychology Quarterly*, **44**(2): 83–92.
Byrne, B.M. 2010. *Structural Equation Modeling for AMOS: Basic concepts, applications, and programming* (2nd edn). New York: Routledge.
Cardon, M.S. 2008. Is passion contagious? The transference of entrepreneurial emotion to employees. *Human Resource Management Review*, **18**(2): 77–86.
Cardon, M.S. and Kirk, C. 2015. Entrepreneurial passion as mediator of the self-efficacy to persistence relationship. *Entrepreneurship Theory and Practice*, **39**(5).
Cardon, M.S., Foo, M.-D., Shepherd, D.A. and Wiklund, J. 2012. Exploring the heart: Entrepreneurial emotion is a hot topic. *Entrepreneurship Theory and Practice*, **36**(1): 1–10.
Cardon, M.S., Grégoire, D.A., Stevens, C.E. and Patel, P.C. 2013. Measuring entrepreneurial passion: Conceptual foundations and scale validation. *Journal of Business Venturing*, **28**(3): 373–96.
Cardon, M.S., Wincent, J. Singh, J. and Drnovšek, M. 2009. The nature and experience of entrepreneurial passion. *Academy of Management Review*, **34**(3): 511–32.
Chang, S.-J., van Witteloostuijn, A. and Eden, L. 2010. From the Editors: Common method variance in international business research. *Journal of International Business Studies*, **41**(2): 178–84.
Chen, C.C., Greene, P.G. and Crick, A. 1998. Does entrepreneurial self-efficacy distinguish entrepreneurs from managers? *Journal of Business Venturing*, **13**(4): 295–316.
Chen, X.P., Yao, X. and Kotha, S. 2009. Entrepreneur passion and preparedness in business plan presentations: A persuasion analysis of venture capitalists' funding decisions. *Academy of Management Journal*, **52**(1): 199–214.
Cornelissen, J.P. and Clarke, J.S. 2010. Imaging and rationalizing opportunities: Inductive reasoning and the creation and justification of new ventures. *Academy of Management Review*, **35**(4): 539–57.
Damasio, A.R. 1994. *Descartes' Error: Emotion, reason and the human brain*. New York: Grossett/Putnam.
Damasio, A. 2005. *Descartes' Error: Emotion, reason, and the human brain* (2nd edn). New York: Penguin Putnam.
D'Andrade, R. 1984. Cultural meaning systems. In R.A. Shweder and R.A. LeVine (eds), *Culture Theory: Essays on mind, self, and emotion*: 88–119. Cambridge: Cambridge University Press.
DeVos, G. 1985. Dimensions of the self in Japanese culture. In A.J. Marsella and F.L.K. Hsu (eds), *Culture and Self: Asian and Western perspectives*: 149–84, London: Tavistock.
Dillman, D.A., Smyth, J.D. and Christian, L.M. 2009. *Internet, Mail, and Mixed-mode Surveys: The tailored design method* (3rd edn). Hoboken, NJ: John Wiley & Sons.
Dolan, R.J. 2002. Emotion, cognition, and behavior. *Science*, **298**(8): 1191–4.
Drnovšek, M., Wincent, J. and Cardon, M.S. 2010. Entrepreneurial self-efficacy and business start-up: developing a multi-dimensional definition. *International Journal of Entrepreneurial Behaviour & Research*, **16**(4): 329–48.
Ekman, P. 1984. Expression and the nature of emotion. In K. Scherer and P. Ekman (eds), *Approaches to Emotion* (319–44). New York: Psychology Press.
Ekman, P. 1992. An argument for basic emotions. *Cognition and Emotion*, **6**(3–4): 169–200.
Ekman, P., Friesen, W.V. and Ellsworth, P. 1972. *Emotion in the Human Face: Guidelines for research and an integration of findings*. New York: Pergamon Press.
Elfenbein, H.A. and Ambady, N. 2002. On the universality and cultural specificity of emotion recognition: A meta-analysis. *Psychological Bulletin*, **128**(2): 203–35.
Estrada, C., Isen, A. and Young, A. 1997. Positive affect facilitates integration of information and decreases anchoring in reasoning among physicians. *Organizational Behavior and Human Decision Processes*, **72**(1): 117–35.
Foo, M.-D., Uy, M.A. and Baron, R.A. 2009. How do feelings influence effort? An empirical study of entrepreneurs' affect and venture effort. *Journal of Applied Psychology*, **94**(4): 1086–94.

Fredrickson, B.L. 1998. What good are positive emotions? *Review of General Psychology*, **2**(3): 300–319.

Frijda, N.H. 1987. Emotion, cognitive structure, and action tendency. *Cognition and Emotion*, **1**(2): 115–43.

Frijda, N.H. 2009. Emotions, individual differences and time course: Reflections. *Cognition and Emotion*, **23**(7): 1444–61.

Frijda, N.H. and Mesquita, B. 1994. The social roles and functions of emotions. In S. Kitayama and H. Marcus (eds), *Emotion and Culture: Empirical studies of mutual inferences*: 51–87. Washington, DC: American Psychological Association.

Gartner, W.B. 1985. A conceptual framework for describing the phenomenon of new venture creation. *Academy of Management Review*, **10**(4): 696–706.

Geertz, C. 1975. On the nature of anthropological understanding: Not extraordinary empathy but readily observable symbolic forms enable the anthropologist to grasp the unarticulated concepts that inform the lives and cultures of other peoples. *American Scientist*, **63**(1): 47–53.

Griffiths, P.E. and Scarantino, A. 2005. Emotions in the wild: The situated perspective on emotion. In P. Robinsons and M. Aydele (eds), *The Cambridge Handbook of Situated Cognition*: 437–53. New York: Cambridge University Press.

Hair, J.F., Anderson, R.E., Tatham, R.L. and Black, W.C. 1998. *Multivariate Analysis*. Englewood Cliffs, NJ: Prentice Hall International.

Hareli, S. and Parkinson, B. 2008. What's social about social emotions? *Journal for the Theory of Social Behaviour*, **38**(2): 131–56.

Hayton, J.C., George, G. and Zahra, S.A. 2002. National culture and entrepreneurship: A review of behavioral research. *Entrepreneurship Theory and Practice*, **26**(4): 33–52.

Hmieleski, K.M. and Baron, R.A. 2008. When does entrepreneurial self-efficacy enhance versus reduce firm performance? *Strategic Entrepreneurship Journal*, **2**(1): 57–72.

Hofstede, G.H. and Hofstede, G. 2001. *Culture's Consequences: Comparing values, behaviors, institutions, and organizations across nations*. Thousand Oaks, CA: Sage Publications.

Isen, A.M. and Baron, R. 1991. Positive affect as a factor in organizational behavior. In B. Staw and L.L. Cummings (eds), *Research in Organizational Behavior*, vol. 13: 1–53. Greenwich, CT: JAI Press.

Izard, C.E. 1978. On the ontogenesis of emotions and emotion–cognition relationship in infancy. In M. Lewis and L. Rosenblum (eds), *The Development of Affect*: 389–413. New York: Plenum Press.

Jiménez-Jiménez, D. and Sanz-Valle, R. 2011. Innovation, organizational learning, and performance. *Journal of Business Research*, **64**(4): 408–17.

Keltner, D. and Haidt, J. 1999. Social functions of emotions at four levels of analysis. *Cognition & Emotion*, **13**(5): 505–21.

Kline, R.B. 2011. *Principles and Practice of Structural Equation Modeling* (3rd edn). New York: Guilford Press.

Krueger, N.F., Reilly, M.D. and Carsrud, A.L. 2000. Competing models of entrepreneurial intentions. *Journal of Business Venturing*, **15**(5/6): 411–32.

Lutz, C. 1988. *Unnatural Emotions: Everyday sentiments on a Micronesian atoll and their challenge to Western theory*. Chicago, IL: Chicago University Press.

Markman, G.D., Balkin, D.B. and Baron, R.A. 2002. Inventors and new venture formation: The effects of general self-efficacy and regretful thinking. *Entrepreneurship Theory and Practice*, **27**(2): 149–65.

Markus, H.R. and Kitayama, S. 1991. Culture and the self: Implications for cognition, emotion, and motivation. *Psychological Review*, **98**(2): 224–53.

McMullen, J.S. and Shepherd, D.A. 2006. Entrepreneurial action and the role of uncertainty in the theory of the entrepreneur. *Academy of Management Review*, **31**(1): 132–52.

Mitchell, R.K., Busenitz, L.W., Lant, T., McDougall, P.P., Morse, E.A. and Smith, J.B. 2002. Toward a theory of entrepreneurial cognition: Rethinking the people side of entrepreneurship research. *Entrepreneurship Theory and Practice*, **27**(2): 93–104.

Mitchell, R.K., Busenitz, L.W., Bird, B., Gaglio, C.M., McMullen, J.S., Morse, E.A.

and Smith, J.B. 2007. The central question in entrepreneurial cognition research 2007. *Entrepreneurship Theory and Practice*, **31**(1): 1–27.
Mitchell, R.K., Randolph-Seng, B. and Mitchell, J.R. 2011. Socially situated cognition: Imagining new opportunities for entrepreneurship research (Dialogue). *Academy of Management Review*, **36**(4): 774–8.
Mitchell, R.K., Smith, J.B., Seawright, K.W. and Morse, E.A. 2000. Cross-cultural cognitions and the venture creation decision. *Academy of Management Journal*, **43**(5): 974–93.
Mitteness, C., Sudek, R. and Cardon, M.S. 2012. Investor characteristics that determine whether perceived passion leads to higher evaluations of funding potential. *Journal of Business Venturing*, **27**(5): 592–606.
Murnieks, C.Y., Mosakowski, E. and Cardon, M.S. 2014. Pathways of passion: Identity centrality, passion, and behavior among entrepreneurs. *Journal of Management*, **40**(6): 1583–606.
Parkinson, B., Fischer, A. and Manstead, A.S.R. 2005. *Emotion in Social Relations: Cultural, group, and interpersonal processes*. New York: Psychology Press.
Pham, M.T. 2004. The logic of feeling. *Journal of Consumer Psychology*, **14**(4): 360–69.
Podsakoff, P.M., MacKenzie, S.B., Lee, J.Y. and Podsakoff, N.P. 2003. Common method biases in behavioral research: a critical review of the literature and recommended remedies. *Journal of Applied Psychology*, **88**(5): 879–903.
Prodan, I. and Drnovšek, M. 2010. Conceptualizing academic–entrepreneurial intentions: An empirical test. *Technovation*, **30**(5–6): 332–47.
Romer, P.M. 2000. Thinking and feeling. *American Economic Review*, **90**(2): 439–43.
Schwarz, N. and Clore, G.L. 1996. Feelings and phenomenal experiences. In E.T. Higgins and A.W. Kruglanski (eds), *Social Psychology: Handbook of basic principles*: 433–65. New York: Guilford Press.
Shane, S. 1992. Why do some societies invent more than others? *Journal of Business Venturing*, **7**(1): 29–46.
Shane, S., Locke, E.A. and Collins, C.J. 2003. Entrepreneurial motivation. *Human Resource Management Review*, **13**(2): 257–79.
Shane, S. and Venkataraman, S. 2000. The promise of entrepreneurship as a field of research. *Academy of Management Review*, **25**(1): 217–26.
Smilor, R.W. 1997. Entrepreneurship: Reflections on a subversive activity. *Journal of Business Venturing*, **12**(5): 341–6.
Smith, E.R. and Semin, G.R. 2004. Socially situated cognition: Cognition in its social context. *Advances in Experimental Social Psychology*, **36**: 53–117.
Stajkovic, A. and Luthans, F. 2001. Differential effects of incentive motivators on work performance. *Academy of Management Journal*, **44**(3): 580–90.
Stryker, S. and Burke, P.J. 2000. The past, present, and future of an identity theory. *Social Psychology Quarterly*, **63**(4): 284–97.
Thomas, Anisya S. and Mueller, Stephen L. 2000. A case for comparative entrepreneurship: Assessing the relevance of culture. *Journal of International Business Studies*, **31**(2): 287–301.
Van Kleef, G.A. 2010. The emerging view of emotion as social information. *Social and Personality Psychology Compass*, **4**(5): 331–43.
Vandenberg, R.J. and Lance, C.E. 2000. A review and synthesis of the measurement invariance literature: Suggestions, practices, and recommendations for organizational research. *Organizational Research Methods*, **3**(1): 4–70.
Vogel, R. and Güttel, W.H. 2012. The dynamic capability view in strategic management: A bibliometric review. *International Journal of Management Reviews*, **15**(4): 426–46.
Zajonc, R.B. 1980. Feelings and thinking: Preferences need no inferences. *American Psychologist*, **35**(2): 151–75.
Zajonc, R.B. 1984. On the primacy of affect. *American Psychologist*, **39**(2): 117–23.

8. Lessons from the neural foundation of entrepreneurial cognition: the case of emotion and motivation

David A. Baucus, Melissa S. Baucus and Ronald K. Mitchell

ABSTRACT

Our objective in this chapter is to illustrate how entrepreneurs' brains are physiologically the same as any other person's brain, but in terms of experiences and knowledge they are different. Using neurophysiology and relevant concepts from neuroscience, we map the physiological processes involved in transmitting visual data from the periphery (environment) to conscious thought and behavior, with affective processes (emotion and motivation) modulating data flows along the way. We explain how cortical fields and subcortical nuclei (key parts of the brain) process and code neural representations, first as simple data points, but then as behaviorally relevant 'percepts' (perceptions) and 'concepts' (conceptualizations) that carry 'affective value' acquired through structures specialized for emotion and motivation. We also explain how these abstract building blocks of thought – percepts, concepts and affective valuations – decouple from external stimuli owing to repeated activation (experience) and come together with real-time data in the default mode network, with emotion and/or motivation enabling the entrepreneur to adapt behavior to a given context. Understanding this 'standard complement' of physiological processes may allow researchers to explain similarities and differences among entrepreneurs and the opportunities they conceive. We offer 15 researchable premises that can be examined with current social-science methodologies and illustrate the implications of our approach (i.e., detailing how the brain generates behavior) for entrepreneurial cognition theory and research.

INTRODUCTION

Are entrepreneurs like anyone else or are they different? Our objective in this chapter is to explain how entrepreneurs' brains are physiologically the same as any other person's brain, but, in terms of experiences and knowledge, they are different. To achieve our objective, however, we need to take readers down the unfamiliar pathway of neurophysiology, from sensation to conscious thought – perception, conception and affective valuation – and into an attendant level of complexity some readers may find uncomfortable.

Our analysis holds implications for both research and teaching within the domain of entrepreneurship study. In tackling the question of whether entrepreneurs are different or the same – through the exploration of neurophysiology – we contribute to entrepreneurial cognition research by enabling future research to better focus on entrepreneurs' experiences and the implications for their perceptions, conceptions and affective valuations (see also McMullen et al., Chapter 9, this volume). When we assert that 'physiologically entrepreneurs' brains are no different than the brains of others,' we implicate current conceptualizations of the entrepreneur. We are therefore constrained to disagree in part with, for example, the 'unique person' conception of the entrepreneur used by Shane and Venkataraman (2000) (which may overemphasize the role of individuals with incomplete experience-based knowledge in contrast to the standardized perfect information/rational man of neoclassical economics); and instead we offer, and support with an extensive analysis of brain physiology, a conception of the entrepreneur that permits the existence of unique life experiences on one hand while emphasizing a 'standard complement' of neurological processes on the other. Thus we argue that a conceptual middle ground exists for examining similarities and differences among entrepreneurs and phenomena such as opportunity conception.

We reason further that if we better understand the physiological workings of the brain, we can target ways to teach entrepreneurship more effectively. Our assertion that 'entrepreneurs' brains are physiologically the same as others' brains but experientially they are different' offers a compelling rationale for efforts expended by society to teach and learn entrepreneurship – especially experientially. We make the argument in this chapter that a better understanding of how episodic memory actually works in the brain (how it is used by humans in general and also by entrepreneurs specifically), and how episodic memory informs conceptual reasoning, might lead us to validate, refine or develop pedagogies that rely first on creating and engaging students in certain entrepreneurial experiences and then using these new memories to conceive opportunities.

Our analysis in this chapter is motivated by the need to resolve what we perceive to be a false dichotomy in the entrepreneurship literature, which erroneously pits nature (entrepreneurial traits) against nurture (entrepreneurial experiences) to explain entrepreneurs' behaviors (see also Baron, Chapter 1, this volume; Bird, Chapter 3, this volume). The deliberate practice stream of research in the expert information processing theory literature helps with this resolution by suggesting that expertise within a domain results from the combination of so-called 'mental hardware' of the brain (e.g., hard coding, which produces learning rates, forgetting rates, speed of processing, and memory), with the 'mental software' of the brain (e.g., soft coding, which contributes knowledge bases and problem-solving processes) (Charness et al., 1996: 53). In this chapter we seek to map the terrain in the brain where soft coding meets hard coding. This mapping task is consistent with and may help to further amplify recent developments in cognitive psychology and entrepreneurial cognition research that integrate action, embodiment and social situation in explanations of entrepreneurial cognition (cf. Mitchell et al., 2011; Smith and Semin, 2004).

Our analysis is further motivated by findings that suggest a paradox of sorts in entrepreneurial cognitions themselves. On one hand, entrepreneurs as a group have been found to have cognitive patterns that are distinct from those of non-entrepreneurs – even across countries and cultures, suggesting a 'universal culture of entrepreneurship' (Mitchell et al., 2000: 988). On the other hand, within the group of entrepreneurs – depending upon culture – entrepreneurs have been found to have cognitive patterns that are distinct from each other (Mitchell et al., 2002). Theory is needed to explain how entrepreneurial cognitions can at once be the same, yet different. Our explanation of 'why physiologically entrepreneurs' brains are no different than the brains of others but experientially they are' offers such a theory.

As suggested in the first paragraph, to accomplish our task we must invite the reader to engage complexity. Yet the bewildering intricacy of the brain remains far too complicated for more than a highly specific analysis of just a few cognitive processes by way of illustration. Thus, in this chapter we utilize the sight-to-feeling pathway (what happens in the split seconds between seeing and feeling) as this illustration, with entrepreneurial affect (emotion and motivation) as a focal point. In the following sections we therefore: (1) provide additional background to frame our research problem and illustration within entrepreneurial cognition; (2) offer definitions for key terms grounded in (structures–functions) physiology and justify our working from clear definitions; (3) introduce several fundamental principles of systems neuroscience that underlie our analysis;

(4) describe the seeing-to-feeling pathway, starting with the ventral visual stream; (5) explain the hierarchical processing of episodic memory, with a link between percepts and concepts; (6) introduce emotion and then motivation to the exploration; and (7) discuss the implications of the analysis for entrepreneurial cognition research.

BACKGROUND

As noted in the Introduction, the research problem that is our objective in this chapter is to explain why, physiologically, entrepreneurs' brains are no different than the brains of others but, experientially, they are. Researchers have addressed some aspects of this problem over the last decade by refining the definition of the entrepreneur to refer to a thinking, feeling human being (Baron, 2004; Grégoire et al., 2011; Mitchell et al., 2002; Mitchell and Shepherd, 2010), not necessarily a rational economic agent, but an actual person whose behavior derives from a standard complement of neurological processes (Baron et al., 2012; Baron and Tang, 2011; Haynie et al., 2010; Tang et al., 2012), and whose choices may be susceptible to cognitive heuristics and biases (Busenitz and Barney, 1997; Keh et al., 2002; Simon and Houghton, 2002). These researchers argue both that these distinctly human attributes affect the entrepreneurial process (Mitchell and Shepherd, 2010), and events – experiences – in this process affect the individual (Morris et al., 2012).

In this regard, emotion has emerged as a 'hot topic' and a proposed umbrella concept, subsuming the entrepreneur's affective evaluations of many types of experiences (Baron, 2008; Cardon et al., 2012). Researchers have, for example, explored the impact of emotions on opportunity evaluation and exploitation (Foo, 2011; Grichnik et al., 2010; Welpe et al., 2012), creativity (Baron and Tang, 2011) and level of effort (Foo et al., 2009). Others have focused on the impact of specific emotions such as optimism (Hmieleski and Baron, 2009), passion (Breugst et al., 2012; Cardon et al., 2005), happiness (Hahn et al., 2012) or grief (Patzelt and Shepherd, 2011; Shepherd et al., 2009), as well as on coping behaviors and emotional intelligence (Foo et al., 2004; Rhee and White, 2007).

Such work has advanced the study of entrepreneurship through the rendering of the entrepreneur as human; but efforts to date to integrate affective evaluations into the entrepreneurial cognition research literature have been described as fragmented and limited (Grichnik et al., 2010). Significant challenges exist in identifying theoretical foundations from which to work (often requiring the use of disparate literatures); and in our observation, researchers have struggled to define core concepts (e.g., cognition, affect, motivation, emotion and stress) as well as relationships among these concepts (see also Carsrud and Brännback, Chapter 1, this volume; Foo et al., Chapter 5, this volume; Grégoire, Chapter 6, this volume). The research challenge has been magnified owing both to the subjectivity of entrepreneurial cognition as the phenomenon of interest, and to the currently unmet need for the use of common terminology: two research complications that risk placing the topic beyond the scope of science (LeDoux, 2012). Thus, for example, it is very difficult if not impossible to study directly the subjective experiences of entrepreneurs' 'feelings' or to distinguish among neural states corresponding to the glut of labels entrepreneurs may apply to introspective assays of their own feelings (e.g., angry versus frustrated; excited versus anxious; fearful versus stressed; or sad versus inept). Without a deeper, more accessible map of how cognitive processes work within humans as physical beings, the emerging research stream risks coming to resemble an amorphous construction of all things neural, built upon a shaky foundation that is an admixture rather than a coherent core.

We believe that research premises can be obtained from the in-depth analysis of the eye-to-brain pathway, to valuation, to behavior transmission physiology. Accordingly, as a specific and detailed illustration of the role of brain physiology in the complex processes involved in entrepreneurial decisions, actions, behaviors and so on, in this chapter we trace the path of episodically acquired data and their emotional/motivational valuation: from photons striking the retina, the brain identifying and categorizing stimuli, the integration into episodic memory imbued with semantic meaning, and the judging of familiar versus novel, toward the production of emotional and/or motivational value, all within a few hundred milliseconds. We therefore seek to contribute to the entrepreneurial cognition literature by offering a neuroscience-based analysis of entrepreneurial cognition, with entrepreneurial cognition defined as 'all of the computations that support the acquisition and comprehension of episodic experiences in the brain, expressly including the affective evaluations of emotion and motivation.' Thus, in this chapter we are concerned with the 'episodic processing' and 'affective evaluation' portions[1] of entrepreneurial cognition, that is, with the neural substrate (e.g., cortical fields and subcortical nuclei) of entrepreneurial experiences and their emotional/motivational valuation. Study of this neural foundation of entrepreneurial cognition can have important benefits.

As Buzsáki (2004: 446) suggests, 'Features of the physical world do not inherently convey whether [for a brain, or for an entrepreneur, for that matter – eds] a situation is familiar or novel, whether a stimulus is pleasant or repellent, etc. These attributes are added to the information conveyed by sensory inputs by a process referred to as cognition.' Thus we argue that it is not the case, as some would argue according to the Cartesian tradition, which ignored the importance of the body and situation (e.g., as did Descartes, Locke, Hume, Kant), that emotions impact cognition as some outside influence that impinges only on the mind as an inner entity of some kind (van Gelder, 1995: 380): an 'isolated mind' (Gallagher, 2009: 35). Rather, we argue that, and seek to better describe how, the neurophysiological foundation of emotion is an essential part of cognition, and is critical for explaining the making of appropriate choices. Thus we argue that the notion of cognition subsumes all the unconscious and conscious processes that turn sights and sounds into neural representations that allow the individual to make sense of the world, form goals and select suitable behaviors for a changing world, and that the study of the physiology informs a more thorough explanation of the entrepreneur and entrepreneurial cognition – especially of emotion and motivation.

Our illustration objective (Figure 8.1), therefore, is that by articulating the computations that take place as environmental data travel from peripheral sensory circuits to structures subserving episodic memory, as well as those computations associated with the affective modulation of neural activity, we can generate research premises for the further advancement of embodied cognition research within the socially situated entrepreneurial cognition research stream (Mitchell et al., 2011). We begin our analysis with some definitions; follow with an introduction to systems neuroscience; then trace the flow of data through the visual 'what' pathway to the inferior temporal cortex (ITC) where percepts (i.e., perceptions) are identified, known as the ventral visual pathway. We then explain how this information projects to the hierarchical medial temporal lobe (MTL) memory system responsible for episodic memory. Following this, we identify and describe the structures in the MTL that selectively gate for affective data, process spatial data, affix a time stamp, transform percepts into concepts and enable novelty detection. The discussion then turns to how the valuation of data occurs through the physiological stress response of emotion and the real-time computation of value-supporting motivation; and we discuss the implications that the physiology of this sight-to-emotion/motivation pathway has for entrepreneurial cognition research.

We therefore argue that the contribution from incorporating neuroscience research into entrepreneurial cognition (and possibly more generally into management) studies will more likely come from informing research questions that can be addressed with familiar and highly developed social-science methodologies, rather than from relying on (for example) fMRI machine output and identifying which parts of the brain are active when the subject does 'x.' We argue this on both practical and theoretical

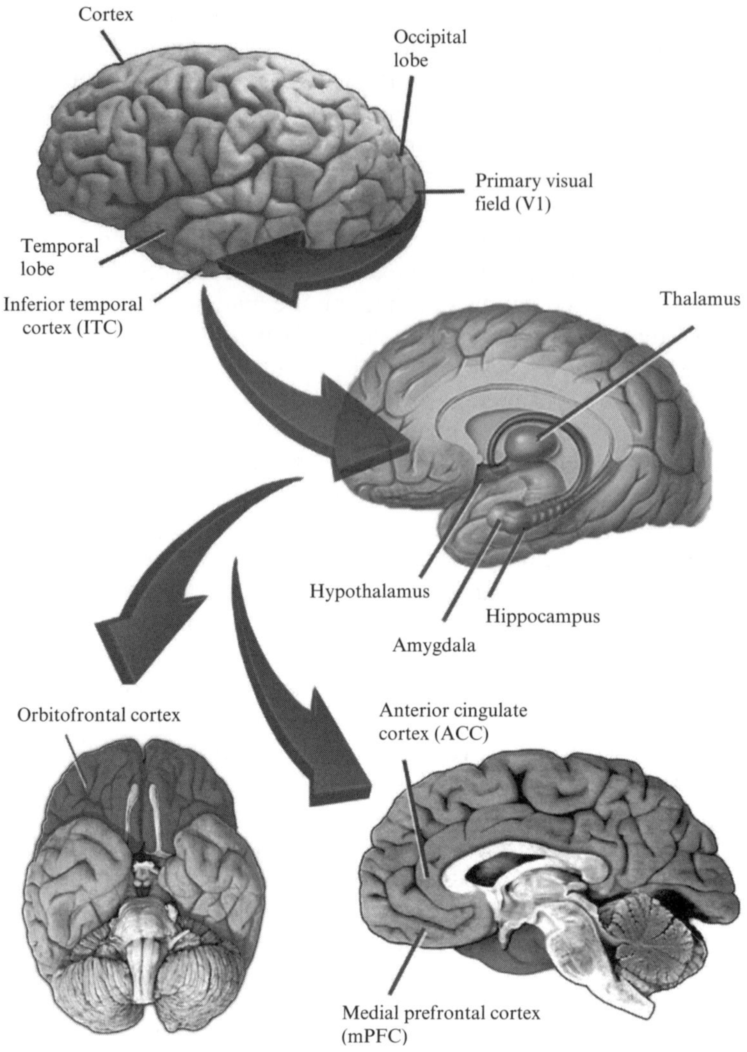

Note: Simplified flow of sensory (visual) data from the periphery to conscious thought: primary visual field (V1) to inferior temporal cortex (ITC); to hippocampus and amygdala (and hypothalamus – triggering a physiological reaction); to orbitofrontal cortex (OFC), medial prefrontal cortex (mPFC) and anterior cingulate cortex (ACC); Cortex ≡ outermost layer of the cerebral hemispheres of the brain.

Source: Adapted from Pyramis Studios, 2007.

Figure 8.1 Simplified flow of visual sensory data from the periphery to conscious thought

grounds. From a pragmatic standpoint, few management (entrepreneurship) researchers presently have access to imaging technologies or the training necessary to conduct the sophisticated statistical analyses required for processing voxel data (e.g., very large samples of regularly spaced data points on a three-dimensional grid). (We note that neuroscience researchers do this kind of research well and we believe it to be sufficient to simply follow their lead to advance entrepreneurship research.) From a theory-building standpoint, we intend to demonstrate how, by tracing the pathway from sight to emotion/motivation and to behavior, likely premises can be isolated for future research attention. In this chapter, we therefore utilize an in-depth analysis of neural structures to generate likely premises for entrepreneurial cognition research.

THE NEED FOR DEFINITIONS

In our observation, some entrepreneurship researchers are beginning to conduct neuroscience research, but perhaps without the advantage of fully grounding their work in analyses of neural structures and functions: the physical 'substrate' or foundation of cognitive processes. We believe studies of neural phenomena, in particular such important areas of study to entrepreneurship as emotions and motivation, with additional grounding in the anatomical and physiological analyses that describe the structure of specific brain areas (i.e., cortical fields and subcortical nuclei) and the functions (i.e., computations) performed by these structures, can assist researchers to explain previously un- or under-explained variance in the impacts of both emotion and motivation on entrepreneurial-cognition-based outcomes (cf. Mitchell et al., 2007). Such a foundation in structures and functions has the potential to free researchers from the constraints of inevitable subjectivity previously attached to explanations of affective phenomena (cf. LeDoux, 2012) and to offer key insights for clearly defining and differentiating among concepts, and for conducting additionally revealing research into the cognitive processes of entrepreneurs. We therefore offer working definitions and abbreviations of the key terminology to be utilized (see Boxes 8.1 and 8.2) as a foundation for the discussion in this chapter.

In the following sections we shall proceed to analyze key portions of the physiological foundation of embodied cognition, that is, 'how the body shapes the mind' (see Gallagher, 2005) to: (1) explain how medial temporal lobe (MTL) memory structures support episodic memory, the rapid coding of one-time experiences (Hafting et al., 2005; Hargreaves et al., 2005; Kondo et al., 2008, 2009; Lavenex and Amaral, 2000; Lavenex

BOX 8.1 DEFINITIONS

Entrepreneur An individual with the same neural structures as anyone else and whose behavior derives from the standard complement of neurological processes. No neurological advantage need exist. This person's adaptive abilities derive from conserved survival circuits that support functions such as the affective valuation of stimuli, prediction and episodic future thought.

Cognition The umbrella concept subsuming all the unconscious and conscious processes, expressly including semantic processing and affective valuations (i.e., emotion and motivation), that turn sights and sounds, mere photons and air-wave fluctuations, into neural representations that allow the individual to make sense of the world, form goals, and select suitable behaviors for an ever-changing world.

Affective valuation The modulation of activity in neural structures from periphery sensory and memory circuits to conscious thought in the DMN, either through direct axonal connection or release of hormones in the body and neurotransmitters in the brain. Valuation derives from the preferential processing of some environmental stimuli (e.g., people, places and things) over others, with prediction errors (e.g., pleasant surprises, missteps or miscues) implicated in inducing the modulation of neural activity. Subsumes emotion and motivation.

Emotion Emotion centers on the amygdala and activation of a visceral stress response through the hypothalamus, sympathetic nervous system and stress hormones: glucocorticoids, epinephrine (a.k.a., adrenaline) and norepinephrine released from the adrenal cortex. An event happens that elicits the coincident neurotransmitter release in the brain and physiological reaction in the body – both modulating information processing. Emotion serves a purpose (function) of tagging behaviorally relevant stimuli or experiences (i.e., beneficial or harmful) that require a response, so 'emotion' may be best defined and measured by a specific stimulus (or an experience) that induces a stress response, the valence (i.e., positive or negative) of the response, and the evoked behavior.

Motivation Motivation refers to a dopamine-modulated state of arousal consonant with anticipation that proceeds and guides reward-seeking (a.k.a. goal-directed or approach) behaviors toward desired outcomes. Appetitive (e.g., food and water) and abstract (e.g., money) stimuli alike acquire subjective reward value as a function of the individual's experiences with these stimuli. The two components of motivation, (1) arousal prior to rewards and (2) reinforcement of reward-seeking behaviors upon the timely delivery of rewards, occur through the operation of the midbrain dopamine (DA) system centered on a distributed network of DA-modulated structures, with special emphasis on the orbitofrontal cortex (OFC), ventral tegmental area (VTA) and ventromedial striatum (a.k.a. nucleus accumbens – NAc). Motivation may be best studied as a specific stimulus with reward value and an evoked behavior.

BOX 8.2 ANATOMICAL ABBREVIATIONS

ACC	anterior cingulate cortex
AMG	amygdala
BLA	basolateral amygdala
CA1 and CA3	hippocampal fields (CA – Cornu Ammonis)
DA	dopamine
DG	dentate gyrus
dlS	dorsolateral striatum
DMN	default mode network
dmS	dorsomedial striatum
dmPFC	dorsomedial prefrontal cortex
ERC	entorhinal cortex
HF	hippocampal formation
HIP	hippocampus
HPA	hypothalamus–pituitary–adrenal axis
ITC	inferior temporal cortex
LEA	lateral entorhinal area
LGN	lateral geniculate nucleus
MDS	midbrain dopamine system
MEA	medial entorhinal area
mPFC	medial prefrontal cortex
MTL	medial temporal lobe (memory system)
NAc	nucleus accumbens (a.k.a. vmS)
OFC	orbitofrontal cortex
PRC	perirhinal cortex
PRPH	perirhinal and parahippocampal cortices
PVN	paraventricular nuclei of the hypothalamus
V1, V2 and V3	visual processing fields
vmPRC	ventromedial prefrontal cortex

et al., 2004; McHugh et al., 2007); (2) describe the amygdala and physiological stress response that weights experiences and stimuli, those slivers of memories entrepreneurs label as emotionally significant (Herman et al., 2005; Kreibig, 2010; Sah et al., 2003; Tsigos and Chrousos, 2002); and

(3) describe the flow of dopamine from the ventral tegmental area (VTA) to the ventromedial striatum (a.k.a. nucleus accumbens – NAc) that supports motivation and value-based decision making (Öngür et al., 2003; Price and Drevets, 2010; Sesack and Grace, 2010; Voorn et al., 2004).

Such study offers a foundation for investigating cognitive processes (e.g., emotion and motivation); and we therefore believe that grounding entrepreneurship studies in neuroscience represents a logical and helpful addition to the ongoing entrepreneurial cognition research effort. To date, entrepreneurship researchers have embraced an agenda taken primarily from cognitive psychology, as illustrated (non-exhaustively) in analyses of cognitive biases, emotion-as-information, expert scripts and venture creation, and effectuation (e.g., Baron, 2008; Busenitz and Barney, 1997; Foo et al., 2009; Grichnik et al., 2010; Mitchell et al., 2002, 2007). Work in neuroscience has the potential to advance the entrepreneurial cognition research agenda by articulating the structures and functions involved in processing information from periphery circuits to conscious thought (in the default mode network – DMN, as discussed below), with emotion and motivation (as focal cases) being shown to modulate the acquisition, processing and storage of information, as well as the selection of behaviors along the way.

KEY PRINCIPLES OF SYSTEMS NEUROSCIENCE

As we illustrate in this chapter, researchers have the opportunity to ground more deeply entrepreneurial-cognition-based explanations in certain established principles of systems neuroscience, expressed as follows: (1) structure drives function; (2) endogenous activity constantly occurs; (3) experience modulates cognition; (4) critical cognitive processes occur offline; and (5) building blocks of cognition exist – that is, far from the periphery, the brain codes for abstract representations more aligned with the meaning and quality of often-seen, familiar and personally relevant percepts than the physical features of things out there (Albright, 2012; Buzsáki, 2004). In the following subsections we offer a deeper explanation of each principle in turn.

Structure Drives Function

The brain is built a certain way. As Wallace and Kerr (2010) note, the anatomy and physiology of individual cells (see Douglas and Martin, 2007; Spruston, 2008) and the structure of the cortex are well described as follows. The cortex is arranged in a laminar, six-layer organization referred to as isocortex (Harris et al., 2011; Treves, 2003). This gray-matter sheath is populated with about 80 percent excitatory pyramidal neurons (the most abundant excitatory cell type in virtually every mammal studied; found in the human cortex, hippocampus and amygdala; and associated with advanced cognitive functions) and 20 percent local inhibitory interneurons (cells thought to stabilize and directly shape the function (computations) of local circuits) (Markram et al., 2004; Spruston, 2008), with excitatory and inhibitory neurons organized as duplications of a stereotypical microcircuit template (Markram et al., 2004). Adaptations of this template are tied to the functions performed by specialized neural structures and the requirements of cognition. We examine the laminar and template structure of cortical fields and subcortical nuclei and their functions at each step in our analysis of the brain's work and workload.

Endogenous Activity Constantly Occurs: No Dormant Starting Condition

Is the brain's workload externally or internally generated? Two views of brain function have existed since the early twentieth century: one posits that the brain is primarily responsive to environmental demands (suggesting an external workload); and the other holds that most brain activity is endogenous (suggesting primarily an internal workload) (Raichle, 2009). The former view has historically influenced neuroimaging studies and may also constitute an extant belief in entrepreneurship, as in an emphasis on studies of overt behaviors in pursuit of opportunities that exist 'out there.' (We note that the discussions (and debates) over entrepreneurial discovery and creation may take some of their shape from such beliefs (cf. Alvarez and Barney, 2007).)

Nonetheless, internal demands are not insignificant. For example, metabolic analyses show that the brain maintains an enormous level of energy consumption, estimated at 20 percent of the body's total oxygen and glucose resources (Raichle, 2010). As much as 80 percent of this energy is consumed through neuron signaling (Sibson et al., 1998), which suggests that it is therefore functionally significant endogenous activity, defined as ongoing and independent of the presence or absence of stimuli or observable behaviors (Snyder and Raichle, 2012). Energy consumption ascribed to environmental demands is estimated at less than 5 percent (Raichle and Gusnard, 2002; Raichle and Mintun, 2006). Hence we assume that the idea of the 'dormant' brain is not credible.

Experience Modulates Cognition

If the brain is always working, but is not necessarily always heavily engaged with the environment, what might be driving the massive internal activity that continuously occurs? Numerous explanations for the existence and function of endogenous brain activity have been posited over the years. Some researchers report that this activity plays a functional role in maintaining an excitatory–inhibitory balance in neural systems (Raichle and Gusnard, 2002; Wehr and Zador, 2003), enabling neurons to detect and respond to sensory stimuli on a millisecond timescale and speed information transfer (Deco et al., 2011; Kayser et al., 2010). Others argue that endogenous activity plays a critical role in maintaining normal brain function, with disruptions of this activity linked to a variety of pathologies and to the degenerative effects of aging (Damoiseaux et al., 2007; Greicius et al., 2004; van den Heuvel and Hulshoff Pol, 2010). Thus both brain balance and brain maintenance consume brain energy; but all of the energy?

Of course, the function of endogenous activity remains the subject of ongoing research, but sufficient evidence exists to also implicate experience in modulating cognition. How experience drives brain activity may also be explained physiologically. Endogenous activity originates in the electrophysiology of neuron spikes and spike correlations (Buzsáki and Draguhn, 2004; He et al., 2008), and this activity is either noise or neural code relaying information through functional networks. Harris (2005) argues that synchronous and repeated activation of neurons during sensory stimulation and behavior enables neuron ensembles (i.e., cell assemblies – groups of neurons that function to sustain a memory trace – Hebb, 1949) to decouple from external events. Thus neurons are thought to maintain activity owing to repeated experiences without sensory stimulation, and (importantly) can be triggered by internal processes alone, for example by thought.

It follows that endogenous brain activity carries on meaningful tasks and, for example, plays important roles in: (1) continuously sharing information (van den Heuvel and Hulshoff Pol, 2010); (2) maintaining ongoing internal representations (abstractions) of the external world; and (3) modulating neural activity during stimulus presentation (Harris, 2005; Kayser et al., 2010). Thus we can see how the brain learns to preferentially respond to and process familiar and relevant stimuli: how experience modulates cognition. We can also draw the inevitable conclusion that there likely is a so-called 'offline' element to cognition.

Cognition Takes Place Offline

Offline cognition matters. A critical discovery in analyses of endogenous activity is that resting-state neuron networks exist, defined as functionally related brain areas with high levels of correlated endogenous activity during rest (Raichle et al., 2001; van den Heuvel and Hulshoff Pol, 2010). Resting-state networks consist of structures with a common function and subserve known processes such as motor function, vision, audition, language and offline cognition in the 'default mode network,' or DMN, a set of brain regions that remain active when an individual is not externally focused, and when the brain is supposedly at rest – that is, offline.

The DMN exhibits heightened neural activity at rest and decreased activation from a 'resting-state' baseline during externally focused tasks: negatively correlated with attention. Thus this endogenous network comprises an offline 'default' mode of cognition to which humans inevitably return when not externally engaged. The DMN subserves, or helps to further or promote, episodic recall and autobiographical memory (Greicius et al., 2003), self-referential thought (Gusnard et al., 2001; Whitfield-Gabrieli et al., 2011), emotion processing (Immordino-Yang et al., 2009), semantic processing (Binder et al., 1999; Wirth et al., 2011), as well as problem solving and episodic future thought (Buckner and Carroll, 2007; Gerlach et al., 2011; Szpunar et al., 2007).

Building Blocks of Thought Exist

At the highest level of processing, the entorhinal cortex and hippocampus developed to support episodic (percepts) and semantic (concepts) memory (Buzsáki and Moser, 2013), while other structures compute affective valuations relevant to emotion and motivation: the amygdala and orbitofrontal cortex respectively. These representations comprise abstractions aligned more with the meaning or quality of that which is often seen, familiar and personally relevant than physical attributes of something out there. In other words, the brain turns dimensionless, man-created concepts (e.g., entrepreneur) into rich representations informed by experience, with meaning derived from statistical regularities (i.e., learned associations) in the environment, and value ascribed to physiological processes, described below. These abstractions comprise the building blocks of thought and, combined with sensory data, ongoing representations of the world in the DMN that support adaptive behaviors. This systems neuroscience characterization of cognitive activity – as ongoing, offline and supported by abstract building blocks (representations) of thought – may align with and refine the concept of 'the mind' employed in other literatures.

Summary

For entrepreneurial cognition researchers, an understanding of the foregoing principles is the first step in laying the foundation for a more exacting

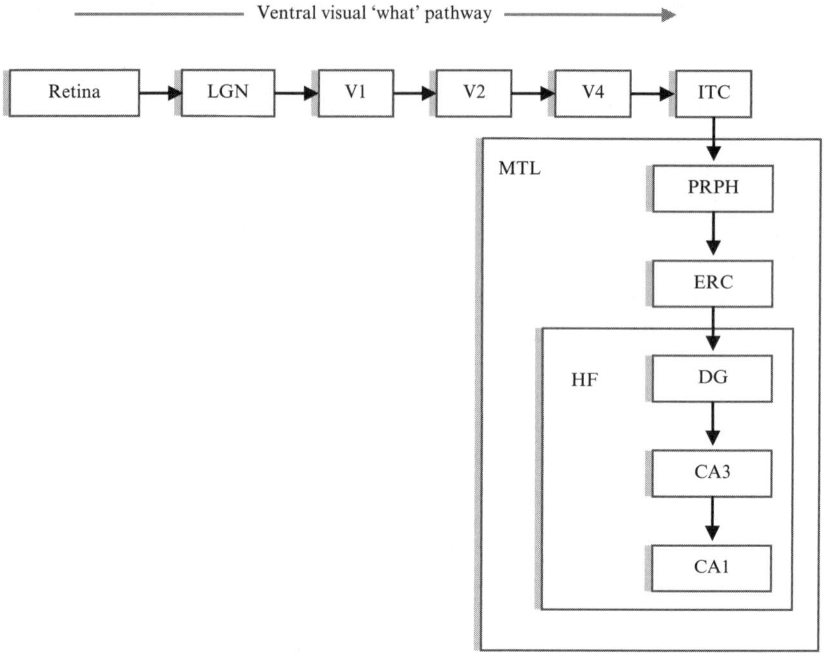

Note: For abbreviations, see Box 8.2.

Figure 8.2 Visual 'what' pathway and medial temporal lobe memory system

physiological explanation of selected entrepreneurial cognition processes/outcomes. The next step in our analysis proceeds with a discussion of the ventral visual 'what' pathway (Figure 8.2) and the distinction between sensation (i.e., of mere photons and sound waves) and perception.[2] The discussion and explanation offered in the next section – and in subsequent sections and subsections – is preceded by a brief explanation or map of the section (which appears in italics), which is offered as a mechanism for the reader to receive a high-level explanation of the material that then appears in detail within that section.

THE VENTRAL VISUAL STREAM

When the individual views the world, visual stimuli enter the brain as mere data points (an orientation and spatial frequency) without meaning. This information undergoes associative processing as it flows through the ventral visual stream (a.k.a. – 'what pathway') and is recognized as behaviorally relevant perceptions (hereafter percepts) in the inferior temporal cortex (ITC). These meaningful percepts are organized categorically (e.g., people, places and things); encompass personally familiar people; are the product of both what was seen (real-time data) and experience-based imagery; and suggest particular behavioral responses (though conscious thought, 'this could be an exciting opportunity' and connections to behavior occur elsewhere). Moreover, the recognition of meaningful percepts can occur (for faces) in under 100 milliseconds.

All sensory data, except olfaction (sense of smell), enter the brain through the thalamus, a structure that gates data transfer to sensory cortices: an initial barrage of excitatory input is rapidly followed by feedforward inhibition that squelches spiking responses in sensory neurons. The thalamus thus increases the temporal precision of firing patterns in sensory cortices and reduces noise: it has a function (Gabarnet et al., 2005). As illustrated in Figure 8.2, we trace the path of visual data from the retina in the eye onward, to clarify the functions performed by other structures in the ventral visual stream, dubbed the 'what' pathway, as well as those in the medial temporal lobe (MTL) memory system.

These functions comprise the initial computations in human cognitive processes, those that transform mere data points (encoded in individual V1, V2, V4 neurons) into high-level percepts recognized and stored in the inferior temporal cortex (ITC), then projected to the MTL (Lavenex and Amaral, 2000) and integrated into episodic memory (i.e., the rich and unique recollection of personal experiences or events from a person's life – Tulving, 1972, 1983), as well as transformed into abstract conceptions of the world – percepts become concepts (Quian Quiroga, 2012). We detail the: (1) computations that take place along the way; (2) content of episodic memory (e.g., percepts and concepts) entrepreneurs may tag as emotionally significant; and, thereby, enable the description of: (3) the essential role of affective valuation in cognition.

We note that episodic memory is defined as memory possessing emotional value: neutral events are not processed and consolidated into long-term memory. Episodic memory also comprises one interpretation of 'experience' from neuroscience. It has the potential to offer new insights into the elements of experience (e.g., scenes, objects, orientations, perspectives, self and other people), so that the entrepreneur may leverage it to envision a business venture. Neuroscientists suggest that the individual flexibly recombines bits and pieces of memory to pre-experience the future through mental simulation, a cognitive trial-and-error process used to select actions with the greatest potential for producing desired outcomes (Addis et al., 2009; Schacter et al., 2008). Thus the ability of the potential entrepreneur to identify and exploit an opportunity may depend on her ability to 'see' the idea and its potential, combined with how she anchors her mental simulation (e.g., a product idea, garage startup concept, desired customer profile, or the benefits of being one's own boss). Of course, not all such starting points lead to success. This is partly due to what happens as sensory data transform into behaviorally relevant percepts, concepts and affective valuations.

From Sensation to Perception

The 'what' pathway is characterized as a hierarchical, recurrent and highly interactive network that processes object quality (Kravitz et al., 2013). Receptive fields, onset latencies and selectivity for complex stimuli increase as data travel from the V1 to the ITC, which is indicative of a hierarchical structure of neural flow. Nonetheless, anatomical data indicate that both feedforward and feedback projections exist among stages; and extrinsic projections to memory, valuation and cognitive systems occur at varied stages (Kravitz et al., 2013). These complex patterns of connectivity help explain the categorical structure of percept data (e.g., clustered by faces, scenes and tools), and the pathways through which top–down influences occur (i.e., how attention is controlled by internal/endogenous factors such as goals).[3]

We therefore offer the following explanation. Neurons in the V1 (the primary visual cortex) have receptive fields selective for minute details of stimuli, with individual cells coding for both the orientation at a point in space and spatial frequency (i.e., sharp edges versus matt surfaces) of stimuli at that location (DiCarlo et al., 2012). Wide diversity exists in the two-dimensional tuning of cells, thereby enabling V1 neurons to respond to the same details of all stimuli, consistent with V1's function as the gateway for visual data into the cortex. The V1, however, does not yield camera-like representations of the world.

Higher-level visual fields (e.g., V2 and V4) perform associative integration functions (e.g., encoding various combinations of input from the V1) to form more complex object representations: they aggregate the output of prior-stage feature detectors, while exhibiting successively larger receptive fields and greater tolerance (i.e., invariance) to variations in size, position and context (Hegdé and Van Essen, 2007; Kravitz et al., 2013; Rust and DiCarlo, 2010). Functional clustering occurs owing to heterogeneous wiring of the 'what' pathway and processing requirements of stimuli. Researchers posit that categories of visual stimuli, requiring large populations of neurons and recurrent processing, cluster in areas affording the necessary connectivity and computing power, as in the fusiform face area (the part of our human visual system specialized in facial and/or object recognition) and scene-selective parahippocampal place area (where scene/place coding and recognition occur) (Kravitz et al., 2013).

More than Meets the Eye: Modulation by Experience

Object recognition occurs in the ITC within about 150 milliseconds, but as Albright (2012: 227) explains: 'It should come as no surprise that what you see is not determined solely by the patterns of light that fall upon your retinae.' Retinal input accounts for only about 20 percent of V1 activity (Muckli and Petro, 2013) and coding in the ITC is posited to be the joint product of current input and repeated experiences. Neurons in the ITC exhibit plasticity into adulthood. Experience-dependent learning steers ITC neurons to respond to familiar stimuli and with greater sparseness: fewer neurons carry better information for directing perceptually guided behavior to well-known objects. It is through recurrent pathways that top–down attentional effects (e.g., the goals of an individual) influence the entire state of the 'what' pathway and filter irrelevant distractors.

Implications for Entrepreneurs

What does this mean for our understanding of entrepreneurial cognition? Stated simply, we learn from this structural account of how signals move from the retina to 'matchmaking' (coding for light patterns + experience) in the ITC, that the structures and functions in the visual 'what' pathway support the transformation of sensations into behaviorally relevant percepts/perceptions, such that what the entrepreneur sees is modulated by experience. Neurons are tuned to familiar people, places and things: object identification. Based on the fundamental structure of the ventral visual system and prior research (Mitchell and Shepherd, 2010) into 'images of opportunity,' we therefore suggest the following premise:

Premise 1: It is likely that entrepreneurs rely on behaviorally relevant percepts – part stimulus-driven and experience-guided imagery of familiar people, places and things – to envision a business venture.

Our analysis of the ventral visual stream aligns with this conjecture, as neurons in the ITC are part stimulus-driven and part experience-tuned to respond to probable things (Albright, 2012). The modulation of perception occurs implicitly, out of habit, and supports refined perceptions (e.g., a doctor's diagnosis of patient symptoms). It enables the craftsperson to leverage perceptual abilities to start a business, but likely hinders her recognition of innovative business opportunities. Our analysis now turns to MTL structures (Figures 8.2 and 8.3) and an analysis of the further hierarchical processing of high-level percepts into integrated episodic memories and concepts.

HOW EPISODIC MEMORIES ARE PROCESSED

Here we describe how percepts from the ITC, as well as all other sensory and endogenously generated data comprising an experience, converge on the medial temporal lobe (MTL) memory system and are integrated into whole episodic memories. Each structure in the MTL performs a specific function, as in gating, spatial processing, distinguishing coups from miscues, associative pattern completion and novelty detection. The types of processing that occur provide insights into the cognitive abilities and composition of episodic memory (experience) that entrepreneurs may use to 'envision' a business opportunity. The MTL is also well recognized as the place at which percepts inform concepts: the nexus between experience and semantic comprehension,

Researchers describe the inferior temporal cortex (ITC) as the pinnacle of the visual 'what' pathway and the long-term repository of visual memory. The ITC massively and reciprocally connects with medial temporal lobe (MTL) memory structures – defined as including the perirhinal–parahippocampal cortices, entorhinal cortex, and hippocampal formation (dentate gyrus, hippocampus (CA3 and CA1) and subiculum (Lavenex and Amaral, 2000)). The MTL supports: (1) rapid recording of one-time episodic experiences; (2) associative learning (plasticity) in cortical systems; as well as (3) memory consolidation and recall. We analyze the flow of data through the MTL, while explicitly linking structures and functions (Figure 8.3).

Affectively Weighted Information Passes through

The perirhinal and parahippocampal cortices (PRPH) provide a powerful illustration of the importance of affective valuation in cognitive processes in the brain. This structure, a pair of cortical fields, gates noisy and otherwise nonessential details (e.g., nameless faces seen on the street), while permitting emotion-laden data to pass through. In other words, high-value (positive or negative) content pertaining to venture

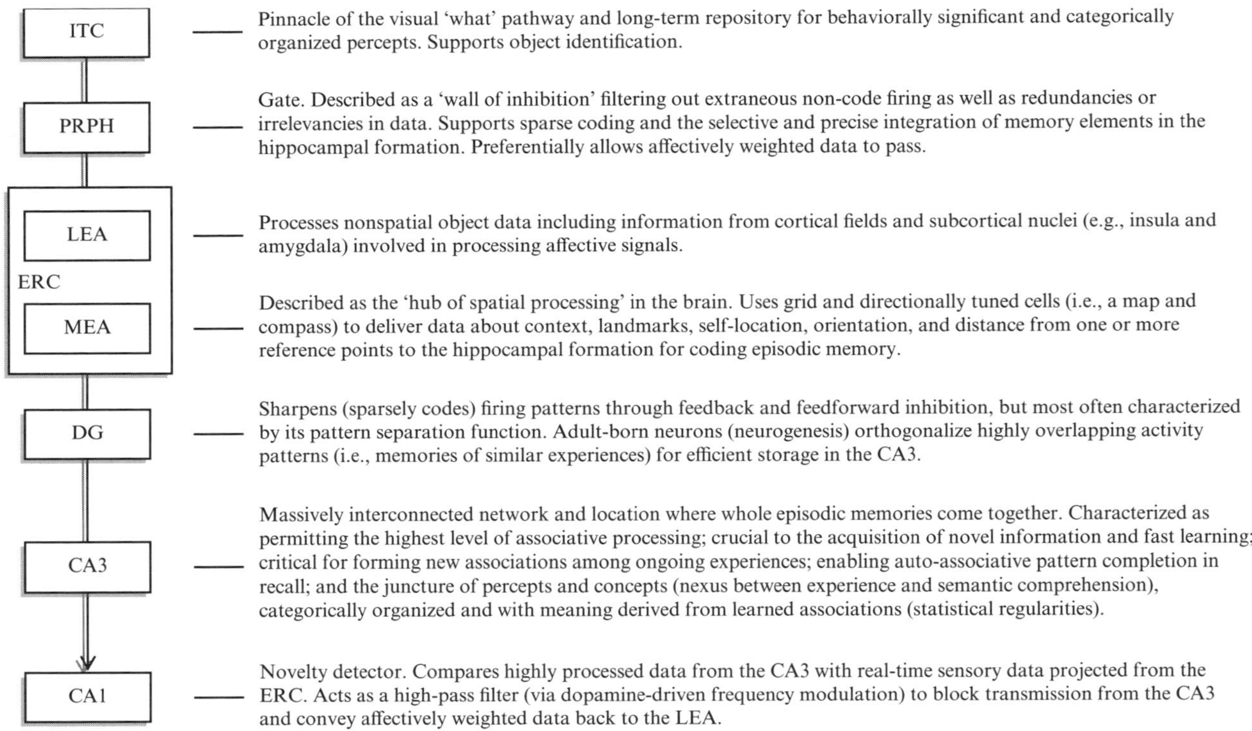

Note: For abbreviations, see Box 8.2.

Figure 8.3 Structures and functions of the medial temporal lobe (MTL)

performance or a promising opportunity gets preferentially processed into episodic memory.

Information from the ITC visual field – along with all other data comprising an experience – converges on the perirhinal and parahippocampal cortices (PRPH), the gateway to the MTL. Together these structures account for nearly two-thirds of the cortical inputs to the entorhinal cortex (ERC) (Insausti et al., 1987), where spatial and nonspatial (e.g., affective) processing occur (Hargreaves et al., 2005). However, anatomists characterize unimodal, polymodal and endogenously generated data as hitting a 'wall of inhibition' at the PRPH, an apt construal of this region's function in the MTL (Curtis and Paré, 2004). This is because the PRPH operates as a gate that inhibits propagation of extraneous data to the ERC and hippocampal formation (strongly implicated in episodic memory), while allowing affectively weighted information to pass. That is, in the PRPH, noise and nonessential details (e.g., nameless faces seen on the street) are deleted, while emotion-laden data pass through.

Animal studies show that stimulation of the ITC excites as many as 42 percent of perirhinal (PRC) neurons compared to only 2 percent of ERC cells, a response characterized as rare and sparse (i.e., stimulus coding by a small percentage of neurons) (Curtis and Paré, 2004). The PRPH likely quashes activity owing to highly spontaneous, non-code firing in cortical circuits (i.e., noise) and redundancies or irrelevancies in data, a function that supports the selective and precise integration of memory elements in the hippocampal formation. Indeed, problems occur when too much information gets into the hippocampus. Failure of inhibitory control has been implicated in overstimulation or epileptic activity in the ERC and hippocampal CA3 and CA1 fields (Curtis and Paré, 2004).

Nevertheless, inhibitory control by the PRPH partially lifts owing to input from the amygdala (strongly implicated in the physiology of emotion) and medial prefrontal cortex (mPFC) (supporting conscious thought, autonomic/emotional control, and flexible/adaptive behavior) to the PRC (see above) and the ERC (Kajiwara et al., 2008). For example, stimulation of the basolateral amygdala (BLA) has been found to activate approximately 15 percent of both PRPH and ERC neurons (Curtis and Paré, 2004). These authors speculate, owing to the accepted roles of the mPFC and BLA in reward-related behavior and emotion (Öngür and Price, 2000), that 'traffic through the rhinal cortices [PRC and ERC] varies depending on the emotional significance of current environmental contingencies' (Curtis and Paré, 2004: 106). From a neuroscience perspective we can see how, due to the PRPH, high-value (positive or negative) content gets preferentially processed into memory (first in the ERC, then in hippocampal formation), illustrating how essential it is to better understand the role of emotion and motivation in cognition generally, and in entrepreneurship specifically. How does the ERC process spatial memory?

Spatial Processing: Map and Compass

The ERC processes contextual (affective and behaviorally relevant) data, but is best known for spatial processing: it acts as a map and compass. Researchers argue the ERC evolved to support spatial navigation (e.g., a mouse foraging for food), but now supports human episodic and semantic memory, with the implication that navigation in physical and mental space relies on the same basic algorithms. The ERC supports landmark-base and path integration (PI) navigation, and these relate to semantic and episodic memory, respectively. We describe the ERC as a map and compass here (a foundation) and develop implications for entrepreneurial cognition in our discussion section.

The entorhinal cortex (ERC) constitutes one of the more interesting structures in the brain owing to its function in orienting the individual to the world, with computations tied to short-term synaptic plasticity – the tendency for the brain to change its connections in response to changing stimuli. Most people know that synaptic plasticity – cellular-level changes in dendritic spine architecture – supports long-term memory (Lamprecht and LeDoux, 2004). Less well known is that short-term plasticity can occur on the dynamic timescale of behavior (i.e., milliseconds) in the performance of neural computations (Abbott and Regehr, 2004; Fujisawa et al., 2008). Short-term plasticity in the ERC supports real-time processing of spatial data, but it may not retain memory for more than minutes or seconds (Almeida et al., 2012).

As shown in Figure 8.2, the entorhinal cortex (ERC) consists of two parts: the lateral entorhinal area (LEA) and the medial entorhinal area (MEA), executing nonspatial and spatial processing respectively (Hargreaves et al., 2005). The LEA processes nonspatial data about context, such as the emotional/behavioral significance of stimuli (i.e., people, places or things), consistent with its inputs from the olfactory and frontal cortices, insular cortex (involved in taste, emotion, homeostasis), and from subcortical amygdala and hypothalamus involved in triggering the physiological stress responses of emotion (de Quervain et al., 2009; Herman et al., 2005; Kreibig, 2010).

The (dorsocaudal) MEA operates as a generalized, context-independent representation of spatial environments, with an internal map of rigid spatial relationships that registers comparable distances across environments and thereby offers a neural metric for distance (Jeffery and Burgess, 2006; Moser and Moser, 2008). Grid cells in layer II code for regularly spaced external locations (Hafting et al., 2005; Moser and Moser, 2008; Sargolini et al., 2006), while cells in layers III, V and VI are directionally tuned based on head direction (Sargolini et al., 2006), with the combination of layers forming an internal map and compass. This structure supports two types of navigation. Grids anchor to landmarks that permit navigation via goal-directed paths, with landmarks functioning as key decision points (i.e., go here, turn there) (Janzen and van Turennout, 2004; Maguire et al., 1997; Rosenbaum et al., 2004). Path integration (PI) navigation uses self-motion cues to track linear and angular movements and continuously update representations of self-location, orientation and distance relative to defined (past) reference points (Etienne and Jeffery, 2004; McNaughton et al., 2006).

In sum, the entorhinal cortex (ERC) processes contextual and spatial information, with the LEA and MEA together deliver data about affective value, landmarks, self-location, orientation and distance from reference points to the hippocampal formation for coding in episodic memory. Thus the entrepreneur may recall specific details from an episodic memory: a scene, where s/he was standing, direction s/he was looking, location of prominent spatial features, objects and other people. S/he may project forward by flexibly combining similar elements of memory to envision a business venture, and this suggests our second premise:

Premise 2: It is likely that entrepreneurs rely on specific details from episodic memory to envision a business venture.

Next, the entorhinal cortex provides the major inputs to the dentate gyrus, which enables the pattern separation skills found in most mammals, including humans (Figure 8.4).

Separating Coups from Miscues

Entrepreneurs acquire episodic experiences daily, with these experiences populated by the same people, places and things. The question arises, 'How does the entrepreneur distinguish this time from last time, or tragedy from triumph?' The dentate gyrus (DG) transforms highly overlapping episodic memories – relying on adult neurogenesis to affix a time stamp – into sparse, orthogonal representations for storage in the hippocampus. The entrepreneur may rely on temporal differences in episodic memories (subtle changes) as signals of opportunity.

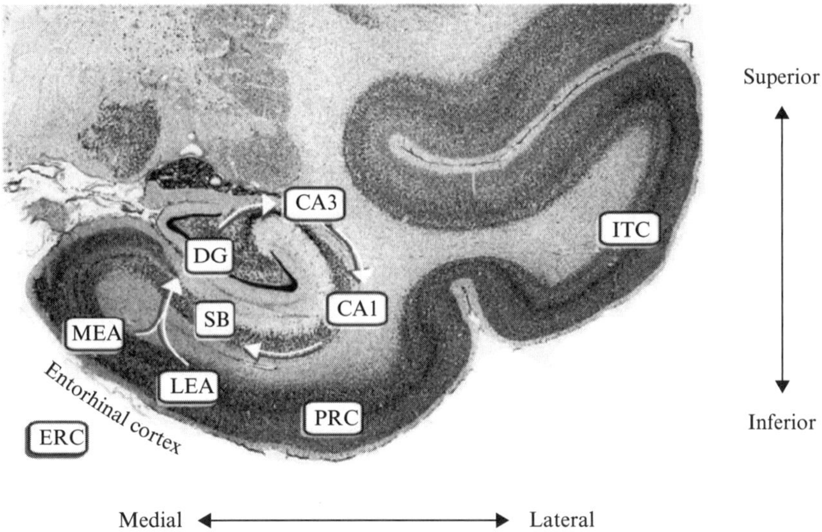

Note: For abbreviations, see Box 8.2.

Source: Adapted from Brainmaps.org.

Figure 8.4 Hippocampal formation

The dentate gyrus (DG) is anatomically notable as the primary entry point into the hippocampus proper (fields CA3 and CA1), where episodic memories are stored (and recalled) and the highest level of associative processing occurs. All DG granule cells (small, densely packed and uniformly sized cells located in the DG granule layer) project to CA3 pyramidal cells (Seress, 2007). In this section we examine the function of the DG as a filter or gate further reducing noise (Hsu, 2007) and a pattern separator for the efficient (i.e., sparse and orthogonal) storage of episodic memories in the extremely small hippocampal region: CA3.

To accomplish this task, the DG functions as a gate by producing both recurrent and feedforward inhibition (Hsu, 2007). For instance, anatomical data show that a higher percentage of inhibitory interneurons exists in the DG compared to the cortex (i.e., 50 percent versus 10 to 20 percent, respectively) (Acsády and Káli, 2007) and, through strong recurrent inhibition, spontaneous non-code firing in the DG is very low (0.1–0.01 spikes/s). DG granule cells are thus extremely quiet (Jung and McNaughton, 1993). When these cells participate in data transfer, interneurons respond very fast to suppress noisy non-code activity in a competitive manner (Acsády and Káli, 2007), so only patterns of ERC activation that are consistently and repetitively presented (i.e., familiar or the focus of attention) pass through to the hippocampal CA3 region (Hsu, 2007).

DG granule cells drive feedforward inhibition in the CA3 region as well. Researchers report that a single granule cell contacts an estimated 15 CA3 pyramid neurons, with each CA3 neuron receiving about 50 excitatory inputs. In contrast, granule cell axons synapse on CA3 inhibitory interneurons four times more frequently than CA3 pyramidal cells (Acsády and Káli, 2007) and CA3 pyramidal cells may have as many as 30000 to 60000 terminals within the CA3 network (Amaral and Witter, 1989; Li et al., 1994). Activation of more inhibitory than excitatory neurons in the massively recurrent CA3 region suggests that DG granule cells may induce a net inhibitory effect in the CA3: they quiet background noise and permit the faithful transmission of information concerning what happened out there (Acsády and Káli, 2007; Amaral et al., 2007; Kondo et al., 2008). Thus the DG dampens the transmission of noise and sharpens firing patterns for storage in the hippocampal CA3.

Nevertheless, the DG is most often characterized by its pattern separation function: it transforms highly overlapping representations (i.e., memories involving common contexts, people and objects) into highly dissimilar, non-overlapping representations (Bakker et al., 2008; McHugh et al., 2007; Treves et al., 2008). Researchers hold that the DG performs this function through adult neurogenesis, the generation of new granular cells throughout the individual's life (Clelland et al., 2009; Deng et al., 2010; Eriksson et al., 1998).

Ninety-five percent of granule cells in the DG are generated during child development and do not divide thereafter, but the DG also contains neural progenitor cells that generate the remaining 5 percent of adult-born neurons (Nakashiba et al., 2012). Young adult-born granule cells are particularly responsive during a critical window after genesis: they may become tuned to events that occur during that time, and participate in the reinstatement of those precise memories (Alme et al., 2010). They orthogonalize memories according to time, in essence, affixing a temporal stamp. Recent evidence suggests that young adult-born DG granule cells contribute to ongoing pattern separation, while older granule cells support the rapid recall of memories through pattern completion (Nakashiba et al., 2012).

In sum, the DG transforms noisy and overlapping inputs into sparse, orthogonal code for storage in hippocampal CA3. It projects familiar data, most likely aligned with emotion-laden remembrances, but distinguishes this time from prior times. We therefore theorize that the entrepreneur depends on the DG to distinguish unsettling miscues from triumphant coups in a common context and thereby learn from successive experiences, and we suggest:

Premise 3: It is likely that entrepreneurs rely on temporal differences in episodic experiences (i.e., between this time and prior times) to envision and conceive a business opportunity or venture.

Pattern separation is essential to learning and memory (Bakker et al., 2008) and is likely critical to entrepreneurial cognition, but, of course, it is not the only thinking skill required. Complete episodic memories come together in the hippocampal CA3, a neural field that permits the highest levels of associative processing and supports perceptual and conceptual reasoning, most likely used by entrepreneurs to envision the future and conceive (i.e., become aware of and semantically comprehend) a new opportunity and/or venture.

All the Pieces Come Together

The hippocampus comprises several neural fields (the CA3, CA1 and subiculum, for our purposes), with the hippocampal CA3 as the place where whole episodic memories are processed and stored. The CA3 is described as highly recursive and supporting auto-associative recall, so a single environmental stimulus, word or thought can enable the entrepreneur to recall a past memory or vision a business opportunity or venture.

It is reasonable to suppose that the entrepreneur experiences an event in a context with physical properties: her location in space (having an orientation and perspective); gaze direction and observed scene being duly noted; spatial relationships among landmarks and salient objects having registered; others moving, doing, saying observed; and affective responses to what happened experienced. These all come together as a coherent representation (i.e., as a whole episodic memory) for storage and recall in the hippocampal CA3 field, an area structured for forming new associations among ongoing experiences and permitting the sparsest code among small ensembles of pyramidal neurons.

The distinctive feature of the CA3 is that pyramidal cells mainly connect to other CA3 pyramidal cells to form a dense, recurrent network (Amaral and Witter, 1989). Any neuron is two or three synapses (junctions through which firing impulses pass) away from about four million other CA3 neurons. Thereby, it can form arbitrary associations among any or all elements of an episodic memory and whole episodic memories can be stored in one unified network (Lavenex and Amaral, 2000; Treves and Rolls, 1994).

The CA3 has been described in computational models as supporting auto-associative pattern completion of whole memories (Marr, 1971; Rolls, 1996; Treves and Rolls, 1994). In other words, a single fragment, degraded or noisy cue can elicit the recall of a whole memory past or present (Leutgeb and Leutgeb, 2007; Marr, 1971; Rolls, 1996). The importance of auto-associative pattern completion cannot be overstated, particularly as it relates to the recall of one-time, single-trial learning experiences (e.g., characterizing entrepreneurial processes), those that do not repeat and for which only partial or vaguely remembered cues are available. Entrepreneurs may also depend on this property to envision (i.e., mentally simulate) a business venture from fragments from memories or ongoing experiences, a potentially critical insight. We therefore suggest:

Premise 4: It is likely that entrepreneurs rely on auto-association in episodic memory (the ability to retrieve a memory from only a tiny sample of itself) to envision a business venture from only fragmentary cues.

Percepts Inform Concepts

The CA3 is also recognized as the juncture between percepts and concepts or, in other words, where the entrepreneur's experiences inform his or her semantic comprehension of a business opportunity. Environmental cues are observed as statistical regularities that attach semantic properties (e.g., belongs to this category) and meaning to concepts: 'the entrepreneur' or 'opportunity' derives meaning from the individual's experiences.

Thus far we have analyzed the hierarchical processing of visual data into high-level percepts from retina to ITC, and explained the generation of sparsely coded, highly personal episodic memories in the CA3. Structures described along the way gated, inhibited, extracted, sharpened and modulated the available data (through experience and affective processes), such that percepts in the ITC and episodic memories in the CA3 emerge from raw sensory data. We have illustrated how neurons become more and more selective and invariantly responsive to that which is familiar, often seen, and personally relevant. We have argued that these percepts and memories constitute abstractions removed from actual sensory stimuli, possessing semantic properties and subjective meaning (Albright, 2012; Quian Quiroga, 2012). Anatomists and computational neuroscientists alike argue the CA3 field resides at the juncture of transformations of percepts into concepts: the nexus between experience and semantic comprehension (Lavenex and Amaral, 2000; Quian Quiroga et al., 2008; Rolls, 1996), and compelling evidence supports the logic.

Quian Quiroga and colleagues, for instance, show that about 14 percent of hippocampal neurons respond selectively and invariantly to objects, landmarks and persons, but the relationship is not a single neuron coding for a specific individual or landmark. These neurons, coined 'concept cells,' commonly respond to a small number of (often related) stimuli and small ensembles of neurons respond to categories of stimuli (e.g., landmarks – Eiffel Tower, Washington Monument or Tower of Pisa) and remain silent otherwise (Mormann et al., 2008; Quian Quiroga, 2012; Quian Quiroga et al., 2005; Suthana and Fried, 2012). Moreover, concept cells that respond selectively to a person's image also respond to the person's written name and synthesized pronunciation of her name, showing independence of internal representations from visual stimuli. Thus it has been suggested that CA3 concept cells and cell ensembles code for category, do not require visual cues, and can be triggered by mere thought (Quian Quiroga, 2012).

The meaning attached to concept cells is thus thought to be derived as follows. Learned associations between stimuli arise from statistical regularities[4] in the observer's environment (e.g., person + place = barista), regularities that prove beneficial for predicting and interpreting future sensory inputs (Albright, 2012). These associations are stored in memory and recalled as needed. Quian Quiroga and colleagues report that a time delay (i.e., latency) exists in the firing of CA3 neurons relative to the ITC. ITC neurons fire at about 100–150 milliseconds after stimulus onset, but hippocampal neurons fire at 300–400 milliseconds, an amount of time substantially and suspiciously greater than required for hierarchical processing through the MTL (Mormann et al., 2008). These authors argue that lateral (associative) processing takes place in the CA3 during this delay and learned associations are used to define semantic properties (e.g., belongs to this category) and meaning of a concept: meaning that likely includes emotional or motivational value. For instance, a business venture may be defined by market or job creation; an entrepreneur by how she thinks, feels and behaves; and 'entrepreneurship' as a concept that may carry positive or negative value for an individual, depending on learned associations.

Concept cells and categorical coding align with the theorized role of the MTL in the consolidation of long-term semantic memories (Bayley and Squire, 2005), adding further credence to the proposed nexus of percepts and concepts. We are left with the conclusion that episodic memory may consist of richly coded percepts – of scenes, objects, orientations, perspectives, self and other people – and concepts – of things familiar, often seen and personally relevant. The implication is that these abstract representations with subjective meaning and value comprise the building blocks of declarative memory (Quian Quiroga, 2012) and offline self-reflective thought about the future in the DMN (our conjecture): not objective physical details of things seen 'out there.' This implication supports arguments in the literature that suggest that 'creation of new ventures is a process by which entrepreneurs come to imagine the opportunity for novel ventures, refine their ideas, and, after an initial investment, justify their ventures to relevant others to gain much-needed support and legitimacy' (Cornelissen and Clarke, 2010: 539; see also Alvarez and Barney, 2007; McMullen and Shepherd, 2006). We therefore suggest:

Premise 5: It is likely that, prior to actual formation, an entrepreneur's reliance on combined episodic memory and self-reflective thought enables a business opportunity or venture to be comprehended as an abstract representation/business concept.

And:

Premise 6: It is likely that interdependencies exist in the entrepreneur's ability to use (1) value-weighted percepts to envision and (2) statistically informed concepts to semantically comprehend a business opportunity or venture – prior to formation.

So where do emotion and motivation enter the picture? What are the neural processes that move cognition from conceptualization to affect and motivation? Enter CA1.

The Detection of Novelty

In this subsection, we introduce a new concept (prediction error ≡ the difference between that which was expected and what just happened) and describe the role of the hippocampal CA1 as a novelty detector that enables the entrepreneur to compare familiar features of the world with real-time data from ongoing experiences. Both emotion and motivation are implicated in the CA1's role in novelty detection.

Our discussion of MTL structures thus far has focused on the refinement of cortical inputs into meaningful sparse code for memory storage. We should also note that the hippocampus participates in both encoding and retrieval, and it dynamically shifts from one to the other on the timescale required to support behavior (Figure 8.5). Researchers describe the hippocampal CA1 field as a novelty detector, rerouting neural activity from memory retrieval to encoding (Duncan et al., 2012). This novelty detection process moves cognition beyond the familiar representations and conceptualizations in the CA3.

With perhaps as few as 4 million neurons based on a 1:3.7 cell count

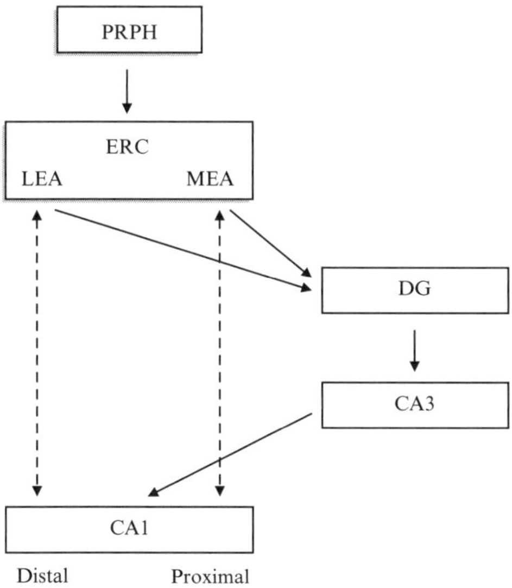

Note: For abbreviations, see Box 8.2.

Figure 8.5 Hippocampal formation wiring diagram

ratio, nonhuman primate to human (Jabès et al., 2011; Seress, 2007), the CA3 temporarily registers and records all the day-to-day events that may become episodic memories over an individual's lifetime. Thus the entrepreneur, for example, relies on this structure to meet a potential client: 'I parked over there; we met over lunch; we talked about production; you requested a proposal for improving efficiency.' She will meet the client again this week and use the pattern completion capacity of the CA3 to seamlessly advance the ongoing agenda. A single cue (e.g., voice on the phone) elicits the recall of details from the prior meeting and she will use episodic memory to form expectations, plan and envision (even script) this week's meeting – an illustration of episodic future thought. Our entrepreneur will continue to rely on memory retrieval as long as experiences match her expectations, but when sensory data (e.g., scowl on the client's face) signal a mismatch, the CA1 will reroute neural activity, that is, engage in novelty detection and encoding functions.

Anatomical data support the CA1's role as a novelty detector or comparator of highly processed, familiar representations of the world transmitted from the CA3 with real-time data from ongoing experiences directly from the ERC (Duncan et al., 2012). The CA1 gates (or compares and allows/

inhibits) the flow of information and switches hippocampal function from retrieval to encoding of novel data, in part defined as unexpected data (i.e., a prediction error) used to update representations about the individual's changing world (Ito and Schuman, 2012; Kumaran and Maguire, 2007).

The CA1 receives two streams of data: one through the ERC–DG–CA3–CA1 pathway as described above, while the other projects data directly from the ERC to the CA1. CA3 sends highly processed, associative information about well-known people, places and things, and constitutes the primary excitatory familiarity signal to the CA1. The ERC in contrast transmits real-time data and (when signaling novelty) can strongly inhibit or regulate activity in the CA1 (Dvorak-Carbone and Schuman, 1999). The CA1 is thus well positioned to detect novelty or prediction error, that is, differences between what was expected and what has just happened. It is at this point that the role of affective valuation enters the picture.

Specifically, affective value plays an essential role in the way CA1 gates information, with affect in this case attached to novel (unexpected) stimuli, as follows. Direct inputs from the medial entorhinal area (MEA) and lateral entorhinal area (LEA) terminate in discrete CA1 fields, a topographical organization that hints at a behaviorally relevant function. MEA fibers send spatial data to proximal CA1 neurons close to the CA3, while LEA axons convey nonspatial data to the distal CA1 cells adjacent to the subiculum (Ito and Schuman, 2012).

Additionally, CA1 neurons are also primary targets for dopamine-releasing projections from the ventral tegmental area (VTA) (Ito and Schuman, 2012; Lisman and Grace, 2005), with dopamine (DA) strongly implicated in signaling behavioral relevance (Ito and Schuman, 2012; Lemon and Manahan-Vaughan, 2006; Lisman and Grace, 2005). Researchers argue the detection of novel data in the CA1 activates VTA neurons, which release dopamine that preferentially impacts LEA inputs, those carrying nonspatial (e.g., object quality and affect) information about what exists out there. DA enhances high-frequency signals (suppresses low-frequency signals) that inhibit CA1 pyramid cells from firing in response to CA3 inputs (Dvorak-Carbone and Schuman, 1999).

Thus CA1 acts as a filter for preferential encoding behaviorally relevant (i.e., novel) information (Ito and Schuman, 2012). The CA1 then transmits real-time data to the amygdala, strongly implicated in triggering the physiological responses of emotion (Fudge et al., 2012), and back to the ERC for further processing. Data streams from the cortex to the ERC and from the CA1 to the ERC are posited to cross paths, with the ERC acting as an interactive hub for incoming and outgoing data (Canto et al., 2008), possibly – in the case of the entrepreneur – prompting her to pause and revisit novel data. Baron (2008: 335), for example, suggests that both dispositional and event-generated affect primes mood-relevant memories, or cues reactions to objects, ideas, people and so on, which in turn have effects on basic cognitive processes such as perceptions, judgments, decisions, memories and so on, thereby influencing such entrepreneurial processes as opportunity recognition and capacity to respond effectively to highly dynamic environments, among other outcomes. We therefore suggest:

Premise 7: The likelihood of entrepreneurs effectively responding to novelty (i.e., prediction error) as an environmental cue to opportunity, and thereafter successfully forming a business venture, is moderated by affective processes, and implicates both emotion (physiological response and label) and motivation.

Our analysis thus turns to emotion and motivation as useful phenomena to illustrate the research potential offered by a neuroscientific rationale for the importance of entrepreneurial emotion and entrepreneurial motivation in entrepreneurial cognition research.

EMOTION

In this section, we argue that emotion may be most productively defined as a stress response – initiated by the sympathetic nervous and neuroendocrine systems – and experienced as a physiological reaction in the body that informs the entrepreneur about what is relevant. The weights and labels attached to stimuli vary across individuals, over time, and with experience. We note that the stress response associated with entrepreneurs' emotions is objectively measurable and can be linked to verbal reports (i.e., labels) entrepreneurs attach to their experiences.

Two lines of reasoning exist in the study of emotion in neuroscience: one focuses on emotions as 'natural kinds' or feeling states that exist in nature independent of the perceiver, while the other explains emotions in terms of structures and functions (see LeDoux, 2012; Lindquist et al., 2012 for reviews). These approaches differ in their grounding in mainstream neuroanatomy and ability to support additive research.

Emotions as Natural and Universal

Natural kinds researchers characterize emotions as discrete feeling states (e.g., fearful, angry, happy or surprised), universally expressed and recognized (in facial expression studies) around the world, conserved across species, and dubbed basic or natural emotions hardwired in the brain (Ekman, 1992a, 1992b, 2011; Izard, 2007). This approach has been widely applied in research, owing to the simplicity of attaching a name to a subjectively defined state; and it may typify the theoretical underpinnings of some studies in entrepreneurship concerning, for example, traits-based entrepreneurship research.[5] Still, problems exist in this neuroscience literature.[6]

In brief, and specifically with respect to emotion, deficiencies in 'natural kinds' research include the lack of agreement about the numbers, names and defining characteristics of basic emotions, as well as their neural structures and functions (see LeDoux, 2012). Little evidence exists to tie discrete emotions to dedicated circuits (but see Vytal and Hamann, 2010), and the idea of emotions as discrete states does not align with the analog character of processes giving rise to emotions, those involving the widespread diffusion of hormones and neuromodulators with long-lasting but imprecise influences on neural systems. As we have explained in the foregoing foundational descriptions, the physiology of emotions is anything but discrete and objectively grasped. Thus 'natural kinds' research remains suspect and unable to sustain additive research, as evidenced by challenges to the most basic question: do natural kind emotions exist? Advocates of this approach must defend their work after 40 years (e.g., Ekman, 2011).

A Structure–Function Account

The second research stream (focusing on structures and functions) dates to the early twentieth century and the discovery that the hypothalamus controls the autonomic nervous system, as well as later evidence that damage to the amygdala can leave subjects with diminished fear and aggression (LeDoux, 2000, 2003a; Sah et al., 2003: historical reviews). Early anatomists established the amygdala and hypothalamus as central to emotion processes (structure); and fear-conditioning studies of the 1970s–1980s began to clarify function. For example, researchers paired an emotionally neutral conditioned stimulus with an aversive unconditioned stimulus, for example a tone (CS) and foot shock (US). After one or more trials, a tone acquired value, was found to signal danger and triggered a physiological reaction capable of provoking a behavior (avoidance) (LeDoux, 2003a). Based on this research, and more recent studies (LeDoux, 2012: extensive review), researchers have therefore argued that the function of emotion is not to produce 'feelings,' but to enable an 'organism to survive and thrive by detecting and responding to challenges and opportunities' (LeDoux, 2012: 654), and which we term 'stressors.'

Emotional triggers

In the structure–function stream, emotions are characterized as a stress response to various stimuli in the individual's life: pleasant or aversive, real-time or temporally distant, and oftentimes the unexpected (e.g., see the previous CA1 discussion). Everyday stressors may include the thought of rising from a warm bed on a cold morning, a number on the bathroom scales, pinched toes in new shoes, attention from a certain other person, the density of traffic on the day's commute, stock market vagaries, gas prices and career choices. Regrets past and worries future (e.g., mortality) enter the person's thoughts as well. Specifically as an entrepreneur, she affectively weights (positive or negative), for example, factors such as the independence, benefits and costs of starting/owning a business, financial investments, debt obligations, monetary gains or losses, customer demands and complaints, hiring or firing decisions, competitor actions, accounting requirements, investor presentations and the possibility of success or failure. Physiological reactions in the body (triggered by the amygdala and related structures) inform the person about what is relevant and steer her through daily decisions (e.g., where to lunch, how to price or whom to trust) and weights are attached to stimuli that vary across individuals, over time and with experience (e.g., rollercoasters lose their thrill value with repetition).

A specialized structure for emotion

The amygdala complex, at the center of emotion processing, consists of about 13 nuclei clustered in anatomical studies as the: (1) basolateral (BLA) complex; (2) cortical group (not a part of our analysis); and (3) centromedial group (Sah et al., 2003; Spampanato et al., 2011). The BLA complex is the primary input site into the amygdala and is posited to assign a valence (i.e., affective value) to stimuli through associative coding of stimuli and physiological data (Cardinal et al., 2002; Sah et al., 2003). The BLA processes data locally and projects to the centromedial nuclei, described as a primary output station (Sah et al., 2003).

Emotion ≡ a physiological response

How does the stress response occur? The amygdala receives sensory data from the thalamus, cortex (e.g., ITC) and MTL structures, while physiological inputs arrive from the hypothalamus and brainstem structures. These data can trigger an automatic stress response: for example, unprocessed sensory data (e.g., firing activity signaling a 98 mph fastball closing in on a batter's head) project directly from the thalamus to the amygdala to trigger an automatic defensive behavior such as ducking (but see Kravitz et al., 2013). Alternatively, cortical areas and MTL structures

(Sah et al., 2003) convey highly processed information about an episodic experience (something 'out there'), as well as endogenous data concerning physiological discomfort (pain), sociopsychological happenings (rejection and uncertainty) or thoughts (starting or failing in business) that knock the individual out of homeostasis (de Quervain et al., 2009; Herman et al., 2005; Kreibig, 2010; McGaugh, 2002; Tsigos and Chrousos, 2002). The amygdala then triggers a peripheral reaction.

Specifically, centromedial nuclei project to the brainstem to activate the sympathetic nervous system, thereby initiating the characteristic physiological stress response: heightened attention, catabolic release of energy from fat cells into the bloodstream, increased blood pressure, galvanic skin response, and the acceleration of cardiac and respiratory output to fuel brain, heart and muscle (Tsigos and Chrousos, 2002). The amygdala also projects to the paraventricular nuclei (PVN) of the hypothalamus to initiate a hormonal cascade through the hypothalamus–pituitary–adrenal (HPA) axis leading to the release of adrenal stress hormones: glucocorticoids, epinephrine and norepinephrine.

This combination of the autonomic nervous system and the hypothalamic neuroendocrine system comprises the efferent limbs through which the brain influences all peripheral stress responses (Tsigos and Chrousos, 2002) and those engaged in classical fear-conditioning studies in which a neutral stimulus (conditioned stimulus (CS)) acquires emotional value when paired with a biologically significant event (unconditioned stimulus (US)) (LeDoux, 2000). For the individual, then, external and internal stimuli (thoughts) produce physiological responses registered in the brain and reflectively interpreted as 'feelings.' The individual then names these feeling as specific emotions, with a time course to the applied labels (Kirkland and Cunningham, 2011). In this manner, emotion arises and a task comes to be viewed as onerous, a co-worker as kind, or thought of becoming an entrepreneur as exciting or terrifying and so forth.

Widespread modulation

Additionally, the amygdala modulates memory and cognition through both indirect and direct means (LaBar and Cabeza, 2006; McGaugh, 2002). For instance, it heightens arousal through interactions with noradrenergic and dopaminergic brainstem and midbrain nuclei, such as the ventral tegmental area (VTA) mentioned earlier (Cardinal et al., 2002). Centromedial nuclei projections and blood-borne epinephrine feedback to the brainstem (de Quervain et al., 2009; Roozendaal et al., 2009) trigger the widespread distribution of excitatory dopamine and norepinephrine through the brain and thereby prime circuits for activation (Sah et al., 2003).

Heightened levels of norepinephrine (plus glucocorticoids and acetylcholine) in the hours after an emotionally charged episode (Paré, 2003) may account for evidence showing that BLA activation strengthens memory by tuning auditory receptor fields to a specific tone; increasing visual stimulus representation, so larger neuron populations store more data; as well as synchronizing frequencies in visual circuits to select behaviorally relevant stimuli (Chavez et al., 2009; Murty et al., 2010). As noted earlier, amygdala activation also modulates the PRPH to selectively convey affectively weighted data to the ERC (Curtis and Paré, 2004) and may elicit frontal-lobe-mediated semantic processing in support of emotional memory encoding (LaBar and Cabeza, 2006; Murty et al., 2010).

Targeted/direct connections

The amygdala also enhances memories of emotionally arousing experiences via direct projections to other brain regions in a narrow window around the time of an emotional event (de Quervain et al., 2009; Roozendaal and McGaugh, 1997; Roozendaal et al., 2009). Researchers report that norepinephrine (widely distributed in the brain) and glucocorticoids (crossing the blood–brain barrier) converge on and activate receptors in the BLA, which then projects to the MTL and cortex (Joëls et al., 2011; Majak and Pitkänen, 2003; McGaugh, 2002; Murty et al., 2010; Paz et al., 2006; Sah et al., 2003). The amygdala receives inputs from MTL structures including rhinal cortices and hippocampus and projects back to these same structures (Majak and Pitkänen, 2003; Sah et al., 2003), creating reciprocal pathways for emotional memory enhancement (Murty et al., 2010).

These reciprocal connections boost and synchronize BLA output and the BLA sends 'massive' projections to MTL structures (specifically the rhinal cortices) that amplify firing rates between MTL structures (Curtis and Paré, 2004; Paz et al., 2006). The amygdala is reported to reduce the threshold for hippocampal long-term potentiation and increase the expression of genes critical in mediating synaptic plasticity. Thus, through these processes, emotion-laden memories are more vividly and longer remembered (LaBar and Cabeza, 2006; McGaugh, 2002; Murty et al., 2010; Paz et al., 2006; Roozendaal et al., 2009). Now (within the structure–function paradigm), having provided a quite detailed (for a social-science chapter) map of the 'structure' of emotion processing, the question of function also can be addressed.

The function of emotion

Neuroscience research leaves little doubt about the modulatory influence of emotion on cognition and memory, but what functions do emotions serve? We offer two answers. First, research on emotion most constructively focuses on the selective processing and valuation of stimuli, rather than a feeling state. Fear studies illustrated the conditioned negative value of stimuli in detecting a threat or danger (LeDoux, 2000). This negative-valence conception broadened to encompass the amygdala's role in positive affect, reward and novelty (Bermudez and Schultz, 2010; Ito and Schuman, 2012), then expanded to recognize the amygdala's role as a relevance detector that responds to stimuli (e.g., pain) deemed important to the organism's wellbeing (Ousdal et al., 2012; Simons et al., 2014). Emotion thus weights events, experiences and outcomes, and thereby influences even daily choices (e.g., where to eat, what to wear, which way to work and when to do a task). In the case of entrepreneurial cognition, it is to be expected that emotion serves to steer the entrepreneur down an unfamiliar path of realizing a business. We thus suggest:

Premise 8: It is likely that stressors and the physiological stress response produced within entrepreneurs (duration and magnitude) prominently describe the process of exploiting a business opportunity (how it 'feels to the entrepreneur').

Premise 9: It is likely the labels entrepreneurs apply to stress responses vary over time and with experience, as well as affect the probability of venture success.

And:

Premise 10: It is likely that the affective weighting of stimuli (emotion) in part explains why entrepreneurs choose one business opportunity or venture over another, as well as paths taken (choices) in exploiting a business opportunity.

We also note that episodic memory refers to rich and unique recollections of personal experiences from a person's life, happenings at a particular time and place, with affective value sufficient to elicit feelings of reexperiencing the past (Tulving, 1972, 1983). Emotions serve the valuable function of delineating self-relevant, self-defining moments that contribute to the formation of the individual's cohesive identity (Conway, 2009; LeDoux, 2003b). To the extent the entrepreneur travels unfamiliar paths, unexpected outcomes will happen; she will remember emotion-laden happenings longer and more vividly than neutral experiences; and those experiences will likely (in part) come to define her as an entrepreneur. We therefore suggest:

Premise 11: It is likely that concepts such as 'entrepreneur' and 'opportunity' as comprehended by the individual are informed by statistical regularities, prediction errors, and affective valuations derived from episodic experiences in the opportunity identification and exploitation processes.

The implication of emotion-driven assistance with venture creation suggests a linkage to entrepreneurial motivation, which we address next.

Motivation

Here we describe the physiology of motivation, including the online valuation of stimuli and how motivation connects with behavior. For entrepreneurs, the midbrain dopamine system supports both the refinement of expert skills /habits and adaptive behavior. It is at this point in our analysis that abstract building blocks of thought – percepts, concepts and affective valuations – come together with real-time data from ongoing experiences in the medial prefrontal cortex (mPFC – part of the default mode network) in support of conscious thought and entrepreneurs' adaptive behaviors – those beyond simple instrumental learning.

The *Merriam-Webster Dictionary* defines motivation as: the act or process of giving someone a reason for doing something. In neuroscience, the term motivation is defined as the dopamine-modulated state of arousal consonant with anticipation that precedes and guides reward-seeking behaviors toward desired outcomes (Düzel et al., 2009; Isoda and Hikosaka, 2011; Schultz, 1997, 2010). In contrast to classical stimulus–stimulus conditioning explanations of emotion in which stimuli acquire value (described above), neuroscience researchers describe motivation as engaging instrumental (i.e., action–outcome or reinforcement) learning (Berridge, 2007; Cardinal et al., 2002; Kelley, 2004; Salamone et al., 2012). Thus the idea asserted in our analysis of motivation is that appetitive and abstract (e.g., money) stimuli do not arrive in the brain with inherent value: these stimuli acquire value through experience (Knutson et al., 2005; Miyapuram et al., 2012; Schultz, 2010; Seymour et al., 2007); are wanted (Berridge, 2007); require effort to obtain (Salamone et al., 2012); and reinforce successful actions leading to consummation. This process is thought to be an evolutionarily old means for coding action–outcome memory that supports adaptive behaviors and – by implicating more recently evolved systems (e.g., medial prefrontal cortex – mPFC) – gives rise to more sophisticated applications of motivation in value-based decision making (Rolls, 2004; Wallis, 2012), effort-related choice (Salamone et al., 2012), goal-directed action (Balleine and O'Doherty, 2010) and choice among immediately available or temporally delayed rewards (Cardinal et al., 2002).[7]

In this section, we describe the distributed network of structures involved

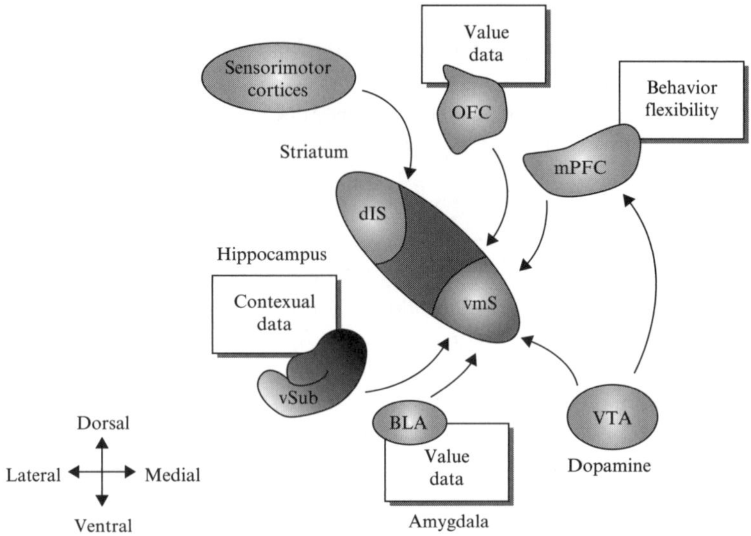

Note: For abbreviations, see Box 8.2.

Figure 8.6 The striatum and midbrain dopamine system

in coding the value of stimuli (amygdala – AMG – and orbitofrontal cortex – OFC) and those involved in mediating behavior (dorsolateral striatum – dlS, ventromedial striatum – vmS, hippocampus – HIP, hippocampal ventral subiculum – vSB, medial prefrontal cortex – mPFC and ventral tegmental area – VTA), as further illustrated in Figure 8.6. We continue to focus our analysis primarily on the processing of visual data from the visual 'what' pathway (see Kravitz et al., 2013). To accomplish this task, we address, first, how value is represented neurologically; following this, we discuss the midbrain dopamine system (MDS) to explain how value is connected to behavior, thereby addressing how habits and expert skills form, adaptive behavior occurs, as well as dopamine's role in motivation.

The Representation of Value in the Brain

The amygdala, as described above, is responsible for coding the value of stimuli. It initiates the characteristic autonomic stress response and hormonal HPA cascade associated with emotion (Tsigos and Chrousos, 2002) and codes stimulus–physiology associations: the neural representation of emotional value. Researchers describe the amygdala as receiving visual data (e.g., from the ITC) and projecting back to the ventral visual stream to orient the individual to emotionally significant stimuli. Researchers implicate the amygdala in the preferential processing and storage of long-lasting (lifetime) memories, and possibly stable preferences, but this structure is thought to lack the flexibility to adjust 'stored' values on the timescale of behavior (Kravitz et al., 2013; Padoa-Schioppa, 2011; Rolls, 2004).

In this subsection we therefore describe the neural computation of motivational value in the orbitofrontal cortex (OFC), which, in contrast to the amygdala, computes the value of stimuli in real time as the individual makes value-based choices (Padoa-Schioppa, 2011). In other words, the OFC relies on real-time value computations rather than 'stored' values to facilitate choice decisions. The OFC, situated on the ventral frontal lobe surface (directly behind the eyes), receives data from most sensory systems and functions as a secondary sensory cortex representing stimulus value, for example, hedonic pleasure (Grabenhorst and Rolls, 2011). Value in the OFC is encoded at the neuron level (as linear increases in firing rates) and constitutes a common unit of measure for comparing disparate choice options (Padoa-Schioppa, 2011: review). This representation is described as subjective (e.g., varying with satiation) and abstract (i.e., simultaneously coding for multiple dimensions on which options are compared, such as identity, hedonic quality, quantity, risk, ambiguity, fairness and temporal delay). OFC representations also display transitivity, range adaptation (e.g., coding choices on a café menu or high-end car showroom) (Padoa-Schioppa, 2011) and some neurons code for prediction error (Grabenhorst and Rolls, 2011). We also note that, while the OFC computes stimuli values, it only weakly connects to motor areas essential for instrumental learning (Carmichael and Price, 1995). This leads us next to consider how instrumental learning occurs: that is, through reinforcement in the midbrain dopamine system, which connects value to behavior.

Connecting Value to Behavior

Researchers describe the midbrain dopamine (DA) system as a broadly distributed 'reinforcement network' enabling the individual to move with purpose toward 'wanted' outcomes (i.e., motivated behavior) (Berridge, 2007; Kelley, 2004). This network, centered on the striatum, serves a central role in connecting stimulus value to behavior: it supports instrumental learning (i.e., habit formation and refinement of expert skills) as well as flexible adaption (Foerde and Shohamy, 2011). How does it work physiologically?

Anatomical studies show that cortical neurons from the OFC and mPFC, as well as limbic structures – basolateral amygdala (BLA) and hippocampus (HIP) (specifically the ventral subiculum – vSB) – project functionally segmented information to distinct regions within the striatum, regions described as specialized by function or neurophysiological processes. The dorsolateral striatum (dlS) predominantly receives sensorimotor information; the ventromedial part takes in limbic (i.e., affective and behaviorally relevant) data; while areas in the middle receive higher-order associational information (Voorn et al., 2004). Thus a dorsolateral to ventromedial gradient exists in the striatum, with researchers implicating dorsolateral (dlS) and dorsomedial (dmS) regions as mediating instrumental learning, with a region-specific shift in coding occurring as learning slows and habits develop and refined skills consolidate (Yin et al., 2009).

Habits and expert skills

Researchers argue the neural representation of habits and expertise are coded as firing patterns in the striatum – like the coding of episodic experiences – and these firing patterns change with practice (Knowlton and Foerde, 2008). Habits and refined skills are expressed in more efficient neural processing and decreased neural activity (i.e., greater fluency and less cognitive effort), as illustrated in the case of a novice painter learning to cut a straight line with a fully loaded paint brush. First attempts require extensive cognitive and perceptual effort as she watches the brush angle, bristle pattern, hand position, movement fluidity and resulting painted line. She talks her way through each brush stroke (e.g., 'watch your elbow'), evidence of the cognitive effort expended. Over time, perceptual representations of experience sharpen (i.e., sparse coding), such that neurons encoding critical details continue to fire; but those coding for less useful information drop out, and the total number of active neurons decreases. Expertise permits the rapid and automatic shifting of attention to successive elements of the task, so less cognitive effort is required and (through practice) expert performance is implemented through the smooth execution of automatic sensorimotor programs (Knowlton and Foerde, 2008).

A gradient thus exists in the acquisition of habits and expertise, with an initial phase of rapid improvement, followed by more gradual improvements as the new firing patterns are consolidated. Actions become automatic, resistant to interference and durable as a region-specific shift in neural activity takes place in the striatum. The associative dmS is preferentially active during training, while the sensorimotor dlS takes over as improvement slows and actions become automatic (Poldrack et al., 2005; Yin et al., 2009).

Adaptive behavior

In contrast, inputs from the OFC, mPFC, BLA and hippocampus (vSB) project heavily to the ventromedial striatum (hereafter nucleus accumbens, NAc): a structure closely related to other DA-innervated limbic areas, defined as a component of the 'extended amygdala' and location in which affect value links to adaptive action (Kelley, 2004; Price and Drevets, 2010). The NAc plays a crucial role in 'why' explanations of behavior (in contrast to prior analyses of 'what' gets coded) and is implicated in the experience of pleasure. The NAc is described as a hedonic hotspot and target of psychostimulants such as cocaine and amphetamine (Berridge and Kringlebach, 2013).

As previously described, the OFC network processes multi-modal sensory data into real-time representations of stimulus value under varying physiological conditions, but does not directly connect to behavior. The mPFC, in contrast, integrates contextual and value data into conscious thought in the DMN (Raichle et al., 2001), modulates autonomic/emotional control (Gabbott et al., 2005), and permits flexible, adaptive control of behavior (Euston et al., 2012).

The mPFC network encompasses areas on the medial (i.e., midline) surface of the frontal cortex and those along the medial edge of the orbital cortex (Price and Drevets, 2010). It exhibits a ventral-to-dorsal functional gradient (Euston et al., 2012) in which affective (valuation data) and highly processed sensory data converge (in support of behavior); and it extensively overlaps with the anterior cingulate cortex (ACC) (Price and Drevets, 2010). The ventromedial prefrontal cortex (vmPFC) receives data from midbrain limbic structures (e.g., OFC, BLA, insula vSB, and ERC) and is posited to mediate physiological–emotional activity, while the dmPFC receives multisensory data from widespread cortical areas and supports motor control (Narayanan and Laubach, 2006) and goal-directed actions (Gabbott et al., 2005). Researchers hypothesize that the vmPFC and dmPFC together enable the individual to adapt emotional and behavioral responses to a given context (Euston et al., 2012), while the ACC has been implicated in conflict monitoring, prediction error detection and the valuation of actions, including the cost of actions (Botvinick et al., 1999; Carter et al., 1998; Quilodran et al., 2008; Rudebeck et al., 2008).

The function of the mPFC in motivation remains the subject of ongoing research, but the nature of neural projections to this structure has prompted speculation. Some researchers posit that the mPFC enables the individual to predict the likely outcomes of actions, good or bad (Alexander and Brown, 2011; Quilodran et al., 2008). Others suggest it enables the individual to infer rules from statistical regularities (i.e., learned associations) in the environment, with adaptive behavior tied to learning novel rules in changing environments (Durstewitz et al., 2010) – a potentially critical insight for entrepreneurs. In the broadest account, the mPFC codes (on a timescale of behavior) for 'rich contextual representations that take into account not only sensory cues but also actions and time' (Hyman et al., 2012: 5086), enables the individual to recall best actions and emotional responses from past experiences (Euston et al., 2012), and allows her to simulate, through cognitive activity in this area of the brain, possible actions to select those with the greatest potential for producing desired outcomes (Addis et al., 2009; Schacter et al., 2008).

In sum, the mPFC maintains ongoing cognitive representations that permit adaptive choices beyond those available in a simple instrumental (i.e., action–outcome) learning model. The individual can adjust her behavior to a boardroom or ballgame – or, using the adaptive processes just described, can peek ahead, compare rewards today to those down the road and make adaptive choices under uncertainty. For example, the motivation to become an entrepreneur, where processes that require choices beyond a simple learning model are strongly implicated (Alvarez and Barney, 2007; Baron, 2008), is therefore likely to be supported by the foregoing neurological systems.

Dopamine's role in motivation

The two components of motivation – arousal prior to reward and reinforcement of reward-seeking behaviors upon the timely delivery of rewards – depend on the circulation of dopamine throughout the MDS (Schultz, 1997). In brief, contextual and value data from widespread cortical fields (OFC and mPFC) and limbic structures (ERC, HIP, BLA) converge on the NAc (nucleus accumbens), which projects to the ventral tegmental area (VTA) (Floresco, 2007; Grace et al., 2007). The VTA, in turn, sends dense dopaminergic projections back to the NAc and other relevant structures – the mPFC for our purposes.

Dopaminergic neurons in the VTA exhibit tonic (irregular pacemaker-like activity) and burst (spike) firing (Grace et al., 2007; Lodge and Grace, 2006). The ongoing tonic release of dopamine from the VTA modulates diffused extracellular DA levels in structures like the mPFC, thereby preparing (arousing) the person to respond to motivation-relevant stimuli (Floresco et al., 2003). Tonic DA activity is punctuated by burst firing to the NAc, a signal that carries information about reward value (specifically better-than-expected rewards). These two DA firing patterns are believed to elicit the hedonic sensations of liking and wanting (Berridge, 2007), facilitate learning from prediction errors (Schultz, 1997), and sustain behaviors sufficient to overcome time- or work-related costs of pursuing delayed rewards (Salamone et al., 2012).

How does dopamine support both instrumental learning and adaptation? Physiological studies suggest the hippocampal vSB (limbic) and mPFC (cortex) compete for control over the NAc, with the vSB steering behavior toward reward-dependent (instrumental learning) and the mPFC driving behavioral flexibility (Floresco, 2007; Goto and Grace, 2008). Increased burst and tonic firing (signaling reward) sustains action–outcome learning (of habits/refined skills), while the cessation of DA to the NAc (when actions no longer yield expected outcomes) induces behavioral flexibility through the mPFC, a component of the DMN (Goto and Grace, 2008).

In summary, our neuroscience explanation of motivation suggests that the values of choice options are computed as real-time, subjective and abstract representations. The MDS supports habit/skill formation, with learning represented as efficient neural processing and decreased neural activity. Adaptive behavior centers on the NAc, as well as the convergence of midbrain limbic and multisensory data in the mPFC, with adaptation (in part) described as learning novel rules (through statistical regularities) in changing environments. Tonic and burst dopamine from the VTA drives both instrumental learning and adaptive behavior, with the entrepreneur's pursuit of a business opportunity or venture influenced by both action–outcome (reward) and higher-order cognitive processes. Thus, with respect to the implications of a neuroscientific analysis of entrepreneurial motivation, we suggest:

Premise 12: It is likely that the individual's choice to become an entrepreneur, a 'why' decision – as well as choices along the way – are a function of affective influences, a combination of:

- *real-time, subjective and abstract computations of the choice option values;*
- *short-term dopamine–reward signaling in instrumental, action–outcome (reward) experiences;*
- *stable (long-term) affective weightings (emotion-based preferences) processed in the self-reflective medial prefrontal cortex (and beyond simple instrumental learning).*

Premise 13: It is likely that the uncertainty entrepreneurs experience in the pursuit of a business opportunity or venture arises as much from within the person (owing to inefficient neural coding) as from the environment (and is resolved as they gain experience/expertise).

Premise 14: It is likely that the processes of pursuing a business opportunity or venture is characterized (in part) as inferring novel rules (about effective action/behaviors) from statistical regularities in unfamiliar environments.

And:

Premise 15: It is likely that the entrepreneur's adaptive abilities in pursuit of a business opportunity or venture arise from self-reflective processes in the DMN (anchored by the mPFC), beyond instrumental learning, and informed by percepts (envisioning), concepts (semantic comprehension) and affective valuations (emotion and motivation processes).

DISCUSSION

In this chapter we have traced the neurophysiological pathway involved in transmitting visual data from the periphery to conscious thought and behavior – with affective processes (emotion and motivation) modulating data flows along the way – with the purpose of illustrating that entrepreneurs' brains are no different than anyone else's, but they are different based on accumulated experiences and knowledge contained in the abstract representations – percepts, concepts and affective valuations that inform conscious thought and behavior. We view our analysis as essential to advancing entrepreneurial cognition as a science, for instance, as in resolving the nature versus nurture dichotomy; mapping hardware to software; explaining similarities and differences among entrepreneurs; and discerning how entrepreneurs conceive (become aware) of opportunities.

We have offered 15 premises in this chapter as researchable assertions informing research into opportunity and venture formation that can be addressed with current social-science methodologies. In Premises 1 through 4, we highlight the importance of experienced-based imagery and episodic memory in envisioning a future business venture; in Premise 5 we connect statistical regularities to the conception/semantic comprehension of business opportunities; and in Premise 6 we emphasize the interdependence of the entrepreneur's abilities to envision and semantically comprehend a business opportunity or venture. In Premises 7 through 10, we propose that novelty (a form of prediction error) and emotion (physiological stress response + label) influence entrepreneurs' choices and venture success; Premise 11 implicates statistical regularities, prediction errors and affective valuations in the person's understanding of terms such as 'entrepreneur' or 'opportunity'; while Premises 12 through 14 highlight the influence of motivation, real-time valuations, instrumental learning, within-subject uncertainty, as well as dopamine signaling on the pursuit of a business opportunity or venture. Premise 15 then suggests that the entrepreneur's adaptive abilities to identify and pursue a business opportunity or venture arise from self-reflective processes in the DMN, beyond instrumental learning, and informed by percepts (envisioning), concepts (semantic comprehension) and affective valuations – the abstract building blocks of thought. As we previously argued herein, this approach stands as an alternative research strategy to reliance on (e.g.) fMRI imaging to identify which parts of the brain are active when the subject does 'x, y or z.'

The Entrepreneur

We now illustrate the implications of our work for theory and research in the science of entrepreneurial cognition, starting with an analogy relating to conceptions of the entrepreneur. In the absence of experience, a necktie is a meaningless word. The formal definition may refer to a band of cloth worn around the neck, but a deeper understanding regarding a necktie arises owing to its path to existence. Imagine, for example, that you draft a line drawing, and pass it to the next person who colors between the lines. One person after another adds shading for dimension, cuts and creases, and folds the paper into an origami necktie. The next individuals emulsify the paper, add a bonding agent and dye, make a fabric, and then cut and sew the material into a desired shape, and the tie is offered for sale in a shop. Then a person buys the tie, so it becomes 'my' tie. He wears it to successful business meetings and it becomes my 'lucky' tie. Through the years the necktie acquires a soup stain, frayed edges and a lifetime of meaning associated with people, places and other statistical regularities in its owner's life. Thus this particular necktie becomes unlike any other necktie and with meaning linked to the man's identity, much more deeply informing than a dictionary conception of a necktie.

We argue that the designation entrepreneur for 'the person' follows a similar route to existence. Here again, a simple definition may lack dimension, while individual experiences breathe life into the conceptual understanding of what it means when a thinking human being becomes an entrepreneur. The term 'entrepreneur' lacks depth of understanding in the absence of experience (cf. Morris et al., 2012), but imagine the following physiological pathway. First, and from unique life experiences, cortical structures and subcortical nuclei gate, inhibit, extract, amplify, sharpen and otherwise modulate sensory data – as in our necktie example. Then sensory circuits tune (through experience) to familiar and personally relevant percepts and percepts merge into emotion-laden episodic memories. Percepts next align with concepts – some distinctive of the entrepreneur – and the concept 'entrepreneur' acquires meaning through statistically predictable regularities in the person's life. This concept carries positive or negative value based on associations. Percepts, concepts and affective valuations come together as abstracted building blocks in ongoing representations of the world in the DMN that shape how she (as a distinctive entrepreneur) thinks, feels and acts: one exemplar or data point.

The challenge for researchers in comprehending the concept of the entrepreneur goes as follows. At one extreme, the entrepreneur is defined as human; all humans have a common complement of neural processes, and (in the absence of anomalies) all entrepreneurs rely on the same physiological processes of cognition, for example, percepts, experience-based imagery, concepts, statistical regularities, prediction errors, physiological stress responses (of emotion), real-time valuations of motivation, instrumental learning and dopamine signaling. At the other extreme all entrepreneurs accumulate unique life experience (small-*N* data) that inform subject-specific variations in percepts, concepts, valuations and so forth. Cognitive processes remain the same, while exemplars exhibit wide intersubject differences – some related to entrepreneurial cognition and others mere noise. What should researchers do?

We argue that entrepreneurial cognition as a science is best served by advancing a common conception of the entrepreneur, a middle ground between 'all the same' and 'all unique,' to identify research questions (and streams) capable of sustaining additive research, those that unravel how experiences, percepts, concepts and affective valuations embody the entrepreneur, as well as explain systematic variance in cognitive abilities (e.g., envisioning, associative processing and conceptual reasoning), including affective processes (e.g., physiological stress response, construal of predictions errors and real-time computations of motivation value) in and among entrepreneur exemplars – as well as relative to other human beings. Our analysis contributes to the advancement of knowledge by refining our understanding of the entrepreneur with common neurological processes and by identifying premises that point to promising questions that may sustain additive research, while perhaps excluding distracting intersubject noise from scientific discussions.

Entrepreneurial Process

Implications exist for how the entrepreneurial process may work as well. Earlier in the chapter we introduced the notion of episodic future thought: how the individual might flexibly recombine bits and pieces of memory to pre-experience the future through simulation, a mental trial-and-error process used to select actions with the greatest potential for producing desired outcomes (Addis et al., 2009; Schacter et al., 2008). Thereby the individual is not locked into the here and now, but can disengage from ongoing sensory stimuli to mentally travel back in time or forward to an envisioned state of the world that does not yet exist: for example, a new opportunity or business venture. Where the potential entrepreneur anchors her envisioned venture remains unknown theoretically, but our analysis suggests that the content of episodic memory (and future thought) may include richly coded percepts of scenes, reference points (landmarks), prominent spatial features, objects, self, self-location, orientation (perspectives) and other people – as well as concepts of familiar, often seen and personally relevant things. Some starting points are likely more promising than others.

Buzsáki and Moser (2013) further explain how the opportunity/venture formation process may work from a neuroscience standpoint. The entorhinal cortex (ERC) and hippocampus (HIP) developed for spatial processing and navigation, with higher-order episodic and semantic memory capabilities layered on top, but relying on the same basic algorithms used in path integration (PI) and landmark-based (map-based) navigation. These authors argue that PI navigation is similar to episodic memory and the exploratory acquisition of new experiences, while semantic memory (e.g., of objects, contexts and events) equates to map-based navigation in a static world, a known environment that can be easily articulated. In other words, navigation in physical and mental space can be viewed as fundamentally the same. Applied as an explanation for a likely entrepreneurial process, the sequence might proceed as follows.

The potential entrepreneur starts out with an ongoing representation of the world in her head, a map-based creation of a known world that can be easily articulated and navigated. She envisions a business venture to meet a need (a business for a missing product or service), a mental simulation that initially remains described as sketchy, essentialized or sparsely populated. This person explores the mental space of starting this business through the acquisition of new and challenging experiences, the fundamental way she learns and adapts. Indeed, the process of identifying an opportunity may require her to gain sufficient experience, form associative connections (in the HIP) to translate percepts into concepts until she reaches a point where she can sufficiently imagine and conceive the opportunity (Cornelissen and Clarke, 2010). Semantic comprehension (again, beyond the scope of our analysis in this chapter) in such a case may equate to the proverbial 'Aha!' moment. Comprehension informs next actions and our potential entrepreneur exploits the opportunity by gaining experiences and familiarity in a new well-articulated terrain of an emerging business: products/services, customers, employees, locations and revenues. This process may include inferring rules from statistical regularities in the new environment (Durstewitz et al., 2010). Key insights exist in this scenario: (1) experience comes before semantic comprehension; and

(2) the heavy lifting of identifying and exploiting a business opportunity may occur offline in the DMN.

Opportunity: The Geometry of Semantic Memory

An intriguing insight from our analysis is that a kind of geometry may exist in semantic memory that parallels the analysis of episodic memory but extends beyond concepts derived from percepts (i.e., personal experiences). Researchers widely hold that semantic memory is organized categorically and hierarchically. This structure may also be pyramidal, with broad concepts encoded by some neurons (i.e., pyramid base), others encoded for associative information (e.g., person + camera = photographer), and rich details of specific percepts encoded by still others (i.e., the peak) (Lin et al., 2005, 2006). As previously noted, concept meaning derives from statistical regularities in the individual's experiences (Quian Quiroga, 2012) and, according to Buzsáki and Moser (2013: 131), 'neuronal mechanisms that evolved to define the spatial relationships among landmarks can also serve to embody associations among [semantic representations of] objects, events and other types of factual information.'

Together, these insights imply that a structure exists in semantic memory (knowledge in the brain that supports thought), maybe resembling a high-dimensional topographical map of sorts, with unknown implications for how entrepreneurs can systematically identify and exploit opportunities. We believe that entrepreneurs in some way rely on category structures, statistical regularities, landmarks and the hierarchy (or pyramid) of broad semantic concepts to specific episodic experiences. Opportunity conception may be theorized to occur through articulating, searching or altering these structures. In this respect, neuroscience may align with metaphor (Cornelissen and Clarke, 2010) in the imagining and enacting of new opportunities and ventures.

Affective Valuation

From a physiological standpoint, affective valuation in either emotion or motivation signals behavioral relevance essential to cognition. Furthermore, the valuation of stimuli – identified, belonging to a category, informed by experience-guided imagery, preferentially selected, familiar, often seen, personally relevant, with semantic meaning derived from statistical regularities, and deemed novel, unexpected, threatening or wanted and so on – appears to be crucial to and inseparable from cognition. We therefore assert, based on our analysis, that affective valuation plays a critical role in all questions of judgment, when no 'correct' answer exists based on external circumstances. In this respect, our neuroscientific analysis makes a contribution to the decades-old question about judgment under uncertainty (cf. Knight, 1921), with even the threshold question in entrepreneurship research, 'Why would anyone choose to become an entrepreneur?' better informed by an affect-implicated answer.

Emotion

Our analysis further suggests (and explains why, from a physiological position) that inherent problems exist in studying emotion as a label attached to introspective assays of a person's feelings. Autonomic and neuroendocrine systems may elicit hormonal or physiological reactions with distinct signatures, but the notion remains speculative. Our research suggests that these systems developed for self-preservation and respond to sociopsychological stressors as they would to a predator, with crude distinctions in physiological signals possible at best. Cues sent to the brain remain vague and subject to interpretation (misinterpretation), so the names applied to 'feelings' likely vary across time with stimuli even flipping in valence: a failed venture may evoke embarrassment and later carry a positive connotation of a crucial learning experience, or an even more negative connotation as a truly terrible experience.

The definition of emotion as a physiological stress response, often induced by prediction errors, affords the opportunity to measure reactions to specific entrepreneurial experiences. Researchers may, for instance, consider using performance sports (biofeedback) technologies to capture real-time physiological data (e.g., heart rate, blood pressure, respiration rate, skin temperature and galvanic skin response data), while potential entrepreneurs pitch business ideas to angel investors or a similar scenario in which prediction errors may occur – and compare these with verbal reports (i.e., labels subjects apply to the stress response). The centrality of prediction errors in emotion research remains unknown and the correlation of physiology to verbal report may inform emotional labeling and the impact on entrepreneurial outcomes.

Motivation

Our analysis of motivation has centered on the link between valuation and behavior, with the midbrain DA system implicated in explanations for learning habits and expert skills, as well as for supporting adaptive behavior. We emphasize that motivation involves the real-time valuation of options critical to judgment-based decisions, like the 'big' choice of becoming an entrepreneur, or the many 'small' choices made in the process. For example, from a motivational standpoint, the individual chooses between rewards today versus those later, with dopamine enabling her to wait (Salamone et al., 2012). The occasion exists to study the valuation of options, costs of actions, and influence of differences in the person's ability to see further down the road (i.e., temporal or psychological distance) on perceived risk or value of delayed rewards. An extensive literature exists in neuroscience concerning how economic and judgment-based decisions occur, and hence we do not proceed beyond this reference to it.

CONCLUSION

We believe value exists in grounding entrepreneurship research in neuroscience, and this chapter has detailed the role of neurophysiology in episodic processing and affective valuation to support the assertion that entrepreneurs' brains are physiologically the same as any other person's brain, but in terms of experiences and knowledge they are different. We have endeavored to illustrate the potential advantage of including discussions of neural structures and functions in entrepreneurial cognition research and offered premises from which beneficial future research may proceed. Still, we concede that our work is neither comprehensive nor definitive, as neuroscience researchers continue to refine our understanding of neurophysiology. We also largely excluded any discussion of semantic processing (a key research stream for investigating knowledge structures and opportunity) owing to the already overwhelming detail for non-neuroscience readers, while we underspecified our descriptions of key neural processes for neuroscientists, a necessary compromise in cross-disciplinary research, in our view. We accept that alternative views exist on key concepts such as emotions (e.g., natural kinds) or the role of context in opportunity and venture formation. Nonetheless, our approach (detailing how the brain generates behavior) stands in contrast to the alternative as previously described: that of taking a man-created word or concept (e.g., emotion) and searching for an explanation, oftentimes outside the individual.

In our view, neuroscience renders the entrepreneur as human and, more importantly, as accessible owing to the explanations possible through its structure–function paradigm; and (we hope) offers a foundation for additive research with well-defined (although cross-disciplinary) concepts and consequent research questions, some prompted through the 'premises' herein. We view the entrepreneur as an ideal research subject given this person's adaptive nature and (as we observe) the affectively treacherous path she follows. The research challenge is then to view the 'entrepreneur' as more than a lifeless moniker dangling like a necktie; to distinguish between the entrepreneur as a research concept (defined as human, with highly developed brain and common neural processes) and exemplar (embodiment of experiences, percepts, concepts and valuations); and to design research questions informed by the research concept to explain variance among exemplars. It is here that we hope researchers will discover a key to unlock previously inaccessible doors to understanding, not the least of which might be in better explanations: of differences among entrepreneurs, franchisees, and nonentrepreneurs; differences between male versus female and young versus older entrepreneurs; distinctions among opportunity conception: recognition, discovery and creation; and the more exact role of embodiment in socially situated entrepreneurial cognition research.

NOTES

1. A more comprehensive definition would include all of the computations that support the acquisition and comprehension of episodic experiences in the brain, expressly including semantic processing. But for the sake of limiting the already almost bewildering level of necessary complexity, semantic processing is not included in this illustration.
2. A parallel dorsal visual stream (a.k.a. 'where' pathway) exists and is associated with motion and location of external objects. While the functional characterization of the dorsal stream has been extensively debated, this material remains beyond the scope of our analysis and an unnecessary complication in our view. We track the steps in the ventral visual stream as a means to clarify the processing of stimuli to perception.
3. Top–down (endogenous) processing is distinguished from bottom–up (exogenous) processing by the source of initiation, with bottom–up processing arising from external stimuli such as a loud noise, a shout, a flash of light and so on; top–down processing arises from the individual's desire to achieve some goal or objective.
4. Statistical regularities refer to observed patterns in the person's environment (observations/data points), as in repeated pairings that inform conceptual meaning. We note that this small-N (maybe $N = 1$) reasoning from unique life experiences likely amplifies unexplainable variance in empirical analyses and thereby complicates efforts to refine our understanding of the 'entrepreneur' as a research concept from noisy data.
5. Mitchell (1994: 25) explains as follows: 'Psychological characteristics-based research focuses on a very broad range of psychological characteristics including (not exhaustively) the need to control and direct, self-confidence, a sense of urgency, good health, comprehensive awareness, realism, superior conceptual ability, needs for status, objectivity in interpersonal relations, emotional stability, attraction to challenge, level of creativity, need for achievement, belief in an internal locus of control (belief in the ability to control the environment through individual actions), risk-taking propensity, and more (Coulton and Udell, 1976; McClelland, 1965; McClelland and Winter, 1969; Rotter, 1966; Welsh and White, 1981).'
6. Problems also exist in the entrepreneurship literature. Mitchell (1994: 25) further explains: 'Examples of subsequently disconfirming research on the most commonly cited

psychological traits: the need for achievement, belief in an internal locus of control, and risk-taking propensity (Sexton and Bowman-Upton, 1991) illustrate the present level of confusion in this research stream. The difficulty arises because the factors that describe entrepreneurs . . . also tend to describe successful people in many areas, such as business, art, music, and education (1991: 9).'

7. To place this discussion in context, some neural processes exist across species and are referred to as evolutionarily conserved (LeDoux, 2012) owing to their utility or function (e.g., survival circuits). The basic notion of action–outcome coding is thought to exist across species (e.g., even simple life forms retreat from toxins), while the human forebrain (cerebrum and subcortical nuclei) enables higher-order value-based decision processes.

REFERENCES

Abbott, L. and Regehr, W. 2004. Synaptic computation. *Nature*, **431**(7010): 796–803.

Acsády L. and Káli, S. 2007. Models, structure, function: The transformation of cortical signals in the dentate gyrus. *Progress in Brain Research*, **163**: 577–99.

Addis, D., Pan, L., Vu, M., Laiser, N. and Schacter, D. 2009. Constructive episodic simulation of the future and the past: Distinct subsystems of a core brain network mediate imagining and remembering. *Neuropsychologia*, **47**(11): 2222–38.

Albright, T. 2012. On the perception of probable things: Neural substrates of associative memory, imagery and perception. *Neuron*, **74**(2): 227–45.

Alexander, W. and Brown, J. 2011. Medial prefrontal cortex as an action–outcome predictor. *Nature Neuroscience*, **14**(10): 1338–44.

Alme, C., Buzzetti, R., Marrone, D., Leutgeb, J., Chawla, M., Schaner, M., Bohanick, J., Khoboko, T., Leutgeb, S., Moser, E., Moser, M., McNaughton, B. and Barnes, C. 2010. Hippocampal granule cells opt for early retirement. *Hippocampus*, **20**(10): 1109–23.

Almeida, L., Idiart, M., Villavicencio, A. and Lisman, J. 2012. Alternating predictive and short-term memory modes of entorhinal grid cells. *Hippocampus*, **22**(8): 1647–51.

Alvarez, S.A. and Barney, J.B. 2007. Discovery and creation: Alternative theories of entrepreneurial action. *Strategic Entrepreneurship Journal*, **1**(1–2): 11–26.

Amaral, D., Scharfman, H. and Lavenex, P. 2007. The dentate gyrus: Fundamental neuroanatomical organization (dentate gyrus for dummies). *Progress in Brain Research*, **163**: 3–22.

Amaral, D. and Witter, M. 1989. The three dimensional organization of the hippocampal formation: A review of anatomical data. *Neuroscience*, **31**(3): 571–91.

Bakker, A., Kirwan, C. Miller, M. and Stark, C. 2008. Pattern separation in the human hippocampal CA3 and dentate gyrus. *Science*, **319**(5870): 1640–42.

Balleine, B. and O'Doherty, J. 2010. Human and rodent homologies in action control: Corticostriatal determinants of goal-directed and habitual action. *Neuropsychopharmacology*, **35**(1): 48–69.

Baron, R.A. 2004. The cognitive perspective: A valuable tool for answering entrepreneurship's basic 'why' questions. *Journal of Business Venturing*, **19**(2): 221–39.

Baron, R.A. 2008. The role of affect in the entrepreneurial process. *Academy of Management Review*, **33**(2): 328–40.

Baron, R.A., Hmieleski, K.M. and Henry, R.A. 2012. Entrepreneurs' dispositional positive affect: The potential benefits and potential costs of being 'up.' *Journal of Business Venturing*, **27**(3): 310–24.

Baron, R.A. and Tang, J. 2011. The role of entrepreneurs in firm-level innovation: Joint effects of positive affect, creativity, and environmental dynamics. *Journal of Business Venturing*, **26**(1): 49–60.

Bayley, P. and Squire, L. 2005. Failure to acquire new semantic knowledge in patients with large medial temporal lobe lesions. *Hippocampus*, **15**(2): 273–80.

Bermudez, M. and Schultz, W. 2010. Reward magnitude coding in primate amygdala neurons. *Journal of Neurophysiology*, **104**(6): 3424–32.

Berridge, K. 2007. The debate over dopamine's role in reward: case for incentive salience. *Psychopharmacology*, **191**(3): 391–431.

Berridge, K. and Kringlebach, M. 2013. Neuroscience of affect: Brain mechanisms of pleasure and displeasure. *Current Opinion in Neurobiology*, **23**(3): 1–10.

Binder, J., Frost, J., Hammeke, T., Bellgowan, P., Rao, S. and Cox, R. 1999. Conceptual processing during the conscious resting state: A functional MRI study. *Journal of Cognitive Neuroscience*, **11**(1): 80–93.

Botvinick, M., Nystrom, L., Fissell, K., Carter, C. and Cohen, J. 1999. Conflict monitoring versus selection-for-action in anterior cingulate cortex. *Nature*, **402**(6758): 179–81.

Breugst, N., Domurath, A., Patzelt, H. and Klaukien, A. 2012. Perceptions of entrepreneurial passion and employees' commitment to entrepreneurial ventures. *Entrepreneurial Theory and Practice*, **36**(1): 171–92.

Buckner, R. and Carroll, D. 2007. Self-projection and the brain. *Trends in Cognitive Sciences*, **11**(2): 49–57.

Busenitz, L.W. and Barney, J.B. 1997. Differences between entrepreneurs and managers in large organizations: Biases and heuristics in strategic decision-making. *Journal of Business Venturing*, **12**(1): 9–30.

Buzsáki, G. 2004. Large-scale recording of neuronal ensembles. *Nature Neuroscience*, **7**(5): 446–51.

Buzsáki, G. and Draguhn, A. 2004. Neuronal oscillations in cortical networks. *Science*, **304**(5679): 1926–9.

Buzsáki, G. and Moser, E. 2013. Memory, navigation and theta rhythm in the hippocampal–entorhinal system. *Nature Neuroscience*, **16**(2): 130–38.

Canto, C., Wouterlood, F. and Witter, M. 2008. What does anatomical organization of entorhinal cortex tell us? *Neural Plasticity*: 1–18.

Cardinal, R., Parkinson, J., Hall, J. and Everitt, B. 2002. Emotion and motivation: The role of the amygdala, ventral striatum, and prefrontal cortex. *Neuroscience and Biobehavioral Reviews*, **26**(3): 321–52.

Cardon, M.S., Foo, M.-D., Shepherd, D.A. and Wiklund, J. 2012. Exploring the heart: Entrepreneurial emotion is a hot topic. *Entrepreneurship Theory and Practice*, **36**(1): 1–10.

Cardon, M.S., Zietsma, C., Saparito, P., Matherne, B.P. and Davis, C. 2005. A tale of passion: New insights into entrepreneurship from a parenthood metaphor. *Journal of Business Venturing*, **20**(1): 23–45.

Carmichael, S. and Price, J. 1995. Sensory and premotor connections of the orbital and medial prefrontal cortex in the macaque monkeys. *Journal of Comparative Neurology*, **363**(4): 642–64.

Carter, C., Braver, T., Barch, D., Botvinick, M., Noll, D. and Cohen, J. 1998. Anterior cingulate cortex, error detection and the online monitoring of performance. *Science*, **280**(5364): 747–9.

Charness, N., Krampe, R. and Mayer, U. 1996. The role of practice and coaching in entrepreneurial skill domains: An international comparison of life-span chess skill acquisition. In K.A. Ericsson (ed.), *The Road to Excellence: The acquisition of expert performance in the arts and sciences, sports, and games*: 51–80. Mahwah, NJ: Lawrence Erlbaum Associates.

Chavez, C., McGaugh, J. and Weinberger, N. 2009. The basolateral amygdala modulates specific sensory memory representations in the cerebral cortex. *Neurobiology of Learning and Memory*, **91**(4): 383–92.

Clelland, C., Choi, M., Romberg, C., Clemenson, G., Fragniere, A., Tyers, P., Jessberger, S., Saksida, L., Barker, R., Gage, F. and Bussey, T. 2009. A functional role for adult hippocampal neurogenesis in spatial pattern separation. *Science*, **325**(5937): 210–13.

Conway, M.A. 2009. Episodic memories. *Neuropsychologia*, **47**(11): 2305–13.

Cornelissen, J.P. and Clarke, J.S. 2010. Imaging and rationalizing opportunities: Inductive reasoning and the creation and justification of new ventures. *Academy of Management Review*, **35**(4): 539–57.

Coulton, R. and Udell, G.G. 1976. The National Science Foundation's Innovation Center: An experiment in training potential entrepreneurs and innovators. *Journal of Small Business Management*, **21**(3): 11–20.

Curtis, M. and Paré, D. 2004. The rhinal cortices: A wall of inhibition between the neocortex and the hippocampus. *Progress in Neurobiology*, **74**(2): 101–10.

Damoiseaux, J., Beckmann, C., Arigita, E., Barkof, F., Scheltens, P., Stam, C., Smith, S. and Rombouts, S. 2007. Reduced resting-state brain activity in the 'default network' in normal aging. *Cerebral Cortex*, **18**(8): 1856–64.

De Quervain, D., Aerni, A., Schelling, G. and Roozendaal, B. 2009. Glucocorticoids and the regulation of memory in health and disease. *Frontiers in Neuroendocrinology*, **30**(3): 359–70.

Deco, G., Buehlmann, A., Masquelier, T. and Hugues, E. 2011. The role of rhythmic neural synchronization in rest and task conditions. *Frontiers in Human Neuroscience*, **5**: 4.

Deng, W., Aimone, J. and Gage, F. 2010. New neurons and new memories: How does adult hippocampal neurogenesis affect learning and memory? *Nature Reviews Neuroscience*, **11**(5): 339–50.

DiCarlo, J., Zoccolan, D. and Rust, N. 2012. How does the brain solve visual object recognition? *Neuron*, **73**(3): 415–34.

Douglas, R. and Martin, K. 2007. The butterfly and the loom. *Brain Research Reviews*, **55**(2): 314–28.

Duncan, K., Ketz, N., Inati, S. and Davachi, L. 2012. Evidence for area CA1 as a match/mismatch detector: A high-resolution fMRI study of the human hippocampus. *Hippocampus*, **22**(3): 389–98.

Durstewitz, D., Vittoz, N., Floresco, S. and Seamans, J. 2010. Abrupt transitions between prefrontal neural ensembles states accompany behavioral transitions during rule learning. *Neuron*, **66**(3): 438–48.

Düzel, E., Bunzeck, N., Masip, M., Witmann, B., Schott, B. and Tobler, P. 2009. Functional imaging of the human dopaminergic midbrain. *Trends in Neurosciences*, **32**(6): 321–8.

Dvorak-Carbone, H. and Schuman, E. 1999. Patterned activity in stratum lacunosum moleculare inhibits CA1 pyramidal neuron firing. *Journal of Neurophysiology*, **82**(6): 3213–22.

Ekman, P. 1992a. An argument for basic emotions. *Cognition and Emotion*, **6**(3–4): 169–200.

Ekman, P. 1992b. Facial expressions of emotion: New findings, new questions. *Psychological Science*, **3**(1): 34–8.

Ekman, P. 2011. What is meant by calling emotions 'basic'? *Emotion Review*, **3**(4): 364–70.

Eriksson, P., Perfilieva, E., Björk-Eriksson, T., Albor, A., Nordborg, C., Peterson, D. and Gage, F. 1998. Neurogenesis in the adult human hippocampus. *Nature Medicine*, **4**(11): 1313–17.

Etienne, A. and Jeffery, K. 2004. Path integration in mammals. *Hippocampus*, **14**(11): 180–92.

Euston, D., Gruber, A. and McNaughton, B. 2012. The role of the medial prefrontal cortex in memory and decision making. *Neuron*, **76**(6): 1057–70.

Floresco, S. 2007. Dopamine regulation of limbic–striatal interplay. *Journal of Psychiatry Neuroscience*, **32**(6): 400–411.

Floresco, S., West, A., Ash, B., Moore, H. and Grace, A. 2003. Afferent modulation of dopamine neuron firing differentially regulates tonic and phasic dopamine transmission. *Nature Neuroscience*, **6**(9): 968–73.

Foerde, K. and Shohamy, D. 2011. The role of the basal ganglia in learning and memory: Insights from Parkinson's disease. *Neurobiology of Learning and Memory*, **96**(4): 624–36.

Foo, M.-D. 2011. Emotions and entrepreneurial opportunity evaluation. *Entrepreneurship Theory and Practice*, **35**(2): 375–93.

Foo, M.-D., Elfenbein, H., Tan, H. and Aik, V. 2004. Emotional intelligence and negotia-

tion: The tension between creating and claiming value. *International Journal of Conflict Management*, **15**(4): 411–29.
Foo, M.-D., Uy, M.A. and Baron, R.A. 2009. How do feelings influence effort? An empirical study of entrepreneurs' affect and venture effort. *Journal of Applied Psychology*, **94**(4): 1086–94.
Fudge, J., Decamp, D. and Becoats, K. 2012. Revisiting the hippocampal–amygdala pathway in primates: Association with immature appearing neurons. *Neuroscience*, **212**: 104–19.
Fujisawa, S., Amarasingham, A., Harrison, M. and Buzsáki, G. 2008. Behavior-dependent short-term assembly dynamics in the medial prefrontal cortex. *Nature Neuroscience*, **11**(7): 823–33.
Gabarnet, L., Jadhav, S., Feldman, D., Carandini, M. and Scanziani, M. 2005. Somatosensory integration controlled by dynamic thalamocortical feed-forward inhibition. *Neuron*, **48**(2): 315–27.
Gabbott, P., Warner, T., Jays, P., Salway, P. and Busby, S. 2005. Prefrontal cortex in the rat: Projections to subcortical autonomic, motor and limbic centers. *Journal of Comparative Neurology*, **492**(2): 145–77.
Gallagher, S. 2005. *How the Body Shapes the Mind*. Oxford: Oxford University Press.
Gallagher, S. 2009. Philosophical antecedents of situated cognition. In P. Robbins and M. Aydede (eds), *The Cambridge Handbook of Situated Cognition*: 35–53. Cambridge: Cambridge University Press.
Gerlach, K., Spreng, R., Gilmore, A. and Schacter, D. 2011. Solving future problems: Default network and executive activity associated with goal-directed mental simulations. *NeuroImage*, **55**(4): 1816–24.
Goto, Y. and Grace, A. 2008. Limbic and cortical information processing in the nucleus accumbens. *Trends in Neuroscience*, **31**(11): 552–8.
Grabenhorst, F. and Rolls, E. 2011. Value, pleasure and choice in the ventral prefrontal cortex. *Trends in Cognitive Neuroscience*, **15**(2): 56–67.
Grace, A., Floresco, S., Goto, Y. and Lodge, D. 2007. Regulation of firing of dopaminergic neurons and control of goal-directed behaviors. *Trends in Neuroscience*, **30**(5): 220–27.
Grégoire, D.A., Corbett, A. and McMullen, J. 2011. The cognitive perspective in entrepreneurship: An agenda for future research. *Journal of Management Studies*, **48**(6): 1443–77.
Greicius, M., Krasnow, B., Reiss, A. and Menon, V. 2003. Functional connectivity in the resting brain: A network analysis of the default mode hypothesis. *Proceedings of the National Association of Science, U.S.*, **100**(1): 253–8.
Greicius, M., Srivastava, G., Reiss, A. and Menon, V. 2004. Default-mode network activity distinguishes Alzheimer's disease from healthy aging: Evidence from functional MRI. *Proceedings of the National Association of Science, U.S.*, **101**(13): 4637–42.
Grichnik, D., Smeja, A. and Welpe, I. 2010. The importance of being emotional: How do emotions affect entrepreneurial opportunity evaluation and exploitation? *Journal of Economic Behavior & Organization*, **76**(1): 15–29.
Gusnard, D., Akbudak, E., Shulman, G. and Raichle, M. 2001. Medial prefrontal cortex and self-referential mental activity: Relation to a default mode of activity. *Proceedings of the National Association of Science, U.S.*, **98**(7): 4259–64.
Hafting, T., Fyhn, M., Molden, S., Moser, M. and Moser, E. 2005. Microstructure of a spatial map in the entorhinal cortex. *Nature*, **436**(7052): 801–6.
Hahn, V., Frese, M., Binnewies, C. and Schmitt, A. 2012. Happy and proactive? The role of hedonic and eudaimonic well-being in business owners' personal initiative. *Entrepreneurship Theory and Practice*, **36**(1): 97–114.
Hargreaves, E., Rao, G., Lee, I. and Knierim, J. 2005. Major dissociation between medial and lateral entorhinal input to dorsal hippocampus. *Science*, **308**(5729): 1792–4.
Harris, K. 2005. Neural signatures of cell assembly organization. *Nature Reviews Neuroscience*, **6**(5): 399–407.
Harris, K., Bartho, P., Chadderton, P., Curto, C., de la Rocha, J., Hollender, L., Itskov, V., Luczak, A., Marguet, S., Renart, A. and Sakata, S. 2011. How do neurons work together? Lessons from the auditory cortex. *Hearing Research*, **271**(1): 37–53.

Haynie, J.M., Shepherd, D., Mosakowski, E. and Early, P.C. 2010. A situated metacognitive model of the entrepreneurial mindset. *Journal of Business Venturing*, **25**(2): 217–29.
He, B., Synder, A., Zempel, J., Smyth, M. and Raichle, M. 2008. Electrophysiological correlates of the brain's intrinsic large scale functional architecture. *Proceedings of the National Academy of Sciences, U.S.*, **105**(41): 16039–44.
Hebb, D. 1949. *The Organization of Behavior: A neuropsychological theory*. New York: Wiley.
Hegdé J. and Van Essen, D. 2007. A comparative study of shape representation in macaque visual areas V2 and V4. *Cerebral Cortex*, **17**(5): 1100–16.
Herman, J., Ostrander, M., Mueller, N. and Figueiredo, H. 2005. Limbic system mechanisms of stress regulation: Hypothamamo–pituitary–adrenocortical axis. *Progress in Neuro-Psychopharmacology and Biological Psychiatry*, **29**(8): 1201–13.
Hmieleski, K.M. and Baron, R.A. 2009. Entrepreneurs' optimism and new venture performance: A social cognitive perspective. *Academy of Management Journal*, **52**(3): 473–88.
Hsu, D. 2007. The dentate gyrus as a filter or gate: A look back and a look ahead. *Progress in Brain Research*, **163**: 601–13.
Hyman, J., Ma, L., Balaguer-Ballester, E., Durstewitz, D. and Seamans, J. 2012. Contextual encoding by ensembles of medial prefrontal cortex neurons. *Proceedings of the National Academy of Sciences, U.S.*, **109**(13): 5086–91.
Immordino-Yang, M., McColl, A., Damasio, H. and Damasio, A. 2009. Neural correlates of admiration and compassion. *Proceedings of the National Academy of Sciences, U.S.*, **106**(19): 8021–6.
Insausti, R., Amaral, D. and Cowan, W. 1987. The entorhinal cortex of the monkey: II, cortical afferents. *Journal of Comparative Neurology*, **264**(3): 356–95.
Isoda, M. and Hikosaka, O. 2011. Cortico-basal ganglia mechanisms for overcoming innate, habitual and motivated behaviors. *European Journal of Neuroscience*, **33**(11): 2058–69.
Ito, H. and Schuman, E. 2012. Functional division of hippocampal area CA1 via modulatory gating of entorhinal cortical inputs. *Hippocampus*, **22**(2): 372–87.
Izard, C. 2007. Basic emotions, natural kinds, emotion schemas and a new paradigm. *Perspectives on Psychological Science*, **2**(3): 260–80.
Jabès, A., Lavenex, P., Amaral, D. and Lavenex, P. 2011. Postnatal development of the hippocampal formation: A stereological study in macaque monkeys. *Journal of Comparative Neurology*, **519**(6): 1051–70.
Janzen, G. and van Turennout, M. 2004. Selective representation of objects relevant for navigation. *Nature Neuroscience*, **7**(6): 673–7.
Jeffery, K. and Burgess, N. 2006. A metric for the cognitive map: Found at last? *Trends in Cognitive Sciences*, **10**(1): 1–3.
Joëls, M., Fernander, G. and Roozendaal, B. 2011. Stress and emotional memory: A matter of timing. *Trends in Cognitive Sciences*, **15**(6): 280–88.
Jung, M.W. and McNaughton, B.L.1993. Spatial selectivity of unit activity in the hippocampal granular layer. *Hippocampus*, **3**(2): 165–82.
Kajiwara, R., Wouterlood, Floris G., Sah, A., Boekel, A., Baks-te-Bulte, L. and Witter, M. 2008. Convergence of entorhinal and CA3 inputs onto pyramidal neurons and interneurons in hippocampal area CA1: An anatomical study in the rat. *Hippocampus*, **18**(3): 266–80.
Kayser, C., Logothetis, N. and Panzen, S. 2010. Millisecond encoding precision of auditory cortex neurons. *Proceedings of the National Academy of Sciences, U.S.*, **107**(39): 16976–81.
Keh, H.T., Foo, M.-D. and Lim, B.C. 2002. Opportunity evaluation under risky conditions: The cognitive processes of entrepreneurs. *Entrepreneurship Theory and Practice*, **27**(2): 125–48.
Kelley, A. 2004. Ventral striatal control of appetitive motivation: Role in ingestive behavior and reward-related learning. *Neuroscience and Biobehavioral Reviews*, **27**(8): 765–76.

Kirkland, T. and Cunningham, W. 2011. Neural basis of affect and emotion. *Cognitive Science*, **2**(6): 656–65.
Knight, F. 2012 (1921). *Risk, Uncertainty and Profit*. New York: Courier Dover Publications.
Knowlton, B. and Foerde, K. 2008. Neural representations of nondeclarative memories. *Current Directions in Psychological Science*, **17**(2): 107–11.
Knutson, B., Taylor, J., Kaufman, M., Peterson, R. and Glover, G. 2005. Distributed neural representation of expected value. *Journal of Neuroscience*, **25**(19): 4806–12.
Kondo, H., Lavenex, P. and Amaral, D. 2008. Intrinsic connections of the macaque monkey hippocampal formation: I. Dentate gyrus. *Journal of Comparative Neurology*, **511**(4): 497–520.
Kondo, H., Lavenex, P. and Amaral, D. 2009. Intrinsic connections of the macaque monkey hippocampal formation: II. CA3 connections. *Journal of Comparative Neurology*, **513**(3): 349–77.
Kravitz, D., Saleem, K., Baker, C., Ungerleider, L. and Miskin, M. 2013. The ventral visual pathway: An expanded neural framework for the processing of object quality. *Trends in Cognitive Sciences*, **17**(1): 26–49.
Kreibig, S. 2010. Autonomic nervous system activity in emotion: A review. *Biological Psychology*, **84**(3): 394–421.
Kumaran, D. and Maguire, E. 2007. Which computational mechanisms operate in the hippocampus during novelty detection? *Hippocampus*, **17**(9): 735–48.
LaBar, K. and Cabeza, R. 2006. Cognitive neuroscience of emotional memory. *Nature Reviews Neuroscience*, **7**(1): 54–64.
Lamprecht, R. and LeDoux, J. 2004. Structural plasticity and memory. *Nature Reviews Neuroscience*, **5**(1): 45–54.
Lavenex, P. and Amaral, D. 2000. Hippocampal–neocortical interactions: A hierarchy of associativity. *Hippocampus*, **10**(4): 420–30.
Lavenex, P., Suzuki, W. and Amaral, D. 2004. Perirhinal and parahippocampal cortices of the macaque monkey: Intrinsic projections and interconnections. *Journal of Comparative Neurology*, **472**(3): 371–94.
LeDoux, J. 2000. Emotion circuits in the brain. *Annual Review of Neuroscience*, **23**: 155–84.
LeDoux, J. 2003a. The emotional brain, fear and the amygdala. *Cellular and Molecular Neurobiology*, **23**(4–5): 727–38.
LeDoux, J. 2003b. *Synaptic Self: How our brains become who we are*. New York: Penguin Group.
LeDoux, J. 2012. Rethinking the emotional brain. *Neuron*, **73**(4): 653–76.
Lemon, N. and Manahan-Vaughan, D. 2006. Dopamine D_1/D_5 receptors gate the acquisition of novel information through hippocampal long-term potentiation and long-term depression. *Journal of Neuroscience*, **26**(29): 7723–9.
Leutgeb, S. and Leutgeb, J.K. 2007. Pattern separation, pattern completion, and new neuronal codes within a continuous CA3 map. *Learning & Memory*, **14**(11): 745–57.
Li, X., Somogyi, P., Ylinen, A. and Buzsáki, G. 1994. The hippocampal CA3 network: An in vivo intracellular labeling study. *Journal of Comparative Neurology*, **339**(2): 181–208.
Lin, L., Osan, R., Shoham, S., Jin, W., Zuo, W. and Tsien, J. 2005. Identification of network-level coding units for real-time representation of episodic experiences in the hippocampus. *Proceedings of the National Academy of Sciences, U. S.*, **102**(17): 6125–30.
Lin, L., Osan, R. and Tsien, J. 2006. Organizing principles of real-time memory encoding: neural clique assemblies and universal neural codes. *Trends in Neurosciences*, **29**: 48–57.
Lindquist, K.A., Wager, T.D., Kober, H., Bliss-Moeau, E. and Barrett, F. 2012. The brain basis of emotion: A meta-analytic review. *Behavioral and Brain Sciences*, **35**(3): 121–43.
Lisman, J. and Grace, A. 2005. The hippocampal–VTA loop: Controlling the entry of information into long-term memory. *Neuron*, **46**(5): 703–13.
Lodge, D. and Grace, A. 2006. The laterodorsal tegmentum is essential for burst firing of ventral tegmental area dopamine neurons. *Proceedings of the National Academy of Sciences*, 103: 5167–72.

Maguire, E., Frackowiak, R. and Frith, C. 1997. Recalling routes around London: Activation of the right hippocampus in taxi drivers. *Journal of Neuroscience*, **17**(18): 7103–10.
Majak, K. and Pitkänen, A. 2003. Activation of the amygdalo-entorhinal pathway in fear-conditioning in rat. *European Journal of Neuroscience*, **18**(6): 1652–9.
Markram, H., Toledo-Rodriquez, M. Yun, W., Gupta, A. Silberberg, G. and Caizhi, W. 2004. Interneurons of the neocortical inhibitory system. *Nature Reviews Neuroscience*, **5**: 793–807.
Marr, D. 1971. Simple memory: A theory for archicortex. *Philosophical Transactions of the Royal Society of London*, **262**(841): 23–81.
McClelland, D.C. 1965. Need achievement and entrepreneurship: A longitudinal study. *Journal of Personality and Social Psychology*, **1**(4): 389–92.
McClelland, D.C. and Winter, D.G. 1969. *Motivating Economic Achievement*. New York: Free Press.
McGaugh, J. 2002. Memory consolidation and the amygdala: A systems perspective. *Trends in Neurosciences*, **25**(9): 456–61.
McHugh, T., Jones, M., Quinn, J., Balthasar, N., Coppari, R., Elmquist, J., Lowell, B., Fanselow, M., Wilson, M. and Tonegawa, S. 2007. Dentate gyrus NMDA receptors mediate rapid pattern separation in the hippocampal network. *Science*, **317**(5834): 94–9.
McMullen, J.S. and Shepherd, D.A. 2006. Entrepreneurial action and the role of uncertainty in the theory of the entrepreneur. *Academy of Management Review*, **31**(1): 132–52.
McNaughton, B., Battaglia, F., Jensen, O., Moser, E. and Moser, M. 2006. Path integration and the neural basis of the 'cognitive map'. *Nature Reviews Neuroscience*, **7**(8): 663–78.
Mitchell, J.R. and Shepherd, D.A. 2010. To thine own self be true: Images of self, images of opportunity, and entrepreneurial action. *Journal of Business Venturing*, **25**(1): 138–54.
Mitchell, R.K. 1994. The composition, classification, and creation of new venture formation expertise. Unpublished doctoral dissertation, University of Utah.
Mitchell, R.K., Busenitz, L.W., Bird, B., Gaglio, C.M., McMullen, J.S., Morse, E.A. and Smith, J.B. 2007. The central question in entrepreneurial cognition research 2007. *Entrepreneurship Theory and Practice*, **31**(1): 1–27.
Mitchell, R.K., Busenitz, L.W., Lant, T., McDougall, P.P., Morse, E.A. and Smith, J.B. 2002. Toward a theory of entrepreneurial cognition: Rethinking the people side of entrepreneurship research. *Entrepreneurship Theory and Practice*, **27**(2): 93–104.
Mitchell, R.K., Smith, B., Seawright, K.W. and Morse, E.A. 2000. Cross-cultural cognitions and the venture creation decision. *Academy of Management Journal*, **43**(5): 974–93.
Mitchell, R.K., Randolph-Seng, B. and Mitchell, J.R. 2011. Socially situated cognition: Imagining new opportunities for entrepreneurship research (Dialogue). *Academy of Management Review*, **36**(4): 774–8.
Miyapuram, K., Tobler, P., Gregorios-Pippas, L. and Schultz, W. 2012. BOLD responses in reward regions to hypothetical and imaginary monetary rewards. *NeuroImage*, **59**(2): 1692–9.
Mormann, F., Kornblith, S., Quian Quiroga, R., Kraskov, A., Cerf, M., Fried, I. and Koch, C. 2008. Latency and selectivity of single neurons indicate hierarchical processing in the human medial temporal lobe. *Journal of Neuroscience*, **28**(36): 8865–72.
Morris, M.H., Kuratko, D.F., Schindehutte, M. and Spivack, A.J. 2012. Framing the entrepreneurial experience. *Entrepreneurship Theory and Practice*, **36**(1): 11–40.
Moser, E. and Moser, M. 2008. A metric for space. *Hippocampus*, **18**(12): 1142–56.
Muckli, L. and Petro, L. 2013. Network interactions: Non-geniculate input to V1. *Current Opinion in Neurobiology*, **23**(2): 1–7.
Murty, V., Ritchey, M., Adcock, R. and LaBar, K. 2010. fMRI studies of successful emotional memory encoding: A quantitative meta-analysis. *Neuropsychologia*, **48**(12): 3459–69.
Nakashiba, T., Cushman, J., Pelkey, K., Renaudineau, S., Buhl, D., McHugh, T., Rodriguez Barrera, V., Chittajallu, R., Iwamoto, K., McBain, C., Fanselow, M. and Tonegawa, S. 2012. Young dentate granule cells mediate pattern separation, whereas old granule cells facilitate pattern completion. *Cell*, **149**(1): 188–201.

Narayanan, N. and Laubach, M. 2006. Top-down control of motor cortex ensembles by dorsomedial prefrontal cortex. *Neuron*, **52**(5): 921–31.

Öngür, D., Ferry, A. and Price, J. 2003. Architectonic subdivision of the human orbital and medial prefrontal cortex. *Journal of Comparative Neurology*, **460**(3): 425–49.

Öngür, D. and Price, J. 2000. The organization of networks within the orbital and medial prefrontal cortex of rats, monkeys and humans. *Cerebral Cortex*, **10**(3): 206–19.

Ousdal, O., Reckless, G., Server, A., Andreassen, O. and Jensen, J. 2012. Effect of relevance on amygdala activation and association with the ventral striatum. *NeuroImage*, **62**(1): 95–101.

Padoa-Schioppa, C. 2011. Neurobiology of economic choice: A good-based model. *Annual Review of Neuroscience*, **34**: 333–59.

Paré, D. 2003. Role of the basolateral amygdala in memory consolidation. *Progress in Neurobiology*, **70**(5): 409–20.

Patzelt, H. and Shepherd, D. 2011. Negative emotions of an entrepreneurial career: Self-employment and regulatory coping behaviors. *Journal of Business Venturing*, **26**(2): 226–38.

Paz, R., Pelletier, J., Bauer, E. and Paré, D. 2006. Emotional enhancement of memory via amygdala-driven facilitation of rhinal interactions. *Nature Neuroscience*, **9**(10): 1321–9.

Poldrack, R., Sabb, F., Foerde, K., Tom, S., Asarnow, R., Bookheimer, S. and Knowlton, B. 2005. The neural correlates of motor skill automaticity. *Journal of Neuroscience*, **25**(22): 5356–64.

Price, J. and Drevets, W. 2010. Neurocircuitry of mood disorders. *Neuropsychopharmacology*, **35**(1): 192–216.

Quian Quiroga, R. 2012. Concept cells: The building blocks of declarative memory functions. *Nature Reviews Neuroscience*, **13**(8): 587–97.

Quian Quiroga, R., Kreiman, G., Koch, C. and Fried, I. 2008. Sparse but not 'grandmother-cell' coding in the medial temporal lobe. *Trends in Cognitive Sciences*, **12**(3): 87–91.

Quian Quiroga, R., Reddy, L., Kreiman, G., Koch, C. and Fried, I. 2005. Invariant visual representation by single neurons in the human brain. *Nature*, **435**(7045): 1102–7.

Quilodran, R., Rothè, M. and Procyk, E. 2008. Behavioral shifts and action valuation in the anterior cingulate cortex. *Neuron*, **57**(2): 314–25.

Raichle, M. 2009. A brief history of human brain mapping. *Trends in Neurosciences*, **32**(2): 118–26.

Raichle, M. 2010. Two views of brain function. *Trends in Cognitive Sciences*, **14**(4): 180–90.

Raichle, M. and Gusnard, D. 2002. Appraising the brain's energy budget. *Proceedings of the National Academy of Sciences, U.S.*, **99**(16): 10237–9.

Raichle, M., MacLeod, A., Snyder, A., Powers, W., Gusnard, D. and Shulman, G. 2001. A default mode of brain function. *Proceedings of the National Academy of Sciences, U.S.*, **98**(2): 676–82.

Raichle, M. and Mintun, M. 2006. Brain work and brain imaging. *Annual Review of Neuroscience*, **29**: 449–76.

Rhee, K. and White, R. 2007. The emotional intelligence of entrepreneurs. *Journal of Small Business & Entrepreneurship*, **20**(4): 409–25.

Rolls, E. 1996. A theory of hippocampal function in memory. *Hippocampus: Special issue on Computational Models of Hippocampal Function in Memory*, **6**(6): 601–20.

Rolls, E. 2004. The functions of the orbitofrontal cortex. *Brain and Cognition*, **55**(1): 11–29.

Roozendaal, B., McEwen, B. and Chattarji, S. 2009. Stress, memory and the amygdala. *Nature Reviews Neuroscience*, **10**(6): 423–33.

Roozendaal, B. and McGaugh, J. 1997. Glucocorticoid enhancement of memory storage involves noradrenergic activation in the basolateral amygdala. *Proceedings of the National Academy of Sciences, U.S.*, **94**(25): 14048–53.

Rosenbaum, R., Ziegler, M., Uinocur, G., Grady, C. and Moscovitch, M. 2004. 'I have often walked down this street before': fMRI studies on the hippocampus and other structures during mental navigation of an old environment. *Hippocampus*, **14**(7): 826–35.

Rotter, J. 1966. Generalized expectancies for internal versus external control of reinforcement. *Psychological Monographs: General and Applied*, **80**(1): 1–28.
Rudebeck, P., Behrens, T., Kennerley, S., Baxter, M., Buckley, M., Walton, M. and Rushworth, M. 2008. Frontal cortex subregions play distinct roles in choices between actions and stimuli. *Journal of Neuroscience*, **28**(51): 13775–85.
Rust, N. and DiCarlo, J. 2010. Selectivity and tolerance ('invariance') both increase as visual information propagates from cortical area V4 to IT. *Journal of Neuroscience*, **30**(39): 12978–95.
Sah, P., Faber, S., Lopez De Armentia, M. and Power, J. 2003. The amygdaloid complex: Anatomy and physiology. *Physiological Reviews*, **83**(3): 803–34.
Salamone, J., Correa, M., Nunes, E., Randall, P. and Pardo, M. 2012. The behavioral pharmacology of effort-related choice behavior: Dopamine, adenosine and beyond. *Journal of the Experimental Analysis of Behavior*, **97**(1): 125–46.
Sargolini, F., Fyhn, M., Hafting, T., McNaughton, B., Witter, M., Moser, M. and Moser, E. 2006. Conjunctive representations of position, direction and velocity in entorhinal cortex. *Science*, **312**(5774): 758–62.
Schacter, D., Addis, D. and Buckner, R. 2008. Episodic simulation of future events: Concepts, data, and applications. *Annals of the New York Academy of Sciences*, **1124**(1): 39–60.
Schultz, W. 1997. A neural substrate of prediction and reward. *Science*, **275**(5306): 1593–9.
Schultz, W. 2010. Subjective neuronal coding of reward: Temporal value discounting and risk. *European Journal of Neuroscience*, **31**(12): 2124–35.
Seress, L. 2007. Comparative anatomy of the hippocampal dentate gyrus in adult and developing rodents, non-human primates and humans. *Progress in Brain Research*, **163**: 23–41.
Sesack, S. and Grace, A. 2010. Cortico-basal ganglia reward network: Microcircuitry. *Neuropsychopharmacology*, **35**(1): 27–47.
Sexton, D.L. and Bowman-Upton, N.B. 1991. *Entrepreneurship: Creativity and growth*. New York: Macmillan Publishing Company.
Seymour, B., Daw, N., Dayan, P., Singer, T. and Dolan, R. 2007. Differential encoding of losses and gains in the human striatum. *Journal of Neuroscience*, **27**(18): 4826–31.
Shane, S. and Venkataraman, S. 2000. The promise of entrepreneurship as a field of research. *Academy of Management Review*, **25**(1): 217–26.
Shepherd, D., Covin, J., and Kuratko, D. 2009. Project failure from corporate entrepreneurship: Managing the grief process. *Journal of Business Venturing*, **24**(6): 588–600.
Sibson, N., Dhankhar, A., Mason, G., Rothman, D., Behar, K. and Shulman, R. 1998. Stoichiometric coupling of brain glucose metabolism and glutamatergic neuronal activity. *Proceedings of the National Academy of Sciences, U.S.*, **95**: 316–21.
Simon, M. and Houghton, S. 2002. The relationship among biases, misperceptions, and the introduction of pioneering products: Examining differences in venture decision contexts. *Entrepreneurship Theory and Practice*, **27**(2): 105–24.
Simons, L., Moulton, E., Linnman, C., Carpino, E., Becerra, L. and Borsook, D. 2014. The human amygdala and pain: Evidence from neuroimaging. *Human Brain Mapping*, **35**(2): 527–38.
Smith, E.R. and Semin, G.R. 2004. Socially situated cognition: Cognition in its social context. *Advances in Experimental Social Psychology*, **36**: 53–117.
Snyder, A. and Raichle, M. 2012. A brief history of the resting state: The Washington University perspective. *NeuroImage*, **62**(2): 902–10.
Spampanato, J., Polepalli, J. and Sah, P. 2011. Interneurons in the basolateral amygdala. *Neuropharmacology*, **60**(5): 765–73.
Spruston, N. 2008. Pyramidal neurons: Dendritic structure and synaptic integration. *Nature Reviews Neuroscience*, **9**(3): 206–21.
Suthana, N. and Fried, I. 2012. Percepts to recollections: Insights from single neuron recordings in the human brain. *Trends in Cognitive Sciences*, **16**(8): 427–36.
Szpunar, K., Watson, J. and McDermott, K. 2007. Neural substrate of envisioning the future. *Proceedings of the National Academy of Sciences, U.S.*, **104**(2): 642–7.

Tang, J., Kacmar, K.M. and Busenitz, L. 2012. Entrepreneurial alertness in the pursuit of new opportunities. *Journal of business Venturing*, **27**(1): 77–94.

Treves, A. 2003. Computational constraints that may have favoured the lamination of sensory cortex. *Journal of Computational Neuroscience*, **14**(3): 271–82.

Treves, A. and Rolls, E. 1994. Computational analysis of the role of the hippocampus in memory. *Hippocampus*, **4**(3): 374–91.

Treves, A., Tashiro, A., Witter, M.P. and Moser, E.I. 2008. What is the mammalian dentate gyrus good for? *Neuroscience*, **154**(4): 1155–72.

Tsigos, C. and Chrousos, G. 2002. Hypothalamic–pituitary–adrenal axis, neuroendocrine factors and stress. *Journal of Psychosomatic Research*, **53**(4): 865–71.

Tulving, E. 1972. Episodic and semantic memory. In E. Tulving and W. Donaldson (eds), *Organization of Memory*: 381–402. New York: Academic Press.

Tulving, E. 1983. *Elements of Episodic Memory*. New York: Oxford University Press.

Van den Heuvel, M. and Hulshoff Pol, H. 2010. Exploring the brain network: A review on resting-state fMRI functional connectivity. *European Neuropsychopharmacology*, **20**(8): 519–34.

Van Gelder, T. 1995. What might cognition be, if not computation? *Journal of Philosophy*, **92**(7): 345–81.

Voorn, P., Vanderschuren, L., Groenewegen, H., Robbins, T. and Pennartz, C. 2004. Putting a spin on the dorsal–ventral divide of the striatum. *Trends in Neurosciences*, **27**(8): 468–74.

Vytal, K. and Hamann, S. 2010. Neuroimaging support for discrete neural correlates of basic emotions: A voxel-based meta-analysis. *Journal of Cognitive Neuroscience*, **22**(12): 2864–85.

Wallace, D. and Kerr, J. 2010. Chasing the cell assembly. *Current Opinion in Neurobiology*, **20**(3): 296–305.

Wallis, J. 2012. Cross-species studies of orbitofrontal cortex and value-based decision-making. *Nature Neuroscience*, **15**(1): 13–19.

Wehr, M. and Zador, A. 2003. Balanced inhibition underlies tuning and sharpens spike timing in auditory cortex. *Nature*, **426**(6965): 442–6.

Welpe, I.M., Spörrle, M., Grichnik, D., Michl, T. and Audretsch, D.B. 2012. Emotions and opportunities: The interplay of opportunity evaluation, fear, joy, and anger as antecedent of entrepreneurial exploitation. *Entrepreneurship Theory and Practice*, **36**(1): 69–96.

Welsh, J.A. and White, J.F. 1981. Converging on characteristics of entrepreneurs. In K.H. Vesper (ed.), *Frontiers of Entrepreneurship Research*: 504–15. Wellesley, MA: Babson Center for Entrepreneurial Studies.

Whitfield-Gabrieli, S., Moran, J., Nieto-Castañón, A., Triantafyllou, R. and Gabrielli, J. 2011. Associations and dissociations between default and self-reference networks in the human brain. *NeuroImage*, **55**(1): 225–32.

Wirth, M., Jann, K., Dierks, T., Federspiel, A., Wiest, R. and Horn, H. 2011. Semantic memory involvement in the default mode network: A functional neuroimaging study using independent component analysis. *NeuroImage*, **54**(4): 3057–66.

Yin, H., Mulcare, S., Hilário, M., Clouse, E., Holloway, T., Davis, M., Hansson, A., Lovinger, D. and Costa, R. 2009. Dynamic reorganization of striatal circuits during the acquisition and consolidation of a skill. *Nature Neuroscience*, **12**(3): 333–41.

9. Entrepreneurial cognition and social cognitive neuroscience

Jeffery S. McMullen, Matthew S. Wood and Leslie E. Palich

ABSTRACT

By offering an alternative to behavioral research methods, neuroscience has transformed cognitive psychology, social psychology, behavioral economics and other disciplines from which entrepreneurial cognition research regularly draws for theoretical insight. In this chapter we examine the potential of neuroscience methods, particularly from cognitive and social cognitive neuroscience, to explain the entrepreneurial phenomenon of forming and successfully implementing opportunity beliefs. After a primer on brain anatomy and neuroscience methods, we partition the entrepreneurial cognition literature into four distinct explanations of the formation and successful implementation of opportunity beliefs and identify the emergence of a fifth explanation that points to the future of neuroscience research. We then review relevant neuroscience literature that could further scholarly understanding of each perspective, and conclude with some general observations about entrepreneurial cognition made possible as a result of this application.

INTRODUCTION

Over the last two decades, neuroscience methods have contributed greatly to our understanding of the brain while providing an alternative to the behavioral research methods that are typically used to study cognition (Lieberman, 2010). As a result, cognitive psychology departments have been transformed by student interest in cognitive neuroscience, demand for neuroscience research, and the dedication of resources and faculty lines to its study (Anderson, 2005). Echoing these developments, social psychology has also become interested in neuroscience's potential for studying social cognition, a field that 'broadly includes the cognitive processes used to understand and store information about other persons including the self, and about interpersonal norms and scripts (or procedures) to navigate efficiently in the social world' (Van Overwalle, 2009: 829). The result has been the field of social cognitive neuroscience (SCN), which uses the tools of neuroscience to study 'the mental mechanisms that create, frame, regulate, and respond to our experience of the social world' (Lieberman, 2010: 143).

The purpose of this chapter is to explore whether neuroscience, particularly in its manifestation as social cognitive neuroscience, has similar potential to transform entrepreneurial cognition, a field concerned with understanding 'the knowledge structures that people use to make assessments, judgments, or decisions involving opportunity evaluation and new venture creation and growth' (Mitchell, Busenitz et al., 2002: 97). With its emphasis on mental mechanisms and reliance on behavioral research methods, entrepreneurial cognition research is likely to confront many of the same obstacles and frustrations that have plagued cognitive psychologists and social psychologists (see Baron, Chapter 1, this volume; Bird, Chapter 3, this volume). Can they be far from following in the footsteps of their contemporaries in psychology who have turned to neuroscience methods to advance their research on cognition?

To answer this question, we begin by introducing entrepreneurial cognition scholars to some of the premises and methods of neuroscience (see also Baucus, Baucus and Mitchell, Chapter 8, this volume). We then seek to organize the entrepreneurial cognition literature around a common research question that is answered differently depending on the explanation of entrepreneurial cognition employed. We identify four established and one emerging explanation, and suggest how neuroscience research might be applied to further entrepreneurial cognition research according to each perspective.

A PRIMER ON BRAIN NEUROSCIENCE

Although brain neuroscience is a relatively new field, over the last two decades researchers across the biological and social sciences have enthusiastically engaged in interdisciplinary attempts to map the functions of the mind on to the structure of the brain. Disciplines including psychology (Ochsner and Lieberman, 2001; Van Overwalle, 2009), economics (Camerer, Loewenstein and Prelec, 2005), and sociology (Franks, 2010; Todorov, Fiske and Prentice, 2011) have sought to examine cognitive neuroscience and social neuroscience. Before examining how their findings may inform research on entrepreneurial cognition, however, a primer on the anatomy of the brain may be in order.

General Brain Anatomy

The brain consists of two nearly mirror-image hemispheres. The left hemisphere receives most of its inputs from, and controls, the right side of the body. In humans it is specialized for language, rule-based reasoning and analytic skills. In contrast, the right hemisphere receives most of its inputs from, and controls, the left side of the body. It is better at visual pattern recognition and more holistic kinds of perception. The two hemispheres are connected by the largest fiber tract in the brain, known as the corpus callosum, which enables a divide-and-conquer approach by which the brain uses its right side to apprehend the big picture and its left side to process details.

The brain is believed to have evolved to handle the difficult cognitive problems that are posed by being a highly social animal (Haxby, 2011). As a result, the inside of the brain where the brain stem meets the spinal cord is the 'oldest' and was coined our 'lizard brain' by Paul MacLean because it is exhibited by reptiles, most primitive vertebrates, and even some invertebrates. This 'reptilian' level boasts the basal ganglia, which mediate instinctive behaviors such as fight-or-flight responses, hunger, defending territory and sexual pursuit. The next part of the brain, known as the limbic system, contains the amygdala, hypothalamus, septum hippocampus, hippocampal complex and cingulate cortex. Because it allows feelings that enable more complex behavior, such as parental nurturing and participation in social dominance hierarchies, the limbic system is believed to have arisen in early mammals. Lastly, we encounter the neo-mammalian part of the brain, known as the neo-cortex, which envelops the limbic system. The larger the neo-cortex, the more folded it tends to be, such that the creases become *sulci* and the areas on the surface become *gyri*. In humans it resembles a wadded up sheet of paper, but in some mammals it is entirely smooth. The neo-cortex allows complex social behavior, intelligence, varying degrees of awareness and, ultimately (in humans), consciousness.

The neo-cortex consists of four major lobes: frontal, parietal, occipital and temporal. The frontal lobe addresses general planning and is found at the front of the brain. The parietal lobe, found at the top back of the brain, handles visual and auditory processing, particularly as they relate to spatial location and navigating through space. The occipital lobe is devoted exclusively to vision and is found in the back of the brain. Finally, the temporal lobe is a tongue-like extension bordering the occipital and parietal lobes. It is known to tackle some auditory signals, some high-order visual functions, like face recognition, and visual processing of object perception.

Social Cognition and the Brain

Few functions that involve behavior as complex as entrepreneurial action exclusively activate one region of the brain, but meta-analysis has revealed a reliable connection between specific functions of the mind and particular regions of the brain. For example, the temporo-parietal junction (TPJ) and the medial prefrontal cortex (mPFC) have been consistently identified as integral to social cognition – the capacity to understand people's behavioral intentions, social beliefs and personality traits (Van Overwalle, 2009). Jenkins and Mitchell (2010) take this a step further, showing that social cognition makes flexible use of different processes as a function of the demands of the social context. Examining the ability to read others' minds – that is, to mentalize – by understanding a wide range of mental states, including not only others' beliefs and knowledge, but also their feelings, desires and preferences, they find that perceivers tend to mentalize more under conditions of uncertainty in which they must generate plausible hypotheses about a target's mental state from ambiguous or otherwise underspecified information (see also Zaki and Ochsner, 2011). Specifically, when mentalizing about different types of mental states (beliefs versus preferences) under uncertainty – for example, making sense of new situations, unfamiliar individuals or ambiguously motivated behavior – the ventral and dorsal aspects of the mPFC responded more than when making unambiguous inferences, regardless of the type of mental state. In contrast, the right TPJ was sensitive to the distinction between beliefs and preferences irrespective of certainty. Studies such as theirs call into question any assumption that social cognition comprises a single mental operation.

RESEARCH METHODS OF SOCIAL COGNITIVE NEUROSCIENCE

Brain Imaging

Various biological and physiological research methodologies have been used to identify and explore brain activations and the relationships that these may have with important social and psychological processes, many of which may be relevant to the study of entrepreneurial cognition. Schultheiss and Stanton (2009) and van Honk (2012) explain how to assay or manipulate hormone levels to index and understand neural functioning, and Lee and Chamberlain (2007) advocate the use of autonomic nervous system (ANS) measures, which can include options such as electrodermal response (EDR, sweat reactions in the skin) and the startle eyeblink response. These reveal shifts in a subject's emotions, motivations, attention and preferences. Still others argue that transcranial magnetic stimulation (TMS, stimulating activity in specific parts of the brain with an electrical current) and facial electromyography (facial EMG, measuring facial muscle activity) offer a unique window into the brain and its functioning (Harmon-Jones and Beer, 2012). But these are only a few of the available options.

Methods such as those mentioned above each offer their own advantages and drawbacks and need to be used where they are best suited to the research task at hand. However, our focus in this chapter will be on the use of brain imaging, which is the most popular neuroscientific study approach in use today. Although the specific details involved in applying these tools are highly technical and require extensive training and lab resource commitments, the general process is relatively simple. Researchers using these technologies are able to determine which regions of the brain are activated in subjects completing an experimental task. These images can then be compared to maps generated by subjects in a control group to discern the unique neural activity that results from the experimental condition.

Neurological imaging and its associated technologies continue to advance rapidly even while the cost of using them (though still substantial) falls. Most researchers employ one of three available methods: positron emission topography (PET); quantitative electro-encephalogram (qEEG), or functional magnetic resonance imaging (fMRI). These are each discussed, in turn, below.

Though still useful to neuroscientific inquiry, PET is the least used and least discussed of the brain-scanning methods available to researchers today. Seen in the context of the rapid developmental pace of neuroimaging technologies, PET is a relatively old technique. Similar to fMRIs, PET does not measure neural activity directly but works instead by sensing the flow of blood to the various regions the brain. These readings serve as a reasonable proxy for neural responses, which naturally determine bloodflow patterns.

The EEG is the oldest of the brain-scanning techniques, first invented by German psychiatrist Hans Berger nearly a century ago (see Berger, 1929). Using advanced signal techniques, it reads directly the electrical activity generated by the activity of populations of neurons inside the brain as measured at the scalp via caps with precisely placed electrodes. When this method is used to index brain activity associated with an experimental task involving specific events (e.g., stimuli or responses), the qEEG response is called an event-related brain potential (ERP).

The qEEG is the direct scanning methodology most often used in studies related to the organizational setting, with applications to advertising research going back at least as far as the early 1980s (e.g., Appel, Weinstein and Weinstein, 1980). For their research on the neuroscience of leadership, Waldman, Balthazard and Peterson (2011: 64) deliberately chose to employ the qEEG for the following reasons.

> The data produced by qEEG can be readily used in conjunction with statistical analysis programs. In addition, qEEG is relatively inexpensive, portable, and non-invasive, and its use involves no health risk. Unlike fMRI and other methods that require people to remain immobilized in an unnatural setting (that is, in a clinical setting within a confined tube), qEEG assessment can be completed while people are comfortably seated and engaged in everyday tasks – including conversations associated with the type of vision communication task [employed in our study].

This justification might suggest that tests using qEEGs are relatively easy to conduct, but they actually require significant preparation (almost one hour per subject), and in many ways the process must be managed in a similar fashion to the conduct of fMRIs and PET scans. For example, participants must be trained to limit their movements as much as possible during the test to prevent 'muscle artifacts' (i.e., brain signals created from the movement and that are unrelated to the experimental treatment being investigated). Even the blinking of eyes will influence EEG data.

Because of its practical importance, the point Waldman and colleagues make about the low cost of EEG neuroimaging methods deserves special mention. The alternatives to the qEEG are extremely expensive. To provide one comparison, fMRI scanners cost around $5 million to purchase and set up and perhaps $500 per hour for personnel and maintenance costs. EEG equipment, on the other hand, costs less than $100,000 to purchase and is relatively inexpensive to maintain. This clearly changes the overall calculus of experimental options.

As the newest of the options available, fMRI is the most popular imaging method in use today. But the fMRI is very different from the qEEG in that it does not measure brain activity directly. Rather, it measures neural responses according to ratios of deoxygenated to oxygenated blood movement in the brain that naturally result from such activity (often called 'blood-oxygenation-level-dependent,' or BOLD, contrast). Johnstone, Kim and Whalen (2009: 314) describe the usefulness of this technology as follows: 'Through some good fortune, oxygenated blood and deoxygenated blood have slightly different magnetic properties; this difference makes it possible to detect the ratio by [means of] a large electromagnet and radio waves.' This allows a researcher to detect brain activations slightly after the fact, though the delay is long enough to spawn criticisms of fMRI and its usefulness in psychological research.

But there are definite drawbacks to using fMRI technology in neuroscience studies. Because such tests must be laboratory based, researchers can find it very difficult to apply this neuroscanning method to the study of actual behavior or naturally occurring situations. For example, it is impossible to test subjects while they are engaged in face-to-face interactions since subjects can only be scanned as they are situated in cramped machines, holding their heads absolutely and unnaturally still. As a result, direct interactions with others are not possible during testing.

The methods of neuroimaging discussed here each have definite strengths and weaknesses. Because they provide direct measures of neural firing, qEEGs present excellent temporal resolution (being able to capture electrical activations less than a millisecond after their occurrence in some cases). At the same time, however, they provide less precise information regarding the origin of the anatomical signals that they receive (spatial resolution only within centimeters). That is, a qEEG senses these signals at the level of the scalp, allowing bioelectric indicators to blur partially as they pass through the brain, skull and scalp. The nature of the technology also prevents the detection of changes that occur within the deeper regions of the brain.

fMRIs and PET scans have their own inherent strengths and limitations. Most notably, they provide only indirect measures of neuronal activity, assessing instead the flow of oxygenated blood to neural tissue that results from this activity. As a result, these methods suffer from lagged temporal resolution (only within seconds), while offering much more precise spatial resolution (within millimeters) (Harmon-Jones and Beer, 2012).

Even with the superior spatial resolution of fMRIs and PET scans, these forms of brain imaging still offer, at best, a crude snapshot of brain activity. As Camerer and colleagues (2005: 12) report, 'Neural processes are thought to occur on a 0.1 millimeter scale in 100 milliseconds (msec), but the spatial and temporal resolution of a typical scanner is only 3 millimeter and several seconds. Multiple trials per subject can be averaged to form composite images, but doing so constrains experimental design.' It is also important to note that the data generated from methods that use blood flows as a proxy for neural activity are fundamentally correlational in nature – that is, they cannot identify the specific regions of the brain that produce observed neuronal responses.

In any case, the technology continues to improve while the cost of equipment declines. Furthermore, researchers are starting to use hybrid techniques to take advantage of the strengths of different methods within the context of a single study. Another important development is the use of 'hyperscanning,' a method by which multiple participants, each in a separate MRI scanner, can interact with each other in the course of an experiment. Since the brain of each subject is being scanned during the analysis, deeper levels of social interaction can be studied (Montague, Berns, Cohen, McClure, Pagnoni, Dhamala, Wiest, Karpov, King, Apple and Fisher, 2002). This approach promises to offer rich insights into the social processes involved in decision making and action, creating new research opportunities and understanding as it is applied to untangling various aspects of entrepreneurial cognition.

APPLICABILITY OF SCN TO ENTREPRENEURIAL COGNITION

Social cognitive neuroscience seeks to understand phenomena in terms of interactions between three levels of analysis: the social level, which is concerned with the motivational and social factors that influence behavior and experience; the cognitive level, which is concerned with the information-processing mechanisms that give rise to social-level phenomena; and the natural level, which is concerned with the brain's mechanisms that instantiate cognitive-level processes. The social cognitive neuroscience approach entails conducting studies and constructing theories that make reference to all three levels and contrasts with traditional social psychological and cognitive neuroscientific research that primarily makes reference to two levels (Ochsner and Lieberman, 2001).

Functions of the Mind, Regions of the Brain, and Selective Inference

Studying the brain activity of entrepreneurs begins with a stimulus–response configuration that captures a particular socio-cognitive function. This function is mapped on to particular regions of the brain through a process of corresponding activation, such that activation of a region of the brain by a participant who is completing a task intended to require for example, planning can be interpreted as evidence that the planning function takes place in the medial prefrontal cortex (mPFC). Currently, however, this inferential process is capable of flowing in only one direction from function to region. The mapping of the brain has not yet progressed far enough to start with images of the activated region of the brain such that the region can then be used to predict with any accuracy what socio-cognitive function the participant is performing.

For some functions of the mind, it is reasonably clear which regions of the brain are activated. For instance, meta-analyses are revealing that the mPFC lights up regularly for social cognition tasks – which are mostly about mentalizing others' beliefs and preferences. This mentalizing is clearly a significant part of entrepreneurship because it addresses intentional causality used in ascribing agency to others – others act because of their beliefs and preferences. Ascribing causality to others is essential for the intense social negotiating involved in the early phases of innovation, as described and discussed by a number of abductive and inductive approaches to problem solving, including design thinking (Brown, 2008) or effectuation (Sarasvathy, 2001; McMullen, 2010). This mentalizing, however, says nothing about cognition regarding engineering-type problems that have to do more with contemplating physical causality (involving rule-based logic or natural laws), which tends to activate different regions of the brain. Fortunately, meta-analyses are beginning to appear that can help to summarize findings across other sub-fields, such as cognitive neuroscience. This should expedite the brain-mapping process and enhance accessibility of neuroscience findings for more applied fields such as entrepreneurship.

For the time being, however, it looks as though the field of entrepreneurial cognition, like cognitive psychology or social psychology, may have more to give than to receive from neuroscience. This is because of what Poldrack (2012) labels selective inference. Based on the psychology literature, researchers supposedly know that a particular task activates some cognitive function of the mind; the question then becomes what region of brain structure will correspondingly be activated. This forward inference from mind function to brain structure works well, but the reverse inference from brain structure to mind function has a way to go before it will be particularly informative. This is because certain regions of the brain (e.g., the anterior cingulate cortex (ACC)) can be activated by numerous functions, providing very little discriminatory information. This is particularly disappointing for entrepreneurship researchers who may have hoped that neuroscience could reveal which regions of the brain were activated by an entrepreneurial task and then use this activated region to infer what is going on in the entrepreneurial mind. Although reverse inferences like this are not yet possible, calls for 'selective inference' indicate that they are not a pipe dream. Neuroscientists are well aware of the problem of reverse inference and have been calling for studies that look for selective connections between mind function and brain structure such that the brain structure is uniquely associated with a particular function of the mind. In its infancy, this initiative is likely to continue to become easier as brain-imaging technologies and computing power improve, allowing for finer-grained images of the brain in real time.

The Need to Identify a Distinctly Entrepreneurial Task

Given selective inference concerns, entrepreneurship cognition researchers may be forced into conducting psychological experiments that are not that different from those that would be conducted without neuroscience except for an additional level of analysis – a hypothesis and results about the region of the brain that is being activated. For this brain information to be more than a novelty, there must be some theoretical expectation, not just about whether the task being conducted activates a certain function, but also about what the mechanism behind that function is. For example, a theory-of-mind task like that recently examined by Bagozzi and colleagues (2013) asks participants to think actively about a problem from their own point of view and from the point of view of another, as well as passively about how they would feel in another's shoes. All three requests involve the same function of the mind – solving someone else's problem – and the task performed clearly activates the problem-solving function, but neuroscience research suggests that different regions of the brain are activated when people think autobiographically versus when they think from another's vantage point versus when they imagine how they would feel in another's shoes.

It may be possible, for example, to use a similar approach to opportunity evaluation tasks (cf. Wood and Williams, 2014) if a clear theoretical explanation can be provided for why the entrepreneurial task is likely to evoke different socio-cognitive functions. McMullen and Shepherd (2006) offer one such possibility by suggesting that individuals distinguish between opportunities for someone versus opportunities for themselves. Their framework of entrepreneurial action, which draws partly from the theory-of-mind literature, could be used to argue that the first-person opportunity in which one believes a situation is an opportunity for himself must have an element of personal commitment that is not shared by a third-person opportunity in which one believes a situation is an opportunity for someone else. Does evaluation of a situation as a third-person opportunity activate the region of the brain associated with taking another's viewpoint? Does the region of the brain associated with personal agency, such as committing to a goal and/or engaging in planning, become activated when evaluating a first-person opportunity but not a third-person opportunity?

Entrepreneurial cognition studies can borrow from psychology theory to predict what tasks elicit particular functions of mind and from neuroscience to predict how these functions will map on to the structure of the brain, but in doing so entrepreneurial cognition studies become conjunctively constrained. These constraints are tolerable if something about the task is clearly and uniquely entrepreneurial. Without a uniquely entrepreneurial quality to the task, however, entrepreneurial cognition scholars run the risk of studying a piece of the entrepreneurial process that is unlikely to be unique to entrepreneurship and therefore possibly redundant with studies that have already been completed in psychology and/or neuroscience. For example, negotiating with investors about equity issues is a task that may be unique or redundant with prior research (e.g., Galinsky and Mussweiler, 2001; Galinsky et al., 2008). If the equity negotiation differs in some theoretically meaningful way from the use of more mundane negotiation tasks, it is possible that a contribution to psychology could be made. Similarly, such a study of equity negotiation may contribute to the neuroscience on empathy and perspective taking (e.g., Decety and Jackson, 2006; Decety and Porges, 2011; Decety et al., 2012) if the nature of the negotiation somehow differs theoretically from Jean Decety's body of research examining the neuroscience of empathy and perspective taking in tasks involving cooperation and competition (e.g., Decety et al., 2004).

Unless entrepreneurship involves a task that requires some unique cognitive function, however, it is unlikely to contribute to either psychology or neuroscience, which often turns to psychology for its theory. A new level of analysis – the biological – does indeed contribute to entrepreneurship by offering an alternative to behavioral research methods, thereby reducing common method bias (Podsakoff et al., 2003). But instead of settling for how neuroscience could make a small contribution to entrepreneurial cognition research, we encourage entrepreneurship researchers to be more ambitious and think about how a social cognitive neuroscience approach to entrepreneurial cognition might shed light on all three levels of analysis, thereby making a contribution to all three fields. In this spirit, we offer an in-depth examination of neuroscience's potential to advance entrepreneurial cognition research.

OPPORTUNITY BELIEFS AS A DISTINCTLY ENTREPRENEURIAL TASK

Mitchell and colleagues (2007) point out that entrepreneurs do not follow the normative/rational model in their thinking and that this creates a need within the field of entrepreneurial cognition to determine what it is that entrepreneurs are following. They ask, 'How do entrepreneurs think, reason, and behave such that they create value and wealth through the identification and implementation of market opportunities?' (Mitchell et al., 2007: 5). This question somewhat unintentionally implies a causal sequence in which entrepreneurs' (1) thinking, reasoning or behavior leads to (2) their identification of a market opportunity that creates value and/or wealth, and/or (3) their implementation of a market opportunity that creates value and/or wealth.[1]

At first, studying these individual links appears reasonable. After all, opportunity identification with its more perceptual emphasis appears to be easily delineated from implementation, which seems to be more behavioral in nature. But a closer examination of the question reveals that, because the opportunity construct specifies its outcome (e.g., market opportunity that creates value and/or wealth), the construct may be identifiable only in retrospect. This makes the concept of opportunity identification dependent on successful implementation. As a result, it does not make sense to ask whether thinking leads to the identification of market opportunities if market opportunities must, by definition, be implemented in order to create value and/or wealth, and the entrepreneur chooses not to do so (McMullen and Shepherd, 2006). Conversely, the question makes implementation equally dependent on identification. For example, how can entrepreneurs implement a market opportunity that creates value and/or wealth if they have not first identified it? Thus, according to the research question articulated above, only situations that are identified as opportunities are implemented, and only those that are implemented successfully enough to create value and/or wealth are ultimately classifiable as 'opportunities.'

This conundrum has led some to suggest that opportunities are empirically problematic because they can only be observed *post hoc* (Dimov, 2011). Such criticism, however, may confound opportunities to try to advance a goal, which are *a priori* opportunities – situations that can be classified as opportunities regardless of outcome, with opportunities to advance a goal, which are *post hoc* opportunities – constructs that can be identified only by their outcome. Consider the following dating analogy, for example. A young man may have the goal of asking a young woman out on a date but lack the opportunity until they are both invited to a party. At that party he may meet her and even have a conversation but lack the courage to ask. This failure does not negate the existence of the opportunity to try to advance a goal, despite his failure to seize it. Therefore identification of an opportunity to try to advance a goal – henceforth 'opportunities to try' – are situations that allow an individual to act on a goal, regardless of outcome or whether implementation takes place at all. All that is needed to establish the objective existence of an opportunity to try are the elements of a goal and the environmental conditions that would allow conversion of that goal into behavior (McMullen, Plummer and Acs, 2007).

Opportunities to try typically involve proximal outcomes (McMullen and Dimov, 2013) and simple tasks whose execution depends solely on the individual who is contemplating them. Because they mostly involve self-focused beliefs about one's own interests and whether an activity will advance those interests, their evaluation can be examined through theories of motivation, perceived efficacy, intent and trying, which are particularly appropriate for asking questions about whether an individual will form the belief that entrepreneurial action is possible (McMullen and Shepherd, 2006; see also Carsrud and Brännback, Chapter 2, this volume). They say little, however, about how customers and other stakeholders might react to those actions.

By contrast, an opportunity to advance a goal such as meeting one's soul mate, for example, is contingent on another's beliefs, preferences and behavior over time. Accurate social inference in this scenario is necessary in order to realize a successful outcome. The focal agent has no way of knowing whether the situation is an opportunity before action proves it to be so. He may believe that the situation is an opportunity to advance the goal of finding his soul mate and ask the girl out. She may say no, refuting the belief. She may say yes, only to say no to a second date or for the young man to realize he was mistaken in his belief such that he chooses not to ask her out for a second date. They may even date for years, marry, and divorce years later. This could refute the belief that this girl is his soul mate. Therefore an opportunity to advance a goal represents a much higher bar than an opportunity to try, which it encompasses, and the height of this bar ultimately depends on the ambition of the goal that one is seeking to advance.

Opportunities to advance a goal are situations that cannot be classified as opportunities *a priori* because they are contingent on outcomes. This is particularly true for entrepreneurial opportunities – situations where new goods, services and organizing methods can be introduced and sold at a price greater than their costs of production (Casson, 1982; McMullen, Plummer and Acs, 2007; Shane and Venkataraman, 2000). As such, entrepreneurial opportunities consist of situations that allow for the successful conversion of a goal (e.g., a desire for financial returns) into an event outcome (e.g., realized profit) through some behavior (e.g., introduction of a product) that is contingent on another party or parties for completion (e.g., investment and/or sale). Thus entrepreneurial opportunities are situations that are both outcome dependent and contingent on others for both the production and reward of task-contingent behavior. They require cooperation, first, to procure the resources (from investors) and labor (from employees) needed to perform the task (e.g., producing a new product), and, second, to receive the reward (e.g., profit from sales to customers) for successfully completing the task.

The need to solicit cooperation requires that entrepreneurs anticipate as well as respond to others' reactions as they seek to realize complex goals. Such anticipations are likely to influence entrepreneurs' beliefs about whether opportunities to try represent opportunities to advance a goal. These beliefs are then supported or refuted by information obtained by the entrepreneur through interactions with stakeholders (investors, employees, suppliers, distributors, bureaucrats etc.) long before a product may even be offered to customers (McMullen, 2014; Wood, McKelvie and Haynie, 2014).

Because entrepreneurs cannot read minds, they must infer stakeholders' beliefs and preferences in hope of soliciting their cooperation. This need for inference introduces uncertainty that precludes opportunities to advance a goal from being classifiable as anything but beliefs until validated through a successful exchange (McMullen, 2010, 2014). In turn, this exchange is an outcome of action that is based on the belief that the opportunity to try is an opportunity to advance a goal. For this reason, it is entirely possible to ask whether certain types of thinking, reasoning or behavior lead individuals 'to identify opportunities to try' without also having to meet the requirement that these opportunities be implemented. The same, however, cannot be said for opportunities to advance a goal.

To the extent that entrepreneurial cognition researchers ask whether particular types of thinking, reasoning and/or behavior lead individuals to classify 'opportunities to try to advance a goal' as 'opportunities to advance a goal,' they are in effect studying the formation of opportunity beliefs. Opportunity beliefs are self-centered mental images or 'theories' about the potential reward for a particular action versus the cost of that action (McMullen and Shepherd, 2006). This suggests that, for entrepreneurs to act on an opportunity idea (Dimov, 2010), they must believe that doing so will result in a successful event outcome, such as profit generation or societal benefits (Lumpkin et al., 2013; Wood et al., 2012). To do this, entrepreneurs judge the possibility of an opportunity not by seeing it directly, but by seeing changes in variables related to action and the corresponding event outcome (Baron and Ensley, 2006). Although this information can be obtained and processed in many ways, social inference – defined as 'a variety of processes invoked as we form representations of the psychological states, traits, and preferences of others' (Lieberman, 2010: 152) – may be particularly crucial because judging the future outcome of acting on a set of circumstances is a subjective and interpretive process (Barreto, 2012; Lachmann, 1970). As such, social inference becomes the mechanism by which perceptions of circumstances and those of potential stakeholders are integrated into visions of what the future might look like and, subsequently, beliefs about the value of hypothetical courses of action (McMullen and Shepherd, 2006; Shackle, 1979; Walsh, 1995). Thus dependency on the complicit feedback and behavior of stakeholders, including customers, investors, employees, suppliers, distributors and so on, is likely to require entrepreneurs to mentalize the beliefs and preferences of observed and imagined stakeholders as they form their opportunity beliefs (Chiles et al., 2010; McMullen, 2010).

Whether these opportunity beliefs are justified, however, may depend on the thinking, reasoning and/or behavior used by the entrepreneur to form them (McMullen, 2014). Are some individuals consistently better at forming opportunity beliefs that correspond with subsequent realities? If so, what role does their thinking, reasoning or behavior play in achieving this convergence? Is the thinking, reasoning or behavior that facilitates successful implementation different from that used either to form opportunity beliefs or to identify opportunities to try? If so, how?

Regardless of the questions asked, the fact remains that entrepreneurial cognition researchers can study entrepreneurial opportunities only in retrospect. To study phenomena in real time, including entrepreneurial opportunities, the field must look to a different 'dependent variable,' be it opportunity beliefs, opportunities to try, or something else entirely. Because there are a number of different theoretical explanations for why some individuals develop beliefs along the lines of 'this is an attractive opportunity for me to take action' (Haynie et al., 2009), while others have little confidence that the same circumstance represents an opportunity for entrepreneurship at all (Dimov, 2007; Wood and McKinley, 2010), it is useful to consider the major 'threads' that run through the entrepreneurial cognition literature as they relate to explaining opportunity beliefs. Accordingly, we now examine the most common ways in which cognition researchers explain the formation and successful implementation of opportunity beliefs and explore the possibility of applying neuroscience techniques to enhance those explanations or develop new ones.

PREDOMINANT EXPLANATIONS OF OPPORTUNITY BELIEF FORMATION

What makes entrepreneurs believe that acting in the face of uncertainty will lead to successful outcomes, such as the realization of profit? Is it because these individuals are 'hard wired' to believe action is the answer when uncertainty reigns? Is it because they have developed a positive attitude about dynamic environments? Is it because they attend to the right information to predict outcomes with accuracy? Maybe it is because they hold domain knowledge that others simply do not have, and this allows them to see things others have overlooked. Examination of the literature to date suggests that studies concerning opportunity beliefs have typically fallen into one of four perspectives, each favoring its own explanation. These perspectives emerge from a composition of two dimensions. The first dimension, or x-axis, addresses whether the formation and successful implementation of opportunity beliefs is triggered primarily by a stimulus that is internal or external to the individual. The second dimension, or y-axis, is concerned with whether differences observed in entrepreneurial thinking, reasoning or behavior are considered to be permanent or temporary phenomena. Together these dimensions form the four quadrants presented in Table 9.1 – internal–permanent (the Trait explanation), external–permanent (the Adaptation explanation), internal–temporary (the Expertise explanation), and external–temporary (the Mindset explanation).

Each explanation flows from research interests unique to each perspective. In the subsections that follow, we briefly outline each perspective, identify the umbrella question(s) asked, highlight some of the research within it, and explore the potential of neuroscience for informing the perspective. Finally, we discuss a fifth, emerging perspective within entrepreneurial cognition that explicitly emphasizes the formation and implementation of opportunity beliefs as a dynamic process.

The Trait Explanation

Early attempts to understand entrepreneurship were squarely focused on the idea that entrepreneurs are people born with certain personality traits – relatively stable dispositions to respond in a certain way across various situations (Caprana and Cervone, 2000; Rauch and Frese, 2007) – that predispose them to entrepreneurship (Cromie, 2000). Despite its intuitive appeal, the traits explanation has received mixed empirical support (see Stewart and Roth, 2001; Zhao and Seibert, 2006; Rauch and Frese, 2007, for reviews). Even though entrepreneurs often exhibit comparatively high ratings on some trait dimensions, such as need for achievement (Brandstätter, 2011; McClelland, 1961/1967), these traits are not exclusively correlated with entrepreneurial action (Brockhaus, 1980; Koh, 1996). At best, this makes them necessary but insufficient for entrepreneurship (Collins, Hanges and Locke, 2004). As a result, Howard Aldrich (1999: 76) concluded that research on traits 'seems to have reached an empirical dead end.' But despite Aldrich's claim, the trait-based explanation has not died. Personality may not be as central to the discussion as it used to be, but enduring traits, such as the Big Five personality

Table 9.1 Explanations of opportunity beliefs and potential application of neuroscience

Trait explanation	Adaptation explanation
Traditional questions: Who is an entrepreneur? What personality traits make entrepreneurs unique? Do particular personality traits encourage individuals to become entrepreneurs? Are these traits responsible for entrepreneurial success?	*Traditional questions:* Which cognitive skills are required to notice opportunities to try to advance a goal? How do those skills affect the cognitive processes that underpin formation of opportunity beliefs and how do variations in those skills relate to successful implementation of those beliefs?
Conceptual logic: Opportunity beliefs emerge because some individuals possess personality traits that compel them to pursue entrepreneurship. Those who do not possess the right personality traits are relegated to traditional employment.	*Conceptual logic:* Opportunity beliefs flow from having the cognitive skills (or innate ability to develop them) needed to respond to the changing requirements of the environment. Such ability to adapt allows for the formation, revision and successful implementation of opportunity beliefs.
Potential of neuroscience: Possible identification of regions of the brain that are activated for people with personality traits that encourage identification of opportunities to try and/or the formation and successful implementation of opportunity beliefs.	*Potential of neuroscience:* Possible identification of regions of the brain that are activated for people who attentively and accurately monitor the physical and social environments.
Expertise explanation	**Mindset explanation**
Traditional questions: What role do knowledge and experience play in opportunity belief formation? How does expertise shape the way entrepreneurs form and implement opportunity beliefs?	*Traditional questions:* Do entrepreneurs have a unique mindset that encourages them to form opportunity beliefs when others do not or cannot? What are the factors that make up the entrepreneurial mindset? What is the relationship between various elements of the entrepreneurial mindset, opportunity beliefs and implementation outcomes?
Conceptual logic: Opportunity beliefs are a function of application of knowledge and expertise developed through formal learning, day-to-day interactions, and previous experiences with business and entrepreneurship.	*Conceptual logic:* Entrepreneurship takes place because individuals use an entrepreneurial mindset (confident, goal-oriented, embracing ambiguity) to develop positive (and sometimes overly optimistic) beliefs of circumstances and what can be done within those circumstances.

Table 9.1 (continued)

Expertise explanation	Mindset explanation
Potential of neuroscience: Mapping brain activity of novices and experts to explore the encouraging and/or discouraging role of expertise in the formation and implementation of opportunity beliefs.	*Potential of neuroscience:* Examining regions of the brain associated with pleasure/pain and approach/avoidance may shed light on the degree to which particular situations temporarily evoke distinct mindsets that influence the formation and implementation of opportunity beliefs.

dimensions, are still considered an important part of predicting entrepreneurship and may yet provide insight into entrepreneurship, especially if studies rely on more carefully constructed samples of entrepreneurs (see, e.g., Zhao, Seibert and Lumpkin's [2010] review).

Although traits studies may provide insight into determining who is likely to become an entrepreneur (Carland, Hoy and Carland, 1988), their implications may be somewhat limited given that, by definition, they cannot be changed. This has led some to question the value of continuing on this path. However, until we are able to provide convergent empirical evidence that either definitively supports or refutes the trait explanation, research in this tradition is likely to continue. If so, it would likely benefit by shifting attention away from asking whether traits matter and toward asking how, why and under what conditions they are likely to matter. Specifically, we need to understand when, how and why traits influence the perception of opportunities to try, the formation of opportunity beliefs, and the successful implementation of them once formed.

To move away from asking 'if' questions and toward asking 'when' and 'why' questions requires researchers to tap into the cognitive processes used to form these opportunity beliefs, which historically has not been empirically possible. Traditional methods, such as self-report scales, can measure how people vary across a range of individual differences, but the use of experiments (Crant, 2000), scenarios (Rauch and Frese, 2007), surveys (Hmieleski and Baron, 2009), and other behavioral approaches do not allow us to look at 'what lies beneath' (Krueger, 2000) – how these traits and the behaviors they are thought to produce are related. Consequently, we have been unable to develop a causal chain that flows from personality traits to cognitions to action. Without these linkages, researchers can measure all of the individual differences to their heart's desire, but payoff will be limited if those measures cannot be readily linked to cognitions.

Consider, for example, the recent search for an entrepreneurial gene (e.g., Nicolaou and Shane, 2014; Shane and Nicolaou, 2013). As in early traits research, findings have not been promising. Even if there is a gene or trait that allows us to predict who will become an entrepreneur, we still have little to no information about why, when, where or how that gene should be expected to influence entrepreneurial cognitions and ultimately entrepreneurial action. However, if a strong correlation between some particular trait and some aspect of entrepreneurial cognition can be established, then neuroscience might reveal the regions of the brain that are activated for people with and without this personality trait as they engage in entrepreneurial thinking, reasoning or behavior. In doing so, researchers may begin to identify the system of neurons that is activated in the entrepreneurial process, compare it to systems that are involved in cognitive functions that are much better mapped, and glean some insight into the psychological and biological mechanisms responsible for translating personality traits into opportunity beliefs.

Applying neuroscience to the trait explanation

Neuroscience may offer insight into the relationship between personality traits and opportunity beliefs. Self-report and other traditional measures can be used to identify populations high in particular traits and compare them to populations low in that same trait as they are presented or asked to imagine opportunities to try to advance a goal – situations expected to encourage or reward use of a particular trait. As both populations conduct various cognitive tasks, such as planning or mentalizing others' possible reactions, neuroscience allows comparisons and contrasts between the regions of the brain activated by the task within each population and provides possible insight into why some who are high in the personality trait did not form an opportunity belief when many others did, or why some who are low in a particular personality trait notice an opportunity to try when others do not. Assuming a difference in brain activity exists across populations according to the degree to which they possess a particular personality trait, there may be reason to believe that this trait plays some role in the situations that one imagines or perceives and how those situations, once encountered, affect the entrepreneur.

Some research along these lines has already been conducted using the Big Five model of personality, which is a taxonomy that consists of extraversion, neuroticism, agreeableness, conscientiousness and openness/intellect. Indeed, 'personality neuroscience' is possible to the extent that regularities in biological functions in the brain are responsible for personality traits that are in turn responsible for regularities in behavior and experiences (DeYoung and Gray, 2009). According to this explanation of entrepreneurial cognition, a particular brain structure could be expected to be more prone to the formation and/or successful implementation of opportunity beliefs because brain structure determines personality, which, in turn, shapes experience and behavior in ways that are amenable to initiating and successfully persisting at entrepreneurial action.

For instance, extraversion may be an outcome associated with heightened activity in the medial orbitofrontal cortex, nucleus accumbens, amygdala and striatum (Canli et al., 2001; Canli, Sivers, Whitfield, Gotlib and Gabrieli, 2002; Cohen, Young, Baek, Kessler and Ranganath, 2005; Deckersbach et al., 2006; Mobbs, Hagan, Azim, Menon and Reiss, 2005). Applying this research to entrepreneurial cognition would suggest that, to the extent that entrepreneurial activities demand sales, marketing, making new contacts and other tasks that would be facilitated by high levels of extraversion, then someone who experiences heightened activity in certain brain regions may require less incentive to engage in these activities than someone who experiences less activity in these regions and therefore experiences less extraversion. Thus brain structure that leads to extraversion would encourage an individual to notice opportunities to try to advance goals that were facilitated by this trait and encourage formation of opportunity beliefs in those instances where this extraversion might facilitate implementation.

Another possible personality trait contributing to the formation and successful implementation of opportunity beliefs is conscientiousness, which reflects the tendency to maintain motivational stability within the individual and to make plans and carry them out in an organized and industrious manner. One biological factor that may be related to conscientiousness is glucose metabolism. Studies indicate that blood glucose – the energy source of the brain – is depleted by acts of self-control and that the extent of this depletion predicts failures of self-control (Gailliot et al., 2007; Gailliot and Baumeister, 2007). Individuals whose metabolism provides their brains with an ample and steady supply of glucose may be higher in conscientiousness, and therefore more successful at implementing opportunity beliefs.

If individual differences in glucose metabolism are indeed involved in conscientiousness, then the question becomes: what brain systems are consuming glucose to fuel acts of self-control? The prefrontal cortex seems likely to be involved, given its central role in planning and voluntary control of behavior and its high consumption of glucose (Gailliot and Baumeister, 2007). An fMRI study (Brown, Manuck, Flory and Hariri, 2006) showed that brain activity in the ventral prefrontal cortex during a response inhibition task was negatively associated with a questionnaire measure of impulsivity that is strongly negatively correlated with conscientiousness (Whiteside and Lynam, 2001). Findings such as these suggest that conscientiousness may discourage impulsivity and thus the formation of opportunity beliefs but encourage their successful implementation once formed owing to its penchant for planning. To the extent that conscientiousness is a personality trait, less glucose is likely to be consumed when exercising the self-control likely to be essential to successful implementation of opportunity beliefs.

These personality traits are only two of many personal characteristics that may be theoretically relevant to entrepreneurial cognition. Research by Zhao and Seibert (2006) or Baron, Hmieleski and Henry (2012) provides other possible candidates or rationales. But, regardless of the operationalization, the logic remains the same: differences in brain activity are likely to manifest as differences in personal characteristics, which in turn influence thinking, reasoning and/or behavior by encouraging the formation and successful implementation of opportunity beliefs intended to make the environment more suited to the individual.

The Adaptation Explanation

A second perspective that has provided insight into the formation and successful implementation of opportunity beliefs is the adaptation approach. This approach builds on the notion that entrepreneurs develop insights into the viability of opportunities to try in part from their interpretation of information signals from their environment (Baron, 2006; Shane, 2000). These interpretive judgments are heavily influenced by one's ability to assimilate information and visualize what that information means for them and others (Dimov, 2010; McMullen, 2010; Mitchell et al., 2000). This means that entrepreneurs must first be attentive to changes in the environment. They must then focus their attention on the most salient aspects of those changes to develop an image of the situation (Shepherd et al., 2007; Mitchell and Shepherd, 2010) that can be tested for veracity (Shepherd, Haynie and McMullen, 2012). To do all of this, entrepreneurs need a set of cognitive skills (Wood, Williams and Grégoire, 2012), and the adaptation perspective advances the idea that these skills are often dynamic. As they develop, one is better able to use information from the environment to develop future-focused visions and conceptualize the viability of turning those visions into reality (cf. Mitchell and Shepherd, 2010; Wood and Williams, 2014).

The adaptation explanation assumes that environmental change rewards individuals who already have particular characteristics or have the capability of developing them. These can be perceptual, such as alertness (Kirzner, 1973) or mindfulness (Corbett and McMullen, 2007), such that an individual can notice opportunities to try, or they can be cognitive, such as cognitive adaptability (Haynie et al., 2012) or the metacognitive ability to learn quickly (Haynie et al., 2010) such that they encourage the formation and successful implementation of opportunity beliefs. Along these lines, prior research has highlighted the use of cognitive scripts (Mitchell et al., 2000), self-regulation processes (Bryant, 2007), and rule-based processing (Wood and Williams, 2014) as somewhat dynamic processes that are central to an entrepreneur's ability to evaluate opportunities. Other researchers have addressed the question of how cognitive skills correlate with measurable outcomes. Categorization of business situations (Palich and Bagby, 1995), persistence in actions that lead to success in business start-up (Gatewood, Shaver and Gartner, 1995), and new-venture performance (Hmieleski and Baron, 2009) are a few of the outcomes that have been associated with certain cognitive skill variables. As a result, we have a fair picture of the skills required to perform certain 'cognitive tasks' (e.g., identifying opportunities) effectively. The problem with this approach is that successful implementation of opportunity beliefs is unlikely to require a single cognitive task even if a single task were all that was needed to form the belief in the first place (which is unlikely to be the case). Consequently, the personal characteristics thought to encourage survival tend to be conceived as cognitive skills or capabilities of a more dynamic nature.

In many cases, for example, the ability to learn quickly may be more important than having all the requisite knowledge, but not all actors possess the ability to learn quickly and, of those who do, only some will be willing to use it, and of those who choose to use it, only some will succeed. Consequently, the configurational nature of the adaptive explanation is likely to rely on explanations that favor lucky matches of supply and demand and evolving capabilities that enable actors to adapt to environmental demands via foresight or the mobility necessary to leave hostile environments. The importance of the environment favors cognitions that heighten environmental sensitivity, such as mindfulness (Corbett and McMullen, 2007) or alertness (Kirzner, 1973), but it is not limited to noticing opportunities to try that promise instantaneously achievable outcomes. Quite the contrary, this perspective's emphasis on dynamic capabilities allows it to speak to the formation of opportunity beliefs whose successful realization may be expected to require long time horizons.

Applying neuroscience to the adaptation explanation

Much as with the trait explanation, the adaptation explanation's reliance on behavioral research methods has limited insight into how dynamic cognitive capabilities might enable the formation and implementation of opportunity beliefs. If opportunity beliefs require the ability to learn quickly or some other dynamic capability such that an entrepreneur perceives that he can change himself, his environment or both, then engagement in thinking about these activities should activate particular regions of the brain, especially those responsible for monitoring the physical and social environments. Neuroscience is equipped to detect such activation.

Attention is one explanatory variable among the many encompassed by the adaptation explanation of entrepreneurial cognition that has received substantial interest among cognitive neuroscientists who have examined its relationship with action. Concepts such as 'attention to action' (Norman and Shallice, 1986) and 'selection for action' (Allport, 1987), for example, are references to the process by which particular cognitive intentions and sensory inputs are selected and coupled with the effector system for the control of action production (Holroyd, Nieuwenhuis, Mars and Coles, 2004). The ACC has been found to play a central role in this process (Posner, Petersen, Fox and Raichle, 1988; Posner and Dehaene, 1994; Posner and DiGirolamo, 1998; Posner and Petersen, 1990). By allowing only particular sources of information access to the output system, the ACC contributes to executive or strategic aspects of motor control and appears to be involved in transforming intentions into actions by selecting actions or action plans that are consistent with task goals. Converging evidence using functional neuroimaging, neuroanatomical, neurophysiological, intracranial stimulation, and lesion studies in humans and animals indicates that a caudal/dorsal area of the ACC is involved in the cognitive control of motor behavior (Bush, Luu and Posner, 2000; Devinsky, Morrell and Vogt, 1995; Dum and Strick, 1991; Goldberg, 1992; Paus, 2001; Picard and Strick, 1996).

Because the ACC is sensitive to incorrect or inappropriate behaviors, one aspect of its control function is error processing and bringing erroneous behaviors into line with desired goals (Holroyd and Coles, 2002; cf. Botvinick, Braver, Barch, Carter and Cohen, 2001). These conclusions have been drawn from observations of the error-related negativity (ERN), a component of the event-related brain potential (ERP) associated with error commission, which appears to be generated in the ACC. To the extent that opportunities to try are a function of changes in the environment that make previous beliefs, assumptions and expectations about what is valuable, desirable, feasible and/or viable incorrect, attention allocation by the ACC would appear to be integral to entrepreneurship. Indeed, Kirzner's (1973) notion of entrepreneurial alertness describes an individual's realization that he or she has erroneously valued means, ends or means–ends frameworks that are no longer optimal owing to various changes in the environment of which the individual was ignorant up until then. Because of the ACC's sensitivity to incorrect or inappropriate behaviors, its control function in processing errors, and its role in bringing erroneous behaviors into line with desired goals, the ACC appears to be fundamental to the perceptual explanations and dynamic cognitive capabilities encompassed by the adaptation approach of entrepreneurial cognition.

The Expertise Explanation

A third explanation of entrepreneurial cognition emphasizes the role of expertise. It is well established that experts are often better task performers than novices, leading researchers to ask what role knowledge and experience play in the formation and successful implementation of opportunity beliefs. The assumption is that entrepreneurs think about a given entrepreneurial task differently when they have extensive experience and knowledge in the domain (cf. Krueger, 2000; Haynie et al., 2009; Mitchell et al., 2003), and that this difference is rewarded by conditions of uncertainty and ambiguity that commonly characterize the novel tasks and environments of entrepreneurship. Following this logic, researchers have proposed that experience (Mitchell et al., 2003), knowledge (Mitchell and Shepherd, 2010) and learning styles (Corbett, 2007) play important roles in opportunity recognition (Baron, 2006) and opportunity evaluation (Wood et al., 2014).

Previous research has found that knowledge improves the odds of opportunity discovery (Ucbasaran, Wright and Westhead, 2003). Haynie and colleagues (2009) document the positive relationship between related knowledge and opportunity attractiveness. Extending this research, Wood and Williams (2014) show that knowledge of both the technology and the opportunity play an important moderating role in opportunity evaluation. In terms of experience, Shepherd (2003) discusses prior failure as influential, and Wood and company (2014) find that prior failure tempers one's view of the attractiveness of opportunity pursuit. In terms of more concrete outcomes, extant research finds links between prior knowledge and firm growth (Cooper, Gimeno-Gascon and Woo, 1994), as well as between knowledge and venture success (Romanelli, 1989). Finally, studies demonstrate that individuals who lack business ownership experience are less likely to gain the support of investors (Wright, Robbie and Ennew, 1997) and are more likely to experience business failure (Song et al., 2008).

Because knowledge and experience appear to play a key role in the formation and implementation of opportunity beliefs, the expertise explanation is closely aligned with the literature that seeks to understand how people become expert entrepreneurs. Mitchell and Chesteen (1995), for instance, advance the idea that expertise is a function of the 'expert scripts' – highly developed and ordered knowledge of a specific area – which people acquire through real-world interactions. These scripts allow experts to process information differently from novices. Relatedly, Corbett (2005) asserts that entrepreneurs' different learning styles (e.g., convergent, assimilation etc.), which ironically tend to be static, influence how entrepreneurs think about opportunities (see also Forbes, Chapter10, this volume). For their part, Holcomb and colleagues (2009: 175) discuss how observing others allows people to form knowledge structures around 'general rules and strategies for approaching new situations.' This is important because the knowledge perspective has been extended by Wood and Williams (2014), who find that entrepreneurs apply their expertise in the evaluation of opportunities by using rule-based thinking and that these rules are developed from formal education, day-to-day experiences and interacting with others (Chaiken and Trope, 1999; Smolensky, 1988). Taken together, these studies highlight the important role that real-world interactions and vicarious experiences play in establishing the expertise that is applied in the formation and successful implementation of opportunity beliefs.

Therefore expertise is important in the cognitive processes that underpin entrepreneurial action because it influences how entrepreneurs interpret new information and form opportunity beliefs about what can and cannot be done under novel circumstances (Mitchell et al., 2003). What is still being investigated, however, is the role that specific kinds of expertise play in determining how experts and novices react to specific entrepreneurial tasks. The field has moved beyond the question of whether expertise matters and is now focused on 'when' and 'why' it matters. As a result, there may be much to learn by examining possible negative influences of particular types of expertise on the formation and/or implementation of opportunity beliefs. For example, effectual logic – using available means to determine possible goals – is believed to be a cognitive strategy that results from experience operating under conditions of uncertainty (Sarasvathy, 2001). Whether these skills are accompanied with metacognitive skills that guide when it is (and is not) appropriate to use effectual logic is less clear (Haynie et al., 2010). In other words, whether experience and expertise also equip entrepreneurs with the prudence and judgment to know when effectual logic is more appropriate than causal logic has not been studied. As a result, it is possible that experience and expertise may not always be strengths; they could also present rigidities that pose weaknesses when implementing opportunity beliefs under more predictable circumstances that lend themselves to rule-based, algorithmic approaches to decision making.

Applying neuroscience to the expertise explanation

The expertise explanation of entrepreneurial cognition emphasizes the importance of prior knowledge in encouraging the formation and successful implementation of opportunity beliefs. Neuroscience has reached a similar conclusion by studying the relationship between expertise and the brain according to the expertise hypothesis and the motor cognition hypothesis, both of which appear to be highly relevant to entrepreneurial expertise.

The expertise hypothesis focuses on the role that expertise plays in making distinctions among objects that novices might overlook and began with researchers examining the fusiform face area (FFA), located in the right lateral part of the mid-fusiform gyrus (Kanwisher et al., 1997; Kanwisher and Yovel, 2006). These early studies suggested that this brain region served the sole purpose of face recognition, but the expertise hypothesis has since questioned this proposition, suggesting that the FFA is a general expertise module specialized for perceptual processes associated with visual individuation (Gauthier et al., 1999, 2000) in domains including birds (Gauthier et al., 2000), cars (Gauthier et al., 2000, 2005; Grill-Spector et al., 2004; Xu, 2005), butterflies (Rhodes et al., 2004), and novel object classes such as greebles (Gauthier et al., 1999). Although researchers have typically sought to rule out performance-based differences by asking participants to identify isolated expertise objects or to remember their location, there have been some recent exceptions. For example, Harley and colleagues (2009) investigated performance differences in FFA activation among expert and novice radiologists as they sought to identify abnormalities in x-ray images. They found that FFA activations were highly correlated with behavioral performance among experts but not among novices. Along these lines, Bilalic, Langer, Ulrich and Grodd (2011) examined the expertise hypothesis in the domain of chess. A favorite among early management science researchers (Simon, 1955), chess requires that players differentiate between objects (Saariluoma, 1995) and therefore involves the same basic recognition process related to individuation of single objects as is entailed in other visual domains. In addition, the objects on the board form complex spatial relationships determined by rules that constrain the way each piece can be moved. Therefore the need to cope simultaneously with multiple objects capable of forming numerous interrelations is not only one of the most difficult aspects of the game of chess, but also it is what makes the game similar to real-life decision situations (Gobet and Simon, 1996). To the extent that the formation and successful implementation of opportunity beliefs require individuals to cope with interrelations with numerous stakeholders, it shares many of the properties of chess, even if the spatial patterns are perhaps more abstract and less visual.

This expertise hypothesis is clearly relevant to the entrepreneurial domain where prior knowledge has been shown to be influential in determining the ideas that entrepreneurs generate and opportunities they notice (Shane, 2000). Indeed, research suggests that entrepreneurs often employ logics, scripts and schemas that differ from those employed by novices (Sarasvathy, 2001; Gaglio and Katz, 2001; Mitchell et al., 2002, 2007). In addition to enhancing entrepreneurs' abilities to see new potential in nuances that novices might overlook and thus form opportunity beliefs, expertise may activate other regions of the brain as well, especially when the entrepreneur is engaged in the application of new technologies. It is this form of learning that is often discussed by studies examining the motor cognition hypothesis mentioned earlier.

The motor cognition hypothesis proposes that action understanding is tied to motor expertise (Gallese et al., 2009), but learning clearly requires understanding of actions not yet in the observer's repertoire. Some scholars have used PET or fMRI to compare brain activation during the observation of simple tool making, such as the prehistoric use of a hammerstone to strike sharp flakes from a cobble core or the more sophisticated technological endeavors of shaping a handaxe. In the former, the parietal and frontal brain regions were involved in sensorimotor coordination, grip selection and 3D shape perception (Stout and Chaminade, 2007), whereas in the latter, the right inferior frontal gyrus (Stout et al., 2008), a region implicated in the hierarchical organization of action (Koechlin and Jubault, 2006), was activated. The increased sophistication of the latter design, which included the previous technology as an input, appears to have required the activation of additional brain regions that move beyond the imitation of mirror neurons and into the realm of inference. To test this possibility, Stout, Passingham, Frith, Apel and Chaminade (2011) crossed observer expertise (Naïve, Trained, Expert) with technological sophistication (Hammerstone, Handaxe) to examine the contribution of resonance and interpretation in understanding actions of varying familiarity and complexity. Regardless of complexity, motor resonance predicted expertise effects in the putative human mirror neuron system (Rizzolatti and Craighero, 2004) and dorsolateral prefrontal cortex (Buccino et al., 2004; Vogt et al., 2007), whereas inference (Saxe and Wexler, 2005) predicted complexity effects in brain regions associated with mental-state attribution, including the medial prefrontal cortex (Frith and Frith, 2006). Findings such as these suggest that higher technical expertise is likely to involve greater inference, and the activation of additional regions associated with social cognition used perhaps to imagine the intent of the designer of the technology. Consequently, neuroscience can be used to show that the formation of some opportunity beliefs is unlikely or even impossible when individuals do not have the prior knowledge or experience necessary to activate parts of the brain that are needed to mentalize the beliefs and preferences required to realize the full potential of the resources and/or data at their disposal.

The Mindset Explanation

The fourth conceptual approach to understanding entrepreneurial cognition is the mindset explanation, which assumes that some individuals take action because they have developed an 'entrepreneurial mindset' (Haynie et al., 2010; McGrath and MacMillan, 2000). A mindset can be thought of as a mental attitude that influences interpretations of incoming information and unique situations. The concept has emerged because entrepreneurs appear to have a rather distinct mental attitude. Specifically, entrepreneurs typically give themselves much higher odds of success than others (Cooper, Woo and Dunkelberg, 1988), are quite optimistic about what they can achieve (Hayward et al., 2006), and are generally very confident of their own abilities (Markman et al., 2002). Thus entrepreneurs seem to have a success-oriented mindset, which influences the confidence they have in their visions of the future (Dimov, 2010). Although this confidence can increase the odds of action, it can also lead to erroneous predictions of outcomes because people who are highly optimistic tend to discount negative information that leads to the pursuit of underperforming opportunities (Hmieleski and Baron, 2009). Thus the formation and successful implementation of opportunity beliefs can be influenced by the mindset that individuals bring to their circumstances.

A number of researchers have used the mindset explanation as justification for investigating a wide range of factors thought to distinguish 'enterprising individuals' (Shane, 2003) from the rest of the population. These factors are typically explored within the context of opportunity pursuit (identification, evaluation, exploitation) and often rest in some way on Ajzen's (1985) concept of 'control beliefs' – an individual's perceptions that he or she has the skills and abilities to complete effectively the task at hand. In that spirit, Gaglio and Katz (2001) argued that entrepreneurial alertness is important because alertness makes individuals more sensitive to profit-potential circumstances. Similarly, Chen, Greene and Crick (1998) document that entrepreneurial self-efficacy is positively related to innovation and risk taking. For their part, Busenitz and Barney (1997) find that entrepreneurs are more confident/overconfident than managers, and this confidence influences decision making. In addition to these studies, some researchers have explored how cognitive adaptability (Haynie, Shepherd and Patzelt, 2012), passion (Cardon et al., 2009), and other factors become part of the entrepreneurial mindset. Taken as a whole, the research to date suggests that the entrepreneurial mindset involves a nexus of factors that coalesce in a way that orients individuals towards vigorously 'seeking and exploiting the best opportunities' (Ucbasaran et al., 2003).

The mindset explanation of entrepreneurial cognition views the interpretive lens through which the environment is perceived as the key discriminator and predictor of entrepreneurial action. Social psychology researchers, however, have shown that much of what is interpreted as evidence for an entrepreneurial mindset may simply be phenomenological effects elicited by engagement in particular tasks or situations. For instance, entrepreneurs have been found to experience a 'narrow view' (Kahneman and Lovallo, 1993) in which they see themselves as immune to odds that they believe are applicable to others (Cooper, Woo and Dunkelberg, 1988; McMullen and DeCastro, 2000). This effect, however, has been shown by Gollwitzer (1990) to be a symptom of an implemental mindset, which is experienced after one commits to goal striving. Before such commitment, the individual experiences a deliberative mindset in which goal-relevant information is evaluated in an objective, detached fashion, but after commitment, a self-serving bias taints the interpretation of goal-relevant information. Thus this 'narrow view' is not uniquely entrepreneurial but rather attributable to motivated information processing. Similarly, some entrepreneurship researchers have suggested that an entrepreneurial mindset is opportunity-seeking and therefore similar to regulatory focus theory's promotion mindset (Brockner, Higgins and Low, 2004), which is evoked by approach-oriented motives, such as situations framed in terms of gains/non-gains (Higgins, 1997). However, empirical research has also shown that prevention focus is capable of encouraging entrepreneurial action if innovation is seen as a way to escape a loss situation (McMullen, Shepherd and Zahra, 2006). This suggests that the entrepreneurial mindset may be better understood as a temporary phenomenological effect evoked by perceptions of situational conditions. These phenomenological cognitive effects then encourage or discourage both the formation and successful implementation of opportunity beliefs.

Applying neuroscience to the mindset explanation

The methodological tools of neuroscience arguably offer more promise for the mindset explanation than for any of the other perspectives of entrepreneurial cognition. Traditional methodological tools, such as surveys and experiments, have captured evidence of various cognitive phenomena associated with a host of task-induced mindsets. To the extent that entrepreneurial action is facilitated by a mindset and this mindset is evoked by the characteristics of particular tasks or environments, it is important to understand how the nature of tasks affects judgment and decision making in action-encumbering or action-facilitating ways that individuals may not be conscious of. Indeed, this explanation of entrepreneurial cognition has the potential to demonstrate that observed differences in entrepreneurial thinking may simply be due to the task and/or environmental demands those individuals are in the midst of facing. Too often, however, researchers of this approach have had to settle for indirect evidence of these mindsets by demonstrating that particular inputs predictably lead to distorted decisional outputs when such orientations are evoked. Consider, for example, the framing effects resulting from a loss orientation in which people make different choices for decisions framed as gains versus losses (Kahneman and Tversky, 1979; Tversky and Kahneman, 1981), or the regulatory focus effects of promotion or prevention orientation observed by Higgins (1997) in which people feel better about a decision when fit between the approach–avoidance-oriented nature of the outcome and the approach–avoidance-oriented nature of the strategy used to attain it match. If people are not conscious of the mindsets they are experiencing, traditional methods such as surveys may not be equipped to study anything other than their effects on various outcomes.

Neuroscience, however, has the potential to allow observation of brain activity such that researchers can witness different regions of the brain being activated when engaging in approach-oriented thinking as opposed to avoidance-oriented thinking (Levin et al., 2012). This could greatly enhance our understanding of the effects of conditions of risk or uncertainty such as those that characterize entrepreneurship and make entrepreneurial judgment and decision making such an important topic for management and economics (Knight, 2012; McMullen and Shepherd, 2006). Both psychologists and neuroscientists have investigated approach and avoidance as potential mechanisms of behavioral impulses and their control.

Approach–avoidance behavior can be automatically triggered by unconscious or consciously controlled processes (Hofmann et al., 2009; Ernst and Fudge, 2009). Indeed, individuals must use the volitional control needed to achieve long-term goals to regulate the immediate impulses to approach positive stimuli and avoid negative stimuli, which are extremely helpful in adjusting to one's environment (Hofmann et al., 2009). Psychologists typically propose dual-process models that consist of two semi-independent systems – an impulsive system responsible for automatic reactions and a reflective system responsible for controlled reactions (e.g., Strack and Deutsch, 2004). Although the striatum and amygdala have been associated primarily with automatic approach reactions toward positive stimuli and automatic avoidance reactions toward negative stimuli, respectively (Ernst and Fudge, 2009; Tomer et al., 2008; O'Doherty et al., 2004; Cunningham et al., 2010; Schlund and Cataldo, 2010), neither subcortical structure processes positive or negative stimuli exclusively (Carretié et al., 2009; Hamann and Mao, 2002). In contrast, neuroscience suggests that controlled approach–avoidance reactions are mediated by the prefrontal cortex (PFC) (Ernst and Fudge, 2009; Miller and Cohen, 2001; Dosenbach et al., 2008), with its dorsolateral parts initiating performance adjustments (Ridderinkhof et al., 2004). Working as a whole, the impulsive system generates the automatic approach–avoidance tendencies of sub-cortical bottom–up activity which is then inhibited and controlled by the prefrontal cortical top–down regulation of the reflective system (Ernst and Fudge, 2009; Bechara, 2005; Ernst et al., 2006). Without both systems, there is either a lack of opportunity for the entrepreneur to notice or a lack of the control of impulse needed for successful implementation of opportunity beliefs once they are formed.

Thus neuroscience methods may be capable of shedding some light on the optimal mix of top–down versus bottom–up information processing when individuals are engaged in goal pursuit. Top–down processing may be highly efficient at noticing opportunities or threats to advancement of focal goals and thus common to the entrepreneurial mindset experienced during the implementation of opportunity beliefs, but it can also blind individuals to even greater opportunities or bigger threats to the advancement of non-focal goals, thereby discouraging the formation of new opportunity beliefs when one is in the midst of goal pursuit. Drawing from neuroscience, entrepreneurial cognition scholars may be able to refine their understanding of the entrepreneurial mindset by clarifying whether the mindset is supposedly making individuals more capable of noticing and exploiting opportunities to advance a goal, which occurs within a given goal structure, or whether it is supposedly making individuals more adept at noticing opportunities to try, which leads to the formation of new opportunity beliefs. Given that environmental conditions may deteriorate in between the formation of an opportunity belief and the successful realization of an event outcome, it would seem imperative that entrepreneurial cognition researchers know whether the entrepreneurial mindset is as functional across goal structures as it is within them. Otherwise, the entrepreneurial mindset runs the risk of allocating scarce resources to the pursuit of goals that environmental changes render suboptimal. If so, the entrepreneurial mindset could actually prevent rather than enable strategic re-optimization during implementation (McMullen and DeCastro, 2000).

COGNITION–ACTION–COGNITION AS A NEW EXPLANATION

In addition to the four explanations outlined above, we introduce a fifth explanation – see Table 9.2 – that can be inferred from some of the other traditions but that has not been made explicit – namely that there is a recursive relationship between action and cognition (Phelps and Hase, 2002). Complexity theory, for example, uses the lens of adaptation to posit that action is 'both the source and the product of cognition' as agents' schemas are actively constructed as the result of perceived regularities in experience (Doolittle, 2000). Likewise, theories of embodied cognition or situated action (e.g., Gibson, 1977) argue that the environment plays a central role in shaping cognitions, such that actions having an impact on the immediate social environment will impact future cognitions. Applied to entrepreneurship, this suggests that, to truly explain entrepreneurial action, we must not only view action as the effect of cognition, but also consider it as the cause of cognition. Entrepreneurship is a process (Shane and Venkataraman, 2000) requiring the completion of a number of tasks (e.g., opportunity recognition, evaluation, business planning, resource procurement etc.). As a task unfolds, the experience of that task will influence how one interacts with the next task. In that way, the cognitive adaptations that occur as the result of one's own action are likely to play a critical role in how the entrepreneurial process unfolds for each individual actor (McMullen and Dimov, 2013).

Entrepreneurial cognition researchers recognize, at least theoretically, the presence of recursive relationships (McMullen and Dimov, 2013). Aldrich and Martinez (2001) assert that entrepreneurs learn during the 'process of constructing their firms, based on feedback from their outcomes.' Wood and McKinley (2010) take a similar position by including a recursive relationship in their model of opportunity production as they assert that the abandonment of opportunity feeds back to influence conceptualizations of future opportunity ideas. For their part, Shepherd, McMullen and Jennings (2007) offer a model of top–down opportunity belief formation that includes recursive matching and updating of cognitive representations of the environment. The common theme appears to be that entrepreneurs are able to use their experiences, as they unfold, to create moment-to-moment images of an envisioned future (Chiles et al., 2010; Shackle, 1958). This perspective serves to focus our attention on the

Table 9.2 The dynamic approach to explaining entrepreneurial cognition

Conceptual logic: Explanations of entrepreneurial cognition must view action as both the effect and the cause of cognition. Thus entrepreneurs use their experiences as they unfold to create moment-to-moment images of an envisioned future from their perspective and those of salient others.
Emergent questions: How does initial action, and the feedback from action, influence the cognitions that underpin future action? How are cognitive processes, like perspective taking, influenced by previous action? How do the changes in those processes influence subsequent action? How does the direction (negative versus positive) of the feedback from initial action influence subsequent cognitions about what can be achieved via future action?

dynamism of entrepreneurship in general and of entrepreneurial cognition in particular.

By taking a dynamic view and thus considering the cognition–action–cognition cycle, an iterative creative process (Chiles et al., 2010) emerges by which entrepreneurs build confidence (Dimov, 2010) in their expectations that taking particular actions will result in anticipated outcomes. This process is facilitated not only by understanding the implications of action for one's self, but also for others (cf. Loewenstein, 1996; McMullen, 2010; Tajima, 2007). Achieving this understanding involves simulating other people's minds (Goldman, 2006) via techniques such as perspective taking, where one considers the world from the viewpoint of others (Galinsky et al., 2008). McMullen (2010) advances the idea that entrepreneurs consider the perspective of others (e.g., second- and third-person perspectives) to conceptualize value propositions associated with the introduction of new products or services. In that regard, the cognition–action–cognition cycle is critical because it occurs both cognitively through imagined interactions with others (e.g., through 'what if' scenarios) and behaviorally through actual interactions (e.g., quizzing others about the envisioned value proposition).

Applying neuroscience to the cognition–action–cognition explanation

This final explanation points to the future of neuroscience research. Although many of the neuroscience studies discussed earlier are likely to contribute to this approach as well, the key differentiator that characterizes this approach is its emphasis on the dynamism of the social mind, which is constantly monitoring the environment for new information about the self, others, and their relationship with natural and artificial objects.

According to this view, the mind is active, constantly informing and being informed by the individual's interaction with the world. As a result, we suspect that the hippocampus, which mediates the transfer of information from short-term to long-term memory and which receives connections from virtually the entire neo-cortex, is likely to be activated in this process. For example, if some color-detecting area of the cortex signals 'gray,' another area signals 'big,' and another signals 'has a trunk,' the neuron in the hippocampus that all three of those neurons project to could be part of an elephant detector or the ever-elusive heffalump that is entrepreneurship (Kilby, 1971). Similarly, a woman might recall something she did recently, like driving home. She may even remember the sequence of events from getting into her car to arriving home, and many of the sights, sounds and perhaps even smells encountered on the trip. If asked, she might be able to draw a picture of some of the places she passed or describe the sounds she heard as those of birds, traffic or airplanes. To conjure up this memory, a mental representation must be formed of both the things that she encountered and the sequence in time on the trip home last night as opposed to some other night. This kind of memory is called episodic memory, which contains not only content (the things remembered) but also context (the particular circumstances of one particular series of events).

This type of mental representation formation is reminiscent of Grégoire and colleagues' (2011) call to examine the antecedents and consequence of various cognitive representations. To this, we might add that neuroscience as applied to this approach to entrepreneurial cognition may also enable examination of how these mental representations are formed and transformed. As methodology and computing improve, the ability to capture real-time neurological data in the field may move beyond science fiction to science fact, and in the process all four of the approaches to entrepreneurial cognition discussed in Table 9.1 are likely to start to resemble the cognition–action–cognition approach often described and discussed in the form of constructs such as sensemaking, comprehension, understanding, embodied cognition and a number of other process-oriented cognitive constructs that are difficult to study empirically.

DISCUSSION

Within the entrepreneurial cognition literature, there have been different approaches to studying entrepreneurial cognition depending on assumptions about whether it is anomalous or ubiquitous, whether it is something the researcher is seeking to predict, explain or describe, and whether it is evoked by the environment, the individual, or some interaction of the two.

Regardless of approach, however, most researchers suggest that variance in action is at least partly attributable to variance in thinking and that manifestation of this thinking in the formation and implementation of opportunity beliefs affects and/or is affected by others. Consequently, entrepreneurship researchers have regularly used a social cognitive lens to gain insight into entrepreneurial cognition.

Whether these socio-cognitive differences are the cause or effect of entrepreneurial action has been unclear. Indeed, Grégoire, Corbett and McMullen (2011: 1444) point out that

> many works are built on the premise that founders and entrepreneurs 'think' differently than other individuals or business executives (e.g., Busenitz and Barney, 1997). But it is far less clear whether this 'cognitive difference' originates from idiosyncratic factors and events that *predate* entrepreneurs' efforts and actions, or from the very experience of entrepreneurship by these individuals (cf. Foo et al., 2009; Sarasvathy, 2008).

This confusion is further exacerbated by the question of whether the 'entrepreneurial' of entrepreneurial cognition refers to a position held – for example, founder (Wasserman, 2008), owner–manager (Hmieleski and Baron, 2009), CEO of a high-growth firm (Penrose, 1959) – or whether it refers to a function fulfilled within the economy – for example, innovator (Schumpeter, 1934), uncertainty bearer (Knight, 2012), market coordinator (Say, 1836), organizational creator (Gartner, 1985) – or both.

Finally, Grégoire and colleagues (2011: 1444) point out that it remains unclear 'whether the "cognitive difference" of entrepreneurs result from tasks and environmental conditions that "reward" individuals with particular "thinking", or from conditions that encourage the expression and/or development of such thinking (cf. Baron, 1998, 2000; Lovallo et al., 2008).' Consequently, they encourage scholars to 'pay attention not only to the consequences of relevant cognitive variables, but also to the origins and development of such variables,' and call and for future research to 'distinguish between different antecedents of entrepreneurship cognition – and to study these distinctions specifically' (Grégoire et al., 2011: 1456).

Combining our brief overview of the SCN's research methods with the four explanations of entrepreneurial action that currently characterize the entrepreneurial cognition literature, it may be possible to detect differences in brain activity related to entrepreneurial activity, to do so in a manner that is consistent with Grégoire and colleagues' recommendations, and to contribute to multiple fields in the process. Do such differences exist, and if so, to what should they be attributed – external sources, internal sources, or the interaction of the two? If entrepreneurial cognition, however conceptualized and operationalized, is believed to be the cause of entrepreneurial action by enabling the formation and implementation of opportunity beliefs, then the question becomes: what antecedent causes this cognition such that it yields action? Just as the nature of this cognition appears to depend significantly on the explanation of entrepreneurial cognition that a scholar uses, so does the antecedent that is identified as the source of this cognition. Therefore designing an experiment that is capable of examining entrepreneurial cognition as a social cognitive neuroscience phenomenon is no easy endeavor, but there may be some structural similarities across explanations that provide some insight into how we might begin such a program.

Revisiting Table 9.1, there is a trait explanation and an adaptation explanation. In both instances, the source of cognitive differences between entrepreneurs and the general population is considered permanent influences on one's thinking. Whether conceived of as traits or the dynamic cognitive skills of the adaptation explanation, these cognitive differences are not believed to vary; instead, they are seen as consistently influencing which data get processed, how such data are likely to be interpreted, and how well the resulting information is used to form and implement opportunity beliefs. In essence, both explanations employ a logic of fit in which the formation and successful implementation of opportunity beliefs result largely from a match between the cognitive requirements of a particular environment and the cognitive abilities of individuals with certain characteristics (be they personality traits or adaptive cognitive skills). Although these explanations differ, they do so in relation to whether the source or trigger of the entrepreneurial process is internal or external. Under the trait explanation, the cognitive difference is generated by the personality trait such that the individual seeks situations that reward use of the trait – fit is sought by the individual. To the extent that the trait is conducive to entrepreneurial action, it is likely to encourage individuals to form and implement opportunity beliefs that will bring their environment into greater alignment with this particular personality trait. In contrast, the adaptation explanation attributes the source or trigger of the entrepreneurial process to the environment. As such, certain environments that are believed to reward entrepreneurial action present potentially lucrative situations for those individuals with the cognitive skills or personal characteristics necessary to notice them, form opportunity beliefs, and successfully implement these opportunities to advance a goal. In the process, the individual willingly seeks to conform to the demands of the environment.

In contrast, the expertise and mindset explanations posit that the cognitive differences that lead to the formation and successful implementation of opportunity beliefs are temporary phenomena that may indeed vary depending on the needs of the individual at any given moment. These explanations differ, however, in regard to whether the source or trigger of the entrepreneurial process is considered internal or external to the individual. The expertise explanation often discusses the formation and successful implementation of opportunity beliefs in terms of goals that are negotiated and co-created with others as a means of producing some desired effect for the parties involved. Consequently, individuals rely on knowledge about who they are, what they know and whom they know in an effort to imagine new ways to use the resources at their disposal to create new value for those involved. Although this process may leverage environmental contingencies, it does not appear to be dependent upon them to get the ball rolling. In contrast, the mindset explanation often discusses entrepreneurial cognition as a process that is triggered by the environment. Opportunities to try are viewed as situations that facilitate goal advancement, but whether they are noticed depends on any one of a host of mindsets or orientations that the individual may be experiencing at any given moment, depending on the motives and goals with which he or she is currently concerned.

Specifying whether entrepreneurial cognition is permanent or temporary and the product of internal or external sources matters for a number of reasons. First, it helps reveal the causal essence of the explanation of entrepreneurial cognition being used by the researcher. In both the trait and adaptation approaches, the explanation lends itself primarily to a between-participants design in which participants who differ in some self-reported or observed individual difference (be it personality trait or adaptive cognitive skill) can be expected to form and/or implement opportunity beliefs better than others who lack this characteristic, either because they search out situations that reward the trait (trait explanation) or they are selected by the situation because of some adaptive cognitive skill they possess (adaptation).

In contrast, the expertise and mindset explanations may be studied using between- or within-participant designs to examine the formation and implementation of opportunity beliefs. Experiments using think-aloud protocols may offer ideal designs for studying these explanations. Because mindsets are often temporary and can be evoked, they lend themselves to experiments using within-participant as well as between-participant designs. Similarly, because expertise is learned, it offers a treatment effect that allows researchers to study differences in brain activity before and after acquiring the knowledge of particular logics, strategies, scripts, schema and so on. Consequently, the expertise approach also lends itself to use of both within- and between-participant experimental design.

With their emphasis on external triggers of the entrepreneurial process, the adaptation and mindset explanations differ from the internally focused trait and expertise explanations. Consequently experimental design of the former is likely to present participants with situations and then observe entrepreneurial cognition as a response to those stimuli, whereas the latter are likely to assign participants goals or evoke motives and then observe entrepreneurial cognition as those participants seek to find or create situations that facilitate the realization of those goals. In the field both scenarios are likely to occur, producing a dynamic and holistic entrepreneurial process. As individuals confront a continuous stream of new information produced by interacting with their social and natural environments, they are likely to engage in a perpetual process of updating their understanding of their situation. Indeed, this dynamism implies that individuals may experience all four explanations on their way to realizing an event outcome as complex as introducing a new product to the market or starting a firm. Therefore it is probably worth noting that none of these explanations of entrepreneurial cognition is likely to explain definitively the formation and implementation of opportunity beliefs, and therefore an experimental program employing good design and multiple perspectives over time may be the prudent researcher's best bet.

CONCLUSION

Just as new technologies are a primary source of innovation and opportunity in entrepreneurship (Schumpeter, 1934; Drucker, 1985), the same might be said for science (Sanders, 2007). Neuroscience did not even exist 20 years ago, but thanks to technological advancements, it has become one of the fastest-growing areas of the biological sciences and a revolutionizing force across the social sciences, challenging assumptions that underlie disciplines ranging from economics to sociology to psychology. Given that entrepreneurial cognition draws on many of these disciplines, the field is unlikely to be immune to neuroscience's transformative impact. At its core, entrepreneurship involves individuals interacting with their social and natural environments such that new possible thoughts, feelings and actions emerge that lead to the creation of new understandings, artifacts, arrangements and relationships that transform individuals, organizations, economies and societies.

This chapter has sought to articulate the nature of the opportunity that neuroscience provides entrepreneurial cognition researchers. In so doing, we have sought to establish a semblance of order to the research that has already been conducted in entrepreneurial cognition such that it might be cross-pollinated with the incredible advancements being made in neuroscience, especially social cognitive neuroscience. Ultimately, however, entrepreneurial cognition researchers will be the judges of whether neuroscience will be embraced by our field. However, if researchers, like entrepreneurs, chase opportunities that promise returns, then we expect that this chapter will not be the last inquiry into the subject. Indeed, we imagine that it will be only a small drop in what will likely become a raging torrent in the years to come.

NOTE

1. Although 1 must precede 2, and 2 must precede 3, sometimes it is unclear in the entrepreneurial cognition literature whether 1 is in its original form when it precedes 3 or whether it has been revised after opportunity identification. We will return to this issue in our discussion of the cognition–action–cognition explanation of entrepreneurial cognition.

REFERENCES

Ajzen, I. 1985. From intentions to actions: A theory of planned behavior. In J. Kuhl and J. Beckmann (eds), *Action Control: From Cognition to Behavior*: 11–39. New York: Springer-Verlag.

Aldrich, H. 1999. *Organizations Evolving*. London: Sage Publications.

Aldrich, H. and Martinez, M.E. 2001. Many are called but few are chosen: An evolutionary perspective for the study of entrepreneurship. *Entrepreneurship Theory and Practice*, **25**(4): 41–56.

Allport, A. 1987. Selection for action: Some behavioral and neurophysiological considerations of attention and action. In H. Heuer and A.F. Sanders (eds), *Perspectives on Perception and Action*: 395–419. Hillsdale, NJ: Erlbaum.

Anderson, J.R. 2005. *Cognitive Psychology and its Implications*. New York: Worth Publishers.

Appel, V., Weinstein, C. and Weinstein, S. 1980. Brain activity responses to magazine and television advertising. *Journal of Advertising Research*, **20**(3): 57–63.

Bagozzi, R.P., Verbeke, W.J., Dietvorst, R.C., Belschak, F.D., van den Berg, W.E. and Rietdijk, W.J. 2013. Theory of mind and empathic explanations of Machiavellianism: A neuroscience perspective. *Journal of Management*, **39**(7): 1760–98.

Barreto, I. 2012. Solving the entrepreneurial puzzle: The role of entrepreneurial interpretation in opportunity formation and related processes. *Journal of Management Studies*, **49**(2): 356–80.

Baron, R.A. 1998. Cognitive mechanisms in entrepreneurship: Why and when enterpreneurs think differently than other people. *Journal of Business Venturing*, **13**(4): 275–94.

Baron, R.A. 2000. Counterfactual thinking and venture formation: The potential effects of thinking about 'what might have been'. *Journal of Business Venturing*, **15**(1): 79–91.

Baron, R.A. 2006. Opportunity recognition as pattern recognition: How entrepreneurs 'connect the dots' to identify new business opportunities. *Academy of Management Perspectives*, **20**(1): 104–19.

Baron, R.A. and Ensley, M.D. 2006. Opportunity recognition as the detection of meaningful patterns: Evidence from comparisons of novice and experienced entrepreneurs. *Management Science*, **52**(9): 1331–44.

Baron, R.A., Hmieleski, K.M. and Henry, R.A. 2012. Entrepreneurs' dispositional positive affect: The potential benefits and potential costs of being 'up.' *Journal of Business Venturing*, **27**(3): 310–24.

Bechara, A. 2005. Decision making, impulse control and loss of willpower to resist drugs: A neurocognitive perspective. *Nature Neuroscience*, **8**(11): 1458–63.
Berger, H. 1929. Über das Elektrenkephalogramm des Menschen. *Archiv für Psychiatrie und Nervenkrankheiten*, **87**: 527–70.
Bilalic, M., Langner, R., Ulrich, R. and Grodd, W. 2011. Many faces of expertise: Fusiform face area in chess experts and novices. *Journal of Neuroscience*, **31**(28): 10206–14.
Botvinick, M.M., Braver, T.S., Barch, D.M., Carter, C.S. and Cohen, J.D. 2001. Conflict monitoring and cognitive control. *Psychological Review*, **108**(3): 624–52.
Brandstätter, H. 2011. Personality aspects of entrepreneurship: A look at five meta-analyses. *Personality and Individual Differences*, **51**(3): 222–30.
Brockhaus, R.H. 1980. Risk taking propensity of entrepreneurs. *Academy of Management Journal*, **23**(3): 509–20.
Brockner, J., Higgins, E.T. and Low, M.B. 2004. Regulatory focus theory and the entrepreneurial process. *Journal of Business Venturing*, **19**(2): 203–20.
Brown, T. 2008. Design thinking. *Harvard Business Review*, **86**(6): 84–92.
Brown, S.M., Manuck, S.B., Flory, J.D. and Hariri, A.R. 2006. Neural basis of individual differences in impulsivity: Contributions of corticolimbic circuits for behavioral arousal and control. *Emotion*, **6**(2): 239–45.
Bryant, P. 2007. Self-regulation and decision heuristics in entrepreneurial opportunity evaluation and exploitation. *Management Decision*, **45**(4): 732–48.
Buccino, G., Vogt, S., Ritzl, A., Fink, G.R., Zilles, K., Freund, H.-J. and Rizzolatti, G. 2004. Neural circuits underlying imitation learning of hand actions: an event-related fMRI study. *Neuron*, **42**(2): 323–34.
Busenitz, L.W. and Barney, J.B. 1997. Differences between entrepreneurs and managers in large organizations: Biases and heuristics in strategic decision-making, *Journal of Business Venturing*, **12**(1): 9–30.
Bush, G., Luu, P. and Posner, M. I. 2000. Cognitive and emotional influences in anterior cingulate cortex. *Trends in Cognitive Sciences*, **4**(6): 215–22.
Camerer, C., Lowenstein, G. and Prelec, D. 2005. Neuroeconomics: How neuroscience can inform economics. *Journal of Economic Literature*, **43**(March): 9–64.
Canli, T., Sivers, H., Whitfield, S.L., Gotlib, I.H. and Gabrieli, J.D. 2002. Amygdala response to happy faces as a function of extraversion. *Science*, **296**(5576): 2191.
Canli, T., Zhao, Z., Desmond, J.E., Kang, E., Gross, J. and Gabrieli, J.D. 2001. An fMRI study of personality influences on brain reactivity to emotional stimuli. *Behavioral Neuroscience*, **115**(1): 33.
Caprana, G.V. and Cervone, C. 2000. *Personality: Determinants, dynamics, and potentials*. New York: Cambridge University Press.
Cardon, M.S., Wincent, J., Singh, J. and Drnovšek, M. 2009. The nature and experience of entrepreneurial passion. *Academy of Management Review*, **34**(3): 511–32.
Carland, J.W., Hoy, F. and Carland, J.A. 1988. Who is an entrepreneur? is a question worth asking. *American Journal of Small Business*, **12**(4): 33–9.
Carretié, L., Ríos, M., De la Gandara, B.S., Tapia, M., Albert, J., López-Martín, S. and Álvarez-Linera, J. 2009. The striatum beyond reward: Caudate responds intensely to unpleasant pictures. *Neuroscience*, **164**(4): 1615–22.
Casson, M. 1982. *The Entrepreneur: An economic theory*. Totowa, NJ: Barnes & Noble.
Chaiken, S. and Trope, Y. 1999. *Dual Process Theories in Social Psychology*. New York: Guilford Press.
Chen, C.C., Greene, P.G. and Crick, A. 1998. Does entrepreneurial self-efficacy distinguish entrepreneurs from managers? *Journal of Business Venturing*, **13**(4): 295–316.
Chiles, T.H., Tuggle, C.S., McMullen, J.S., Bierman, L. and Greening, D.W. 2010. Dynamic creation: Extending the radical Austrian approach to entrepreneurship. *Organization Studies*, **31**(1): 7–46.
Cohen, M.X., Young, J., Baek, J.M., Kessler, C. and Ranganath, C. 2005. Individual differences in extraversion and dopamine genetics predict neural reward responses. *Cognitive Brain Research*, **25**(3): 851–61.

Collins, C.J., Hanges, P.J. and Locke, E.A. 2004. The relationship of achievement motivation to entrepreneurial behavior: A meta-analysis. *Human Performance*, **17**(1): 95–117.
Cooper, A., Gimeno-Gascon, J. and Woo, C. 1994. Initial human and financial capital as predictors of new venture performances. *Journal of Business Venturing*, **9**(5): 371–95.
Cooper, A., Woo, C. and Dunkelberg, W. 1988. Entrepreneurs' perceived chances for success. *Journal of Business Venturing*, **3**(2): 97–108.
Corbett, A.C. 2005. Experiential learning within the process of opportunity identification and exploitation. *Entrepreneurship Theory and Practice*, **29**(4): 473–91.
Corbett, A. (2007). Learning asymmetries and the discovery of entrepreneurial opportunities. *Journal of Business Venturing*, **22**: 97–18.
Corbett, A.C. and McMullen, J.S. 2007. Perceiving and shaping new venture opportunities through mindful practice. In A. Zacharakis and R. Spinnelli (eds), *Process of Entrepreneurship*: 43–64. Westport, CT: Praeger Publishers.
Crant, J.M. 2000. Proactive behavior in organizations. *Journal of Management*, **26**(3): 435–62.
Cromie, S. 2000. Assessing entrepreneurial inclinations: Some approaches and empirical evidence. *European Journal of Work & Organizational Psychology*, **9**(1): 7–30.
Cunningham, W.A., Arbuckle, N.L., Jahn, A., Mowrer, S.M. and Abduljalil, A.M. 2010. Aspects of neuroticism and the amygdala: Chronic tuning from motivational styles. *Neuropsychologia*, **48**(12): 3399–404.
Decety, J. and Jackson, P.L. 2006. A social–neuroscience perspective on empathy. *Current Directions in Psychological Science*, **15**(2): 54–8.
Decety, J., Jackson, P.L., Sommerville, J.A., Chaminade, T. and Meltzoff, A.N. 2004. The neural bases of cooperation and competition: An fMRI investigation. *Neuroimage*, **23**(2): 744–51.
Decety, J., Norman, G.J., Berntson, G.G. and Cacioppo, J.T. 2012. A neurobehavioral evolutionary perspective on the mechanisms underlying empathy. *Progress in Neurobiology*, **98**(1): 38–48.
Decety, J. and Porges, E.C. 2011. Imagining being the agent of actions that carry different moral consequences: An fMRI study. *Neuropsychologia*, **49**(11): 2994–3001.
Deckersbach, T., Miller, K.K., Klibanski, A., Fischman, A., Dougherty, D.D., Blais, M.A. and Rauch, S.L. 2006. Regional cerebral brain metabolism correlates of neuroticism and extraversion. *Depression and Anxiety*, **23**(3): 133–8.
Devinsky, O., Morrell, M.J. and Vogt, B.A. 1995. Contributions of anterior cingulate cortex to behaviour. *Brain*, **118**(1): 279–306.
DeYoung, C.G. and Gray, J.R. 2009. Personality neuroscience: Explaining individual differences in affect, behavior, and cognition. In P.J. Corr and G. Matthews (eds), *The Cambridge Handbook of Personality Psychology*: 323–46. Cambridge, UK: Cambridge University Press.
Dimov, D. 2007. From opportunity insight to opportunity intention: The importance of person–situation learning match. *Entrepreneurship Theory and Practice*, **31**(4): 561–83.
Dimov, D. 2010. Nascent entrepreneurs and venture emergence: Opportunity confidence, human capital, and early planning. *Journal of Management Studies*, **47**(6): 1123–53.
Dimov, D. 2011. Grappling with the unbearable elusiveness of entrepreneurial opportunities. *Entrepreneurship Theory and Practice*, **35**(1): 57–81.
Doolittle, P.E. 2000. Complex constructivism: A theoretical model of complexity and cognition: Draft. http://www.tandl.vt.edu/doolittle/research/complex1.html. Accessed on December 1, 2000.
Dosenbach, N.U., Fair, D.A., Cohen, A.L., Schlaggar, B.L. and Petersen, S.E. 2008. A dual-networks architecture of top-down control. *Trends in Cognitive Sciences*, **12**(3): 99–105.
Drucker, P.F. 1985. The discipline of innovation. *Harvard Business Review*, **63**(3): 67–72.
Dum, R.P. and Strick, P.L. 1991. The origin of corticospinal projections from the premotor areas in the frontal lobe. *Journal of Neuroscience*, **11**(3): 667–89.
Ernst, M. and Fudge, J.L. 2009. A developmental neurobiological model of motivated

behavior: anatomy, connectivity and ontogeny of the triadic nodes. *Neuroscience & Biobehavioral Reviews*, **33**(3): 367–82.
Ernst, M., Pine, D.S. and Hardin, M. 2006. Triadic model of the neurobiology of motivated behavior in adolescence. *Psychological Medicine*, **36**(3): 299–312.
Foo, M.-D., Uy, M.A. and Baron, R.A. 2009. How do feelings influence effort? An empirical study of entrepreneurs' affect and venture effort. *Journal of Applied Psychology*, **94**(4): 1086–94.
Franks, D.D. 2010. *Neurosociology: The nexus between neuroscience and social psychology*. New York: Springer.
Frith, C.D. and Frith, U. 2006. The neural basis of mentalizing. *Neuron*, **50**(4): 531–4.
Gaglio, C.M. and Katz, J. 2001. The psychological basis of opportunity identification: Entrepreneurial alertness. *Small Business Economics*, **16**(2): 95–111.
Gailliot, M.T. and Baumeister, R.F. 2007. The physiology of willpower: Linking blood glucose to self-control. *Personality and Social Psychology Review*, **11**(4): 303–27.
Gailliot, M.T., Baumeister, R.F., DeWall, C.N., Maner, J.K., Plant, E.A., Tice, D.M. and Schmeichel, B.J. 2007. Self-control relies on glucose as a limited energy source: Willpower is more than a metaphor. *Journal of Personality and Social Psychology*, **92**(2): 325–36.
Galinsky, A.D., Maddux, W.W., Gilin, D. and White, J.B. 2008. Why it pays to get inside the head of your opponent: The differential effects of perspective taking and empathy in negotiations. *Psychological Science*, **19**(4): 378–84.
Galinsky, A.D. and Mussweiler, T. 2001. First offers as anchors: The role of perspective-taking and negotiator focus. *Journal of Personality and Social Psychology*, **81**(4): 657–69.
Gallese, V., Rochat, M., Cossu, G. and Sinigaglia, C. 2009. Motor cognition and its role in the phylogeny and ontogeny of action understanding. *Developmental Psychology*, **45**(1): 103–13.
Gartner, W.B. 1985. A conceptual framework for describing the phenomenon of new venture creation. *Academy of Management Review*, **10**(4): 696–706.
Gatewood, E.J., Shaver, K.G. and Gartner, W.B. 1995. A longitudinal study of cognitive factors influencing start-up behaviors and success at venture creation. *Journal of Business Venturing*, **10**(5): 371–91.
Gauthier, I., Curby, K.M., Skudlarski, P. and Epstein, R.A. 2005. Individual differences in FFA activity suggest independent processing at different spatial scales. *Cognitive, Affective, & Behavioral Neuroscience*, **5**(2): 222–34.
Gauthier, I., Skudlarski, P., Gore, J.C. and Anderson, A.W. 2000. Expertise for cars and birds recruits brain areas involved in face recognition. *Nature Neuroscience*, **3**(2): 191–7.
Gauthier, I., Tarr, M.J., Anderson, A.W., Skudlarski, P. and Gore, J.C. 1999. Activation of the middle fusiform 'face area' increases with expertise in recognizing novel objects. *Nature Neuroscience*, **2**(6): 568–73.
Gibson, J.J. 1977. The theory of affordances. In R.E. Shaw and J. Bransford (eds), *Perceiving, Acting and Knowing*: 67–82. Hillsdale, NJ: Erlbaum.
Gobet, F. and Simon, H.A. 1996. Templates in chess memory: a mechanism for recalling several boards. *Cognitive Psychology*, **31**(1): 1–40.
Goldberg, G. 1992. Premotor systems, attention to action and behavioural choice. In J. Kien, C.R. McCrohan and W. Winslow (eds), *Neurobiology of Motor Programme Selection: New approaches to the study of behavioural choice*: 225–49. Oxford: Pergamon Press.
Goldman, A.I. 2006. *Simulating Minds: The philosophy, psychology, and neuroscience of mindreading*. New York: Oxford University Press.
Gollwitzer, P.M. 1990. Action phases and mind-sets. In E.T. Higgins and R.M. Sorrentino (eds), *Handbook of Motivation and Cognition: Foundations of social behavior*: 53–92. New York: Guilford Press.
Grégoire, D.A., Corbett, A. and McMullen, J. 2011. The cognitive perspective in entrepreneurship: An agenda for future research. *Journal of Management Studies*, **48**(6): 1443–77.
Grill-Spector, K., Knouf, N. and Kanwisher, N. 2004. The fusiform face area subserves face perception, not generic within-category identification. *Nature Neuroscience*, **7**(5): 555–62.

Hamann, S. and Mao, H. 2002. Positive and negative emotional verbal stimuli elicit activity in the left amygdala. *Neuroreport*, **13**(1): 15–19.

Harley, E.M., Pope, W.B., Villablanca, J.P., Mumford, J., Suh, R., Mazziotta, J.C., Enzmann, D. and Engel, S. A. 2009. Engagement of fusiform cortex and disengagement of lateral occipital cortex in the acquisition of radiological expertise. *Cerebral Cortex*, **19**(11): 2746–54.

Harmon-Jones, E. and Beer, J.S. 2012. Introduction to social and personal neuroscience methods. In E. Harmon-Jones and J.S. Beer (eds), *Methods in Social Neuroscience*: 1–9. New York: Guilford Press.

Haxby, J.V. 2011. Social neuroscience and the representation of others. In A. Todorov, S. Fiske and D. Prentice (eds), *Social Neuroscience: Toward understanding the underpinnings of the social mind*. Oxford Scholarship Online.

Haynie, J.M., Shepherd, D. and McMullen J. 2009. An opportunity for me? The role of resources in opportunity evaluation decisions. *Journal of Management Studies*, **46**(3): 337–61.

Haynie, J.M., Shepherd, D., Mosakowski, E. and Early, P.C. 2010. A situated metacognitive model of the entrepreneurial mindset. *Journal of Business Venturing*, **25**(2): 217–29.

Haynie, J.M., Shepherd, D.A. and Patzelt, H. 2012. Cognitive adaptability and an entrepreneurial task: The role of metacognitive ability and feedback. *Entrepreneurship Theory and Practice*, **36**(2): 237–65.

Hayward, M.L.A., Shepherd, D.A. and Griffin, D. 2006. A hubris theory of entrepreneurship. *Management Science*, **52**(2): 160–72.

Higgins, E.T. 1997. Beyond pleasure and pain. *American Psychologist*, **52**(2): 1280–300.

Hmieleski, K.M. and Baron, R.A. 2009. Entrepreneurs' optimism and new venture performance: A social cognitive perspective. *Academy of Management Journal*, **52**(3): 473–88.

Hofmann, W., Friese, M. and Strack, F. 2009. Impulse and self-control from a dual-systems perspective. *Perspectives on Psychological Science*, **4**(2): 162–76.

Holcomb, T.R., Ireland, R.D., Holmes, R.M. and Hitt, M.A. 2009. Architecture of entrepreneurial learning: Exploring the link among heuristics, knowledge, and action. *Entrepreneurship Theory and Practice*, **33**(1): 173–98.

Holroyd, C.B. and Coles, M.G. 2002. The neural basis of human error processing: Reinforcement learning, dopamine, and the error-related negativity. *Psychological Review*, **109**(4): 679–708.

Holroyd, C.B., Nieuwenhuis, S., Mars, R.B. and Coles, M.G.H. 2004. Anterior cingulate cortex, selection for action, and error processing. In M.I. Posner (ed.), *Cognitive Neuroscience of Attention*: 219–31. New York: Guilford Press.

Jenkins, A.C. and Mitchell, J.P. 2010. Mentalizing under uncertainty: Dissociated neural responses to ambiguous and unambiguous mental state inferences. *Cerebral Cortex*, **20**(2): 404–10.

Johnstone, T., Kim, M.J. and Whalen, P.J. 2009. Functional magnetic resonance imaging in the affective and social neurosciences. In E. Harmon-Jones and J.S. Beer (eds), *Methods in Social Neuroscience*: 313–36. New York: Guilford Press.

Kahneman, D. and Lovallo, D. 1993. Timid choices and bold forecasts: A cognitive perspective on risk taking. *Management Science*, **39**(1): 17–31.

Kahneman, D. and Tversky, A. 1979. Prospect theory: An analysis of decision under risk. *Econometrica: Journal of the Econometric Society*, **47**(2): 263–91.

Kanwisher, N., McDermott, J. and Chun, M.M. 1997. The fusiform face area: A module in human extrastriate cortex specialized for face perception. *Journal of Neuroscience*, **17**(11): 4302–11.

Kanwisher, N. and Yovel, G. 2006. The fusiform face area: A cortical region specialized for the perception of faces. *Philosophical Transactions of the Royal Society B: Biological Sciences*, **361**(1476): 2109–28.

Kilby, P. 1971. *Entrepreneurship and Economic Development*. New York: Free Press.

Kirzner, I.M. 1973. *Competition and Entrepreneurship*. Chicago, IL: University of Chicago Press.

Knight, F. 2012 (1921). *Risk, Uncertainty and Profit*. New York: Courier Dover Publications.
Koechlin, E. and Jubault, T. 2006. Broca's area and the hierarchical organization of human behavior. *Neuron*, **50**(6): 963–74.
Koh, H.C. 1996. Testing hypotheses of entrepreneurial characteristics: A study of Hong Kong MBA students. *Journal of Managerial Psychology*, **11**(3): 12–25.
Krueger, N.F. 2000. The cognitive infrastructure of opportunity emergence. *Entrepreneurship Theory and Practice*, Spring: 5–23.
Lachmann, L.M. 1970. *The Legacy of Max Weber*. Berkeley, CA: Glendessary.
Lee, L. and Chamberlain, L. 2007. Neuroimaging and psychophysiological measurement in organizational research: An agenda for research in organizational cognitive neuroscience. In C. Senior and J.R. Butler (eds), *The Social Cognitive Neuroscience of Organizations*: 18–42. Boston, MA: Blackwell Publishing.
Levin, I.P., Xue, G., Weller, J.A., Reimann, M., Lauriola, M. and Bechara, A. 2012. A neuropsychological approach to understanding risk-taking for potential gains and losses. *Frontiers in Neuroscience*, **6**: Article 15.
Lieberman, M.D. 2010. Social cognitive neuroscience. In S.T. Fiske, D.T. Gilbert and G. Lindzey (eds), *The Handbook of Social Psychology* (5th edn): 143–93. New York: McGraw-Hill.
Loewenstein, G. 1996. Out of control: Visceral influences on behavior. *Organizational Behavior & Human Decision Processes*, **65**(3): 272–92.
Lovallo, D., Powell, T.C., Fox, C.R. and Teece, D.J. 2008. Psychological foundations of strategic management. *Strategic Management Journal*. Special issue, http://strategicmanagement.net/pdfs/SMJ_Special_Issues_Call_for_Papers.pdf. Document accessed January 25, 2010.
Lumpkin, G.T., Moss, T.W., Gras, D.M., Kato, S. and Amezcua, A.S. 2013. Entrepreneurial processes in social contexts: how are they different, if at all? *Small Business Economics*, **40**(3): 761–83.
Markman, G.D., Balkin, D.B. and Baron, R.A. 2002. Inventors and new venture formation: The effects of general self-efficacy and regretful thinking. *Entrepreneurship Theory and Practice*, **27**(2): 149–65.
McClelland, D.C. 1961. *The Achieving Society*. Princeton, NJ: D. Van Nostrand Company. Reprinted 1967, New York: Free Press.
McGrath, R.G. and MacMillan, I.C. 2000. *The Entrepreneurial Mindset: Strategies for continuously creating opportunity in an age of uncertainty*. Boston, MA: Harvard Business Press.
McMullen, J.S. 2010. Perspective taking and the heterogeneity of the entrepreneurial imagination. In R. Koppl (ed.), *Advances in Austrian Economics*: 113–44. Greenwich, CT: JAI Press.
McMullen, J.S. 2014. Entrepreneurial judgment as empathic accuracy: A sequential decision-making approach to entrepreneurial action. *Journal of Institutional Economics*. In press.
McMullen, J.S. and DeCastro, J. 2000. Entrepreneurial decision making and risk assessment in the midst of environmental uncertainty: The Everest disaster. *Academy of Management Proceedings*, **2000**(1): F1–F6.
McMullen, J.S. and Dimov, D. 2013. Time and the entrepreneurial journey: The problems and promise of studying entrepreneurship as a process. *Journal of Management Studies*, **50**(8): 1481–512.
McMullen, J.S., Plummer, L.A. and Acs, Z.J. 2007. What is an entrepreneurial opportunity? *Small Business Economics*, **28**(4): 273–83.
McMullen, J.S. and Shepherd, D.A. 2006. Entrepreneurial action and the role of uncertainty in the theory of the entrepreneur. *Academy of Management Review*, **31**(1): 132–52.
McMullen, J.S., Shepherd, D. and Zahra, S. 2006. Regulatory focus and executives' intentions to commit their firms to entrepreneurial action. In A. Zacharakis et al. (eds), *Frontiers of Entrepreneurship Research*: 661–74. Babson Park, MA: Babson College.
Miller, E.K. and Cohen, J.D. 2001. An integrative theory of prefrontal cortex function. *Annual Review of Neuroscience*, **24**(1): 167–202.

Mitchell, J.R. and Shepherd, D.A. 2010. To thine own self be true: Images of self, images of opportunity, and entrepreneurial action. *Journal of Business Venturing*, **25**(1): 138–54.
Mitchell, R.K., Busenitz, L.W., Bird, B., Gaglio, C.M., McMullen, J.S., Morse, E.A. and Smith, J.B. 2007. The central question in entrepreneurial cognition research 2007. *Entrepreneurship Theory and Practice*, **31**(1): 1–27.
Mitchell, R.K., Busenitz, L.W., Lant, T., McDougall, P.P., Morse, E.A. and Smith, J.B. 2002. Toward a theory of entrepreneurial cognition: Rethinking the people side of entrepreneurship research. *Entrepreneurship Theory and Practice*, **27**(2): 93–104.
Mitchell, R.K. and Chesteen, S. 1995. Enhancing entrepreneurial expertise: Experiential pedagogy and the new venture expert script. *Simulation and Gaming*, **26**(3): 288–306.
Mitchell, R.K., Morse, E.A. and Sharma, P. 2003. The transacting cognitions of non-family employees in the family business setting. *Journal of Business Venturing*, **18**(4): 533–51.
Mitchell, R.K., Smith, J.B., Seawright, K.W. and Morse, E.A. 2000. Cross-cultural cognitions and the venture creation decision. *Academy of Management Journal*, **43**(5): 974–93.
Mitchell, R.K., Smith, J.B., Morse, E.A., Seawright, K.W., Peredo, A.M. and McKenzie, B. 2002. Are entrepreneurial cognitions universal? Assessing entrepreneurial cognitions across cultures. *Entrepreneurship Theory and Practice*, **26**(4): 9–32.
Mobbs, D., Hagan, C.C., Azim, E., Menon, V. and Reiss, A.L. 2005. Personality predicts activity in reward and emotional regions associated with humor. *Proceedings of the National Academy of Sciences of the United States of America*, **102**(45): 16502–6.
Montague, P., Berns, G.S., Cohen, J.D., McClure, S.M., Pagnoni, G., Dhamala, M., Wiest, M.C., Karpov, I., King, R.D., Apple, N. and Fisher, R.E. 2002. Hyperscanning: Simultaneous fMRI during linked social interactions. *Neuroimage*, **16**(4): 1159–64.
Nicolaou, N. and Shane, S. 2014. Biology, neuroscience, and entrepreneurship. *Journal of Management Inquiry*, **23**(1): 98–100.
Norman, D.A. and Shallice, T. 1986. Attention to action: Willed and automatic control of behavior. In R.J. Davidson, G.E. Schwartz and D. Shapiro (eds), *Consciousness and Self-Regulation (Vol. 4): Advances in Research and Theory* (1–18). New York: Plenum Press.
Ochsner, K.N. and Lieberman, M.D. 2001. The emergence of social cognitive neuroscience. *American Psychologist*, **56**(9): 717–34.
O'Doherty, J., Dayan, P., Schultz, J., Deichmann, R., Friston, K. and Dolan, R.J. 2004. Dissociable roles of ventral and dorsal striatum in instrumental conditioning. *Science*, **304**(5669): 452–4.
Palich, L.E. and Bagby, D.R. 1995. Using cognitive theory to explain entrepreneurial risk-taking: Challenging conventional wisdom. *Journal of Business Venturing*, **10**(6): 425–38.
Paus, T.S. 2001. Primate anterior cingulate cortex: Where motor control, drive and cognition interface. *Nature Reviews Neuroscience*, **2**(6): 417–24.
Penrose, E.T. 1959. *The Theory of the Growth of the Firm*. Oxford: Basil Blackwell.
Phelps, R. and Hase, S. 2002. Complexity and action research: Exploring the theoretical and methodological connection. *Educational Action Research*, **10**(3): 507–24.
Picard, N. and Strick, P.L. 1996. Motor areas of the medial wall: A review of their location and functional activation. *Cerebral Cortex*, **6**(3): 342–53.
Podsakoff, P.M., MacKenzie, S.B., Lee, J.Y. and Podsakoff, N.P. 2003. Common method biases in behavioral research: a critical review of the literature and recommended remedies. *Journal of Applied Psychology*, **88**(5): 879–903.
Poldrack, R.A. 2012. The future of fMRI in cognitive neuroscience. *Neuroimage*, **62**(2): 1216–20.
Posner, M.I. and Dehaene, S. 1994. Attentional networks. *Trends in Neuroscience*, **17**(2): 75–9.
Posner, M.I., and DiGirolamo, G.J. 1998. Executive attention: Conflict, target detection, and cognitive control. In R. Parasuraman (ed.), *The Attentive Brain*: 401–23. Cambridge, MA: MIT Press.
Posner, M.I. and Petersen, S.E. 1990. The attention system of the human brain. *Annual Review of Neuroscience*, **13**: 25–42.

Posner, M.I., Petersen, S.E., Fox, P.T. and Raichle, M.E. 1988. Localization of cognitive operations in the human brain. *Science*, **240**(4859): 1627–31.

Rauch, A. and Frese, M. 2007. Let's put the person back into entrepreneurship research: A meta-analysis on the relationship between business owner's personality traits, business creation, and success. *European Journal of Work and Organizational Psychology*, **16**(4): 353–85.

Rhodes, G., Byatt, G., Michie, P.T. and Puce, A. 2004. Is the fusiform face area specialized for faces, individuation, or expert individuation? *Journal of Cognitive Neuroscience*, **16**(2): 189–203.

Ridderinkhof, K.R., Ullsperger, M., Crone, E.A. and Nieuwenhuis, S. 2004. The role of the medial frontal cortex in cognitive control. *Science Signaling*, **306**(5695): 443–7.

Rizzolatti, G. and Craighero, L. 2004. The mirror-neuron system. *Annual Review of Neuroscience*, **27**: 169–92.

Romanelli, E. 1989. Organization birth and population variety: A community perspective on origins. In L. Cummings and B. Staw (eds), *Research in Organization Behavior*, vol. 1: 211–46. Greenwich, CT: JAI Press.

Saariluoma, P. 1995. *Chess Players' Thinking: A cognitive psychological approach*. London: Routledge.

Sanders, M. 2007. Scientific paradigms, entrepreneurial opportunities, and cycles in economic growth. *Small Business Economics*, **28**(4): 339–54.

Sarasvathy, S.D. 2001. Causation and effectuation: Toward a theoretical shift from economic inevitability to entrepreneurial contingency. *Academy of Management Review*, **26**(2): 243–63.

Sarasvathy, S.D. 2008. *Effectuation: Elements of entrepreneurial expertise*. Cheltenham, UK and Northampton, MA, USA: Edward Elgar Publishing.

Saxe, R., and Wexler, A. 2005. Making sense of another mind: The role of the right temporo-parietal junction. *Neuropsychologia*, **43**(10): 1391–9.

Say, J.-B. 1836. *A Treatise on Political Economy: Or the production, distribution, and consumption of wealth*. Boston, MA: Grigg & Elliot.

Schlund, M.W. and Cataldo, M.F. 2010. Amygdala involvement in human avoidance, escape and approach behavior. *Neuroimage*, **53**(2): 769–76.

Schultheiss, O.C. and Stanton, S.J. 2009. Assessment of salivary hormones. In E. Harmon-Jones and J.S. Beer (eds), *Methods in Social Neuroscience*: 17–44. New York: Guilford Press.

Schumpeter, J.A. 1934. *The Theory of Economic Development: An inquiry into profits, capital, credit, interest, and the business cycle*. Cambridge, MA: Harvard University Press.

Shackle, G.L.S. 1958. *Time in Economics*. Amsterdam: North Holland.

Shackle, G.L.S. 1979. *Imagination and the Nature of Choice*. Edinburgh: Edinburgh University Press.

Shane, S. 2000. Prior knowledge and the discovery of entrepreneurial opportunities. *Organization Science*, **11**(4): 448–69.

Shane, S. 2003. *A General Theory of Entrepreneurship: The individual–opportunity nexus*. Cheltenham, UK and Northampton, MA, USA: Edward Elgar.

Shane, S. and Nicolaou, N. 2013. The genetics of entrepreneurial performance. *International Small Business Journal*, **31**(5): 473–95.

Shane, S. and Venkataraman, S. 2000. The promise of entrepreneurship as a field of research. *Academy of Management Review*, **25**(1): 217–26.

Shepherd, D.A. 2003. Learning from business failure: Propositions of grief recovery for the self-employed. *Academy of Management Review*, **28**(2): 318–28.

Shepherd, D.A., Haynie, J.M. and McMullen, J.S. 2012. Confirmatory search as a useful heuristic? Testing the veracity of entrepreneurial conjectures. *Journal of Business Venturing*, **27**(6): 637–51.

Shepherd, D.A., McMullen, J.S. and Jennings, P.D. 2007. The formation of opportunity beliefs: Overcoming ignorance and reducing doubt. *Strategic Entrepreneurship Journal*, **1**(1–2): 75–96.

Simon, H.A. 1955. A behavioral model of rational choice. *Quarterly Journal of Economics*, **69**(1): 99–118.
Smolensky, P. 1988. On the proper treatment of connectionism. *Behavioral and Brain Sciences*, **11**(1): 1–23.
Song, M., Podoynitsyna, K., Van Der Bij, H. and Halman, J.I. 2008. Success factors in New Ventures: A meta analysis. *Journal of Product Innovation Management*, **25**(1): 7–27.
Stewart, W.H. and Roth, P.L. 2001. Risk propensity differences between entrepreneurs and managers: A meta-analytic review. *Journal of Applied Psychology*, **86**(1): 145–53.
Stout, D. and Chaminade, T. 2007. The evolutionary neuroscience of tool making. *Neuropsychologia*, **45**(5): 1091–100.
Stout, D., Passingham, R., Frith, C., Apel, J. and Chaminade, T. 2011. Technology, expertise and social cognition in human evolution. *European Journal of Neuroscience*, **33**(7): 1328–38.
Stout, D., Toth, N., Schick, K.D. and Chaminade, T. 2008. Neural correlates of early Stone Age tool-making: Technology, language and cognition in human evolution. *Philosophical Transactions of the Royal Society B: Biological Sciences*, **363**(1499): 1939–49.
Strack, F. and Deutsch, R. 2004. Reflective and impulsive determinants of social behavior. *Personality and Social Psychology Review*, **8**(3): 220–47.
Tajima, K. 2007. The theory of institutions and collective action in Adam Smith's Theory of Moral Sentiments. *Journal of Socio-Economics*, **36**(4): 578–94.
Todorov, A., Fiske, S. and Prentice, D. (eds). 2011. *Social Neuroscience: Toward understanding the underpinnings of the social mind.* Oxford Scholarship Online.
Tomer, R., Goldstein, R.Z., Wang, G.J., Wong, C. and Volkow, N.D. 2008. Incentive motivation is associated with striatal dopamine asymmetry. *Biological Psychology*, **77**(1): 98–101.
Tversky, A. and Kahneman, D. 1981. The framing of decisions and the psychology of choice. *Science*, **211**(4481): 453–8.
Ucbasaran, D., Wright, M. and Westhead, P. 2003. A longitudinal study of habitual entrepreneurs: Starters and acquirers. *Entrepreneurship and Regional Development*, **15**(3): 207–28.
Van Honk, J. 2012. Neuroendocrine manipulation of the sexually dimorphic human social brain. In E. Harmon-Jones and J.S. Beer (eds), *Methods in Social Neuroscience*: 17–44. New York: Guilford Press.
Van Overwalle, F. 2009. Social cognition and the brain: A meta-analysis. *Human Brain Mapping*, **30**(3): 829–58.
Vogt, S., Buccino, G., Wohlschlager, A.M., Canessa, N., Shah, N.J., Zilles, K., Eickhoff, S.B., Freund, H.-J., Rizzolatti, G. and Fink, G.R. 2007. Prefrontal involvement in imitation learning of hand actions: Effects of practice and expertise. *NeuroImage*, **37**(4): 1371–83.
Waldman, D.A., Balthazard, P.A. and Peterson, S.J. 2011. Leadership and neuroscience: Can we revolutionize the way that inspirational leaders are identified and developed? *Academy of Management Perspectives*, **25**(1): 60–74.
Walsh, J.P. 1995. Managerial and organizational cognition: Notes from a trip down memory lane. *Organization Science*, **6**(3): 280–321.
Wasserman, N. 2008. The founder's dilemma. *Harvard Business Review*, **86**(2): 102–9.
Whiteside, S.P. and Lynam, D.R. 2001. The five factor model and impulsivity: Using a structural model of personality to understand impulsivity. *Personality and Individual Differences*, **30**(4): 669–89.
Wood, M.S. and McKinley, W. 2010. The production of entrepreneurial opportunity: A constructivist perspective. *Strategic Entrepreneurship Journal*, **4**(1): 66–84.
Wood, M.S., McKelvie, A. and Haynie, J.M. 2014. Making it personal: Opportunity individuation and the shaping of opportunity beliefs. *Journal of Business Venturing*, **29**(2): 252–72.
Wood, M.S. and Williams, D. 2014. Opportunity evaluation as rule-based decision making. *Journal of Management Studies*, **51**(4): 573–602.

Wood, M.S., Williams, D.W. and Grégoire, D.A. 2012. The road to riches? A model of the cognitive processes and inflection points underpinning entrepreneurial action. In J.A. Katz and A.C. Corbett (eds), *Entrepreneurial Action: Advances in entrepreneurship, firm emergence and growth*, vol. 14: 207–52. Bingley, UK: Emerald.

Wright, M., Robbie, K. and Ennew, C. 1997. Venture capitalists and serial entrepreneurs. *Journal of Business Venturing*, **12**(3): 227–49.

Xu, Y. 2005. Revisiting the role of the fusiform face area in visual expertise. *Cerebral Cortex*, **15**(8): 1234–42.

Zaki, J. and Ochsner, K. 2011. Reintegrating accuracy into the study of social cognition. *Psychological Inquiry*, **22**(3): 159–82.

Zhao, H. and Seibert, S.E. 2006. The Big Five personality dimensions and entrepreneurial status: A meta-analytical review. *Journal of Applied Psychology*, **91**(2): 259–71.

Zhao, H., Seibert, S.E. and Lumpkin, G.T. 2010. The relationship of personality to entrepreneurial intentions and performance: A meta-analytic review. *Journal of Management*, **36**(2): 381–404.

10. The infrastructure of entrepreneurial learning

Daniel P. Forbes

ABSTRACT

The quantity and quality of entrepreneurial activity in a society are critically influenced by the extent to which people in that society possess knowledge relevant to the practice of entrepreneurship. In most times and places in history, the acquisition of entrepreneurial knowledge has been a relatively slow process, and a person's ability to acquire it has often been restricted by his or her geographic location and social network. Over the past several decades, however, there has emerged a rich set of resources that help people acquire entrepreneurship-related knowledge. These new resources are characterized by (1) the large-scale codification of entrepreneurial knowledge through the development of books, periodicals, blog posts, podcasts, videos and other media that distill portions of what practicing entrepreneurs and others have learned, and (2) the formation of networks, markets and other social structures devoted specifically to the exchange of this knowledge. Although many of these resources are familiar to entrepreneurship scholars, their collective emergence and impact remain theoretically underappreciated. In this chapter, I propose a new way of thinking about these resources, and I consider their implications for future research. In particular, I propose that future scholars should pay more careful attention to the content of the knowledge people exchange through these mechanisms and to the potential for these mechanisms to shape the way people think about whether and how to start new ventures.[1]

INTRODUCTION

When Steve Jobs died in October 2011, the event sent ripples of reflection through many communities of people interested in entrepreneurship. Accordingly, much was written about Jobs's life, his management career, and the evolution of his flagship company, Apple, and its path-breaking products. Less has been said about how the larger world of entrepreneurship had changed over the course of Jobs's career. But on this front, too, there is much to reflect on. When Jobs founded Apple in 1975, for example, there were comparatively few resources available to help nascent entrepreneurs learn about the process of launching a new venture. Jobs had the help of some informal networks, such as the 'Homebrew Computer Club,' which had taken root in the Palo Alto area, but relatively few college courses or guidebooks were available to help him, and no 'Startup Weekend' events or 'live-tweeted' panel discussions of venture-funding alternatives were available to inform his efforts. Today, by contrast, aspiring entrepreneurs have access to an extraordinary range of learning resources. Some of these resources address general challenges of entrepreneurship, such as marketing and resource mobilization, while others address more specific issues. For example, there are books and articles written specifically for people who want to open a food truck or an 'app'-based business (e.g., Mayyasi, 2013). And there are online communities that enable subgroups of entrepreneurs – such as mothers with young children or military veterans – to exchange knowledge that is especially relevant to their experiences (Feld, 2012).

As part of the larger 'entrepreneurial revolution' that has been written about at some length (e.g., Audretsch, 2007; Bornstein, 2007), many of these learning resources are well recognized in society at large and among entrepreneurship scholars in particular. At the same time, however, scholars have only just begun to fully appreciate the cumulative significance of these resources. There are scholarly literatures devoted to the study of specific types of resources, such as entrepreneurial education programs (e.g., Kuratko, 2005). In practice, however, many entrepreneurs access information from multiple resources simultaneously (Ozgen and Baron, 2007). And in the future the set of available learning resources is likely to expand and to exert more pervasive effects on whether and how people engage in entrepreneurial activity. As a consequence, it behooves scholars to understand more fully the evolving array of resources from which people learn about entrepreneurship and to consider more carefully their implications for entrepreneurial behavior (see also Baron, Chapter 1, this volume; Bird, Chapter 3, this volume).

My goal in this chapter is to call attention to these learning resources as an important phenomenon worthy of more focused scholarly attention, to help position them in the context of what we know about cognition and entrepreneurship, and to consider their implications for future research. I begin by calling attention to the fact that entrepreneurs need knowledge that enables them not only to recognize opportunities – a process that has attracted the lion's share of scholarly attention in recent years – but to evaluate and exploit them as well (Markman,

2007; Shane and Venkataraman, 2000). In particular, I observe, entrepreneurs need knowledge relevant to the practical demands of creating a new venture and managing it during its earliest stages. In addition, I draw on Van de Ven's (1993) concept of a societal-level 'entrepreneurial infrastructure' to propose a shift in our research focus away from differences in the knowledge possessed by individual entrepreneurs and towards differences in the knowledge possessed by large groups of people within and across societies. Consistent with this approach, I go on to document the emergence of a rich array of learning resources, and I contend that these learning resources comprise a kind of subsidiary infrastructure – an 'infrastructure of entrepreneurial learning' – within the larger infrastructure of entrepreneurship. I conclude by considering some implications of these ideas for the study of entrepreneurial cognition.

KNOWLEDGE AND LEARNING IN NEW VENTURES

The idea that organizations acquire and use knowledge to enhance their effectiveness has a long history in the management literature (Argote and Ingram, 2000; Easterby-Smith and Lyles, 2011). The acquisition of knowledge that is useful or potentially useful to an organization, moreover, has been characterized as 'organizational learning' (Huber, 1991). Much of what organizations learn is derived from their own experience as they perform their tasks (Argote and Miron-Spektor, 2011), but they can also learn from sources outside their own boundaries, a process that has been characterized as 'knowledge transfer' (Argote and Ingram, 2000). Much of what has been written about organizational learning and knowledge transfer has focused on relatively established firms. In a notable exception, however, Huber (1991) described a process of knowledge transfer that occurs during the creation of a new venture, which he called 'congenital learning.' He explained the process this way:

> Organizations do not begin their lives with clean slates. The individuals or organizations that create new organizations have *knowledge about the new organization's initial environment and about the processes the organization can use to carry out its creator's intentions, and they make this knowledge available to the new organization's members*. [italics added] . . . There invariably exists a time interval between when an organization is first conceived of and the rather arbitrarily defined birth event . . . During this interval the organization's founders employ vicarious learning, grafting, or searching to learn on behalf of the organization-to-be. Thus an organization's *congenital knowledge* is a combination of the knowledge inherited at its conception and the additional knowledge acquired prior to its birth. (p. 91)

Thus Huber calls attention to the knowledge that informs the way founders create their ventures as well as the fact that they impart knowledge to their ventures.

Entrepreneurship scholars, for their part, have also discussed knowledge and learning in connection with new venture creation (Harrison and Leitch, 2005; Markman, 2007). Unlike Huber, however, they have tended to focus primarily on those elements of knowledge that enable a person to recognize an entrepreneurial opportunity (e.g., Corbett, 2007; Dimov, 2007; Shepherd and DeTienne, 2005). For example, Shane (2000) called attention to the extent to which an individual's capacity to recognize an entrepreneurial opportunity is a function of his or her 'idiosyncratic prior knowledge.' Relatedly, it is common for entrepreneurship scholars to speak of startup formation as a process through which people seek to 'commercialize knowledge' (e.g., Mueller, 2006). In these cases, the link to opportunity recognition may or may not be made explicit, but the knowledge referred to generally serves as a source of entrepreneurial opportunity (Audretsch and Keilbach, 2007).

Without disputing the critical role knowledge plays in opportunity recognition, I wish to focus here on elements of knowledge that enable people to act more effectively during other stages of the entrepreneurial process as well, including knowledge that enables people to more effectively evaluate and exploit opportunities (Markman, 2007). In particular, I am interested in the knowledge that enables people to create a new independent firm and manage it during its earliest stages (Baron and Shane, 2008). This kind of knowledge may still include some knowledge relevant to opportunity recognition, but it extends well beyond that and overlaps more fully with the kinds of practical knowledge implied in Huber's reference to 'the processes the organization can use to carry out its creator's intentions.' For the sake of simplicity, the range of knowledge I refer to encompasses the set of topics that are commonly covered in a standard introductory entrepreneurship textbook. This includes such topics as market research, venture financing and attracting and managing human capital (e.g., Allen, 2006; Baron and Shane, 2008). Knowledge of these topics often takes the form of tacit knowledge possessed by people having direct or indirect experience of venture creation (Davidsson and Honig, 2003; Minniti and Bygrave, 2001). However, as I explain in subsequent sections, knowledge about various aspects of new firm creation and management is increasingly being codified – that is, transmitted in formal, symbolic language (Edmondson et al., 2003; see also Clarke and Cornelissen, Chapter 11, this volume). And it is being transferred in a wider variety of ways and to a wider range of people than has traditionally been the case.

Past research on knowledge in entrepreneurship has also tended to focus on individual differences in knowledge and on the implications of those differences (e.g., Baum et al., 2007; Alvarez and Busenitz, 2001). My focus here, however, is on knowledge differences as they are manifested across large groups of people, such as those that may exist between groups of people within or across societies, or within a society over time. It is important to consider what large groups of people know about entrepreneurship, because the knowledge they collectively possess is likely to influence both the quantity and the quality of the new ventures launched in a society (Busenitz et al., 2000; Levie and Autio, 2008). Consider, for example, the choices made by the potential entrepreneurs in a society about whether to start a new venture. Sarasvathy (2004) observes that in any society there is likely to be a subset of people 'who want to become entrepreneurs but do not,' owing to one or more barriers that may exist (p. 707). Among the barriers that inhibit people from creating a new venture is a lack of knowledge about how to evaluate and exploit an entrepreneurial opportunity.

Clearly, knowledge is but one of many factors that influence entrepreneurial activity in a society. In an effort to enumerate these factors, Van de Ven (1993) proposed the concept of an 'infrastructure for entrepreneurship.' The infrastructure he referred to encompassed the set of actors and entities within a given society that both enabled and constrained entrepreneurial activity. 'Popular folklore notwithstanding, the process of entrepreneurship is a collective achievement' (p. 211), he wrote, which depended not only on the efforts of individual entrepreneurs, but on 'many other public- and private-sector actors who perform critical functions to develop and commercialize a new technology' (p. 214). As he characterized it, the infrastructure for entrepreneurship comprised three key components: (1) 'proprietary functions'; (2) 'resource endowments'; and (3) 'institutional arrangements.' Proprietary functions refers to the firms directly involved in commercializing an innovation as well as their immediate buyers and suppliers – in short, those actors that fall within a traditional strategic definition of an industry environment (Hitt et al., 2010). Resource endowments refers to the set of actors and entities that provide resources critical to the development of an innovation. Institutional arrangements refers to the authorities and structures that govern and legitimize collective action. In subsequent formulations, he added 'market consumption' as a fourth component, distinguishing consumer demand and the cultural norms affecting it from other proprietary functions (e.g. Van de Ven, 2005). Figure 10.1 provides a graphical representation of the infrastructure.

Within this framework, the availability of knowledge relevant to the creation and management of new ventures can be regarded as a 'resource endowment' that exists at the level of societies, not firms or individuals.

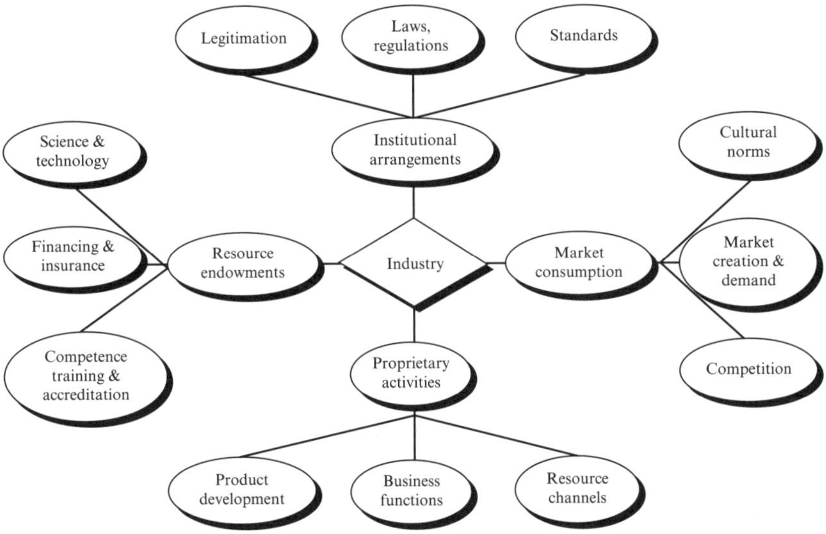

Source: Van de Ven (2005), Van de Ven et al. (1999), adapted from Van de Ven and Garud (1989).

Figure 10.1 The infrastructure for entrepreneurship

For example, Van de Ven called attention to the extent to which people in a society are generally educated as well as the extent to which they possess 'specific skills related to [a particular] innovation' (p. 216). In societies with richer endowments along these dimensions, he contended, actors would be more likely to collectively construct the environmental niches that give rise to the emergence of new organizational populations and, relatedly, to instances of organizational founding and technological innovation (pp. 212–14).

The infrastructure framework provides a useful tool for understanding the learning resources that have emerged over the last several decades, because it prompts us to regard those resources not as discrete fragments of activity but rather as part of a coherent set of societal endowments that shape entrepreneurial activity. Taken together, these learning resources comprise a kind of subsidiary infrastructure within the larger infrastructure of entrepreneurship Van de Ven identified: an 'infrastructure of entrepreneurial learning.' More specifically, this learning infrastructure represents a subset of what Van de Ven characterized as the 'resource endowments' component of the larger entrepreneurial infrastructure insofar as it affects the competence of the human resources available in a society. But because these elements of infrastructure are today considerably deeper and more elaborate than they were when this framework was introduced, they merit more explicit and detailed consideration than they were given at that time. In the sections that follow, I elaborate some key elements of the contemporary infrastructure of entrepreneurial learning, and I consider their implications for entrepreneurial activity. But first, in order to underscore the distinctiveness of these new knowledge resources, I offer a more general discussion of how entrepreneurs acquire new knowledge.

HOW DO PEOPLE ACQUIRE ENTREPRENEURIAL KNOWLEDGE?

In most times and places in history, people have acquired entrepreneurial knowledge in two main ways: directly, by engaging in entrepreneurial activity themselves; or indirectly, by closely associating with others who are so engaged. The direct acquisition of entrepreneurial knowledge is best illustrated by the case of serial entrepreneurs (e.g., Westhead et al., 2005). Minniti and Bygrave (2001) propose that people who repeatedly engage in entrepreneurship acquire a 'stock of knowledge accumulated on the basis of past experiences' (p. 5) and that this knowledge informs their subsequent entrepreneurial efforts. Often – although not always – this knowledge improves the effectiveness with which they carry out specific venture creation activities, and this in turn accounts for the performance advantages that some studies have found to accrue to serial entrepreneurs (e.g., Ucbasaran et al., 2008). Indirect knowledge acquisition, on the other hand, has been documented by the literatures on role modeling and family or early-life influences on entrepreneurial activity (e.g., Carroll and Mosakowski, 1987; Van Auken et al., 2006). These literatures have shown that people can acquire entrepreneurial knowledge by communicating with and closely observing the behaviors of people engaged in entrepreneurship. Having a social relationship with an entrepreneur may affect a person's entrepreneurship-related choices in a variety of ways. For example, the experience of having an entrepreneur as a parent may affect the values or emotions a person associates with entrepreneurial activity (see also Foo, Murnieks and Chan, Chapter 5, this volume; Grégoire, Chapter 6, this volume). But such a relationship often involves a transfer of knowledge as well (Shaver and Scott, 1991), as parents (or other associates) are likely to explain and model various elements of the entrepreneurial activities they are engaged in.

These traditional modes of entrepreneurial knowledge acquisition are powerful, and they remain important modes of knowledge transfer today. But to the extent we are interested in understanding how entrepreneurial knowledge might disseminate within large groups of people, it is clear that these modes of knowledge transfer have some significant limitations. For example, the direct acquisition of experience requires people to actually engage in entrepreneurial activity and to bear the associated risks and costs. And the indirect mode of learning from others has traditionally depended on the formation of relationships between people who are geographically proximate (e.g., between parents and children, or between entrepreneurs and their employees). In many cases, too, the formation of these relational ties is further dependent on pure chance – as in the case of a child being born to entrepreneurial parents – or on an individual's social status (Ruef et al., 2003). Finally, much of the knowledge acquired in these ways is tacit, rather than explicit, a feature that tends to slow its dissemination. 'Because tacit knowledge is bound to the senses, personal experience and bodily movement,' Von Krogh and colleagues (2000) observe, its transfer 'requires close physical proximity while the work is being done' (p. 83).

Over the last several decades, these traditional modes of entrepreneurial knowledge acquisition have been supplemented by new learning resources. These new resources are characterized by (1) the large-scale codification of entrepreneurial knowledge through the development of books, periodicals, blog posts, podcasts, videos and other media that distill portions of what practicing entrepreneurs and others have learned through their direct and indirect experiences with entrepreneurial activity, and (2) the deliberate formation of networks, markets and other social structures devoted specifically to the exchange of this knowledge. Table 10.1 presents some examples of these kinds of learning resources.

THE EMERGENCE OF NEW LEARNING RESOURCES

The resources shown in Table 10.1 did not exist on anything approaching their current scale when Van de Ven (1993) offered his original formulation of the 'resource endowments' component of an entrepreneurial infrastructure some 20 years ago. Accordingly, Van de Ven confined his attention to people's levels of general education and to their possession of knowledge related to particular innovations. As he put it, 'new technologies mean that new competencies are required to perform essential tasks, be they related to research, manufacturing or marketing' (p. 216). This characterization of the relevant human resource endowments was appropriate to the learning resources that were generally available at that time, and it reflected the fact that, for much of the twentieth century, the

Table 10.1 Examples of entrepreneurial learning resources that have emerged or grown significantly since 1980

Type	Description	Examples
Formal educational offerings	These include degree programs and courses available through colleges and universities	• Undergraduate and graduate business degrees reflecting an emphasis or concentration on entrepreneurship • Entrepreneurship-focused courses within business degree programs • Coursework in non-business disciplines (e.g., design, engineering) that incorporates significant entrepreneurial content
Periodicals	These include periodicals that include entrepreneurship-related content, either entirely or frequently	• *Harvard Business Review* • *Stanford Social Innovation Review* • *Inc.* magazine • *Fast Company* magazine
Books	These include books that transmit entrepreneurial knowledge	• *Startup Owner's Manual* by Steve Blank and Bob Dorf • *How to Change the World* by David Bornstein • *Launching New Ventures* by Kathleen Allen • *Business Model Generation* by Alex Osterwalder and Yves Pigneur
Online resources	These include sites that provide original content, such as blogs, and those that serve as portals to other content	• Steve Blank (steveblank.com) • Stanford Entrepreneurship Corner (ecorner.stanford.edu) • Fred Wilson (www.avc.com)
Social media communities	Groups that exchange knowledge virtually using social media technologies	• Twitter • LinkedIn groups
Offline communities	Groups that meet in person and provide networking, training and support	• Startup Weekend • Meetup.com groups • MIT Enterprise Forum

set of knowledge relevant to innovative activity was understood to include a mix of general knowledge, business-related knowledge and technical knowledge specific to the domains of activity being undertaken. Thus Van de Ven identified some specific mechanisms through which those kinds of knowledge could be transferred, including educational training programs at colleges and universities, corporate recruitment and training processes, and knowledge sharing via conferences and trade publications. Today, on the other hand, the elements of knowledge available to potential and existing entrepreneurs are much more specifically targeted and much more widely available. For example, an entrepreneur today can find guidance related to identifying and evaluating new business opportunities (Mullins, 2010), developing and refining various elements of a startup business model (Mullins and Komisar, 2006), soliciting startup funding (DeClercq et al., 2007), and cultivating a regional startup community (Feld, 2012).

Universities have played a critical role in the codification and exchange of entrepreneurial knowledge by fostering the development of scholarly research and formal educational programs devoted to the study of entrepreneurship. Scholarly research on entrepreneurship has grown dramatically over the last three decades, for example, during which time scholars have established entrepreneurship as an academic field (Aldrich, 2012). Key developments in that process have included the publication of books and new scholarly journals, the development of academic conferences and the establishment of doctoral programs, all of which foster the generation and exchange of new knowledge (Katz, 2003). Although much academic research does not produce knowledge that is immediately applicable to the practice of entrepreneurship, it often informs the development of formal educational programs in entrepreneurship and the textbooks used in those programs (Baron and Shane, 2008, pp. 22–8). In addition, it generates frameworks and insights that are sometimes directly adapted and translated for broader audiences. Examples of this translation process include Noam Wasserman's (2012) book *The Founder's Dilemma*, which condenses for a popular audience Wasserman's own academic research on the tradeoffs associated with the pursuit of alternative entrepreneurial goals. Formal educational programs in entrepreneurship, moreover, have themselves increased in number and sophistication. Some of the oldest programs have targeted students in elementary and secondary schools, including Junior Achievement and NFTE-sponsored programs, but the greatest growth in recent decades has been in undergraduate- and masters-level programs (Katz, 2003; Kuratko, 2005). Today, as a result, students in many countries around the world can study entrepreneurship at every level of schooling. Entrepreneurial education is also increasingly integrated into the higher-level study of subjects traditionally taught outside of business schools, such as design and engineering (Thorp and Goldstein, 2010).

Beyond universities, too, there has emerged a variety of networks, markets and other social structures through which people exchange knowledge about the practice of entrepreneurship. These activities are more diffuse and less comprehensively documented than university-based activities, but they involve many more people than university-based activities. Among these activities are the publication of books and periodicals by for-profit publishers. For example, *The Lean Startup*, a book written by entrepreneur Eric Ries and published by a subsidiary of Random House, offers to entrepreneurs a set of techniques for managing the product development process in a new venture. Similarly, periodicals like *Fast Company* (launched in 1995) and *Inc.* (launched in 1979) convey entrepreneurial knowledge codified by journalists, former entrepreneurs and others to hundreds of thousands of readers.

The largest category of non-university activity, however, involves the development and maintenance of online communities, social media networks and offline communities. These communities and networks comprise people who create, co-create and exchange knowledge about entrepreneurship, often through patterns of interaction that resemble those of 'open source' innovation communities (Von Hippel and Von Krogh, 2003). Accordingly, these social structures tend to evolve informally through the accumulated efforts of many individual participants. However, more formal elements of network structure are sometimes established by enterprising individuals (such as Steve Blank, who operates an elaborate and influential website at http://steveblank.com), nonprofit organizations (such as the Kauffman Foundation) or firms (such as Meetup.com, an online networking site).

Sometimes such networks are affiliated with schools and universities (as in the case of the MIT Enterprise Forum, for example), but many others have no such affiliation and are rooted instead in social networks defined by geography, demography or subjects of interest. For example, a search of the Meetup.com site based on the terms 'Chicago entrepreneurs' reveals over 50 groups that feature labels such as: 'Startup Evanston,' the 'Founding Moms Exchange,' the 'Boostrappers Breakfast Group' and the 'Chicagoland Conscious Capitalists.' Each of these groups includes between 100 and 800 members. Groups and networks can leverage in-person mechanisms of knowledge exchange, as in the case of Meetup groups or the 'Startup Weekend' series of venture launch workshops (http://startupweekend.org), as well as virtual mechanisms, such as Internet blogs and social media technologies like Twitter. Twitter, for example, enables networked individuals to exchange short, ready-access linkages to websites, blog postings, articles and other content that contains elements of entrepreneurial knowledge. For example, a search of Twitter in March 2014 for the term 'entrepreneurship' surfaced a link offered by the Kauffman Foundation to a blog post authored by Zen Chu, an entrepreneur in residence at MIT (Chu, 2014). The post provided information on 'meaningful metrics' appropriate for measuring return on investment in healthcare startups and contained several links to additional sites with supporting information.

To some extent these modes expand the reach of traditional modes of knowledge transfer, for example by enabling larger numbers of people to more easily acquire direct and indirect experience of entrepreneurial activity. But they go beyond traditional modes of knowledge transfer insofar as they represent the large-scale codification of knowledge and insofar as they enable people to rapidly form relationships with large numbers of others from whom they might learn. These developments, in turn, are likely to significantly expand the ability of existing and potential entrepreneurs to acquire knowledge relevant to the decisions they face. Figure 10.2 illustrates this point while also highlighting the role that existing and former entrepreneurs play in feeding knowledge back to the infrastructure of entrepreneurial learning.

In characterizing activities such as those shown in Table 10.1 as modes of knowledge transfer, I hasten to add several important caveats. First, I do not mean to suggest that all – or even most – of the knowledge that entrepreneurs learn from experience can be readily codified and disseminated. As in any complex domain, much important knowledge cannot be effectively codified, and not all knowledge that can be codified is readily exchanged (Nonaka and von Krogh, 2011). Accordingly, there are important limits to what entrepreneurs can effectively learn from these

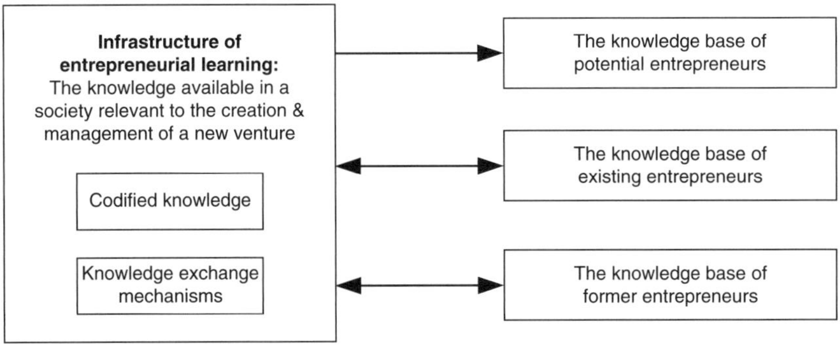

Figure 10.2 The infrastructure of entrepreneurial learning and its implications

resources. Second, the knowledge needs of entrepreneurs vary considerably across contexts, and they are liable to change and expand over time as organizations and their environments evolve (Aldrich and Ruef, 2006). Thus the fact that knowledge is available does not mean it will continue to be relevant indefinitely or that it will be interpreted and applied with due regard to its contextual limitations. Third, we cannot assume that the members of a society will have equal access to the infrastructure of entrepreneurial learning. Rather, access is likely to vary significantly across groups within a society. Finally, the validity of the knowledge offered through these learning resources varies widely. (I return to these last two points shortly.) For all of these reasons, there will inevitably remain significant gaps between the codified knowledge available through the mechanisms identified here and the demands of entrepreneurial activity. At the same time, the collective emergence of these mechanisms represents a clear 'sea change' in the extent to which knowledge about the creation and management of new ventures is available within societies. Accordingly, it behooves us to ask how the availability of this knowledge is affecting the way entrepreneurs think and make decisions.

IMPLICATIONS FOR THE STUDY OF ENTREPRENEURIAL COGNITION

Research on entrepreneurial cognition has traditionally focused on understanding relatively general differences in how people process information and in how those differences affect people's entrepreneurial capacities. For example, scholars have explored the extent to which people rely on biases and heuristics in making decisions, or the extent to which they rely on effectual or causal reasoning processes (Grégoire et al., 2011). These lines of inquiry can help explain larger patterns of behavior on the part of individual entrepreneurs and their firms, but they are less well suited to explaining differences in entrepreneurial behavior across societies. To answer such questions, we need to pay more attention to the societal-level resources that shape people's entrepreneurial abilities. The availability of knowledge resources specific to the practice of entrepreneurship is likely to enhance the quality and quantity of entrepreneurship in a society by enabling the members of that society to make better decisions about whether, when and how to launch new ventures. This, in turn, opens up several important opportunities for scholars interested in entrepreneurial cognition.

At the most basic level, it would be useful for scholars to begin to characterize more clearly and comprehensively the various kinds of entrepreneurially relevant knowledge being exchanged in societies. Some categories of knowledge encompassed by the infrastructure of entrepreneurial learning are relatively well developed in a theoretical sense, such as ideas pertaining to opportunity recognition. But the infrastructure also codifies and disseminates knowledge about a variety of context-specific problems entrepreneurs encounter in the course of building and managing new ventures. For example, if an entrepreneur wants to know how to assemble a board of directors, or how to negotiate a term sheet with a venture capitalist, there are articles and books specifically devoted to those issues. This kind of knowledge is not well characterized in the existing literature on entrepreneurial cognition. In other words, as scholars we lack a robust set of terms and frameworks for describing and discussing some of the more applied forms of knowledge that entrepreneurs work with.

In developing such tools, it may help to turn to theories of socially situated cognition, which emphasize the physical and social environments in which knowledge is used as well as the technologies with which knowledge is accessed. For example, Smith and Semin (2007) contend that 'theory should reflect the way that cognition is socially enabled and distributed through communication,' observing that 'cognitive processes draw not only on resources in the social environment but also on technical equipment (monitors, readouts, maps) into which considerable knowledge is downloaded' (p. 134). Thus studies of entrepreneurial cognition should be expanded beyond their traditional focus on how entrepreneurs think: going forward, we need to know more about *what* entrepreneurs think about and about how they label, access and exchange knowledge in their communications with others.

Second, scholars should begin to critically examine the quality and validity of the knowledge resources that comprise the infrastructure of entrepreneurial learning. Although some of the knowledge made available is based on the insights of past research, much of it is not. Rather, the knowledge offered is often based on the accumulated experiences of individuals. To some extent, the highly distributed and grounded nature of this knowledge is part of its strength; in aggregate, it represents a form of knowledge that differs from scholarly knowledge but which is also valid and appropriate to the purposes it serves (Van de Ven and Johnson, 2006). But not all of it is valid, of course, and some of the knowledge available is in fact contradictory. For example, in his book *The Lean Startup*, Eric Ries advocates an approach to product development centered on the concept of a 'minimum viable product' (MVP). But in a recent blog post, former entrepreneur Bill Aulet warns of 'our dangerous obsession with the MVP' (Aulet, 2014). This raises the question, 'Which view is correct?' Or might it depend on the context – and, if so, then how?

Scholars can help answer questions like these by leveraging their distinctive skills to help clarify which knowledge claims are more and less valid. Scholars can also help identify patterns of cognitive bias in knowledge exchange. For example, past research has shown that there are times when practitioners' own articulations of what they know differ considerably from the intuitions they tend to employ in practice (e.g., Zacharakis and Meyer, 1998). Such discrepancies – between articulated and actual knowledge – suggest that knowledge resources based entirely on practitioners' own articulations of what they know may be misleading to actual and aspiring entrepreneurs. Scholars can call attention to such discrepancies, therefore, and look for ways to more accurately distill and articulate key elements of entrepreneurial expertise (Baron and Ensley, 2006).

Finally, there is a need for scholars to help determine how and to what extent these resources shape entrepreneurial activity within societies. For example, can variations in entrepreneurial activity be traced to differences in the extent to which specific groups of people (across societies or within a given society) have access to the infrastructure of entrepreneurial learning? Alternatively, do people interpret or utilize that learning infrastructure differently based on differences in their institutional/cultural contexts (Elsbach et al., 2005; Mitchell et al., 2000)? Studies have begun to examine the effectiveness of formal entrepreneurship education in connection with questions like these, but more work is needed, especially with regard to the implications of the less-formal knowledge diffusion mechanisms outlined above. Social cognitive theory provides a basis for theorizing about how people can learn about entrepreneurship from less-formal knowledge resources insofar as it has sought to explain how mass communication technologies facilitate vicarious learning and verbal persuasion (Bandura, 2002).

In exploring these and related questions, moreover, scholars will need to bear in mind the tensions that can exist between the quality and quantity of a society's entrepreneurial activity. A society in which people make high-quality entrepreneurial decisions, after all, will be one in which many people make well-considered decisions *not* to start new ventures. This is not to deny the fact that all entrepreneurial activity entails risk, but it is to acknowledge that, as Gilbert and Eyring (2010) have put it, 'not all risks are created equal.' Thus, to the extent that newly available knowledge resources succeed in helping people to make well-informed decisions not to become entrepreneurs, they can contribute to the quality of a society's entrepreneurial activity while simultaneously helping to constrain certain measures of its quantity.

CONCLUSION

For people in the contemporary world with interests in entrepreneurship, the infrastructure of entrepreneurial learning has become so seemingly ubiquitous and its constituent elements so intimately familiar that it may seem superfluous or unnecessary for scholars to turn their attention to it. But while scholars have actively sought to understand various pieces of this infrastructure, they have yet to develop a robust understanding of how those pieces fit together and how they are collectively shaping the way people engage in entrepreneurial activity in societies around the world.

To some extent, this oversight reflects a general tendency on the part of recent entrepreneurship research to emphasize lower levels of analysis – such as individuals, teams and firms – as opposed to higher levels, such as industries and environments (Busenitz et al., 2003; Forbes and Kirsch, 2011). That tendency, in turn, reflects in part the ascendance of the cognitive perspective in entrepreneurship, which has tended to focus scholars' attention on psychological processes as they unfold within individuals and groups (Grégoire et al., 2011). Clearly, individual cognition plays a critical role in the launch of new ventures, given the limited organizational infrastructure such firms generally possess. Increasingly, however, entrepreneurs are able to draw upon an extensive set of cognitive resources from their external environments that are specifically geared to their own, context-specific entrepreneurial challenges. These resources codify and transmit the insights of a large, distributed network of peers, experts and advisors. As such, they represent 'scaffolds' that entrepreneurs can use to help them scale the cognitive challenges of their work (Semin et al., 2012). As this scaffolding evolves and plays an increasingly important role in shaping entrepreneurial behavior, it behooves scholars interested in entrepreneurial cognition to pay more direct attention to the knowledge that scaffolding contains and the ways that knowledge is used. In this chapter, I have sought to make it easier for scholars to theorize about this emerging scaffolding by linking it to some more established ideas pertaining to the role of knowledge in entrepreneurship. By building on those ideas, scholars can help existing theories of entrepreneurial cognition to adapt to the rapidly evolving world of contemporary entrepreneurial practice.

Finally, scholars can benefit from integrating these ideas with newer ideas, such as theories of socially situated cognition. The approach I take here is broadly consistent with a 'situated cognition' perspective on entrepreneurship insofar as it calls attention to the ways in which people's entrepreneurial behaviors are enabled and constrained by the knowledge-related resource endowments that exist within their societies (Mitchell et al., 2007). As with more traditional approaches to the study of entrepreneurial cognition, this approach still conceives of entrepreneurial activity as a set of behaviors critically shaped by the knowledge and beliefs people hold. At the same time, however, it incorporates elements of the institutional and evolutionary perspectives insofar as it seeks to situate discrete instances of entrepreneurial activity within a larger macroenvironmental context that changes over time.

NOTE

1. I am grateful to Michael Cummings, Yuqing Ren, Andy Van de Ven and the editors, among others, for their input.

REFERENCES

Aldrich, H. 2012. The emergence of entrepreneurship as an academic field. A personal essay on institutional entrepreneurship. *Research Policy*, **41**(7): 1240–48.

Aldrich, H. and Ruef, M. 2006. *Organizations Evolving* (2nd edn). London: Sage.

Allen, K. 2006. *Launching new ventures: An entrepreneurial approach*. Boston, MA: Houghton-Mifflin.

Alvarez, S. and Busenitz, L. 2001. The entrepreneurship of resource-based theory. *Journal of Management*, **27**(6): 755–75.

Argote, L. and Ingram, P. 2000. Knowledge transfer: A basis for competitive advantage in firms. *Organizational Behavior & Human Decision Processes*, **82**(1): 150–69.

Argote, L. and Miron-Spektor, E. 2011. Organizational learning: From experience to knowledge. *Organization Science*, **22**(5): 1123–37.

Audretsch, D. 2007. *The Entrepreneurial Society*. Oxford: Oxford University Press.

Audretsch, D. and Keilbach, M. 2007. The theory of knowledge spillover entrepreneurship. *Journal of Management Studies*, **44**(7): 1242–54.

Aulet, B. 2014. Our dangerous obsession with the MVP. Blog post dated March 1 at http://techcrunch.com/2014/03/01/our-dangerous-obsession-with-the-mvp/.

Bandura, A. 2002. Social cognitive theory of mass communication. In J. Bryant and D. Zillman (eds), *Media Effects: Advances in theory and research*: 121–54. Mahwah, NJ: Lawrence Erlbaum Associates.

Baron, R.A. and Ensley, M.D. 2006. Opportunity recognition as the detection of meaningful patterns: Evidence from comparisons of novice and experienced entrepreneurs. *Management Science*, **52**(9): 1331–44.

Baron, R.A. and Shane, S. 2008. *Entrepreneurship: A process perspective* (2nd edn). Cengage.

Baum, J., Frese, M. and Baron, R. (eds). 2007. *The Psychology of Entrepreneurship*. Mahwah, NJ: Lawrence Erlbaum Associates.

Bornstein, D. 2007. *How to Change the World: Social entrepreneurship and the power of new ideas*. Oxford: Oxford University Press.

Busenitz, L.W., Gomez, C. and Spencer, J.W. 2000. Country institutional profiles: Unlocking entrepreneurial phenomena. *Academy of Management Journal*, **43**(5): 994–1003.

Busenitz, L., West, G.P., Shepherd, D., Nelson, T., Chandler, G. and Zacharakis, A. 2003. Entrepreneurship research in emergence: Past trends and future directions. *Journal of Management*, **29**(3): 285–308.

Carroll, G. and Mosakowski, E. 1987. The career dynamics of self-employment. *Administrative Science Quarterly*, **32**(4): 570–89.

Chu, Z. 2014. Why healthcare entrepreneurs should focus on meaningful metrics, http://www.entrepreneurship.org/eMed/eMed-Blog/2014/March/Why-healthcare-entrepreneurs-should-focus-on-meaningful-metrics.aspx. Document accessed on March 7, 2014.

Corbett, A.C. 2007. Learning asymmetries and the discovery of entrepreneurial opportunities. *Journal of Business Venturing*, **22**(1): 97–118.

Davidsson, P. and Honig, B. 2003. The role of social and human capital among nascent entrepreneurs. *Journal of Business Venturing*, **18**(3): 301–31.

DeClercq, D., Fried, V., Lehtonen, O. and Sapienza, H. 2007. An entrepreneur's guide to the venture capital galaxy. *Academy of Management Perspectives*, **20**(3): 90–112.

Dimov, D. 2007. From opportunity insight to opportunity intention: The importance of person–situation learning match. *Entrepreneurship Theory and Practice*, **31**(4): 561–83.

Easterby-Smith, M. and Lyles, M. (eds). 2011. *The Handbook of Organizational Learning & Knowledge Management* (2nd edn). New York: John Wiley & Sons.

Edmondson, A., Winslow, A., Bohmer, R. and Pisano, G. 2003. Learning how and learning what: Effects of tacit and codified knowledge on performance improvement following technology adoption. *Decision Sciences*, **34**(2): 197–223.

Elsbach, K.D., Barr, P.S. and Hargadon, A.B. 2005. Identifying situated cognition in organizations. *Organization Science*, **16**(4): 422–33.

Feld, B. 2012. *Startup Communities: Building an entrepreneurial ecosystem in your city*. New York: Wiley.

Forbes, D. and Kirsch, D. 2011. The study of emerging industries: Recognizing and responding to some central problems. *Journal of Business Venturing*, **26**(5): 589–602.

Gilbert, C. and Eyring, M. 2010. Beating the odds when you launch a new venture. *Harvard Business Review*, **88**(5): 67–73.

Grégoire, D.A., Corbett, A. and McMullen, J. 2011. The cognitive perspective in entrepreneurship: An agenda for future research. *Journal of Management Studies*, **48**(6): 1443–77.

Harrison, R. and Leitch, C. 2005. Entrepreneurial learning: Researching the interface between learning and the entrepreneurial context. *Entrepreneurship Theory and Practice*, **29**(4): 351–71.

Hitt, M.A., Ireland, R.D. and Hoskisson, R.E. 2010. *Strategic Management* (9th edn). South-Western Cengage.

Huber, G. 1991. Organizational learning: The contributing processes and literatures. *Organization Science*, **2**(1): 88–115.

Katz, J.A. 2003. The chronology and intellectual trajectory of American entrepreneurship education: 1876–1999. *Journal of Business Venturing*, **18**(2): 283–300.

Kuratko, D. 2005. The emergence of entrepreneurship education: Development, trends and challenges. *Entrepreneurship Theory and Practice*, **29**(5): 577–97.

Levie, J. and Autio, E. 2008. A theoretical grounding and test of the GEM model. *Small Business Economics*, **31**(3): 235–63.

Markman, G. 2007. Entrepreneurs' competencies. In J. Baum et al. (eds), *The Psychology of Entrepreneurship*: 19–40. Mahwah, NJ: Lawrence Erlbaum Associates.

Mayyasi, A. 2013. Food truck economics, http://blog.priceonomics.com/post/45352687467/food-truck-economics. Document accessed on March 5, 2014.

Minniti, M. and Bygrave, W. 2001. A dynamic model of entrepreneurial learning. *Entrepreneurship Theory and Practice*, **25**(3): 5–16.

Mitchell, R.K., Busenitz, L.W., Bird, B., Gaglio, C.M., McMullen, J.S., Morse, E.A. and Smith, J.B. 2007. The central question in entrepreneurial cognition research 2007. *Entrepreneurship Theory and Practice*, **31**(1): 1–27.

Mitchell, R.K., Smith, J.B., Seawright, K.W. and Morse, E.A. 2000. Cross-cultural cognitions and the venture creation decision. *Academy of Management Journal*, **43**(5): 974–93.

Mueller, P. 2006. Exploring the knowledge filter: How entrepreneurship and university–industry relationships drive economic growth. *Research Policy*, **35**(10): 1499–508.

Mullins, J. 2010. *The New Business Road Test: What entrepreneurs and executives should do before writing a business plan*. New York: Prentice-Hall.

Mullins, J. and Komisar, R. 2006. *Getting to Plan B*. Cambridge, MA: Harvard Business School Press.

Nonaka, I. and Von Krogh, G. 2011. Tacit knowledge and knowledge conversion: Controversy and advancement in organizational knowledge creation theory. *Organization Science*, **20**(3): 635–52.
Ozgen, E. and Baron, R. 2007. Social sources of information in opportunity recognition: Effects of mentors, industry networks and professional forums. *Journal of Business Venturing*, **22**(2): 174–92.
Ruef, M., Aldrich, H. and Carter, N. 2003. The structure of founding teams: Homophily, strong ties, and isolation among U.S. entrepreneurs. *American Sociological Review*, **68**(2): 195–225.
Sarasvathy, S.D. 2004. The questions we ask and the questions we care about. *Journal of Business Venturing*, **19**(5): 707–17.
Semin, G., Garrido, M. and Palma, T. 2012. Socially situated cognition: Recasting social cognition as an emergent phenomenon. In S.T. Fiske and N.C. Macrae (eds), *The Sage Handbook of Social Cognition*: 138–64. Thousand Oaks, CA: Sage Publications.
Shane, S. 2000. Prior knowledge and the discovery of entrepreneurial opportunities. *Organization Science*, **11**: 448–69.
Shane, S. and Venkataraman, S. 2000. The promise of entrepreneurship as a field of research. *Academy of Management Review*, **25**(1): 217–26.
Shaver, K. and Scott, L. 1991. Person, process, choice: The psychology of new venture creation. *Entrepreneurship Theory and Practice*, Winter: 23–45.
Shepherd, D.A., and DeTienne, D.R. 2005. Prior knowledge, potential financial reward, and opportunity identification. *Entrepreneurship Theory and Practice*, **29**(1): 91–112.
Smith, E.R. and Semin, G.R. 2007. Situated social cognition. *Current Directions in Psychological Science*, **16**(3): 132–5.
Thorp, H. and Goldstein, B. 2010. *Engines of Innovation: The entrepreneurial university in the twenty-first century*. Chapel Hill, NC: University of North Carolina Press.
Ucbasaran, D., Alsos, G., Westhead, P. and Wright, M. 2008. Habitual entrepreneurs. *Foundations & Trends in Entrepreneurship*, **4**(4): 309–450.
Van Auken, H., Fry, F. and Stephens, P. 2006. The influence of role models on entrepreneurial intentions. *Journal of Developmental Entrepreneurship*, **11**(2): 157–67.
Van de Ven, A. 1993. The development of an infrastructure for entrepreneurship. *Journal of Business Venturing*, **8**(3): 211–30.
Van de Ven, A. 2005. Running in packs to develop knowledge-intensive technologies. *MIS Quarterly*, **29**(2): 365–78.
Van de Ven, A.H. and Garud, R. 1989. A framework for understanding the emergence of new industries. *Research on Technological Innovation Management and Policy*, vol. 4. Greenwich, CT: JAI Press.
Van de Ven, A. and Johnson, P. 2006. Knowledge for theory and practice. *Academy of Management Review*, **31**(4): 802–21.
Van de Ven, A.H., Polley, D.E., Garud, R. and Venkatsaraman, S. 1999. *The Innovation Journey*, New York: Oxford University Press, p. 154.
Von Hippel, E. and Von Krogh, G. 2003. Open source software and the 'private collective' innovation model: Issues for organization science. *Organization Science*, **14**(2): 209–23.
Von Krogh, G., Ichiro, K. and Nonaka, I. 2000. *Enabling Knowledge Creation: How to unlock the mystery of tact knowledge and release of the power of innovation*. Oxford: Oxford University Press.
Wasserman, N. 2012. *The Founder's Dilemmas: Anticipating and avoiding the pitfalls that can sink a startup*. Princeton, NJ: Princeton University Press.
Westhead, P., Ucbasaran, D. and Wright, M. 2005. Decisions, actions and performance: Do novice, serial and portfolio entrepreneurs differ? *Journal of Small Business Management*, **43**(4): 393–417.
Zacharakis, A. and Meyer, G.D. 1998. A lack of insight: Do venture capitalists really understand their own decision process? *Journal of Business Venturing*, **13**(1): 57–76.

11. How language shapes thought: new vistas for entrepreneurship research

Jean S. Clarke and Joep P. Cornelissen

ABSTRACT

Entrepreneurial cognition research, reflecting classic assumptions within cognitive science more generally, has traditionally conceptualized cognition as static, disembodied and individual. In this chapter, we outline a more dynamic, embodied and social perspective of entrepreneurial cognition. This perspective focuses in particular on the formative role of language in shaping the ideas of entrepreneurs and their attempts to gain a broader understanding and recognition of a new venture from relevant stakeholders and resource providers. We describe the theoretical underpinnings of this approach, conceptualize key constructs and outline a number of emergent areas for future research.

INTRODUCTION

Like many fields that have developed from the 'cognitive revolution' of the 1960s and 1970s, the field of entrepreneurial cognition has traditionally relied heavily on theories that describe individual actors as constructing, activating and applying abstract and disembodied symbolic representations (such as schemas, prototypes etc.) (Gibbs, 2006). There is, however, a growing body of work on socially situated cognition, embodied cognitive science and cognitive linguistics that offers a promising alternative allowing us, as entrepreneurship scholars, to conceive of a new cognitive agenda in entrepreneurship (Mitchell et al., 2011). In this chapter we draw on these literatures and elaborate an alternative approach to entrepreneurial cognition, one that highlights the role of language and communication in entrepreneurial cognition.

This approach starts from the position that language, as both a central modality for thought and as *the* medium for social interaction, influences what entrepreneurs are able to cognitively imagine and infer as possible venture opportunities. The consequence of this premise is that ventures are in effect 'created' or 'made' in and through language, as opposed to a disembodied view that sees opportunity identification or creation as resulting from conscious internal cognitive processing. Instead, conscious thoughts about, for example, venture opportunities emerge in the act of speaking, when individuals use language to create and sustain an image of the reality they are facing and through which they make their situations rationally accountable to themselves and others (Cornelissen and Clarke, 2010).

Language is therefore, we argue, best thought of not as the packaged communicative output of a separate 'internal' cognitive process but rather as having a direct and formative role in the construction of meaning. This broad conceptualization of entrepreneurship provides, we suggest, a processual framework for how entrepreneurs talk novel ventures into being for themselves as well as for others whom they seek to convince. With language being typically used in social settings (as opposed to a fully internalized dialogue in one's head), it also presents a linkage between internal, self-conscious sensemaking and social sensegiving, or the conveying of meaning to others, such as potential stakeholders of a venture.

In this way, then, the perspective in this chapter provides not only a more embodied perspective by focusing on the modality of language as an embodied act of conceptualization, but also a perspective that is socially situated, in emphasizing how language use in context may influence others to come on board and support a burgeoning venture. In other words, language matters because it provides the rationale for both the entrepreneur to develop and commercialize a venture as well as for key constituencies to invest in and support the venture, enabling the venture in turn to evolve, grow and persist. In this chapter we summarize and position this perspective on entrepreneurial cognition, and will do so by drawing on and extending our earlier papers (Cornelissen and Clarke, 2010; Clarke and Cornelissen, 2011). We will unfold our ideas at two levels: first we elaborate the dynamic and active interrelation of language and thought, labeled here as 'thinking-for-speaking' (Slobin, 1996); and second, we outline the important role of language as a key mediating mechanism or device in influencing the cognitions of others, such as investors and other prospective stakeholders of a venture.

TRADITIONAL AND CONTEMPORARY VIEWS ON ENTREPRENEURIAL COGNITION

Entrepreneurial cognition has traditionally been conceptualized as an internal cognitive process where entrepreneurs perceptually notice and bracket breaks in their experience, and through repeated experiences then develop richer and more specific cognitive representations and mental models of their environment (Baron and Ensley, 2006; Korunka, Frank, Luegar and Mugler, 2003; Krueger, Reilly and Carsrud, 2000; Mitchell et al., 2000). Such representations and mental models are assumed to represent the world around them, and as such it is assumed that these representations provide entrepreneurs with a basis for cognitive inferences about possible venture opportunities (e.g., Baron and Ensley, 2006). The ability of individuals to make sense of new circumstances in relation to their accumulated experiences then depends on the mental capacity of individuals to draw generalized abstractions across variations in experience and to learn from exceptions, which, in turn, with sufficient regularity, can become the basis for new or updated mental models or representations.

This traditional approach thus suggests that entrepreneurs learn by cognitively detecting regularities in their direct environments and compressing them into much less detailed cognitive representations or mental models, which then guide their perceptions, inferences and behavior. As such, it provides a take on how cognition exhibits regularity in how individuals form thoughts by relating cues to existing cognitive representations or mental models, but it has, using the words of Turner (2001: 145), proved 'nearly useless for explaining how a new schema can arise *before* it is manifest in our regular experience.' In other words, it cannot account directly for how entrepreneurs creatively and inductively infer new venture opportunities, which may in effect break with past experiences and existing mental models (Cornelissen and Clarke, 2010).

A further corollary of the traditional cognitive perspective is that entrepreneurs are cast as cognitive and disembodied agents, who in some senses are seen to dream up ventures outside of a social context and apart from interacting with others. As such, when entrepreneurs label and articulate their ideas to others, it is assumed that they are simply externalizing or expressing 'some neutral, definitive and ready-made sense of events produced through a process such as noticing what the world is like and then putting it into words' (Edwards, 1997: 144). Their language use, accordingly, is seen as revealing cognitive interpretations and inferences, rather than playing a formative role in the construction of novel ideas. In our recent work (Cornelissen and Clarke, 2010) we stress an alternative perspective that considers entrepreneurial cognition as a sensemaking process. Sensemaking combines language and thought in a dynamic and recursive manner that allows us to account for how entrepreneurs make sense of markets or industries, create and design opportunities for a venture and rationalize such opportunities to others.

The perspective that we elaborate is closely allied with the emerging focus in social psychology and entrepreneurship on socially situated cognition (Smith and Semin, 2004; Mitchell et al., 2011). Socially situated cognition offers a direct critique of the classic computational view of cognition and its base assumptions around stable representations in the heads of individuals that are 'offline,' abstract and disembodied. The approach instead advances a research agenda based on four core ideas (Smith and Semin, 2004): (1) Cognition is linked to action and adaptive behaviours of individuals in a social context – as opposed to seeing cognition as automatic, fundamental and separate from any particular context of action. (2) Cognition is embodied. Individuals draw on their bodies, sensorimotor capacities and specific embodied modalities such as language and gestures to make sense of their environments – as opposed to seeing cognition as amodal, abstract and purely mental representations that are 'stored' and 'retrieved' in their heads. (3) Cognition is situated. Individuals think and act in real-world social and physical environments, which implies that cognition is not stable nor simply a property of abstract representations in individuals' heads but is shaped and influenced by such environments and may also evolve as a result of interactions with others. (4) Cognition is distributed spatially, socially and temporally, across places, people and objects (Hutchins, 1995). Many tasks such as setting up a new venture require the collaborative operation of a group, which means that an entrepreneur must persuade others (such as business partners, advisors and investors) to come on board around a shared reality that facilitates the joint effort and coordination required to give a venture a chance of succeeding. In such situations, cognition is distributed and found in collaborative communication rather than in any one single individual's head (Smith and Semin, 2007).

Our approach shares many of the same assumptions, and we subscribe to the overall importance of advancing this research agenda within entrepreneurship (Mitchell et al., 2011). Smith and Semin (2004) offer the perspective of social situated cognition as an umbrella or 'integrative' perspective that is able to bring together and accommodate a range of studies across social and experimental psychology, cognitive science, cognitive linguistics, artificial intelligence and specific areas of anthropology and sociology. We think this exercise of joining structurally disconnected but otherwise coherent areas of research is important, also in terms of advancing a much broader shift in thinking across the social sciences. At the same time, the breadth of this exercise also requires a certain degree of depth and focus around key constructs, such as embodiment or language, that might provide the 'bridge' (Smith and Semin, 2006) to join these various insights and approaches into a coherent whole. One such construct is language, which connects with the four main themes of socially situated cognition, and provides a way of linking the individual entrepreneur to the act of imagining and developing a venture in a specific social environment. In the remainder of this chapter we unfold our theoretical approach to language and communication in entrepreneurship.

THINKING-FOR-SPEAKING, OR HOW LANGUAGE SHAPES ENTREPRENEURIAL COGNITION

The starting point for our theoretical approach is that language is not just a code for communication or simply an outward representation of thought, but is inseparably involved with processes of thinking and reasoning (Cornelissen, 2013; Cornelissen and Clarke, 2010). In other words, when an entrepreneur makes sense of a novel venture to himself and others, it 'is as much a matter of thinking that is acted out conversationally in the world as it is a matter of knowledge and technique applied to the world' (Weick et al., 2005: 412). We therefore stress the importance of embedding entrepreneurs in a social context, and recognizing the role of that social environment in the creation and justification of opportunities for ventures. Consistent with this approach, we argued in an earlier paper that, 'whilst the inner thoughts and imaginations of entrepreneurs matter, they are not spoken or even necessarily speakable,' and we should therefore direct our gaze, as researchers, 'to the point where ideas take form in the stream of an individual's experience with external speech reconfiguring ideas to fit the demands of spoken language' (Cornelissen and Clarke, 2010: 542). The linguist Slobin (1996) labels this 'thinking-for-speaking,' which refers to how individuals organize their thinking to meet the demands of linguistic encoding online, during acts of speaking with others. The term 'thinking-for-speaking' turns a familiar pairing of terms – thought and language – into a more dynamic expression, drawing attention to the kinds of mental processes that occur during the act of formulating speech. As Slobin notes (1996: 71), 'whatever else language may do in human thought and action, it surely directs us to attend – while speaking – to the dimensions of experience that are enshrined in grammatical categories.' Within this process, thought and language are intimately and dynamically connected at the point where individuals verbalize their ideas, and, while communicating, articulate them in their speech to others (Cornelissen and Clarke, 2010).

This view has recently gained a great deal of currency across the social sciences, including in research on embodied and socially situated cognition. It in effect harks back to the famous Sapir–Whorff hypothesis, the basic idea that the semantic structure of a language shapes or limits the ways in which a speaker forms conceptions of the world. While strong versions of the hypothesis have been refuted, there is a wealth of evidence in cognitive science and cognitive linguistics that language may not directly determine what gets thought (in a one-on-one fashion) but it shapes what gets thought and inferred by individuals in a social setting (e.g., Boroditsky, 2011). In the context of entrepreneurship, one important consequence of this interrelation is that, although basic perceptions and thoughts of entrepreneurs may exist outside of language, language use is formative in the scoping and articulation of an incipient idea (see, e.g., Pinker, 2010). And since entrepreneurs 'operate at the edge of what they do not know' (Hill and Levenhagen, 1995: 1057), they must seek to make equivocal events non-equivocal by constructing a new 'vision' of the business environment. Specifically, we argued in our earlier work that they will use either analogical or metaphorical reasoning to inductively produce an understanding for a novel venture opportunity (Cornelissen and Clarke, 2010). In both instances, specific experiences or observations are discursively extended and generalized to a new situation. Analogies are discursively produced extensions of prior entrepreneurial experiences and observations of industries to a novel venture situation that familiarizes the new situation, reduces uncertainty and supports further inferences (e.g., Ward, 2004). Metaphors are discursively produced extensions of experiences and cultural frames outside of a specific entrepreneurial or professional context (and, hence, are not literal, analogical extensions of past experiences or observations) that similarly allow entrepreneurs to make sense of new or unfamiliar situations and produce links to action by virtue of the inferences for action that they evoke (Cornelissen and Clarke, 2010).

What the use of analogies or metaphors signifies is that language provides a tool, both linguistically and cognitively, to pair, rearrange and recombine words and grammatical constructions. Given that words and grammatical constructions will reference certain basic cognitive categories, such pairings and combinations may present a novel conceptual image or representation of a venture opportunity (Cornelissen and Clarke, 2010). Outside of entrepreneurship, people in general have also been found to activate domains of knowledge related to their embodied experiences (i.e., human motor actions in the sense of bodily movements, the bodily manipulation of physical objects and the felt experience of bodily engagement with objects) as an inductive basis for reasoning about abstract ideas such as future states or unfamiliar situations (e.g., Gibbs, 2006; Grady, 1997; Lakoff and Johnson, 1999). Such grounding in embodied experiences provides inductive reasoning with a 'human scale' in the sense that it acquires a 'direct and experiential basis' (Gibbs, 2006: 117) and can as such be easily understood (Lakoff and Johnson, 1999).

As a tool, language may thus be an important device that entrepreneurs use to direct their thinking processes about a new idea and the venture that they develop. It may give them a sense of control and a 'grasp' of what the venture may be all about, which then in effect allows them to regulate their thoughts and emotions, and motivate themselves to make the venture a reality (see also Carsrud and Brännback, Chapter 2, this volume; Randolph-Seng, Williams and Hayek, Chapter 4, this volume). In this way, then, language may not be 'revealing of' deeper or inner thoughts, as traditional accounts of entrepreneurial cognition assume, but it may directly mediate thoughts and emotions. It may, as mentioned, be the direct trigger, or cause, for a new idea. The evolving language of an entrepreneur also gives them a means to structure and control their thinking and to motivate themselves, with such language also harboring a degree of commitment and in some instances fostering the ever-stronger conviction that the venture idea is worth pursuing (Cornelissen, 2013).

What this means is that language in general does not simply express previously formed mental concepts (Cornelissen and Clarke, 2010). Indeed, in this more general sense, bodily practices, including language, social interaction and gesturing, are part and parcel of the very activity in which concepts and conceptualizations are formed (e.g., Gibbs, 2006; Cornelissen et al., 2012). As Lakoff and Johnson (1999: 19) claim, 'the peculiar nature of our bodies shapes our very possibilities for conceptualization and categorization.' When this realization is extended to the domain of entrepreneurship, it offers the opportunity to move away from a more static information-processing or computational model of cognition and toward an explanation of entrepreneurial cognition as ultimately linked to embodied actions (Varela et al., 1991; see also Sarasvathy, Chapter 14, this volume).

THE SOCIAL DIMENSION: LANGUAGE AND COMMUNICATION ABOUT NEW VENTURES

When entrepreneurs create a novel venture, they do not simply imagine or design a venture in isolation but do so in interaction with relevant others (employees, investors) whose acceptance, support and resources are critical to the success of the venture. In other words, there is a close connection between inductive thought and language in entrepreneurial thinking-for-speaking and communication about novel ventures. In the early stages of entrepreneurial venture formation, for example, entrepreneurs will draw on their available knowledge and experiences to make sense of the novel venture, but will also begin to construct stories and accounts that coherently address questions about who they are, why they are qualified, what they want to do and why they think they will succeed. Again, analogies and metaphors play a useful role here in that they encode and articulate novel ideas that entrepreneurs may have in terms of words and larger frames that, when well chosen, are already largely familiar to listeners or recipients (Cornelissen and Clarke, 2010).

In this respect, entrepreneurs also often use embodied metaphors around physical movement or the physical manipulation of ventures as objects (e.g., 'leverage client base' 'build market awareness' 'expand market share' 'acquire market acceptance') (Martens et al., 2007: 1118) in their accounts toward (prospective) investors at the time of their initial public offering. These metaphors, by themselves or as part of a larger story, cue an image of an (embodied) operator who physically manipulates a venture and hence can control and direct the likelihood of its success. Such metaphorical scenes are 'structured by a SOURCE–PATH–GOAL schema in the time domain' whereby the initial state is characterized as a source, the final state as a destination or goal and the events as locations on a path towards the destination (Lakoff, 1987: 285–6). The embodied nature of these metaphors connects with, and hence reinforces, images of entrepreneurial agency and control for potential investors.

Another important aspect of this process is that language, as part of communication in context, presents an important mediating mechanism between individuals, and specifically in terms of how they understand each other's intentions as a situation unfolds. In contexts of communication and social interaction, individuals need to be able to model each other's evolving and contingent intentions and goals, based on perceptible cues, but also primarily by attending to one another's communicative actions and displays (e.g., Sperber and Wilson, 1995). As we cannot read each other's minds directly, this means that 'no mind can influence another except via mediating structure' (Hutchins and Hazlehurst, 1995: 4). Externalized speech, once spoken, features as an important mediating mechanism in that it allows individuals to attribute intentions and thoughts to one another, without actually being privy to what someone is effectively thinking. Individuals, while communicating, also do not need to share or even have access to the same knowledge about the venture. Instead, in ongoing processes of communication, individuals generally exploit the built-up 'common ground' between them as a resource for understanding and for deriving pragmatic inferences (Clark, 1996). As an entrepreneur gets feedback from stakeholders who display whether they understand and accept the intentions for the venture, it allows them to validate that understanding or to correct it, and this in turn has consequences for the way in which the understanding of the venture evolves.

Baker et al.'s (2003: 264) study of entrepreneurial start-ups provides an example of this process of design through, and as a result of, communication. Their study shows that the creation and implementation of new start-ups was initially and to a degree 'psychological or driven by internal needs' but also significantly 'driven by exogenous demands by external resource providers for founders to provide accounts that make their firms appear like legitimate investment opportunities, suppliers or customers.' The result was an improvisational process with the design and implementation of these ventures emerging at least in part from verbal interactions with resource providers. Baker et al. (2003) illustrate this process in their observation that, in interactions with employees, entrepreneurs added analogies or metaphors (e.g., of the venture organization as a 'family') that they had 'made up on the fly to make their fledgling firms seem comfortable and normal – that is, legitimate – to potential employees.' However, such social constructions 'became part of employee expectations and the emerging culture of the organization after the people [had] joined the firm' (Baker et al., 2003: 263).

One consequence of this is that the conceptual pacts that are agreed on by entrepreneurs and stakeholders, in this case seeing the culture of a novel venture as a family, have a strong claim on mutual understanding and any subsequent interactions. A fruitful avenue for further empirical research would be to explore how, as a result of verbal interactions between entrepreneurs and stakeholders, common ground is built up, which in turn provides an agreement about the scope and design of the venture. How common ground accumulates and binds the participants to a course of action has been hinted at in prior research (Baker et al., 2003) but has so far not yet been systematically explored. Common ground, or mutual knowledge, is the set of presuppositions that entrepreneurs and relevant others, as a result of their ongoing sensemaking and interaction with others, take to be true – and believe their interaction partners also take to be true (Clark, 1996).

Entrepreneurs obviously face the challenge of building up common ground with other stakeholders and resource providers. Such common ground is, however, essential as it provides the basis for common understanding, and facilitates coordination and interaction. Common ground is built up gradually, as a result of interaction between the entrepreneur and others as they together establish, revise and maintain a common referential understanding, which itself is mutually calibrated at each step of an interaction's progression (Clark, 1996). Common ground around key labels or descriptions of the venture also affords economy of expression in that the greater the common ground between an entrepreneur and stakeholders and resource providers, the less effort it takes to make themselves understood. In essence, what this means is that, first of all, entrepreneurs need to tune their language to the understanding of others in social contexts, as a basis for common ground to be enabled and built up. This process of perspective taking is often known as audience design (Smith and Semin, 2004), the process of constructing an utterance and using specific forms of language to reflect the speaker's beliefs about the listener's knowledge (Clark, 1996). The greater the ability and effort of an entrepreneur to put themselves in the shoes of others, the more likely it is that they will be able to make themselves understood. Of course, attempts at such perspective taking are not always successful. Part of this is due to the tendency of entrepreneurs, as well as people in general, to overestimate the transparency of their intentions. Entrepreneurs may also overestimate how well they communicate or may simply be too persuaded of their own ideas to feel that they have to adapt their utterances for a specific audience.

A second point is that, providing entrepreneurs are sufficiently responsive to their audience, we would expect common ground to be built up gradually and in time to contract around key terms or phrases that describe the venture and its essential features. At the start, entrepreneurs and the stakeholders or resource providers that they are speaking to are likely to use more complex phrases, iterate between various utterances, and their language may also demonstrate disfluencies and aborted descriptions of the venture (cf. Clark, 1996). They may also explicitly mark uncertainty in their descriptions using hedges ('sort of') or by regular interruptions or pauses in speech. Over the course of repeated interactions, however, their descriptions gradually converge around key labels and phrases, speech becomes more fluid and as a result the entrepreneur and their stakeholders come to share understanding. They also become more certain that they both share the conceptualization of the venture, as the key referent, with their descriptions (which are in reality labels rather than full-scale descriptions) becoming shorter and more iconic. In other words, there are thus systematic relationships that we would expect to observe between the expansion of common ground between entrepreneurs and stakeholders, and linguistic features of their communication.

The process of building up common ground is, we think, a useful analytical framework for further research on entrepreneurial cognition. It starts with a recognition of the social contexts in which cognition and communication take place. It also highlights the evolving nature of language and communication, with initial utterances seeding the context for the next, and with language and thought being dynamically interrelated. In this respect, a particular utterance may, for example, trigger a thought in a business advisor and his response in turn may affect the thinking and subsequent word choice of an entrepreneur. Common ground also usefully highlights how, in repeatedly referring in the same way to a venture, entrepreneurs and stakeholders may create conceptual pacts about the essential way of describing the venture. These conceptual pacts may be created as a result of lengthy and extensive step-by-step interactions leading to an increasingly refined common understanding. The process may also be shortened considerably, we think, when entrepreneurs use well-chosen analogies or metaphors, as part of their language, that may ease and facilitate understanding and form a basis for joint commitments and actions (Cornelissen and Clarke, 2010). In effect, such analogies or metaphors may then feature as a boundary object that the different actors relate to and that provides a basis for coordinated actions in pursuit of common goals. The notion of common ground furthermore raises intriguing questions as to whether its accumulation may make it hard to abort a chosen path or trajectory for a venture even when it fails to gain traction in the marketplace. If so, this would suggest that a key capability of entrepreneurs is to be particularly strong on perspective taking, so that they are able to reframe and rethink their venture as a result of interactions with stakeholders and in a way that keeps the path for a venture at least initially open and ambiguous and without yet any specific or strong commitments.

NEW VISTAS FOR RESEARCH

The approach to entrepreneurial cognition offers a fruitful alternative to existing entrepreneurial cognition research in that it sees language not simply as an extension or representation of past cognitive experiences but as actually formative of thought. Hence it is also a resource that individual entrepreneurs use to create or produce understanding and that they adapt to local circumstances and changing conditions (Edwards, 1997). This approach incorporates both psychological and social elements because, when entrepreneurs 'make sense' of their ideas they simultaneously construct accounts that help 'give sense' and justify a new venture to relevant others (Cornelissen and Clarke, 2010). Consequently, when entrepreneurs create a novel venture, they do not simply imagine or design a venture in isolation (i.e., outside of a social context) but do so in interaction with relevant others (employees, investors) whose acceptance, support and resources are critical to the success of the venture. The theoretical and empirical advantage of a shift from abstracted cognitive models or socially shared language to specific discursive formulations is that it retains, but reworks, the explanatory power of prior cognitive experience and socially accepted and legitimate frames as elements that entrepreneurs discursively draw on to make sense of the world, while managing to deal with the specific detail, contingency and functionality of how they do so on particular occasions. Centering on the role of language and communication and its interdependencies with individual and group cognition is one of the fundamental aspects of current psychological and psycholinguistic research on socially situated and embodied cognition (e.g., Gibbs, 2006; Pinker, 2010). Thus, in erecting and advancing a socially situated and embodied cognition agenda in the field of entrepreneurial cognition, we believe that one of the most fundamental and promising lines of inquiry concerns a focus on language and communication-based mechanisms that enable and support the creation of individual and joint understanding.

This perspective offers a number of promising directions for research, first and foremost the importance of an empirical examination of speech and communication in entrepreneurship research. Future research attention should be directed to specific features of entrepreneurs' use of language, such as metaphor and analogies in contexts of natural interaction where individuals are trying to give sense and direct the interpretations of others regarding a new venture. Within the field of cognitive linguistics, procedures are available that allow researchers to move in a systematic and grounded way from the identification of metaphors to interpretations of the use of those metaphors and thereby place their claims about the significance and use of metaphorical language and thought in sensegiving on a firm and explicit empirical footing (e.g., Bowdle and Gentner, 2005; Pragglejaz Group, 2007; Goldberg, 1995). Another area for further research involves a detailed examination of language-based interactions between entrepreneurs and stakeholders with a view to seeing how common ground is built up. Here, researchers may draw directly on the analytical tools developed by Clark and colleagues (Clark, 1996) such as a focus on utterances and descriptions, as the main unit of analysis, in an attempt to see how such descriptions may be retained or changed over time (as in the Baker et al. (2003) study). The phenomenon that entrepreneurship researchers may be particularly focused on in such analyses is the linguistic notion of lexical entrainment, by which entrepreneurs and stakeholders over time reuse the same expressions in an increasingly laconic form. Depending on the context, this could be an analogy or metaphor, idiom, or a compact (and therefore iconic) description of the venture. We also believe that there is merit in detailed studies of the bodily actions that entrepreneurs produce alongside their speech in acts of sensegiving, particularly in the early stages of the venture creation process. Based on recent insights from the field of embodied cognition, bodily practices, especially hand gestures, are an integral part of speech and communication and they play a key role in the formation of understanding of both speaker and perceiver (e.g., Gibbs, 2006; Cornelissen et al., 2012). This research agenda can be readily connected to techniques from cognitive linguistics and embodied cognition that offer rigorous procedures for the identification and analysis of bodily movements and gesture and their use in the context of entrepreneurial sensemaking (McNeill, 2005; Kendon, 2004; Gibbs, 2006).

The suggestions we have made so far all relate to methods and research topics at a micro-level of analysis. Such analyses are not without their challenges. For one, such methods require that researchers gain access and immerse themselves in a context, so that they are better able to infer from words and expressions, as well as from other observations (such as symbolic gestures), the basic meanings that are being conveyed and agreed upon by entrepreneurs and their stakeholders. This challenge is similar to how, in general, qualitative researchers make inferences about processes of sensemaking on the basis of an extended engagement in the field. Another challenge involves the use of protocols from linguistics, or at least a basic familiarity with some linguistic concepts and terms. As many entrepreneurship researchers are not trained linguists (and we ourselves are not), this may at first glance seem difficult and potentially even deter some from going in this direction. Yet many analytical protocols such as the one mentioned for identifying and analyzing metaphors are quite accessible, and are increasingly becoming mainstreamed as part of the broader research agenda on socially situated cognition across the social sciences.

Besides analyses at a micro-level, we believe that there are also significant opportunities to extend the ideas and arguments presented in this chapter into more macro-level analyses. Examples are studies of the effect of entrepreneurial narratives (Martens et al., 2007) and of gesture usage and bodily movements (Chen et al., 2009) on investor persuasion. As demonstrated by these studies, a key challenge for such research is to find valid and robust measures of language or gesture usage. In both cases the measures were quite abstract and contained many more specific linguistic categories. For example, the Chen et al. (2009) study focused on a composite factor of bodily movements, including gestures, as conveying an entrepreneur's passion during his or her pitch to investors. Yet there are likely to be significant differences at the level of what gestures are being used by an entrepreneur, ranging from simply beats or marking gestures (repetitive hand movements) to symbolic or iconic gestures (that symbolize the basic idea). By the same token there may be differences in the persuasiveness of a pitch depending on the type of gesture used as well as the synchronicity with verbal speech. As in this specific example, there is real potential in importing measures from areas such as cognitive science and linguistics (McNeill, 2005) to study aspects of the entrepreneurial process and their influence on the overall success of a venture in larger samples of entrepreneurs.

In closing, in this chapter we have outlined a theoretical perspective on entrepreneurship that recasts traditional understandings of entrepreneurial cognition. We have described the basic contours of this perspective and have highlighted key questions and some of the emerging areas of research associated with the overall approach. In doing this, we have also suggested a number of ways in which researchers interested in entrepreneurial cognition may help advance what we think is an important research agenda for our field.

REFERENCES

Baker, T., Miner, A.S. and Eesley, D.T. 2003. Improvising firms: Bricolage, account giving and improvisational competencies in the founding process. *Research Policy*, **32**(2): 255–76.

Baron, R.A. and Ensley, M.D. 2006. Opportunity recognition as the detection of meaningful patterns: Evidence from comparisons of novice and experienced entrepreneurs. *Management Science*, **52**(9): 1331–44.

Boroditsky, L. 2011. How language shapes thought. *Scientific American*, February: 62–5.

Bowdle, B.F. and Gentner, G. 2005. The career of metaphor. *Psychological Review*, **112**(1): 193–216.

Chen, X.P., Yao, X. and Kotha, S. 2009. Entrepreneur passion and preparedness in business plan presentations: A persuasion analysis of venture capitalists' funding decisions. *Academy of Management Journal*, **52**(1): 199–214.

Clark, H.H. 1996. *Using Language*. Cambridge: Cambridge University Press.

Clarke, J.S. and Cornelissen, J.P. 2011. Language, communication and socially situated cognition in entrepreneurship. *Academy of Management Review*, **36**(4): 776–8.

Cornelissen, J.P. 2013. Portrait of an entrepreneur: Vincent van Gogh, Steve Jobs, and the entrepreneurial imagination. *Academy of Management Review*, **38**(4): 700–709.

Cornelissen, J.P. and Clarke, J.S. 2010. Imaging and rationalizing opportunities: Inductive reasoning and the creation and justification of new ventures. *Academy of Management Review*, **35**(4): 539–57.

Cornelissen, J., Clarke, J.S. and Cienki, A. 2012. Sensegiving in entrepreneurial contexts: The use of metaphors in speech and gesture to gain and sustain support for novel ventures. *International Small Business Journal*, **30**(3): 213–41.

Edwards, D. 1997. *Discourse and Cognition*. London: Sage.

Gibbs, R.W. Jr. 2006. *Embodiment and Cognitive Science*. Cambridge: Cambridge University Press.

Goldberg, A.E. 1995. *Constructions: A construction grammar approach to argument structure*. Chicago, IL: University of Chicago Press.

Grady, J. 1997. Foundations of meaning: Primary metaphors and primary scenes. Unpublished doctoral dissertation, University of California, Berkeley.

Hill, R. and Levenhagen, M. 1995. Metaphors and mental models: Sensemaking and sensegiving in innovative and entrepreneurial activities. *Journal of Management*, **21**(6): 1057–74.

Hutchins, E. 1995. *Cognition in the Wild*. Cambridge, MA: MIT Press.

Hutchins, E. and Hazlehurst, B. 1995. How to invent a shared lexicon: The emergence of shared form–meaning mappings in interaction. In E. Goody (ed.), *Social Intelligence and Interaction: Expressions and implications of the social bias in human intelligence*: 53–67. Cambridge, UK: Cambridge University Press.

Kendon, A. 2004. *Gesture: Visible action as utterance*. Cambridge, UK: Cambridge University Press.

Korunka, C., Frank, H., Luegar, M. and Mugler, J. 2003. The entrepreneurial personality in the context of resources, environment, and the startup process: A configurational approach. *Entrepreneurship Theory and Practice*, **28**(1): 23–42.

Krueger, N.F., Reilly, M.D. and Carsrud, A.L. 2000. Competing models of entrepreneurial intentions. *Journal of Business Venturing*, **15**(5/6): 411–32.

Lakoff, G. 1987. *Women, Fire, and Dangerous Things: What categories reveal about the mind*. Chicago, IL: University of Chicago Press.

Lakoff, G. and Johnson, M. 1999. *Philosophy in the Flesh: The embodied mind and its challenge to Western thought*. New York: Basic Books.

Martens, M.L., Jennings, J.E. and Jennings, P.D. 2007. Do the stories they tell get them the money they need? The role of entrepreneurial narratives in resource acquisition. *Academy of Management Journal*, **50**(5): 1107–32.

Mitchell, R.K., Randolph-Seng, B. and Mitchell, J.R. 2011. Socially situated cognition: Imagining new opportunities for entrepreneurship research (Dialogue). *Academy of Management Review*, **36**(4): 774–8.

Mitchell, R.K., Smith, J.B., Seawright, K.W. and Morse, E.A. 2000. Cross-cultural cognitions and the venture creation decision. *Academy of Management Journal*, **43**(5): 974–93.

McNeill, D. 2005. *Gesture and Thought*. Chicago, IL: University of Chicago Press.

Pinker, S. 2010. The cognitive niche: Coevolution of intelligence, sociality, and language. *Proceedings of the National Academy of Sciences*, **107**(Supplement 2): 8993–9.

Pragglejaz Group. 2007. MIP: A method for identifying metaphorically used words in discourse. *Metaphor and Symbol*, **22**(1): 1–39.

Slobin, D. 1996. From 'thought and language' to 'thinking for speaking'. In J.J. Gumperz and S.C. Levinson (eds), *Rethinking Linguistic Relativity*: 70–96. Cambridge, UK: Cambridge University Press.

Smith, E.R. and Semin, G.R. 2004. Socially situated cognition: Cognition in its social context. *Advances in Experimental Social Psychology*, **36**: 53–117.

Smith, E.R. and Semin, G.R. 2006. Socially situated cognition as a bridge. In P.A.M. van Lange (ed.), *Bridging Social Psychology: Benefits of trans-disciplinary approaches*: 145–50. Mahwah, NJ: Laurence Erlbaum.

Smith, E.R. and Semin, G.R. 2007. Situated social cognition. *Current Directions in Psychological Science*, **16**(3): 132–5.

Sperber, D. and Wilson, D. 1995. *Relevance: Communication and cognition* (2nd edn). Oxford/Cambridge: Blackwell Publishers.

Turner, M. 2001. *Cognitive Dimensions of Social Science*. Oxford: Oxford University Press.

Varela, F., Thompson, E. and Rosch, E. 1991. *The Embodied Mind: Cognitive science and human experience*. Cambridge, MA: MIT Press.

Ward, T.B. 2004. Cognition, creativity, and entrepreneurship. *Journal of Business Venturing*, **19**(2): 173–88.

Weick, K.E., Sutcliffe, K.M. and Obstfeld, D. 2005. Organizing and the process of sensemaking. *Organization Science*, **16**(4): 409–21.

12. Thinking big from the start: entrepreneurial growth cognitions

Andrew C. Corbett

ABSTRACT

It is argued that the predominant cognitive perspective among entrepreneurs and related stakeholders is to focus on start-ups and that this mental representation affects the development of the high- and rapid-growth ventures that policy makers hope for their programs and regions. With a desire in most localities to drive economic growth, we need to think differently about venture growth at inception. An argument is made here for a research agenda that focuses on cognition as one way to make this happen. This chapter develops the concept of 'entrepreneurial growth cognitions' – the mental representations of how we develop rapid-, big-growth-oriented firms right from the start – as a mechanism for stimulating scholars to investigate this nascent and fruitful area of research. The chapter shows why this line of inquiry is important for both scholars and practitioners, and provides examples of methodological approaches, sample designs and research questions.

INTRODUCTION

If you had to choose between apportioning more of society's means toward (1) giving birth to more babies or (2) raising smart, independent, well-adjusted children, which would you choose? This is the question and memorable analogy my colleague Dan Isenberg raises while urging policy makers and other stakeholders across the globe to refocus their energies away from start-ups and toward growth (Isenberg, 2012). Whereas it is clear that the birth of new ventures is necessary, there appears to be an overemphasis by economic development agents in cities and countries across the globe on 'starting up businesses' with little accompanying focus on developing the high-growth businesses that deliver the economic benefits policy makers and government officials desire. This is despite evidence that suggests that fewer high-growth starts are more beneficial to society than hundreds of small operations that might employ the same amount of people as the few large ones (Isenberg, 2012).

While Isenberg and others (Grégoire, Corbett and McMullen, 2011) implore scholars and practitioners to consider how we think about entrepreneurship and growth, the extant literature shows that research examining new ventures growth and cognition is virtually non-existent. In fact Grégoire et al. show that very few articles within the entrepreneurial cognition domain examine growth as a dependent variable. It is argued here that there is a perfect storm of forces that appear to be dampening the development of high-growth starts around the globe: (1) scholars are neglecting the intersection of entrepreneurial cognition and growth; (2) policy makers are over-focused on 'starts' as opposed to growth; thus (3) entrepreneurs are starting many firms with low growth potential.

Part of the goal of research is to build new knowledge that both scholars and practitioners can effectively use in their field of endeavor. As such, the aim of this chapter is to create a pathway for new thinking about the nexus of research between entrepreneurial cognition and growth. Focusing on start-ups is fine but incomplete. As noted recently in *The Economist*, technology and lessening of other barriers has made launching new businesses 'fairly easy,' but getting them to thrive, grow and scale up is not (Siegele, 2014). The overarching point is that we appear to know quite well how to start businesses. I am not advocating for fewer small start-ups but instead want to highlight the need to focus on how we can also develop larger, growth-oriented business. Practitioners (both entrepreneurs and policy makers) need to think differently about growth from the start, and academics need to assist them by exploring different avenues of research within this domain. This chapter will bring together these two disparate subfields of entrepreneurship (entrepreneurial cognition and venture growth) to build an agenda for research on entrepreneurial growth cognitions.

The chapter continues with a brief discussion of the literature on entrepreneurial growth and then highlights the paucity of research that crosses entrepreneurial cognition and growth. An argument for the need to change the scripts or mental models of entrepreneurs and others away from 'starting' to 'growth' supports the thesis that all stakeholders need to think not simply about growth – but growing rapidly and big – from the inception of the venture. Those involved with economic development want growth, but not the type of growth generally referred to in the 'new ventures growth literature.' The chapter concludes by offering suggestions for future research questions and methodological designs for entrepreneurial growth cognitions. Economies around the globe need more high-growth ventures and an important step in that direction requires a change in thinking at the start-up stage. This change in thinking needs to begin with scholarly inquiry into entrepreneurial growth cognitions.

NEW VENTURE GROWTH

The literature on new venture growth is vast, but concerns about a lack of understanding or development of the concept are just as great (Gilbert, McDougall and Audretsch, 2006; McKelvie and Wiklund, 2010; Wiklund, Patzelt and Shepherd, 2009), and, as noted above, there is little work that crosses over to cognition (Grégoire et al., 2011). Despite great interest from a variety of stakeholders – researchers, government officials, entrepreneurs and other service providers who support new venture development – we still have not advanced our understanding of growth very far (Leitch, Hill and Neergaard, 2010). Conceptual development has not progressed (Wiklund et al., 2009), and this stagnation may be due to an over emphasis on growth outcomes as opposed to understanding the process. McKelvie and Wiklund (2010) explain that we need to better understand how firms grow as opposed to just measuring how much they grow.

Gilbert at al. (2006) first raised the concern of 'how' in their extensive review of new venture growth literature. After exploring the canon of work on why ventures grow, these authors argue for more research that examines the 'how' and the 'why' of venture growth. Gilbert and her colleagues suggest that we need to better understand whether this growth comes from internal or external means, and also address a need to parse out growth that comes from domestic versus international markets. Wiklund has been arguing for a process approach for some time (Wiklund and Shepherd, 2003; Wiklund et al., 2009; McKelvie and Wiklund, 2010), demonstrating the importance of the growth process and its underlying components (resources, opportunities etc.). Despite this, he and his colleagues still conclude that we are in the nascent stage of understanding entrepreneurial growth (McKelvie and Wiklund, 2010).

Leitch et al. (2010) posit that new venture growth will continue to struggle theoretically until we (1) close the gap between practitioners and researchers regarding what constitutes growth, (2) further investigate growth as a process, and (3) encourage researchers to examine new perspectives on the phenomenon. Achtenhagen, Naldi and Melin (2010) argue that researchers and policy makers need a common, explicit understanding of what they mean by the term growth. Part of this lack of understanding appears to evolve from the point that the new venture growth literature is – rightly – focused on the firm level, while policy makers are generally more concerned about economic growth on a broader level. This is where the fundamental misunderstanding comes from: new venture growth researchers focus on the growth outcomes, growth process, and the outcomes of growth for individual firms (McKelvie and Wiklund, 2010), but growing more small firms does not bring what policy makers desire (Isenberg and Brown, 2014). The job creation and regional growth that policy makers yearn for does not come from incubating more small firms. In fact, there is no evidence to show that growing more new ventures leads to economic growth (Shane, 2009). It appears to be just the opposite, with research showing that a few firms that grow fast from the start account for job creation and economic growth (Isenberg and Brown, 2014). So, part of the disconnect is that researchers and policy makers need to be clear about what they mean when they say 'growth.' While a focus on cognition will not alleviate all of the current concerns within the new ventures growth literature, it can contribute to bringing each of these stakeholders closer to a common understanding. A scholarly research focus on the crossroads of cognition and new venture growth can bring researchers and policy makers together to develop an understanding of what is needed to build an ecosystem that grows some ventures fast and big from the start to augment the many other small-growth ventures.

Cognition and Growth

There is a paucity of research that includes models with growth and some cognitive element (see also Sarasvathy, Chapter 14, this volume). As such, foundational understanding of issues such as whether entrepreneurs think differently about growth or environmental conditions drive entrepreneurs to think and act with growth intentions (Baron, 1998, 2000) still eludes us. Grégoire et al. (2011) illustrated this dearth of work when their review of 30+ years of entrepreneurial cognition research showed only ten works that combined cognition and growth.

There was little diversity in these few works in the 30-year study as nearly all of these articles had growth as their dependent variable. One set of papers within this group of ten articles examines the effect that some individual-level cognition orientation (managerial versus entrepreneurial heuristic; rational versus intuitive thinking; local versus a global mindset) has on growth or the ability to stimulate growth (Sadler-Smith, 2004; Wright, Hoskisson, Busenitz and Dial, 2000). A few others examine cognitive dimensions of social capital and its relationship to the growth of high-technology, science-based or international new ventures (Frese, Krauss, Keith, Escher, Grabarkiewics, Luneng, Heers, Unger and Friedich, 2007; Partanen, Moller, Weterlund, Rajala and Rajala, 2008; Presutti, Boari and Fratocchi, 2007). Only two moved away from the primary direct relationship of an orientation or thinking style and growth by also incorporating the impact of an environmental or competitive condition (Dutta and Thornhill, 2008; Lau and Busenitz, 2001).

While this examination of growth as an outcome of a few cognitive mechanisms has nudged the field forward, we need to think more broadly about growth if the desired outcome is policy to drive economic growth (Shane, 2009) and to think about and develop more high-growth starts from inception (Isenberg, 2012). Recall, it appears that part of the reason why policy makers focus on start-ups as opposed to growth is that there is a lack of understanding of what growth means (Achtenhagen et al., 2010). The argument here is that, beyond growing new ventures, all stakeholders need complementary 'entrepreneurial growth cognitions': the mental representations of how we develop rapid-, big-growth-oriented firms right from the start. If not, policy makers will continue to end up with a great many small firms instead of the big job-creating ones that make an impact (Isenberg, 2012). We have become proficient at starting up new businesses, but not so skilled at developing the high-growth ventures (Siegele, 2014) necessary to anchor the ecosystems that policy makers desire in their individual economic regions.

ENTREPRENEURIAL GROWTH COGNITIONS

To drive this desired growth, researchers need to expand the thinking of new venture growth as decision making of entrepreneurs about how and why they grow (Gilbert et al., 2006) to all forms of cognition. Entrepreneurial cognition is action-oriented, embodied, situated and distributed (Mitchell, Randolph-Seng and Mitchell, 2011), and the call here is for researchers to develop a line of inquiry into entrepreneurial growth cognition to both expand our understanding of the phenomenon and help all stakeholders (entrepreneurs, government officials, investors and other policy makers) drive the growth they desire. We need to examine (1) the manner in which we mentally represent the action we need to take for growth (action-oriented); (2) how our actions shape future cognitions about growth (embodied); (3) what influence the environment and context have on both thoughts and opportunities for growth (situated); and (4) how growth cognitions are exchanged, shared, developed or co-created between actors in the entrepreneurship ecosystem (distributed).

One place to start could be to simply examine the scripts of entrepreneurs. Scripts are mental roadmaps developed from experience to make decisions (Abelson, 1981). Over time, individuals develop domain-specific knowledge structures or mental schemas (Lord and Maher, 1990) that develop an ability for performance due to continued mapping of ordered mental steps pertinent to a particular action, activity or field of interest (Read, 1987). Exporting the concept of scripts to the entrepreneurship literature is not unprecedented as they have been used to show the stability of mental models of entrepreneurs across culture (Mitchell, Smith, Seawright and Morse, 2000), how they influence corporate entrepreneurs (Corbett and Hmieleski, 2007), and their relationship to entrepreneurial failure (Corbett, Neck and DeTienne, 2007).

However, bringing the concept of scripts to bear on fast-growth starts could help us clarify the confusion between researchers and practitioners (Achtenhagen et al., 2010) as well the need to drive more high-growth ventures (Isenberg, 2012).

Beyond scripts, there is a wide array of opportunities for researchers to examine the nexus of entrepreneurial cognition and growth. As Grégoire et al. (2011) note, there is room for research that goes beyond the current entrepreneurship cognition literature, which appears overly focused on consequences (the implications of the effect of a cognitive variable: if you think a certain way, you will achieve some entrepreneurial effect). Within the domain of entrepreneurial growth cognition there is room to examine not just consequences, but antecedents and representations (the characteristics of a cognitive variable). As such, growth research that uses cognitive variables across teams, firms, industries, cities, nations and other contexts is also likely to move scholarship forward.

AN OUTLINE FOR ENTREPRENEURIAL GROWTH COGNITION RESEARCH

Developing an agenda for entrepreneurial growth cognition begins with a foundational understanding of the activities required by entrepreneurs who want to grow their ventures. In order to stimulate scholars to generate their own line of inquiry, I first outline below all of the divergent tasks that a 'growth entrepreneur' needs to do to scale the business. This is followed by a sampling of ideas about how to design a study to examine entrepreneurial growth cognition. Finally, building on prior work with colleagues on entrepreneurship cognition (Grégoire et al., 2011), a section on foundational research questions for entrepreneurial growth cognition is followed by others rooted in psychology, economics and sociology.

Tasks of the Growth Entrepreneur

As noted above, research that has examined the intersection of entrepreneurship and cognition tends to focus on the effect that a cognitive variable has on an entrepreneurial outcome (opportunity identification, intention, performance etc.). As we change the vector to growth, an entirely new set of research opportunities opens up for scholars. If we want to spur growth from the start and truly understand it, researchers need to move to issues that affect businesses that scale as the venture transitions from just the founders and the founding team to one that employs scores of people, generates millions in revenue, and truly drives economic development in their region. By investigating the tasks and cognitions of the high-growth entrepreneur, we can inform the practice of entrepreneurship going forward.

This begins with a rethinking of what entrepreneurs do when they are focused not on starting but on growing. We all know that entrepreneurs 'have to do it all,' especially at the beginning. In the ventures that do not scale, this can continue throughout the life of the business. Now, let's think about all of the jobs a growth entrepreneur needs to do that are likely different from those of the entrepreneur who builds a small-business one- or two-person venture that does not scale up. With respect to financial capital, almost all start-ups require only family, friends and their own savings to fund their venture. In fact, according to the Global Entrepreneurship Monitor, only 2 percent of entrepreneurs get what might be termed professional investment (Kelley et al., 2012). So those who seek growth need to think quite differently. Someone who is starting any new business needs to think about marketing and sales, but those focused on building a growth venture need to build a formidable (and larger) sales and marketing operation. A design where the entrepreneur does everything is different from one in which an entrepreneur needs to build an organization. Entrepreneurs who build ventures that scale also need to manage a larger and more diverse set of people. High-growth entrepreneurs need to find, develop and manage more supplier and distributor relationships than smaller start-ups with only a handful of employees.

Methodological Approach

If you accept my assumption that growth entrepreneurs have different tasks, then my argument that they need to think differently about their venture follows logically. We need entrepreneurial growth cognition research! But how should we go about studying entrepreneurial growth cognitions? What designs or approaches might best help us uncover how they think? Similarly, in order to inform practice, what designs and approaches may give us insights into how they *should* think?

No one methodological design is best. The intent, of course, is to design the best approach, given what it aims to uncover. With this understanding,

I would suggest designs that allow one to analyze the words and thoughts of entrepreneurs and other stakeholders who help them achieve early, fast growth. Could survey items that examine intentions, self-efficacy, growth and other constructs help us understand the cognition behind this type of growth? Certainly, but I would argue that they might be better augmented by designs that examine the actual thoughts, discussions and decisions that individuals make about growth.

Interviews – structured or semi-structured – with successful entrepreneurs, nascent entrepreneurs working on their ventures in incubators or accelerators, incubator/accelerator directors and staff, angels, venture capitalists, government officials and other interested stakeholders would work well. Designs of this form would allow us to examine verbal protocols to find differences, similarities and relationships between various constructs within the entrepreneurial growth cognition domain.

In addition to interviews, cleverly crafted experiments would be useful in providing a window into the cognition of high-growth entrepreneurs. Controlled situations such as conjoint experiments would allow researchers to uncover the different environments, contexts and factors that affect how entrepreneurs think about growth.

With so few prior studies in the entrepreneurship literature examining the intersection of cognition and growth, entrepreneurial growth cognition research is in its nascent stages. In sum, at this time the field would gain greatly from interview and experimental designs that let us unpack exactly what entrepreneurs think about growth and how our research can inform practice. Other approaches will of course bring value, but the important issue is to match the appropriate design to the question under investigation.

RESEARCH QUESTIONS

What questions should we consider? My thesis is that we need more work on entrepreneurial growth cognition because the programs of policy makers across the globe are leading us to plenty of firms with low growth potential and too few firms that deliver the impact these officials desire (Isenberg, 2012). The motivation for this entire line of work comes from these two concerns and the fact that there is a scarcity of research in the area (Grégoire et al., 2011).

Below I offer some questions linked to base disciplines, but I first catalog some fundamental questions that could be approached from multiple theoretical perspectives. Entrepreneurship researchers who examine either cognition or growth could tackle any of the following questions using a theoretical lens of psychology, economics, sociology or a multitude of other perspectives:

- How do entrepreneurs and, perhaps more importantly, nascent entrepreneurs think about high growth and rapid growth?
- How do stakeholders surrounding nascent entrepreneurs – advisors, directors of accelerators, incubators and other 'start-up' organizations – think about high growth and rapid growth?
- What variables can be seen as cognitive antecedents to a high-/rapid-growth intention?
- What cognitive variables lead to high/rapid growth?
- How does an individual's experience (prior work or entrepreneurial experience, education, family etc.) influence the development of a high-/rapid-growth intention?
- How does an individual's experience (prior work or entrepreneurial experience, education, family etc.) impact the ability to perceive high-/rapid-growth opportunities in the market?
- How does dealing with an uncertain future state influence decisions about high-/rapid-growth opportunities?

These questions provide a foundation for the domain of entrepreneurial growth cognitions. However, they can be seen as just an *entrée* to a richer set of disciplinary-based questions that could lead to both theoretical advances and better practice. There are a number of anchors for entrepreneurial cognition work, but none as large and strong as psychology. As such, I begin with some suggestions for entrepreneurial growth cognition research questions with a psychological base.

Psychology and Entrepreneurial Growth Cognitions

Prior research demonstrates that the discipline of psychology provides the richest terrain for research on entrepreneurship cognition (Grégoire et al., 2011), so it is only natural that entrepreneurial growth cognition would also find fertile ground here. While some of the fundamental questions above could be designed to identify differences in variables between growth-focused entrepreneurs and others, there is room for work that examines the antecedents of these variables, and other interaction effects.

From the primary 'differences perspective' we could ask: do nascent entrepreneurs who are focused on developing large-growth companies rely upon cognitive styles, abilities, knowledge or heuristics that differ from those of lifestyle entrepreneurs? Moving beyond this, however, we can see more nuanced questions. For instance, what individual factors shape the acquisition and/or development of cognitive aptitude, abilities, knowledge or representations that are distinctly focused on entrepreneurial growth? If we look outside of the individual to his or her environment, we can ask: what environmental or contextual factors shape the acquisition and/or development of cognitive aptitude, abilities, knowledge or representations that are distinctly focused on entrepreneurial growth? Similarly, what organizational factors (capabilities and competencies) allow the entrepreneur to envision and mentally represent the system, structure and organizational design necessary for growth? How does prior success or failure moderate these factors and their relationship to growth? And, of course, we could move most of these questions to team, firm, industry or national levels as well – for instance, can team cognition lead to growth through the development of more opportunities, more products, multiple business models, or accelerated international development? Moving up from the individual level of analysis brings not just more complex questions, but questions that are pertinent to the policy makers who work across all of these levels as they try to incubate and develop more high-growth firms.

Economics and Entrepreneurial Growth Cognitions

With the recent rise of interest in behavioral economics, it is not surprising that cognitive variables have appeared in studies at the intersection of entrepreneurship and economics. As Grégoire et al. (2011: 1463) note, 'Austrians (e.g., Shackle, 1979; Lachmann, 1976), neo-institutionalists (e.g., North, 2005), evolutionary economists (e.g., Witt, 1998; 2000), growth theorists (e.g., Audretsch and Keilbach, 2007) and entrepreneurship scholars (e.g., Sarasvathy, 2001; McMullen and Shepherd, 2006; Chiles et al., 2009)' are all investigating within this domain. These streams of research have all examined the links between cognitive variables and the economics of entrepreneurial action. What is interesting is that, given economists' general concern for performance and growth, little work has focused on the intersection of cognition and entrepreneurial growth.

This provides an opportunity to assist practitioners by investigating questions at the interface of economics, growth and entrepreneurial cognition. The recent strong influence of behavioral economics makes one wonder: what role does cognition play in creating information asymmetries regarding those who look to build growth companies and those who stay small? Relatedly, are there specific growth cognitions that allow some to see large-scale opportunities where others do not? Are there specific knowledge or capabilities at play here? For instance, is scientific, technical or supply-side knowledge more or less important than knowledge of the market in isolating growth opportunities? What role does cognition play in creating information asymmetries at the industry level and hence competitive advantage and growth? Again, by changing the level of analysis, growth cognition across levels of analysis could include questions such as: does it matter whether knowledge about a growth opportunity, growth sectors or growth industries is concentrated or distributed, within or across levels of cognition? If it is concentrated in the entrepreneur, does it alter his or her cognition more than it would if it were distributed across members of a group, firm or industry?

Sociology and Entrepreneurial Growth Cognitions

In my review with colleagues (Grégoire et al., 2011), we found two primary sets of sociology-based works: work on cognitive legitimacy framed by institutional theory (e.g., Aldrich and Fiol, 1994; Busenitz et al., 2000; Shepherd and Zacharakis, 2003); and scholarship that uses network theory to examine the cognition of social capital (e.g., De Carolis and Saparito, 2006; Lee and Jones, 2008; Ozgen and Baron, 2007). Taken together with other works in the canon of entrepreneurship cognition, we can see that work in this area is often motivated by more than just monetary returns. The social entrepreneurship literature is perhaps the best evidence of this trend.

All of this provides great fodder for my extension of work in entrepreneurial growth cognitions. Within today's growing environment of social, environmental, sustainable and other 'specialized' forms of entrepreneurship, what are the implications for growth? Questions abound, such as what are the interactions between societal norms and individual representations of entrepreneurial growth? Especially with a recent preference for buying local and supporting small businesses, do some nascent entrepreneurs eschew growth? For example, how do social institutions influence entrepreneurial perception, and what effect does this perception have on building a rapid-, high-growth entrepreneurial venture? How do entrepreneurs think about society's perceptions of business and success, and do negative images of 'big business' and 'unethical corporations' affect an entrepreneur's thoughts about growth? More broadly, how does today's interest in social, environmental and sustainable entrepreneurship affect growth? Do these different emphases allow for greater growth? Or might entrepreneurs with these boundary conditions miss growth opportunities? Researchers might also consider social networks and ask questions such as: what impact do social networks have on how entrepreneurs think about growth? In other words, does the proliferation and ease with which we can connect allow nascent entrepreneurs to think differently about growth opportunities?

Summary

There certainly are other avenues for entrepreneurial growth cognition research outside of psychology, economics and sociology, but I believe the agenda put forward above from these disciplines and the fundamental questions I started with provide a bounty of work for scholars to examine. In some respects I categorized these questions for simplicity, but there is clear overlap and interplay between many of them. It is tough to disentangle many of these questions completely and set them in just one of these theoretical classifications. Clearly, however, classification is not the goal here.

CONCLUSIONS

The goal of this work was to outline a blueprint for researchers to investigate the important but nascent area of entrepreneurial growth cognitions. By outlining the current landscape, the practitioner need, and questions for scholars to investigate, this work helps chart a course for all stakeholders to overcome the perfect storm referred to in the introduction. As the community of scholars takes up the agenda put forward here, the effect should be that more entrepreneurs and policy makers think big, and we will see, as a result, an increase in ventures that can scale up and grow fast from the start.

REFERENCES

Abelson, R.P. 1981. The psychological status of the script concept. *American Psychologist*, **36**(7): 715–29.

Achtenhagen, L., Naldi, L. and Melin, L. 2010. Business growth: Do practitioners and scholars really talk about the same thing? *Entrepreneurship Theory and Practice*, **34**(2): 289–316.

Aldrich, H.E. and Fiol, C.M. 1994. Fools rush in? The institutional context of industry creation. *Academy of Management Review*, **19**(4): 645–70.

Audretsch, D. and Keilbach, M. 2007. The theory of knowledge spillover entrepreneurship. *Journal of Management Studies*, **44**(7): 1242–54.

Baron, R.A. 1998. Cognitive mechanisms in entrepreneurship: Why and when enterpreneurs think differently than other people. *Journal of Business Venturing*, **13**(4): 275–94.

Baron, R.A. 2000. Counterfactual thinking and venture formation: The potential effects of thinking about 'what might have been'. *Journal of Business Venturing*, **15**(1): 79–91.

Busenitz, L.W., Gomez, C. and Spencer, J.W. 2000. Country institutional profiles: Unlocking entrepreneurial phenomena. *Academy of Management Journal*, **43**(5): 994–1003.

Chiles, T.H., Tuggle, C.S., McMullen, J.S., Bierman, L. and Greening, D.W. 2010. Dynamic creation: Extending the radical Austrian approach to entrepreneurship. *Organization Studies*, **31**(1): 7–46.

Corbett, A.C. and Hmieleski, K.M. 2007. The conflicting cognitions of corporate entrepreneurs. *Entrepreneurship Theory and Practice*, **31**(1): 105–24.
Corbett, A.C., Neck, H.D. and DeTienne, D.R. 2007. How corporate entrepreneurs learn from fledgling initiatives: Entrepreneurial cognition and the development of a termination script. *Entrepreneurship: Theory and Practice*, **31**(6): 829–52.
De Carolis, D.M. and Saparito, P. 2006. Social capital, cognition, and entrepreneurial opportunities: A theoretical framework. *Entrepreneurship Theory and Practice*, **30**(1): 41–56.
Dutta, D.K. and Thornhill, S. 2008. The evolution of growth intentions: Toward a cognition-based model. *Journal of Business Venturing*, **23**(3): 307–32.
Frese, M., Krauss, S., Keith, N., Escher, S., Grabarkiewicz, R., Luneng, S., Heers, C., Unger, J. and Friedich, C. 2007. Business owners' action planning and its relationship to business success in three African countries. *Journal of Applied Psychology*, **92**(6): 1481–98.
Gilbert, B.A., McDougall, P. and Audretsch, D.B. 2006. New venture growth: A review and extension. *Journal of Management*, **32**(6): 926–50.
Grégoire, D.A., Corbett, A. and McMullen, J. 2011. The cognitive perspective in entrepreneurship: An agenda for future research. *Journal of Management Studies*, **48**(6): 1443–77.
Isenberg, D. 2012. Focus entrepreneurship policy on scale-up, not start-up. *Harvard Business Review Blog Network*, **29**(3–4): November; http://blogs.hbr.org/cs/2012/11/focus_entrepreneurship_policy.html.
Isenberg, D. and Brown, R. 2014. For a booming economy, bet on high growth firms, not small business. *Harvard Business Review Blog Network*, February; http://blogs.hbr.org/2014/02/for-a-booming-economy-bet-on-high-growth-firms-not-small-businesses/.
Kelley, D., Ali, A., Rogoff, E., Brush, C., Corbett, A., Majbouri, M. and Hechavarria, D. 2012. *Global Entrepreneurship Monitor (GEM) 2011 United States report*. Boston, MA: Babson College.
Lachmann, L. 1976. 'From Mises to Shackle: an essay on Austrian economics and the kaleidic society'. *Journal of Economic Literature*, **14**: 54–62.
Lau, C. and Busenitz, L. 2001. Growth intentions of entrepreneurs in a transitional economy: The People's Republic of China. *Entrepreneurship Theory and Practice*, **26**(1): 5–20.
Lee, R. and Jones, O. 2008. Networks, communication and learning during business start-up. *International Small Business Journal*, **26**(5): 559–94.
Leitch, C., Hill, F. and Neergaard, H. 2010. Entrepreneurial and business growth and the quest for a 'comprehensive theory': Tilting at windmills? *Entrepreneurship Theory and Practice*, **34**(2): 249–60.
Lord, R.G. and Maher, K.J. 1990. Alternative information-processing models and their implication for theory, research, and practice. *Academy of Management Journal*, **15**(1): 9–28.
McKelvie, A. and Wiklund, J. 2010. Advancing firm growth research: A focus on growth mode instead of growth rate. *Entrepreneurship Theory and Practice*, **34**(2): 261–88.
McMullen, J.S. and Shepherd, D.A. 2006. Entrepreneurial action and the role of uncertainty in the theory of the entrepreneur. *Academy of Management Review*, **31**(1): 132–52.
Mitchell, R.K., Smith, J.B., Seawright, K.W. and Morse, E.A. 2000. Cross-cultural cognitions and the venture creation decision. *Academy of Management Journal*, **43**(5): 974–93.
Mitchell, R.K., Randolph-Seng, B. and Mitchell, J.R. 2011. Socially situated cognition: Imagining new opportunities for entrepreneurship research (Dialogue). *Academy of Management Review*, **36**(4): 774–8.
North, D. (2005). *Understanding the Process of Economic Change*. Princeton, NJ: Princeton University Press.
Ozgen, E. and Baron, R. 2007. Social sources of information in opportunity recognition: Effects of mentors, industry networks and professional forums. *Journal of Business Venturing*, **22**(2): 174–92.
Partanen, J., Moller, K., Westerlund, M., Rajala, R. and Rajala, A. 2008. Social capital in the growth of science-and-technology-based SMEs. *Industrial Marketing Management*, **37**(5): 513–22.

Presutti, M., Boari, C. and Fratocchi, L. 2007. Knowledge acquisition and the foreign development of high-tech start-ups: A social capital approach. *International Business Review*, **16**(1): 23–46.
Read, S.J. 1987. Constructing causal scenarios: A knowledge structure approach to causal reasoning. *Journal of Personality and Social Psychology*, **52**(2): 288–302.
Sadler-Smith, E. 2004. Cognitive style and the management of small and medium sized enterprises. *Organization Studies*, **25**(2): 155–81.
Sarasvathy, S.D. 2001. Causation and effectuation: Toward a theoretical shift from economic inevitability to entrepreneurial contingency. *Academy of Management Review*, **26**(2): 243–63.
Shackle, G.L.S. 1979. *Imagination and the Nature of Choice*. Edinburgh: Edinburgh University Press.
Shane, S. 2009. Why encouraging more people to become entrepreneurs is bad public policy. *Small Business Economics*, **33**(2): 141–9.
Shepherd, D. and Zacharakis, A. 2003. A new venture's cognitive legitimacy: An assessment by customers. *Journal of Small Business Management*, **41**(2): 148–67.
Siegele, L. 2014. Launching a start-up has become fairly easy, but what follows is back-breaking work, *The Economist*, January 18.
Wiklund, J., Patzelt, H. and Shepherd, D.A. 2009. Building an integrative model of small business growth. *Small Business Economics*, **32**(4): 351–74.
Wiklund, J. and Shepherd, D. 2003. Aspiring for, and achieving growth: the moderating role of resources and opportunities. *Journal of Management Studies*, **40**(8): 1919–41.
Witt, U. 1998. Imagination and leadership – the neglected dimension of an evolutionary theory of the firm. *Journal of Economic Behavior and Organization*, **35**: 161–77.
Witt, U. 2000. Changing cognitive frames – changing organizational forms: an entrepreneurial theory of organizational development. *Industrial and Corporate Change*, **9**: 733–55.
Wright, M., Hoskisson, R.E., Busenitz, L.W. and Dial, J. 2000. Entrepreneurial growth through privatization. *Academy of Management Review*, **25**(3): 591–601.

13. Simulating socially situated cognition in exchange creation

Ronald K. Mitchell, J. Robert Mitchell, Miles A. Zachary and Michael R. Ryan

ABSTRACT

In this chapter, using an agent-based-model simulation, we model the socially situated cognition notion that dynamism in cognition results from the moment-to-moment interaction of an entrepreneur's inner environment and outer environment, using exchange formation as the relevant outcome. We present a fundamental model of exchange: individuals create works for sale to others, as the basis for the simulation. We utilize data from an experiment to represent resource uncertainty and relational uncertainty perceptions as an entrepreneur's 'inner environment.' We utilize assumptions developed from the literature to model the outer environment as represented by the 'entry' and 'exit' rates for each iteration (tick) in the simulation. We observe that both inner and outer environment interactions can be effectively modeled as predicted by socially situated cognition entrepreneurship theory; and that – as suggested by Simon (1981) – inner environments form a foundation whereby variations in a dynamic outer environment have more impact on exchange creation than would be expected when looking at variations in the inner environment alone. The implications of this simulation for theory building in socially situated cognition entrepreneurship research are explored.

INTRODUCTION

> A man, viewed as a behaving system, is quite simple. The apparent complexity of his behavior over time is largely a reflection of the complexity of the environment in which he finds himself . . . there are only a few 'intrinsic' characteristics of the inner environment of thinking man that limit the adaptation of thought to the shape of the problem environment.
>
> (Simon, 1981: 65, 66)

This intriguing premise – that the complexity of behavior results more from external factors than from internal ones, as individuals interact with environments – is at the core of our investigation in this chapter. In the field of entrepreneurial cognition, the theory of socially situated cognition (Mitchell et al., 2011b; cf. Smith and Semin, 2004) provides a theoretical rationale for such an examination. This theory suggests a means whereby entrepreneurial action, which has previously been viewed to occur as part of a 'complex system' (e.g., Minniti, 2004: 642), might actually be viewed as the interaction of two systems, only one of which is complex: inner (simple) and outer (complex). Herein we suggest: (1) an internal system – consisting of the actions governed by mental and physical responses to relational and resource uncertainty; and (2) an external system – consisting of the impacts of the socially situated (and hence dynamic and widely distributed) cognitive environment of exchange creation. In social psychology this socially situated view (that action-oriented, embodied, socially situated and distributed cognitions comprise the essence of a dynamic versus a static view of thinking humanity [Mitchell et al., 2011b]) is important to the advancement of entrepreneurial cognition research because it can help to untangle the main elements of exchange creation (e.g., the internal and external systems) one from another.

Helpfully, as applied herein, Simon's notion of 'inner' and 'outer' environments (alluded to in the aforementioned quotes) encompasses a totality of the conceptual space within which situated cognition exists – where the inner environment reflects the cognitive and affective aspects of the entrepreneur-as-individual and the outer environment reflects the conditions in which the individual operates (see also Foo, Murnieks and Chan, Chapter 5, this volume; Grégoire, Chapter 6, this volume). In this way, Simon's (1981) assertion is an ideal fulcrum point from which to examine what we see as the fundamental assertion of socially situated cognition entrepreneurship research: that it is the intensive moment-to-moment interaction of inner with outer environments that better explains cognitively derived results (Mitchell et al., 2011b; Smith and Conrey, 2009: 459). Our research question, as it applies the socially situated cognition view to entrepreneurship research, is: to what degree is exchange creation simple versus complex? We therefore investigate the extent to which, proportionately, inner environment and outer environment account for complexity in the exchange creation process.

To explore this question in detail, in this chapter we: (1) introduce into the discussion a parsimonious cognitive structure for entrepreneurial action (Mitchell et al., 2012) – to represent Simon's (1981) notion that inner environments are simple; (2) suggest the properties of this system that may be derived from socially situated cognition; (3) explore and illustrate, through a simulation, how the variations in both inner (internal) and outer (external) environments as suggested by socially situated cognition theory might be expected to affect outcomes and, specifically, how variations in perceptions of uncertainty (the inner environment) impact the frequency of new exchanges within various outer environments (ranging from dynamic to static); (4) discuss the theory-building implications of the simulation results for socially situated cognition-based entrepreneurial cognition research; and, somewhat incidentally, (5) examine a bit more fully the premise that simplicity in the cognitive system leads us to conclude that any complexity that arises over time is 'largely a reflection of the complexity of the environment' (Simon, 1981: 65).

In doing this, we hope to make four contributions to the entrepreneurial cognition literature. First, we seek to make more explicit the relative roles, in entrepreneurial action, of entrepreneurs' inner and outer environments. Second, we hope to demonstrate some of the dynamics in entrepreneurial action: through use of a computer simulation to model the effects of interactions between inner and outer environments and thereby generate a deeper understanding of the socially situated entrepreneurial cognition model (see also Sarasvathy, Chapter 14, this volume). Third, we propose the idea that a more fine-grained interpretation of previous work (e.g., Minniti's [2004] helpful assertion that entrepreneurship is a complex system) provides better theory for further research in a variety of streams – for example, entrepreneurial action (McMullen and Shepherd, 2006), opportunity creation (Alvarez and Barney, 2007), economic development (Minitti, 2004) and so on. And fourth, we seek to illustrate the benefit of simulation approaches in the exploration of socially situated entrepreneurial cognition theory.

A PARSIMONIOUS COGNITIVE STRUCTURE FOR ENTREPRENEURIAL ACTION

We assert that there is inherent value possible in human relationships; and that not all inherent value in human economic relationships just 'emerges' – that for a variety of reasons, at least some of the potential or 'un-emerged' new value encounters value-preventing obstacles, and is therefore dormant. On this assumption, an entrepreneur functions as 'the economic actor who identifies the obstacles to value emergence, acts to reduce them, and thereby brings dormant value into existence' (Mitchell, Morse and Sharma, 2003: 536; Mitchell, 2005: 195). In this section we describe how a simple model of exchange can yield a basic characterization of an entrepreneur's 'inner environment.' This characterization can then be used to: (1) identify the fundamental obstacles to exchange emergence faced by entrepreneurs; (2) point to the cognitive structures that entrepreneurs use to generate obstacle-reducing action; and thereby (3) create new value (by bringing new exchanges into existence that would otherwise not have occurred without their intervention). The argument develops as follows.

We have suggested in prior research, first, that the structure of human-relationship-based value creation takes the form of quite standard exchanges involving only three elements: (1) an individual (exchange creator), who creates (2) a work, which is offered for sale to (3) others (Mitchell, 2001; Mitchell et al., 2003); and second, that, depending upon levels of both 'relational' and 'resource' uncertainty present in the exchange environment, the actual number of completed exchanges will vary (Mitchell, Mitchell, Mitchell and Alvarez, 2012: 91). Viewed in light of these assertions and findings, the basic function of the entrepreneur is to act to alter the elements of the exchange (individual, works, others) in ways that reduce uncertainty obstacles, and result in the creation of new value through the emergence of new exchanges. As further explained later in this chapter, we argue that it is in this process of action that complexity arises primarily due to the nature of the environment encountered, that is, the dynamic environment – easier adaptation through action; stable environment – more difficult adaptation through action; static environment – difficult adaptation through action (cf. Mitchell et al., 2011a).

It is therefore of interest to seek a better understanding of the cognitive underpinnings of actions that entrepreneurs engage in to reduce uncertainty and thereby enable exchange. As mentioned earlier, Simon (1981: 65) argues that human beings, 'viewed as a behaving system, are quite simple. [That] the apparent complexity of behavior over time is largely a reflection of the complexity of the environment.' Thus it is possible to argue that the latent structure (Merton, 1968) of an entrepreneurial action will also be relatively simple, focused around the three suggested elements of exchange; and that complexity will be likely to arise at the interface of inner and outer environments, where the required adaptive action that arises from thinking and problem-solving behavior occurs.

Helpfully, the basic model of exchange suggested above (Mitchell, 2001: 43; Mitchell et al., 2003: 536; Mitchell et al., 2012: 107, 108) specifies three types of cognition that follow from the simple structure of exchange. As explained next, these three basic types of cognitions, which arise in frequent exchange behavior (Williamson, 1985: 31), enable the action that can reduce the uncertainty obstacles created by the existence of individuals, others and work. For example, planning cognitions enable entrepreneurs to 'put the future at the service of the present' (Bernstein, 1996: 1) and thereby to lower the uncertainty obstacles otherwise imposed by the bounded rationality introduced by the 'individual' (Williamson, 1985: 31). Likewise, promise cognitions enable entrepreneurs to increase trust in exchange relationships, and thereby reduce the obstacles posed by the potential for opportunism introduced by 'others' (Mitchell, 2001, 2003, 2005; Williamson, 1985: 31). Finally, competition cognitions enable entrepreneurs to produce a work that is competitive within the marketplace, thereby reducing the uncertainty obstacles posed by the specificity that is introduced by a 'work' (Mitchell, 2001, 2003, 2005; Williamson, 1985: 31).

We therefore argue that each of these three types of cognitions results in supple and varied combinations of action, and thereby produces much of the complexity we often observe in entrepreneurship. That is, because planning, promise and competition cognitions are applied within potential exchanges to remove the uncertainty obstacles that stem from the bounded rationality, opportunism and asset specificity that can be encountered in the almost limitless variety of social situations that arise in a dynamic exchange environment (Mitchell, 2001; cf. Williamson, 1985: 31), a virtually unlimited number of entrepreneurial thoughts and behaviors is expected to result. Hence complexity arises.

Thus, according to socially situated cognition theory (Mitchell et al., 2011b; Smith and Semin, 2004), when action occurs (i.e., when an exchange is in process) it can be anticipated to be cognitively dynamic, and there likely will be the following thinking conditions present: (1) as action occurs (2) the embodied cognitions that, in part, motivate that action, will be, (3) widely distributed, (4) within a socially situated context. That is, an individual entrepreneur/social actor might therefore be expected to have some motivation that arises from an embodied source (e.g., physiological needs, safety needs, socioeconomic needs; cf. Maslow, 1943) that is impacted by that person's perceptions and knowledge of their social/economic situation in their potential exchange relationships with others and the work (Fiske and Taylor, 1984; Hayek, 1937), as that situation is in turn shaped and defined by the larger meaning/institutional environment (Minniti, 2004). Hence we argue that, in the dynamics of exchange, socially situated cognition theory will help to explain the source of the complexity that arises as a potential entrepreneur encounters the obstacles to exchange that are manifest at the socially situated interface of inner and outer environments: as individuals create works for others in a varyingly dynamic environment.

Thus we suggest that planning, promise and competition cognition-based dynamic action is linked to the person/environment interface through responses to uncertainty as represented in Figure 13.1 (see also Mitchell, 2001: 43). Action based in planning cognitions serves the inner or thinking and feeling environment of an entrepreneur by diminishing

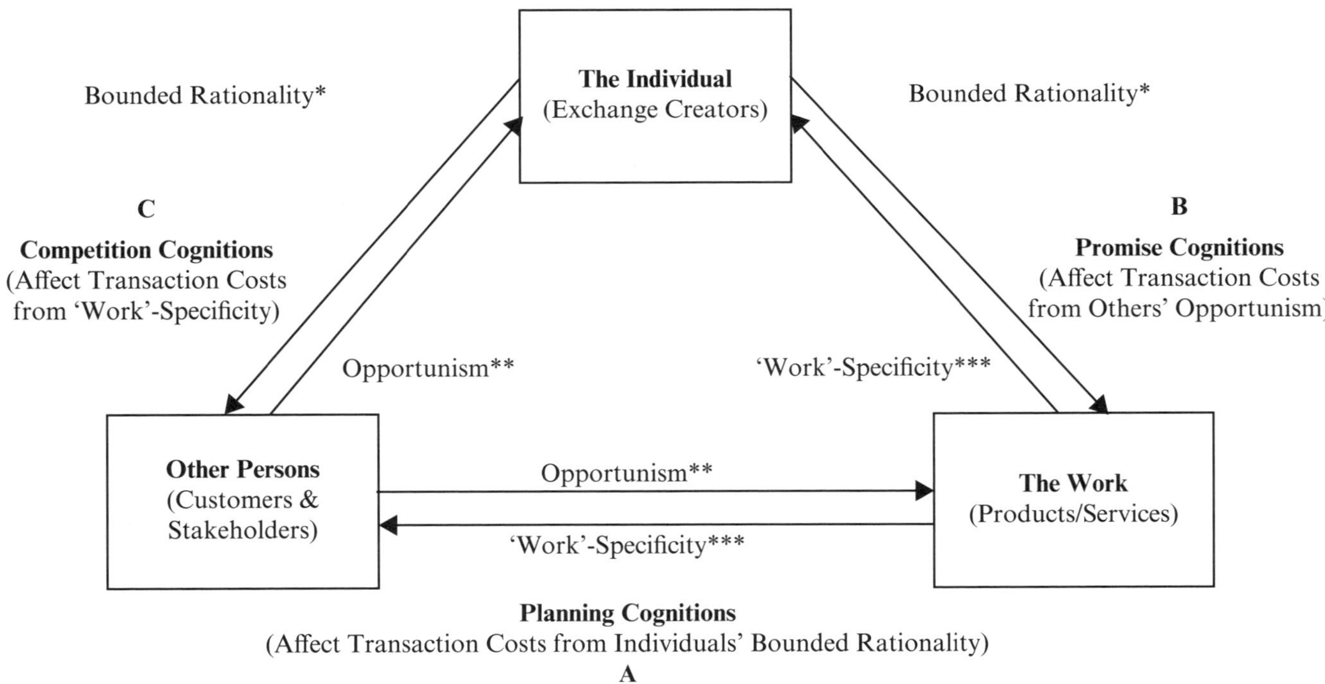

Note: * A basis of environmental uncertainty; ** a basis of relational uncertainty; *** a basis of resource uncertainty.

Source: Adapted from Mitchell (2001: 43), based on Gardner (1993) and Williamson (1985).

Figure 13.1 The basic dynamics of exchange

uncertainty that stems from bounded rationality to enable the infinitely variable problem-solving behavior required as the outer or social environment is encountered. Likewise, action based in promise and competition cognitions also serves the inner or thinking and feeling environment of an entrepreneur by diminishing relational uncertainty that stems from potential opportunism, and resource uncertainty that stems from specificity requirements respecting the outer environment, once again, to enable entrepreneurs to engage in the infinitely variable problem-solving behavior required as a dynamic outer or social environment is encountered.

On the basis of this simple cognitive structure (to represent the inner environment of entrepreneurs), and the application of socially situated cognition theory (to explain the variety and complexity that results as inner environment encounters outer environment), we next offer a parsimonious set of representative properties of the system that may be derived from socially situated exchange-cognition theory, to support a simulation-based analytical model.

THE PROPERTIES OF A REPRESENTATIVE MODEL

> Without a theory sophisticated enough to permit modeling or simulations, the whole process of entrepreneurial research slows down.
>
> MacMillan and Katz (1992: 6)

As will be seen, a very small but crucial addition to the properties of current complexity theory will be necessary for entrepreneurship to be characterized and simulated as both a simple (e.g., inner environment) and a complex (e.g., outer environment) system. Minniti (2004) cites Arthur et al. (1997) to argue that entrepreneurship as a complex system exhibits 'at least five crucial properties' (of a complex system): '(1) a large number of heterogeneous agents interact *locally* in a variety of unpredictable forms; (2) no single agent can control, exploit or plan all opportunities or interactions in the *specified space*; (3) learning and evolving heterogeneous agents cause a continual process of adaptation and *change*; (4) perpetual novelty in the action *landscape* creates new niches in the ecology of the system; and (5) the system is unlikely to be near a *global* optimum and there exist out-of-equilibrium dynamics with either zero or many equilibria' (2004: 643, emphasis added). We do not disagree that such properties of complexity are to be found in observed entrepreneurial action, especially in terms of outer environments (as emphasized in the quotation). However, we seek to refine this conceptualization using socially situated entrepreneurial cognition theory to suggest that, when seen as dynamic cognition, and contrary to common assumptions, entrepreneurship (while still complex in its outer environment) is nonetheless based on a more simple inner-environment-based system at its actual core than previously has been thought.

Accordingly, with respect to the five properties previously suggested (Minniti, 2004: 643), and based on the socially situated-cognition argument in the preceding section – which asserts a simple-structure account of the objectives and cognitions of entrepreneurs – we may therefore relax the two parts of Property (1). We do so by having developed theoretically, as an extension of prior research, an underlying and predictable latent-structure form of socially situated interaction: the exchange (Figure 13.1). Specifically, by positioning a consistent and uniform conceptualization of exchange at the center of entrepreneurship (i.e., individuals, creating works for others), we can eliminate relatively intractable variance from 'a large number or *heterogeneous* agents interact locally in a variety of *unpredictable* forms' and replace it with the relatively tractable variance form, 'theoretically *homogenous* agents interact locally in *predictable* form.' By these two revisions – from heterogeneous to homogeneous, and from unpredictable to predictable – we are enabled to take account of, and to simulate, exchange as a fundamentally simple inner system that is situated in the more complex outer environment. As we work to demonstrate in the following sections, it is this combination of 'inner' and 'outer' that matters, with the outer environment generating the complexity; and the situated, inner environment giving that complexity relevance and meaning. The essence of entrepreneurship is thus captured by the phenomenon of exchange formation; while the essence of exchange formation – its relevance and meaning – is encapsulated within the inner system of response to uncertainty. So, whereas entrepreneurship may appear to be complex to an outside observer, we argue that the essence of entrepreneurship nevertheless lies in the simple system.

Nowak (2004) argues that 'it is possible to explain [parsimoniously] complex psychological and social phenomena with very simple models if these models are dynamic . . . without sacrificing depth of understanding,' and that 'computer simulations have proven to be especially useful for investigating emergent properties in simple models' (2004: 183). In the next sections we therefore report our investigation along these lines.

A SIMULATION MODEL

To examine variations in outcomes of entrepreneurship as a system (specifically: variations in the number of new exchanges) due to variation of both inner and outer environments, we developed a simple agent-based model that simulates exchange completion (an individual connects with a work, and/or others) under varying cognitive and affective aspects of the individual (inner environment) and varying environmental conditions in which the individual operates (outer environment). Again, our objective is to create a clearer picture of how a socially situated cognition view applied within a new-exchange-occurrence framework, within a representative set of inner and outer environment boundaries, can enable theory development through an examination of the contrast suggested by Simon (1981): outer environments generate more of the complexity in a given situation than do inner environments and inner environments represent the simple system, which we argue is at the core of entrepreneurship. In doing so, we seek to explain how the intensive moment-to-moment interaction of inner with outer environments can explain cognitively derived results (Smith and Conrey, 2009: 459). That is, we seek to highlight the simplicity with which inner environments can explain differences in the frequency of new exchanges in the face of complexity in outer environments. In doing so, we also seek to demonstrate the utility of simulation methods to entrepreneurial cognition research, especially regarding the examination of cognitive dynamism as suggested by socially situated cognition.

In this section, after briefly describing our research logic, we: (1) discuss simulations in general; (2) explain our simulation selection process; (3) explain our rationale for representative assumptions about the outer environment; (4) explain our rationale for the data selected to represent the inner environment; (5) summarize the theoretical assumptions developed for the 12 representative cases simulated as included in the research model (see Figure 13.2, which illustrates the socially situated cognition simulation process research logic and assumptions); (6) describe the results of the simulation: 24000 runs of 30000 events per run, for each of the 12 conditions theorized; and (7) based upon a trial-and-error examination of runs and events (8000 to 100000 and 20000 to 60000 events (ticks) respectively, with a 24000 and 30000 combination selected to be both representative and parsimonious), report the results of a statistical examination of the representative data sets derived from the simulation results, which compares the likely impacts of both inner and outer environments as modeled in the simulation.

Thus, in this simulation research we followed the logic shown in three sections in Figure 13.2: (1) initial conditions; (2) simulation interactions; and (3) outcomes. As illustrated in Figure 13.2, and further described later in this chapter, the initial conditions of the simulation consisted of assumptions about the outer environment that were modeled in the simulation; but the initial conditions for the inner environment were not assumptions, but rather were experimentally derived data that were input

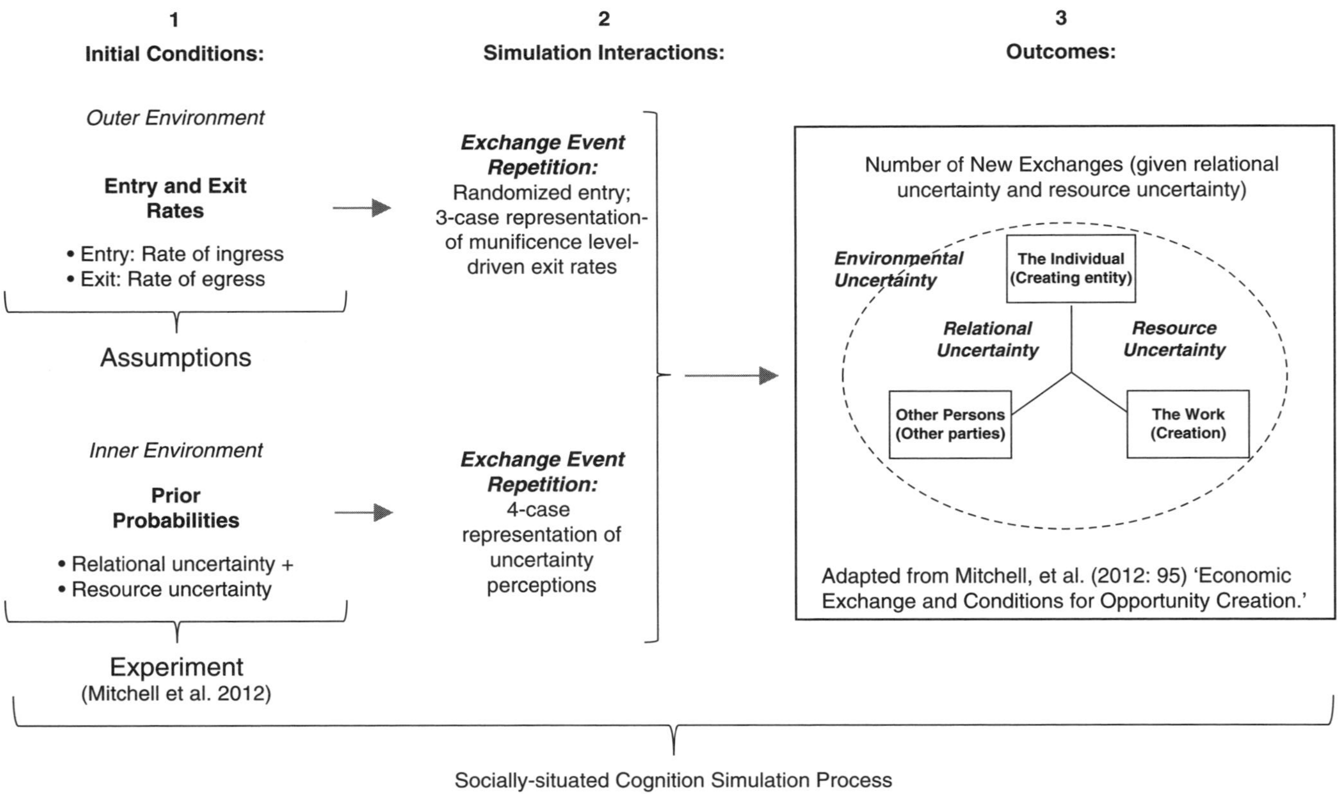

Figure 13.2 A socially situated cognition model of exchange

to represent the inner environment. The simulation interactions consisted of 12 exchange-event repetition types composed of three exit rate levels (outer environment) at four uncertainty perception levels (inner environment). The outcomes are the number of new exchanges completed given the foregoing assumptions, conditions, data and interactions. The nature of simulation research and the selections of the simulation type and of the assumptions and data for the initial conditions are described next.

Simulations

The term 'simulation' loosely refers to an imitation of reality in which conditions underlying such a reality can be imposed to constrict or constrain the behavior of agents operating within its domain, according to theory. A variety of simulation-based research methodologies exists. Laboratory experiments can involve participative simulation (e.g., Mitchell et al., 2012) wherein researchers reproduce a representative situation, implement system controls, facilitate participant interaction in the simulation, and observe the results. Other types of simulation include field study simulations,[1] wherein researchers present realistic scenarios to field participants (e.g., Fredrickson and Mitchell, 1984), participant interview simulations of individual-, firm-, or industry-level decision making or cognitions (e.g., Reger and Huff, 1993), or context-control simulations where the setting of a phenomenon is somewhat controlled in quasi-experiments or natural experiments (e.g., Venkatraman and Zaheer, 1990). In such cases, simulation participants are asked to make decisions within the framework specified by the researcher while their behavior is cataloged and scrutinized.

While the preceding types of simulation have a rich history in organizational research, computer simulation is gaining considerable acclaim. In a computer simulation, computer software is used 'to model. . ."real world" processes, systems, or events' (Davis, Eisenhardt and Bingham, 2007: 481; Law and Kelton, 1991). Researchers construct a virtual representation of theoretical linkages among constructs and test these relationships repeatedly, altering the conditions of the simulation as they proceed, in order to generate a clearer picture of the simulated process or outcome. It is this type of simulation that we have utilized in this research.

Simulation Selection

While there are many types of computer simulations, we focused on those that would assist us to develop a simulation that could properly represent a market environment characterized by multiple actors and participants engaging in the simple exchanges consistent with the theoretical properties that bound the analysis, as previously discussed. Thus we chose to build an agent-based model.

Agent-based modeling is a special type of computer simulation in which autonomous agents interact interdependently (Macy and Willer, 2002). Agents follow simple rules, yet their actions are adaptive and reflect path-dependence, helping to evolve the simulation outcomes over time, which is represented by 'clicks' (or 'tries'). Agent-based models are particularly designed to examine the behavior of agents within a larger system also marked by simple rules, and they have been specifically suggested to help scholars extract complex causal relationships in entrepreneurship research (MacMillan and Katz, 1992; McKelvey, 2004). Thus we suggest that agent-based modeling is an appropriate research technique through which entrepreneurial cognition scholars may model the dynamic cognitive events that characterize the social situation of entrepreneurs attempting to create work for others, in the face of uncertainty, in an outer environment with varying levels of dynamism.

By simulating exchange creation in an agent-based model, we were able to incorporate four essential components of agent-based modeling: (1) various agents; (2) decision-making heuristics; (3) an interactive topology; and (4) non-agent environmental conditions. Three sets of 'agents' were therefore developed based on a cognitive theory of exchange creation (Mitchell et al., 2012): individuals; others; and works as the basic elements of exchange. In the simulation, prior to an exchange, each element exists/operates autonomously; yet when initially linked with one other element, this dyad forms the initial stages of a path-dependent complete exchange (i.e., where all three elements must combine for the exchange to be completed). The linkage process was guided by the representations of both outer and inner environments in the simulation.

To model the outer environment, non-agent environmental conditions were developed using extant literature, representing conditions under which exchange could reasonably be expected to take place. Specifically, the shape of the problem (outer) environment was represented by the rate of agent entry and exit, which are controlled and allowed to vary based on theoretical expectations. Then, following a Bayesian simulation design method, we developed the agent decision-making heuristics (the constraints of the inner environment) using the results of a laboratory experiment conducted by Mitchell and colleagues (2012). Results of the experiment informed our expectations for individual agent actions; in particular, the cognitive receptivity of 'agents' to combination is expected to be a function of relational uncertainty and resource uncertainty. Thus, based on previous empirical results, we provided initial probability distributions to realistically simulate perceived exchange opportunities from the

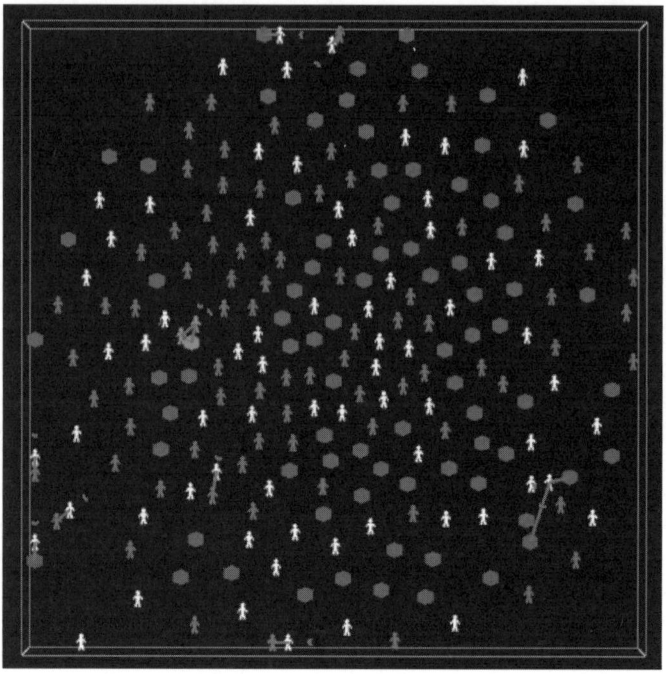

Figure 13.3 Screenshot of agent-based model in NetLogo 4.1.3

perspective of the exchange creator, wherein individuals can create works for other persons.

Our model was constructed in NetLogo 4.1.3, a computer program specifically designed for multidisciplinary agent-based modeling (Wilensky, 1999). An example of the visual output of this simulation is shown in Figure 13.3.

A more detailed explanation of the exact derivation of both the outer environment assumptions and inner environment data follows.

Outer Environment Assumptions

As noted earlier, the (effects of the) outer environment were represented in the simulation by the rate of agent entry and exit. We explain these assumptions for each, in turn.

Entry

In any given marketplace, the number of individual entrepreneurs entering the market varies according to certain conditions, the most important of which are unknown/unknowable, and are therefore by default assumed to be random. However, it also can be expected that any marketplace will be populated by some set of existing agents attempting to exchange; and it is reasonable to suppose that even a random level of entry will nevertheless be shaped (to some extent at least) by the preexistence of these agents in the market space. So, to begin each simulation run, we first established initial conditions under which the simulation would operate. That is, in each run of the simulation an initial number of individuals, others and works (agents) was introduced (15 of each based on trial-and-error exploratory runs of various initial numbers of elements present [for each: low 6 to high 60], to discover the relevant range). Then, consistent with the other properties of the simulation (i.e., random entry and exit criteria), the number of individuals, others and works in the market space changed over time as the simulation progressed.

Theoretically, it is also reasonable to suppose that the resource scarcity and tight niche packing arguments of the population ecologists will be likely to be helpful in explaining, and justifying as reasonable, the rate of entry into the market space: that entry will be slowed by, for example, crowding (or density) within the marketplace (cf. Carroll and Hannan, 1989). Accordingly, the rate of market entry for each new agent was programmed to represent marketplace crowding as a function of the number of exchanges completed: as exchanges increased, the rate of entry decreased. The equation constructed to simulate the rate of market entry for new exchange-creating individuals entering a variously crowded market space is as follows:

$$\textit{Rate of Individual Entry} = \alpha_0 + X_i \sim \textit{Pois}(\lambda_i)$$

$$\lambda_i = \frac{1}{1+n}$$

where n is the number of exchanges present in the marketplace, and α_0 is the initial value.[2] This representation of the rate of individual entry is reasonable to expect since, for example, individuals connected with works and others are not as active *and* they consume resources (the works and others that are still in the market space and not available to connection-seeking individuals); and therefore there will not be room for other individuals, others and works to enter the market space. Accordingly, this representative rate of entry was calculated by a random number generated using a Poisson distribution: where, as the number of exchanges increased,

Table 13.1 Complete exit rate assumptions and logic (stated as a maximum value, Downtime is defined as: a metric used to determine when an agent in a particular state will be compelled to exit the exchange space)

Agent	Condition	Calculation of maximum	Minimum time	Maximum time
Individual Others Works	No linkage	Max-Downtime + 1	Max-Downtime + 1	Max-Downtime + 1
Individual Others	Individual Other Link	Others-Linked-Max-Downtime + 1	Max-Downtime + 1	Others-Linked-Max-Downtime + 1
Individual Works	Individual Work Link	Work-Linked-Max-Downtime + 1	Max-Downtime + 1	Work-Linked-Max-Downtime + 1
Individual Others Works	Completed Exchange	Others-Linked-Max-Downtime + 1	1 tick (exchange formed in first round)	Others-Linked-Max-Downtime + 1

Table 13.1 (continued)

Values used for downtime variables:

Variable name	Assigned value	Agents minimum TTL[1]	Agents maximum TTL[2]
Max-Downtime	0	2 ticks (0, 1)	10 ticks (0 − 9)
Max-Downtime	8	10 ticks (0 – 9)	18 ticks (0 – 9)+(0 – 9) − 1
Max-Downtime	17	19 ticks (0 – 18)	28 ticks (0 – 18)+(0 – 9) − 1
Max-Downtime	26	28 ticks (0 – 27)	37 ticks (0 – 27)+(0 – 9) − 1
Max-Downtime	35	37 ticks (0 – 36)	46 ticks (0 – 36)+(0 – 9) − 1
Max-Downtime	44	46 ticks (0 – 45)	55 ticks (0 – 45)+(0 – 9) − 1
Others-Linked-Max-Downtime	8	Based on Max-Downtime value	
Work-Linked-Max-Downtime	8	(see times above)	

Notes:

1. Agent's Minimum Total Time to Live (TTL) assumes that an exchange does not occur within the specified number of ticks.
2. Agent's Maximum Total Time to Live (TTL) is dependent on linkage occurring no later than the Max-Downtime. The first set of numbers is the time available to secure the first linkage. The second set is the time available to complete the exchange (form second linkage with complementary agent).

the mean (λ) decreased, moving the distribution inward and constricting the number generated to a lower number, thus reducing the rate of market entry, as the theory we have proposed would dictate.

Exit

Consistent with theory (e.g., Mitchell et al., 2011a), the market exit rate was used to represent the level of market dynamism present in the variably hostile market space that it is reasonable to assume is experienced in some degree by most entrepreneurs attempting to start up a new stream of exchanges. Helpfully, it has recently been found that, in the face of uncertainty, agents in 'more dynamic environments make less erratic strategic decisions'; and that 'erratic strategic decisions increase with environmental hostility, but do so more when environmental dynamism is low than when it is high' (2011: 694). Thus, in the relatively hostile new-exchange market space, we can expect that individuals who must immediately complete an exchange (and if they do not, must exit) exist in a highly dynamic outer environment. We therefore represented high environmental dynamism in the simulation with a speedy exit rate (after 1 tick); medium dynamism with a moderate exit rate (after 9 ticks); and low dynamism with a very long exit rate (18 ticks).

Thus a market exit rate was programmed for each combination set of agents (individuals, others and works) as follows: unlinked agents, individual–other-linked agents, individual–work-linked agents, and agents that had combined to form a completed exchange (see Table 13.1 for the exit decision logic). It was necessary to program multiple rates of market exit given that, in practice, individuals and opportunities have different rates of market exit depending on market-space dynamism: the extent to which an exchange develops toward completion as constrained by, for example, the ecology of the outer environment (e.g., the operation of variation, selection and retention processes). For instance, an individual 'still' seeking an 'other' (e.g., a customer) *and* working to create a 'work' (e.g., a product or service) after a given time (number of ticks), that is, an exchange creator who is therefore lagging in the path-dependent process, must exit sooner than an individual who has already paired with an opportunity (a work) and is therefore further along in the process than the unlinked individual, because path-dependent selection has not occurred, and therefore retention is dubious.

Inner Environment Data

As previously noted, following a Bayesian simulation design, we used prior information collected in the experiment reported by Mitchell

Table 13.2 Results of analysis of covariance for opportunity creation[a]

a: Results of ANCOVA for top and bottom quartile for relational uncertainty[a]

Covariate	Economic exchange		
	F	Group Mean	η^2
Group task motivation (control)	1.62		0.06
Knowledge of other group members (control)	1.71		0.06
Hypothesis 1[b]	25.95 **		0.50
High condition		20.60 (1.50)	
Low condition		32.03 (1.61)	
$R^2 = .54$			

Notes:
a. We estimated marginal group means with standard errors in parentheses
b. For the high condition, n = 16 groups; for the low condition, n = 14 groups
* p < .05; ** p < .001

b: Results of ANCOVA for top and bottom quartile for resource uncertainty[a]

Covariate	Economic exchange		
	F	Group Mean	η^2
Group task motivation (control)	2.28		0.08
Knowledge of other group members (control)	6.24 *		0.18
Hypothesis 1[b]	12.82 **		0.31
High condition		21.40 (1.62)	
Low condition		29.41 (1.52)	
$R^2 = .49$			

Notes:
a. We estimated marginal group means with standard errors in parentheses
b. For the high condition, n = 15 groups; for the low condition, n = 17 groups
*p < .05; ** p < .001

Source: Adapted from Mitchell et al. (2012).

and colleagues (2012) to inform an individual's inner environment: the propensity to link with others and works based upon their perceived levels of relational and resource uncertainty. To accomplish this, we used the manipulation checks in the experiment (2012) reflecting relational and resource uncertainty to examine the proportionate occurrence of exchanges between low and high uncertainty quartiles in each case. We used the reported series of analyses of covariance to compute and compare the mean levels of exchange completion for low and high quartiles respectively, with appropriate controls (Table 13.2). Based on this analysis, we were able to approximate the prior probabilities of participants' ability to establish connections with others and works as: 64.3 percent for relational uncertainty and 72.76 percent for resource uncertainty. These probabilities were obtained by dividing in each case the mean of the high quartile by the mean of the low quartile (relational uncertainty = 20.596/32.033 = 0.643; resource uncertainty = 21.401/29.411 = 0.7276). We therefore used the mean scores as the basis for representing relational and resource uncertainty (the inner environments) within the simulation. We have reason to suppose that these prior probabilities are fairly robust, given that the explained variance (R^2) of the ANCOVA analysis = 0.54 (for relational uncertainty), and 0.49 (for resource uncertainty). Using the counts underlying these percentages, we programmed into the simulation an individual's likelihood of forming exchange linkages.

Theoretically Representative Cases

In the simulation we conceptualized four prototypical inner-environment entrepreneurial 'individuals,' distinguished one from the other by their level of resource and relational uncertainty (respectively): high–high; high–low; low–high; and low–low (HH, HL, LH, LL). We ran 24000

Table 13.3 Results of simulation (percentage of exchanges completed)

Trial	Resource uncertainty	Relational uncertainty	Time to exit	Mean %	S.D.	Max	Min	Median	Mode
1.	High	High	Long	3.81	0.11	4.3	3.39	3.81	3.82
2.	High	High	Medium	5.65	0.13	6.17	5.17	5.64	5.63
3.	High	High	Short	10.16	0.16	10.75	9.58	10.16	10.14
4.	High	Low	Long	5.89	0.13	6.40	5.35	5.89	5.90
5.	High	Low	Medium	8.41	0.15	9.02	7.78	8.41	8.41
6.	High	Low	Short	14.02	0.18	14.76	13.24	14.02	14.02
7.	Low	High	Long	5.21	0.12	5.74	4.75	5.21	5.23
8.	Low	High	Medium	7.53	0.15	8.13	6.94	7.53	7.56
9.	Low	High	Short	12.82	0.17	13.55	12.18	12.82	12.82
10.	Low	Low	Long	8.06	0.15	8.7	7.5	8.06	8.07
11.	Low	Low	Medium	11.27	0.17	12.03	10.48	11.27	11.27
12.	Low	Low	Short	18.34	0.2	19.15	17.58	18.34	18.31

Note: 30000 ticks each rep – 24000 reps: Time to exit: Long = 18 ticks; Medium = 9 ticks; Short = 1 tick

simulations of 30 000 ticks each, for every one of these prototypes in each of three levels of environmental dynamism (3 × 24 000 = 72 000 runs for each prototype). As previously noted, these three levels are a function of assumed exit rates to reflect the shape of the outer (external) environment.

Thus 12 theoretically representative cases were simulated according to the foregoing assumptions, initial values and inner-environment probability data. Descriptive statistics for each of these 12 cases are presented in Table 13.3.

Simulation Results

Figure 13.3 displays a screenshot of the two-dimensional agent-based model interface in NetLogo 4.1.3. Action in the simulation involves the components of an exchange moving randomly as time (ticks) moves forward. Individuals are represented by the 'person' figure colored white; others are represented by the 'person' figure colored dark gray; and works are represented by the 'box' shapes colored light gray. As the agents (each of the components) move, individuals identified via programmed selection begin to initiate exchanges. When an individual either links with an 'other' or creates 'work,' the connection between individual and other or between individual and work is graphically represented by a gray arrow connecting the individual and other or work component. When an individual has both linked with an other and created a work, the connection closes, a completed triangle forms, and all the components turn gray (this allows us to better see when an exchange has occurred). In Figure 13.3, a completed exchange can be observed in the left-central portion.

Table 13.4 Percentage of exchanges completed over time under differing levels of resource and relational uncertainty (based on random observations)

Resource uncertainty	Relational uncertainty	Exit time: after		
		Short time: 1 tick	Medium time: 12 ticks	Long time: 38 ticks
High	High	0.096	0.056	0.018
High	Low	0.134	0.082	0.038
Low	High	0.122	0.073	0.026
Low	Low	0.178	0.109	0.039

Note: N = 668 (compare Table 13.3: N = 24 000, total ticks each = 30 000)

Statistical Examination of Results

To understand the output of the simulations in a statistical sense (e.g., to examine questions such as the extent to which there are possibly significant relationships among inner-environment variables (resource uncertainty and relational uncertainty: low versus high) and outer-environment variables (time to exit: short versus long)), we generated random data using a normal distribution of the means and standard deviations of simulation results (i.e., the agents' behavior: p(exch) and tick exit times). Specifically, we created four distinct random data sets that represent the distinct inner environments (HH, HL, LH, LL) regressed with exit times as the independent variable, and probability of exchange (p(exch)) as the dependent variable, which then allowed us to examine modeled statistics that represent the exchange behavior emergent in the simulation. The descriptive statistics utilized are presented in Table 13.4.

As further explained, we also introduced a quadratic term into the regression equation to satisfy two conditions, one of which was observed in the larger simulation – an asymmetric tendency over very long runs (e.g., 24 000 runs at 30 000 ticks each: Table 13.3); and the other to address the impossibility of a negative probability (i.e., situations where the relationship between time and exchanges crosses the X-axis) that emerged in the regression. Note that the 'long' exit time reported in Table 13.4 was further extended from 18 ticks (Table 13.3) to 38 ticks (Table 13.4) to align with the inflection point produced through use of the quadratic term in the analysis, to sharpen the asymptotic relationship observed in the long simulation runs (Table 13.3).

To avoid the possibility of a Type II error (i.e., based on the number of paths and effect size, we would not overestimate), we conducted a power analysis to identify the minimum number of observations required to detect a small effect size. Thus we generated 668 random observations from a normal distribution using the mean and standard deviations reported for each designated downtime[3] (exit) simulated (i.e., 1, 9, and 18; Table 13.3). The four different levels of resource and relational uncertainty (HH, HL, LH, LL) were then categorically regressed on to linear and quadratic exit-rate terms, producing a curvilinear regression equation for each level and type of uncertainty. These regression equations are plotted in Figure 13.4.

Based on regression estimates, we are able to take into account the quantification of our asymptotic observations from the very long-run simulations through use of a quadratic term to sharpen the relationship for purposes of statistical examination. This then allows us to credibly represent the relationship between time in a market (represented by exit

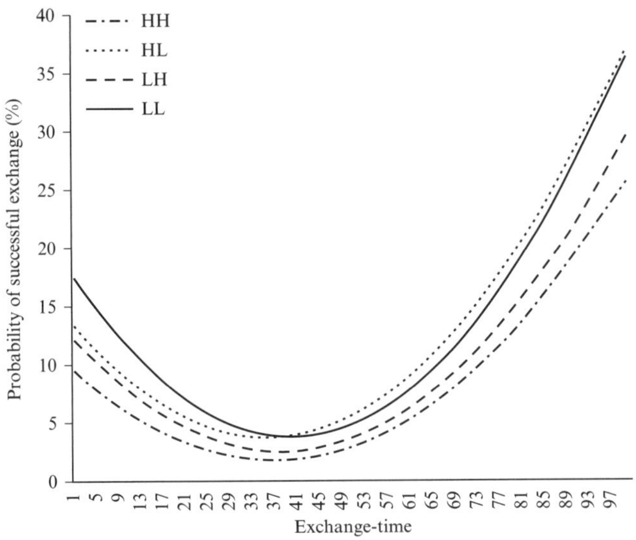

Notes:

- HH = high resource uncertainty, high relational uncertainty.
- HL = high resource uncertainty, low relational uncertainty.
- LH = low resource uncertainty, high relational uncertainty.
- LL = low resource uncertainty, low relational uncertainty.

Regression equations:
% of exchanges completed (HH) = 10.026 − 0.444(Time) + 0.006($Time^2$)
% of exchanges completed (HL) = 13.928 − 0.570(Time) + 0.008($Time^2$)
% of exchanges completed (LH) = 12.717 − 0.533(Time) + 0.007($Time^2$)
% of exchanges completed (LL) = 18.229 − 0.719(Time) + 0.009($Time^2$)

Figure 13.4 Probability of successful exchange under different levels of uncertainty and time allowed for exchange

rate) and the probability of exchange as curvilinear with the probability of exchanging decreasing quickly and remaining low for a period of time, followed by an increase in the probability of completing an exchange.

These results suggest that individuals are best served by entering a market, locating the requisite components, and quickly completing an exchange. This is consistent with the logic derived from the literature as previously noted, that, following market inception, finite resources and relationships are quickly claimed, leaving little to facilitate exchanges for others over time. However, after a certain period of time, it appears that individuals with the means to remain in the market (or also late entrants) are able to operate given the existing resources and relationships, and

Table 13.5 Means, ratios and ANOVA differences between: the percentage of exchanges completed over two times (short = 1 tick; long = 38 ticks), under differing levels of resource and relational uncertainty

Resource uncertainty	Relational uncertainty	Short: 1 tick exit time	Long: 18 tick exit time	Outer environment		
		Mean % completed	Mean % completed	Ratio	*N = 668*	*F* value
Low	Low	17.815	3.865	4.532x	334	51 000**
High	High	9.585	1.815	5.281x	334	30 000**
Inner	Ratio	1.828x	2.130x			
environment	*N = 668*	334	334			
	F value	17 000**	56 059**			

Note: ** $p < 0.01$.

complete their exchange. Individuals with greater access to 'patient' capital are therefore better able to bide their time and complete satisfactory exchanges. One example in practice might be the case of family businesses that, with their propensity to hold larger stores of longer-term capital, are often in a better position to capitalize on fluctuating markets that are unattractive to more short-term-oriented firms (Zellweger, 2007).

Finally, we conducted an ANOVA analysis of the means produced in the four-dataset-case statistical examination. Specifically, we were interested in the extent to which – based on extensive simulation of the influence of both inner and outer environments on individuals with varying inner-environment perceptions of resource and relational uncertainty (HH versus LL) – Simon's (1981) assertion concerning the relative impact of the outer environment versus the inner environment might hold. As reported in Table 13.5, in a comparison of the significant relationships found, it appears that both inner and outer environments matter, but that the relative impact of the outer environment is greater (cf. Dean and Meyer, 1996). That is, regardless of outer-environment dynamism, the percentage of exchanges completed is between 1.828 and 2.130 times greater when inner-environment uncertainty perception is low (versus high). Likewise, in dynamic outer environments, regardless of inner-environment uncertainty perception, the percentage of exchanges completed is between 4.532 and 5.281 times greater when exit time is short (versus long). In this way, there is likely to be a significantly greater impact of the dynamic outer environment (4.532 and 5.281 times, respectively for LL and HH cases) than of the inner (1.828 and 2.130 times, respectively for short and long cases).[4] These results thus support the notion that, in a new-exchange-occurrence framework, it is the intensive moment-to-moment interaction of inner with outer environments that can explain cognitively derived, and socially situated, results (Smith and Conrey, 2009: 459).

DISCUSSION

> What is remarkable looking at epidemiology, criminology, paleontology, or entrepreneurship is the tremendous wealth of techniques and models used in each field. What is present in the first three fields, and sadly missing in entrepreneurship, are bedrock theoretical models and established empirical links between theories . . . [and] experimentation of modeling using key variables in research settings . . .
>
> (MacMillan and Katz, 1992: 5)

In the foregoing quotation, MacMillan and Katz (1992) call for 'bedrock theoretical models' and 'modeling' to be introduced into the field of entrepreneurship research. Based upon a socially situated cognition model of exchange, we have demonstrated in this chapter some of the recent progress that has been possible in this regard. In the following subsections we address the implications of our theorizing for: action, embodiment, distribution, and social situation, as the essence of a dynamic versus a static view of thinking humanity – and specifically of thinking entrepreneurs.

Entrepreneurial Action

The study of entrepreneurial action has been gaining increased prominence in recent years, specifically as it concerns the behaviors that occur as entrepreneurs work to create, discover and/or exploit opportunities (Alvarez and Barney, 2007; McMullen and Shepherd, 2006; Venkataraman, 1997). The problem, of course, is that, while certain actions may have pervasive long-term consequences, such action is nevertheless fleeting, and as such is subject to the difficulties of data obscurity bemoaned by MacMillan and Katz (1992). In short, entrepreneurial action has tractability problems.

In the case of the cognitions that underlie the creation of new exchanges, there is an expectation that – as with most event-specific cognition – the fleeting nature of such dynamic cognition will necessitate research procedures that treat these actions as essentially unobservable; and therefore tractable only by virtue of their representation by surrogate indicators (Posner, 1973). Further, researchers in the field of entrepreneurship are presently evaluating the extent to which the cognitions that underlie opportunity creation (e.g., as represented in opportunity creation events) may also be quite rare (Alvarez, Young and Woolley, 2011). In this chapter we have developed and demonstrated the idea that entrepreneurial cognition theory is sufficiently robust to support both an experiment and a follow-on simulation grounded in the real-world results of that experiment, and to examine in greater depth the socially situated nature of specific entrepreneurial actions: the step-by-step creation of new exchanges.

Our study highlights the symbiotic value added to both laboratory/field experiments and computer simulation methods in the examination of entrepreneurial action. From a Bayesian perspective, we expected that using prior information generated on a smaller scale in an experiment should inform posterior estimates derived in a simulation. Our agent-based modeling approach to simulating entrepreneurial action demonstrates how this could be accomplished. Using the prior probabilities of linkage formation among individuals, others and works, we were able to develop a simulation that incorporated a more accurate measure of the likelihood, in simulation terms, that agents would link; and in opportunity creation terms, that an exchange would occur.

The results of the simulation suggest that, under different uncertainty perception conditions, individuals who differ in their propensity to create exchanges will nevertheless be constrained or enabled by an outer environment. Theoretically speaking, the results of this simulation suggest the action orientation of entrepreneurial cognition, and the important influence of motivation and time pressure – two of the three primary tenets of action orientation in the socially situated cognition model: 'motivation shapes action,' 'time pressure shapes cognition,' and 'mental representations are action-oriented' (Smith and Conrey, 2009: 456–7).

Motivation

Within the simulation, variations in the motivation–action linkage were represented by variations in the resource and relational uncertainty profiles that represented four prototypical inner environments (HH, HL, LH, LL, respectively). Socially situated cognition theory suggests that the motivation component of action orientation depends to a great extent on uncertainty perceptions (e.g., willingness to commit errors of commission or omission – or vice versa – depending on self-regulatory focus [e.g., Crowe and Higgins, 1997] – or need for belonging as it affects sensitivity to rejection [e.g., Pickett, Gardner and Knowles, 2004]). One theory-building possibility suggested by our results is therefore the notion that the black box of 'uncertainty,' as it is being pursued within entrepreneurial cognition research, would benefit from theories (e.g., self-regulatory focus theory) that further seek to explain the inner environment of entrepreneurs in terms of motivation, as uncertainty might shape action-oriented cognition.

Time pressure

The notion of time pressure may, possibly, be introduced into action-oriented cognition by either inner-environmental or outer-environmental demands. In our simulation we represented time pressure in terms of time-to-exit variations – an outer-environment condition. However, in addition to following prior research that frames time pressure as an outer-environment-based variable (cf. Mitchell and Shepherd, 2010), we found little theory to suggest that time pressure would derive from an uncertainty-based inner environment. In this sense, there is a compelling case that time pressure is a key feature of environmental dynamism, which was an important notion in our conceptualization and representation of the outer (external) environment. As our results demonstrate, time pressure (as we can argue represented by exit-rate variations) played a critical role in the explanation of reasons for variations in exchange-creation outcomes. We think that the influence of this time-pressure notion on the conceptualization of action orientation in entrepreneurial cognition offers a compelling reason to suggest time as a crucial construct in the better explanation of exchange creation. The results of our simulation offer substantial theoretical encouragement for the importance of time-focused, action-oriented entrepreneurship research. In this regard, Smith and Conrey (2009: 457) note additionally that 'an emphasis on action as the goal of cognitive activity suggests the importance of the body – the vehicle of all action – as a constraint on cognition.' We therefore turn next to the implications of embodiment for socially situated entrepreneurial cognition research.

Embodied Entrepreneurial Cognition

Embodiment, in the socially situated cognition sense, is defined as 'how the body shapes the mind' (Robbins and Aydede, 2009: 3; Mitchell et al., 2011b: 774). In the case of entrepreneurial cognition, we have argued that it is uncertainty (e.g., opportunity or threat concerning resources and relationships) that, embodied through the perceptions of individuals (i.e., their inner environment), shapes the disposition of the entrepreneurial mind toward exchange. Thus our 'agent' individual (whom we endowed with probabilistic qualities that simulated its representation of the world as grounded in that agent's capacity to sense the 'exchange world of interest' in our theory building) was shaped in its actions by these embodied uncertainty-focused attributes. This is of interest in theorizing, because much of the entrepreneurial action literature focuses on uncertainty as the primary basis for entrepreneurial action (McMullen and Shepherd, 2006); and we suggest that the influence of the embodiment–action fusion is important in understanding entrepreneurial cognition.

While the results of our simulation did not suggest that embodiment (i.e., the inner environment as we have positioned it) was the larger source of variance (when compared to the outer environment), they do suggest that variations in the inner environment as they concern uncertainty perceptions are nevertheless a significant and consistent source of variation in new-exchange-creation outcomes. The question that emerges from our analysis and this line of reasoning is: To what extent does embodiment (e.g., the inner environment) enable or constrain the initiation of exchange behavior? In our simulation we assumed initial conditions and 'entry'; but we did not model the type of refusal behavior that might entirely preempt exchange (cf. Mitchell, 2001). In this sense, embodiment might play a much larger role than we can fully anticipate through viewing new exchange probabilities through the lens of our simulation results.

The foregoing limitation concerning the influence of embodiment on exchange formation outcomes leads us to discuss (but briefly) the potential research avenues available through recent developments in entrepreneurship-focused neuroscience (see Baucus, Baucus and Mitchell, Chapter 8, this volume; McMullen, Wood and Palich, Chapter 9, this volume). For example, as Baucus et al. (Chapter 8) suggest, cognition becomes the umbrella concept subsuming all the unconscious and conscious processes that transform sights and sounds into neural representations, which then allow the individual to make sense of the world, form goals, and select suitable behaviors within a dynamic outer environment. Because sensory functions of the body are essential, in combination with the brain, to convey, for example, familiarity versus novelty in a situation or pleasure versus displeasure with a given stimulus, it is necessary for sensory inputs to be merged with stored information through embodied cognition processes (Buzsáki, 2004: 446). Thus the functioning of cognition as embodiment suggests, for example, that, rather than viewing emotions as impacting cognition, we should view them as an essential part of cognition and in fact critical for making appropriate choices (Damasio, 2005).

So when, in response to uncertainty in the outer environment, an individual has a fear reaction (e.g., fight or flight), it is reasonable to expect that – as seen through an embodied-cognition lens, and as reported in our results – such an individual will have variability in responses to resource and relational uncertainty, which will be expected, in turn, to impact the likelihood of successful engagement of others and works within the market environment. From a theory-development standpoint, we can therefore (based on this logic) agree with the suggestions by Baucus et al. (Chapter 8, this volume) that studies of neural phenomena in entrepreneurship (e.g., entrepreneurial affect [Baron, 2008], entrepreneurial effort [Foo, Uy and Baron, 2009], etc.) are better grounded 'in the anatomical and physiological analyses that describe the structure of specific brain areas (i.e., cortical fields and subcortical nuclei) and the functions (i.e., computations) performed by these structures' (Baucus et al., Chapter 8, this volume, p. 261). Such techniques as diffusion-tensor imaging or resting-state magnetic resonance imaging (rs-fMRI) studies can – by mapping the functional systems through which information flows and cognition occurs – reduce observational subjectivity (cf. LeDoux, 2012) and offer key insights for more clearly differentiating among the cognitive processes of entrepreneurs.

Distributed Entrepreneurial Cognition

The idea that thinking occurs beyond a given individual – that 'the boundaries of cognitive systems lie outside the envelope of individual organisms encompassing features of the physical and social environment' (Clark and Chalmers, 1998; Wilson, 2004) – suggests that 'the mind leaks out into the world, and [that] cognitive activity is distributed across individuals and situations' (Robbins and Aydede, 2009: 7–8). This is the idea behind the notion of distributed cognition. In this view, 'mind is much more a matter of what we do within environmental and social possibilities and bounds' (van Gelder, 1995: 280). Fortunately, dynamic systems theory tools have been suggested to help to model how various elements of a cognitive system change in relation to one another over time, 'because those state changes depend as much on changes in the external environment as on changes in the internal one' (Robbins and Aydede, 2009: 8).

It is within this theoretical context – that of distributed entrepreneurial cognition – that we have constructed the simulation reported in this chapter. As reported earlier, both outer and inner environment are implicated in the creation of new exchanges. What might this mean for theory building within the entrepreneurial cognition literature?

First, we suggest that, by effectively modeling the key entrepreneurial process of exchange creation, we have reason to suggest the importance and viability of tracking cognitive causal processes that are both internal to, and which also cross the boundary of, the individual organism (Robbins and Aydede, 2009: 8). From a theory-building standpoint, we have provided the basis for hypotheses that investigate such boundary crossing in the case of living and breathing entrepreneurs.

And, second, we suggest that, in the results of our simulation may be seen how cognition may often be 'enabled by information-processing loops that pass through the outside world as well as the mind' (Smith and Conrey, 2009: 461). This demonstration has theoretical relevance to broader research questions that propose entrepreneurship as symbol manipulation (e.g., Cornelissen and Clarke, 2010). Cornelissen and Clarke suggest a 'a more integrative understanding that embeds individual entrepreneurs within their social contexts,' and better permits them to 'imagine the opportunity for novel ventures, refine their ideas, and, after an initial investment, justify their ventures to relevant others to gain much-needed support and legitimacy' (2010: 540). They suggest 'linguistics and discourse analysis for analyzing shifts and changes in how entrepreneurs inductively reason about novel ventures.' Our research provides a theoretical rationale for hypotheses that argue for such analytical methods by demonstrating how the looped interaction of individuals with a dynamic outer (external) environment where factors external to the mind – such as the distributed nature of meaning or language in that environment – ultimately influence entrepreneurial outcomes (see Clarke and Cornelissen, Chapter 11, this volume).

Hence we wonder whether entrepreneurs, whose cognitive maps underestimate the role of distributed cognition on eventual outcomes, may encounter exchange-creation results that are suboptimal. Research that connects inner and outer environments based on the notion of distributed entrepreneurial cognition appears to have substantial promise in better understanding such limitations.

Socially Situated Entrepreneurial Cognition

It is becoming well established that dynamic cognition research eschews the 'boxology notion of, for example, input–process–output as it applies to entrepreneurship research' (Mitchell et al., 2011b). Instead, socially situated cognition research 'rejects this picture of autonomous, context-free inner processes in favor of a view of an organism as involved in intensive moment-to-moment interaction with its environment' (Smith and Conrey, 2009: 458–9). Our simulation was conducted based on these socially situated assumptions.

One of the primary observations that we can make, considering the results we have reported in this chapter, is that the socially situated entrepreneurial cognition model is important to the further exploration of the creation of new exchanges as it applies to such mechanisms in entrepreneurship research as opportunity recognition, discovery and creation (Sarasvathy, Dew, Velamuri and Venkataraman, 2003). This is because the conceptual structuring of the dynamic view – for example, where the social situation is a looping interface between inner and outer environments – can be expedited by having the point of departure offered by the effective simulation of both uncertainty (inner environment) and dynamism (outer environment) as they iterate.

Another observation we make is that it appears – again subject to data, assumptions and a simplifying theoretical rationale – that the impact of the environment in the looping dynamism of exchange-creating cognition is not linear. We suggest, based on our simulation observations, that, over very long exchange-time periods, the relationship between exchange-time and percentage of exchanges completed may be asymptotic; and we illustrate this using likely results of our statistical examination of the descriptive statistics produced in extensive simulations (24000 runs or 30000 ticks each, for each of 12 prototypical cases). The hypothesized existence of an asymptotic relationship between outer environment and exchange creation may be an important and useful dimension of the 'social situation' as further theory is proposed to explain opportunity formation.

Limitations

While it has been possible to model certain socially situated entrepreneurial cognitions related to exchange creation – through use of a simulation fed by both empirical (inner-environment) and likely-assumption-based (outer-environment) data, it has not been possible to accomplish this without encountering limitations imposed by the nature of the research methodology and the theories applied. In the following paragraphs, we identify the primary limitations we have identified and explain the steps we have taken to mitigate their effects in this research. These include assumptions-based limitations, the disadvantages coincident to the use of computer simulations, results-interpretation limitations, and issues related to the selection of the simulation itself.

Assumptions-based limitations

By definition, entrepreneurial cognition – even socially situated cognition that implicates the external environment – requires actual people/entrepreneurs for it to occur. Thus the simulation results that we report (in the sense that we have made entry and exit assumptions with respect to the outer environment) are somewhat more distant from real organisms than they would be had we also had (for example) experimental data available as inputs to the outer-environment conditions, as we did in the case of the inputs representing human inner environments. However, we also note that, as weaknesses go, the availability of data from 'real people' to represent the inner environment is a substantial improvement over most simulations, which are not enriched in this manner. And, therefore, we do not view the utilization of non-organism assumptions to represent the outer environment to be a severe limitation, given that economic or other statistics are often used to represent exchange environments in other research (e.g., Dean and Meyer, 1996).

Disadvantages of computer simulations

Primarily, computer simulations suffer from problems of limited external validity. Given that researchers must often distil complex latent constructs into simple and tractable quantitative representations, the applicability and generalizability of the findings generated by simulation research can fail to accurately mirror reality (Davis et al., 2007). As a solution to this problem, Davis and colleagues (2007) suggest that, following the creation of a simulation, scholars should verify findings with empirical data when possible. This helps to extend the external validity of simulated findings to real-world contexts. This is, we believe, an area for fruitful future research.

Interpretation of results

Untangling assumptions from results in simulations is often also a challenge. It is important in such research to demonstrate how, and especially in what manner, the results are not spurious and simply a reflection of inputs, but rather, are useful observations for such purposes as theory building. Of course, we readily admit that – as in all empirical research – the results are the reflection of the inputs; but, we assert, only to the extent that they constrain the nature of the interactions (e.g., that the prior probabilities of interactions – based upon experimentally obtained data – do in fact reflect expectations for the impact of resource and relational uncertainty on the likelihood of agents in the model achieving linkage with another exchange element). Additionally, once the interactions begin to occur as the simulation proceeds, with randomness, with empirics, with initial values, and with entry and exit assumptions as a beginning point, it is not unreasonable to expect that the unknown territory describing 'how things will play out' is not strictly 'determined,' but is, rather, somewhat revealing, at least.

Computer simulations have several additional advantages that could be leveraged by entrepreneurship scholars. First and foremost, computer simulations are particularly suited for developing theoretical insights under conditions of high internal validity (Cook and Campbell, 1979; Davis et al., 2007). Given that in a computer simulation the researcher controls every facet of the simulated process or outcome, computer simulations are highly tractable and are thus able to generate insight into both simple and complex interrelated causal relationships (Ganco and Agarwal, 2009). Second, computer simulations can be conducted at a fraction of the cost of other experimental methods. Laboratory and field simulation techniques require considerable resource outlays and extensive research procedures for the effective and appropriate management of human participants. Given the increasingly stringent requirements of institutional review boards at most major research institutions, for example, the time and effort expended to develop and institute an effective laboratory or field simulation can be daunting. Moreover, to collect longitudinal data, researchers are required to conduct multiple iterations of the simulation, introducing problems found in similar research designs using human participants (e.g., maturation, attrition, fatigue, historical bias, etc.). As an alternative, computer simulations are readily manipulated to impose new underlying assumptions or conditions without obscuring the observations and can run countless times with few of the limitations of human-simulations research.

Simulation selection issues

The serviceability of agent-based modeling methods to model outcomes is also open to discussion. This is because of the limitation in simulation methods to produce 'final' outcomes (in the sense that statistical tests of probability samples may be asserted with respect to some population of interest). Rather, simulation outcomes are multifinal, or possibly equifinal. Agent-based modeling has been suggested as a serviceable means for investigating and abstracting from multifinal and equifinal outcomes (McKelvey, 2004). Multifinality refers to an event in which different solution possibilities exist from a single niche problem. In an opposite scenario, equifinality refers to a system in which multiple initial circumstances lead to the same outcome (Gresov and Drazin, 1997; Payne, 2006; von Bertalanffy, 1968). Despite the ubiquitous nature of both equifinality and multifinality in entrepreneurial phenomena, organizational scholars have seldom investigated them specifically, often due to the problem of data obscurity (cf. Macmillan and Katz, 1992).

However, understanding the implications of alternative starting conditions and solutions for inherently obscure entrepreneurial phenomena holds considerable value for entrepreneurship scholars and entrepreneurs alike. Our simulation demonstrates how, under certain specified conditions in the model, we were able to evaluate variance across assumptions (12 prototypical cases) and to generate and thereby assess a variety of outcomes (i.e., the multifinality phenomenon). However, we have also been able to demonstrate how, by varying assumption and initial condition values of a simulation across a great many runs, we have been able to see if and how agents reach similar outcomes given vastly different origins (the equifinality phenomenon).

CONCLUSION

We began this chapter by claiming that cognition is dynamic: action-oriented, embodied, distributed and socially situated. We also advanced the notion that complexity arises, as suggested by Simon (1981), from the interaction of inner and outer environments. This introduction led us to consider a question relevant to entrepreneurial cognition research, which we suggested socially situated cognition theory might assist to better explain: to what degree is exchange creation simple versus complex?

Conventional wisdom suggests that it is complex – depending on such a wide variety of factors and elements that the latent structure is virtually impossible to conceptualize in one simple model (Minniti, 2004). Socially situated entrepreneurial cognition research offers a different account: the outer environment generates the complexity, and the situated, inner environment gives that complexity relevance and meaning. While not reducing the actual number or extent of variables, it nevertheless permits a simple-system viewpoint to be conceptualized: one that explains most of the complexity in terms of categories of latent variables (Merton, 1968): inner environment/uncertainty perception levels; and outer environment: levels of dynamism. Based on this theoretical structure, the boundaries of these assertions have been examined through our simulation; we have found that – as Simon says – both inner and outer environments matter a lot, with the inner environment offering simplicity in providing relevance and meaning; and the outer environment introducing complexity. This supports the socially situated cognition assertion that cognition is dynamic precisely because of the moment-by-moment interactions of inner and outer environments. We therefore conclude that our simulation and theory-building analysis has substantial explanatory power in untangling the observed complexity in at least the exchange-creation process of entrepreneurship.

NOTES

1. For a comprehensive overview of field simulations, see Snow and Thomas (1994).
2. Thus α_0 represents the number of individual entries allowed by the researchers: each tick in addition to the number of individual entries randomly generated by the Poisson distribution. Hence this term allowed the researchers to force more individuals to enter the market to the extent required by theory. As noted, the initial value determined to

represent a relevant range was 15 agentic elements for each of individuals, works and others in the simulation.

3. Downtime in a simulation is a variable that is set by the experimenter. It provides an indication of the maximum life of a particular agent in a particular state. We use the downtime feature to simulate exit dynamism. Thus downtime is a metric used to detect the actual time (ticks) that an agent has been engaged in a particular state. This appears in our simulation as 'Values used for downtime variables,' according to the logic as follows: If actual time (at the conclusion of the tick) is greater than downtime, the agent is compelled to exit. Actual time starts with time = 0, therefore if the downtime = 0, then an agent will have 2 (0 and 1) ticks to 'live in the simulation.' The 0 tick is the initial run. At the end of that run, as it is not greater than downtime, the actual time will be incremented by 1 and it will then go through the second tick. At the end of that tick, actual time is now 1 which is greater than downtime (0) and the agent will be compelled to exit.
4. According to this analysis, the outer environment has two and one half times the effect on exchange completion than the inner environment (e.g., 4.532/1.828 or 5.281/2.130 = 2.48:1). This relationship can be seen to hold for the very large case simulations as well (see means in Table 13.3).

REFERENCES

Alvarez, S.A. and Barney, J.B. 2007. Discovery and creation: Alternative theories of entrepreneurial action. *Strategic Entrepreneurship Journal*, **1**(1–2): 11–26.

Alvarez, S., Young, S.L. and Woolley, J.L. 2011. The creation of the world's deadliest catch. *Frontiers of Entrepreneurship Research*, **31**(15): Article 3.

Arthur, W.B., Durlauf, S. and Lane, D.A. 1997. Introduction. In W.B. Arthur, S. Durlauf and D.A. Lane (eds), *The Economy as an Evolving Complex System II*: 1–14. Reading, MA: Addison-Wesley.

Baron, R.A. 2008. The role of affect in the entrepreneurial process. *Academy of Management Review*, **33**(2): 328–40.

Bernstein, P.L. 1996. *Against the Gods: The remarkable story of risk*. New York: John Wiley & Sons.

Buzsáki, G. 2004. Large-scale recording of neuronal ensembles. *Nature Neuroscience*, **7**(5): 446–51.

Carroll, G.R. and Hannan, M.T. 1989. Density delay in the evolution of organizational populations: A model and five empirical tests. *Administrative Science Quarterly*, **34**(3): 411–30.

Clark, A. and Chalmers, D. 1998. The extended mind. *Analysis*, **58**(1): 10–23.

Cook, T.D. and Campbell, D.T. 1979. *Quasi-experimentation: Design and analysis issues for field settings*. Chicago, IL: Rand McNally.

Cornelissen, J.P. and Clarke, J.S. 2010. Imaging and rationalizing opportunities: Inductive reasoning and the creation and justification of new ventures. *Academy of Management Review*, **35**(4): 539–57.

Cornelissen, J.P. and Clarke, J.S. 2010. Imaging and rationalizing opportunities: Inductive reasoning and the creation and justification of new ventures. *Academy of Management Review*, **35**(4): 539–57.

Crowe, E. and Higgins, E.T. 1997. Regulatory focus and strategic inclinations: Promotion and prevention in decision-making. *Organizational Behavior & Human Decision Processes*, **69**: 117–32.

Damasio, A.R. 1994. *Descartes' Error: Emotion, reason and the human brain*. New York: Grossett/Putnam.

Damasio, A. 2005. *Descartes' Error: Emotion, reason, and the human brain* (2nd edn). New York: Penguin Putnam.

Davis, J.P., Eisenhardt, K.M. and Bingham, C.B. 2007. Developing theory through simulation methods. *Academy of Management Review*, **32**(2): 480–99.
Dean, T.J. and Meyer, G.D. 1996. Industry environments and new venture formations in the United States: A conceptual and empirical analysis of demand determinants. *Journal of Business Venturing*, **11**(2): 107–32.
Fiske, S.T. and Taylor, S.E. 1984. *Social Cognition*. Reading, MA: Addison-Wesley.
Foo, M.-D., Uy, M.A. and Baron, R.A. 2009. How do feelings influence effort? An empirical study of entrepreneurs' affect and venture effort. *Journal of Applied Psychology*, **94**(4): 1086–94.
Fredrickson, J.W. and Mitchell, T.R. 1984. Strategic decision processes: Extensions, observations, future directions. *Academy of Management Journal*, **27**(2): 399–423.
Ganco, M. and Agarwal, R. 2009. Performance differentials between diversifying entrants and entrepreneurial startups: A complexity approach. *Academy of Management Review*, **34**(2): 228–52.
Gresov, C. and Drazin, R. 1997. Equifinality: Functional equivalence in organization design. *Academy of Management Review*, **22**(2): 403–28.
Hayek, F.A. von (1937). Economics and knowledge. *Economica*, n.s., **4**: 33–54.
Law, A.M. and Kelton, D.W. 1991. *Simulation Modeling and Analysis* (2nd edn). New York: McGraw-Hill.
LeDoux, J. 2012. Rethinking the emotional brain. *Neuron*, **73**(4): 653–76.
MacMillan, I.C. and Katz, J.A. 1992. Idiosyncratic milieus of entrepreneurial research: The need for comprehensive theories. *Journal of Business Venturing*, **7**(1): 8.
Macy, M.W. and Willer, R. 2002. From factors to actors: Computational sociology and agent-based modeling. *Annual Review of Sociology*, **28**(1): 143–66.
Maslow, A.H. 1943. A theory of human motivation. *Psychological Review*, **50**(4): 370–96.
McKelvey, B. 2004. Toward a complexity science of entrepreneurship. *Journal of Business Venturing*, **19**(3): 313–41.
McMullen, J.S. and Shepherd, D.A. 2006. Entrepreneurial action and the role of uncertainty in the theory of the entrepreneur. *Academy of Management Review*, **31**(1): 132–52.
Merton, R.K. 1968. *Social Theory and Social Structure*. New York: The Free Press.
Minniti, M. 2004. Entrepreneurial alertness and asymmetric information in a spin-glass model. *Journal of Business Venturing*, **19**(5): 637–58.
Mitchell, J.R., Mitchell, R.K., Mitchell, B.T. and Alvarez, S. 2012. Opportunity creation, underlying conditions and economic exchange. In A.C. Corbett and J.A. Katz (eds), *Entrepreneurial Action: Advances in entrepreneurship, firm emergence and growth*, vol. 14: 89–123. Emerald Group Publishing Limited.
Mitchell, J.R. and Shepherd, D.A. 2010. To thine own self be true: Images of self, images of opportunity, and entrepreneurial action. *Journal of Business Venturing*, **25**(1): 138–54.
Mitchell, J.R., Shepherd, D.A. and Sharfman, M.P. 2011a. Erratic strategic decision: When and why managers are inconsistent in strategic decision making. *Strategic Management Journal*, **32**(7): 683–704.
Mitchell, R.K. 2001. *Transaction Cognition Theory and High Performance Economic Results*. A research monograph summarizing the development of transaction cognition theory at multiple levels of analysis. International Centre for Venture Expertise, University of Victoria.
Mitchell, R.K. 2003. A transaction cognition theory of global entrepreneurship. In J.A. Katz and D. Shepherd (eds), *Cognitive Approaches to Entrepreneurship Research: Advances in Entrepreneurship, Firm Emergence and Growth*, vol. 6: 183–231. Greenwich, CT: JAI Press.
Mitchell, R.K. 2005. Tuning up the global value creation engine: The road to excellence in international entrepreneurship education. In J.A. Katz and D. Shepherd (eds), *Cognitive Approaches to Entrepreneurship Research: Advances in entrepreneurship, firm emergence and growth*, vol. 8: 185–248. Greenwich, CT: JAI Press.
Mitchell, R.K., Morse, E.A. and Sharma, P. 2003. The transacting cognitions of non-family employees in the family business setting. *Journal of Business Venturing*, **18**(4): 533–51.

Mitchell, R.K., Randolph-Seng, B. and Mitchell, J.R. 2011b. Socially situated cognition: Imagining new opportunities for entrepreneurship research (Dialogue). *Academy of Management Review*, **36**(4): 774–8.
Nowak, A. 2004. Dynamical minimalism: Why less is more in psychology. *Personality and Social Psychology Review*, **8**(2): 183–92.
Payne, G.T. 2006. Examining configurations and firm performance in a suboptimal equifinality context. *Organization Science*, **17**(6): 756–70.
Pickett, C.L., Gardner, W.L. and Knowles, M. 2004. Getting a cue: The need to belong and enhanced sensitivity to social cues. *Personality & Social Psychology Bulletin*, **30**(9): 1095–107.
Posner, M.I. 1973. *Cognitions: An introduction*. Glenview, IL: Scott, Foresman & Company.
Reger, R.K. and Huff, A.S. 1993. Strategic groups: A cognitive perspective. *Strategic Management Journal*, **14**(2): 103–23.
Robbins, P. and Aydede, M. 2009. A short primer on situated cognition. In P. Robbins and M. Aydede (eds), *The Cambridge Handbook of Situated Cognition*: ch. 1. Cambridge, UK: Cambridge University Press.
Sarasvathy, S.D., Dew, N., Velamuri, S.R. and Venkataraman, S. 2003. Three views of entrepreneurial opportunity. In J.A. Zoltan and B. David (eds), *Handbook of Entrepreneurship Research*: 141–160. London: Kluwer Academic Publishers.
Simon, H.A. 1981. *The Sciences of the Artificial*. Cambridge, MA: The MIT Press. Further editions in 1982 and 1996.
Smith, E.R. and Conrey, F.R. 2009. The social context of cognition. In P. Robbins and M. Aydede (eds), *The Cambridge Handbook of Situated Cognition*. Cambridge, UK: Cambridge University Press.
Smith, E.R. and Semin, G.R. 2004. Socially situated cognition: Cognition in its social context. *Advances in Experimental Social Psychology*, **36**: 53–117.
Snow, C.C. and Thomas, J.B. 1994. Field research methods in strategic management: Contributions to theory building and testing. *Journal of Management Studies*, **31**(4): 457–80.
Van Gelder, T. 1995. What might cognition be, if not computation? *Journal of Philosophy*, **92**(7): 345–81.
Venkataraman, S. 1997. The distinctive domain of entrepreneurship research. In J.A. Katz (ed.), *Entrepreneurial Action: Advances in entrepreneurship, firm emergence and growth*, vol. 3: 119–38. Greenwich, CT: JAI Press.
Venkatraman, N. and Zaheer, A. 1990. Electronic integration and strategic advantage: A quasi-experimental study in the insurance industry. *Information Systems Research*, **1**(4): 377–93.
Von Bertalanffy, L. 1968. *General System Theory*. New York: Braziller.
Wilensky, U. 1999. *NetLogo*. http://ccl.northwestern.edu/netlogo/. Center for Connected Learning and Computer-Based Modeling, Northwestern University: Evanston, IL.
Williamson, O.E. 1985. *The Economic Institutions of Capitalism*. New York: The Free Press.
Wilson, R.A. 2004. *Boundaries of the Mind*. Cambridge, UK: Cambridge University Press.
Zellweger, T. 2007. Time horizon, costs of equity capital, and generic investment strategies of firms. *Family Business Review*, **20**(1): 1–15.

14. The whole deal: models, metaphors and mechanisms in entrepreneurial cognition

Saras D. Sarasvathy

ABSTRACT

The recognition that cognition is embodied within entrepreneurs as unique people sets the stage for an examination of both formal and informal models as tools for entrepreneurial cognition research. To minimize the loss in fidelity, as researchers construct models to represent more complex phenomena, models must be: simple, parsimonious, observable and interesting. Of the four model types that emerge from model building, informal non-essentialist models (i.e., constructive models) have the advantage that they do not overspecify definitions (simplicity and parsimony); and they consist of heuristics that embody 'even-if' rather than 'as-if' conceptions of situations and behaviors (observability and interest). In this chapter, after the foregoing discussion of model-building veracity, the theory of effectuation is introduced as a constructive model, characteristic of a science of the artificial: where thinking-based heuristics are utilized to unpack its behavioral assumptions into embodied even-if conceptualizations that explain entrepreneurship in cognitive terms.

INTRODUCTION

In Terry Bisson's story *They're Made of Meat*, an extraterrestrial alien honcho is informed that the sentient life forms they'd come searching for on Earth were made of meat (Bisson, 1991). At first, the honcho doesn't comprehend. Thinking meat? There are attempts at denial. 'Thinking meat' is obviously an oxymoron. Perhaps the Earthlings were only part meat? Perhaps the Earthlings only go through a temporary meat stage? But ultimately it sinks in. The planet the aliens had identified as having sentient life on account of non-meat radio waves and non-meat machines is entirely filled with thinking meat. 'Conscious meat. Loving meat. Dreaming meat.' The alien honcho cannot escape the truth: 'The meat is the whole deal!'

Indeed it is. And of course, we are those aliens. In the last 50 or

60 years, we have come to realize that if we wish to understand thinking – cognition – seriously, we cannot study it separately from its 'meat,' or, more elegantly, from its embodiment (Lakoff and Johnson, 1980; Varela, Thompson and Rosch, 1991).

The importance of this insight is hard to overstate. It immediately tells us that language, a central representative medium for cognition, should reflect this embodiment as well. Our language-use bears out this intuition. For example, we say someone decided to 'trust their *gut*' or that they 'knew it in their *bones*' or that they 'dug in their *heels*' and insisted on 'a fresh *approach* to the problem' which of course required 'a *leap* of faith'. As Lakoff and Johnson demonstrated, orientation metaphors and ontological metaphors are based on the way our bodies relate to space, time and objects (Lakoff and Johnson, 1980).

Or, more generally, to situations. Even before the importance of embodiment for cognition came to be widely recognized, it had been known for a very long time that language use could tell us a great deal about how people think about situations (Boring, 1953). Perhaps the earliest application of this idea was in the analysis of narrated dreams. William James also used introspected reports to build the theory of psychology (James, 1950 (1890)), and in AI, Simon and his colleagues used verbal protocols with great success to build the first programs – models – of human cognition (Newell and Simon, 1972).

But the fact that language reflects not just cognition, but *embodied* cognition, tells us something more, namely, that such conceptualizations need to take embodiment into account. In the context of entrepreneurship, this is to recognize that every representation of an entrepreneurial situation, whether venture or deal or institutional change, always includes the fact that it contains this *particular* entrepreneur. Not just a model of a generalized disembodied ideal of an entrepreneur, but the particular entrepreneur under consideration. Our models, formal or otherwise, have to find ways to work with or around the facts that (a) ignoring particularity has significant consequences and (b) such particulars themselves are situation-dependent and change over time.

MODELS

The first requirement for a good model of anything is that it should not multiply explanatory units beyond necessity. The second requirement qualifies this necessity: we have enough explanatory variables when we judge the model is able to describe the phenomenon of interest only in terms of its explanatory units. A model is not a dictionary. For example, in his justly famous dictionary, Samuel Johnson could define a fishing net as 'a reticulated fabric decussated at regular intervals.' He had the privilege of being able to use any set of English words to define some English words. A model of the English language, however, would be more like defining every word in terms of a much smaller subset, say, Basic English. Hence a model is more like an act of translation into a different language. As with all good translation, this involves hard choices on what constitutes an affordable loss in fidelity.

The first requirement: Occam's Razor, and the second requirement: closure, have to do with parsimony. The third requirement is a sanity check: it requires that the explanatory units be observable – preferably, measurable – in some way. The fourth and final requirement is that the model should tell us something we did not already know. Or, less stringently, it should tell us something we were not aware we knew. The emotional satisfaction that comes from working with a good model is often a consequence of this fourth requirement.

There are several ways in which we can build models to achieve these four requirements. We can fix a couple of handles on the large vase of possibilities by classifying models as either formal or informal. Another set of handles is affixed by classifying models as either essentialist or non-essentialist. This gives four different ways to pick up the vase:

- formal essentialist models;
- formal non-essentialist models;
- informal essentialist models (constitutive models); and
- informal non-essentialist models (constructive models).

Since my interest here is primarily in informal approaches, I will discuss formal models as a single group.

Formal Models

Formal models, by which I mean mathematical models, are models that characterize the explanatory units in terms of mathematics. For example, in Simon's model of altruism, the basic explanatory unit is 'docility,' which he defines as 'the tendency to depend on suggestions, recommendations, persuasion, and information obtained through social channels as a major basis for choice' (Simon, 1993: 156). The definition is not part of his formal model. In his formal model, docility is characterized as a non-negative real-valued function. Such characterizations are the key difference between formal and informal models. Formal models translate the problem to be studied into the language of mathematics. Informal models choose a less drastic strategy, as we will see later.

Formal models tend to be conscientious about all four requirements of good model building. They also tend to interpret the fourth requirement – that a model tell us something we did not already know – as a call to make true predictions. Technically, all formal models are tautological, so every prediction is a teasing out of the information already contained in the axioms. For example, it is obvious that Arrow's Impossibility theorem is not a prediction but a deductive consequence of his four axioms or conditions on social welfare functions (MacKay, 1980). But it is less obvious that exactly the same is true of all the formal models in neoclassical economics. This means that the base assumptions become very important since they encode the 'predictions' that will be generated. Consequently, formal model building is really about having the judgment to pick the 'right' set of assumptions. All else follows from deduction. However, there is no denying the exhilaration in seeing a minimalist formal model generate one unexpected deduction after another, much like watching a magician produce a never-ending ream of ribbon from an apparently empty hand.

But this satisfaction often comes at the cost of considerable distortion. This distortion is especially problematic when it is embodiment that is eliminated in the translation process. For example, a formal economic analysis of an entrepreneur's decision to hire a lawyer or start a new venture starts with the implicit assumption that an entrepreneur's decision can be discussed independently of the entrepreneur making the decision. This may be called the Disembodiment Assumption. What it seeks to do is to de-emphasize the role of the subject in the discussion. The formalist is interested in the entrepreneurial decision, not in entrepreneurs *per se*, and certainly not in any particular entrepreneur. Note that the last sentence has a subject ('the formalist'), who bears a certain relationship ('is interested in') to the object ('the entrepreneurial decision'). Were the sentence disembodied, it might read:

> It is the entrepreneurial decision that is of interest, not entrepreneurs *per se*.

The above sentence assumes that the reader has no interest in knowing just who is interested in the entrepreneurial decision. It has stripped the sentence of its body, so to speak. Though stylistic guides roundly condemn such disembodied sentences as jargon, obscurantist claptrap, newspeak, academese, their proliferation suggests that for all the cries of outrage, such sentences have their uses. Disembodiment creates a sense of objectivity. Consider the last sentence. I could have written:

> I think disembodiment creates a sense of objectivity.

But had I done so, it would have registered as a personal belief, lacking the weighty truth of the same sentence with the personal reference removed.

Similarly, formal models that get rid of the 'meat' also feel: more objective; more universal; more useful. For example, an economics of serial entrepreneurship simply feels more useful than an economics of, say, Milton Hershey.

In a peculiar twist, once the Disembodiment Assumption has stripped the situation of its body, the decision, now wearing a generic uniform called 'entrepreneurial,' acquires a scarecrow body of sorts. The decision-scarecrow has arms that can choose among options, legs that can clamber up and down demand and supply curves, and a head that cannot distinguish between a situation with an unknown probability measure and a situation in which a probability measure cannot be defined or does not exist. We no longer speak of an entrepreneur making a decision; instead we speak of 'entrepreneurial decisions.'

I do not mean to imply that formal models of entrepreneurial cognition are useless. Far from it. As Paul Krugman (1994) explained in an MIT lecture on developmental economics, the well-designed formal model is persuasive when it is right and useful when it is wrong.

To get a taste of how quantitative formal models look at a problem in entrepreneurial cognition, consider the problem of modeling serial entrepreneurship. Some entrepreneurs start only one firm. Others start one firm after another. Any sample of entrepreneurs – generated using industry, region or any other sampling criteria – is associated with a frequency distribution showing how many entrepreneurs start just one firm, two firms, three firms, four firms and so on. The problem is to develop a model that can generate this curve.

At the micro-level, the decision is quite complex. Whether an entrepreneur will choose to start multiple firms could depend on a great many factors, ranging from personality, family background, life events and skill sets through economic incentives and access to social resources to the region's history and outcomes of prior attempts by both this entrepreneur as well as other entrepreneurs.

The Disembodiment Assumption allows economics to equate firms with entrepreneurs. However, in the context of serial entrepreneurship, this convenient equation breaks down, since an entrepreneur may start multiple firms. Depending on how entrepreneurial success and failure are defined, the entrepreneurial failure rate may not be the firm failure rate. Also, the survival statistics of successive ventures are neither independent nor identically distributed because they were all started by the 'same' entrepreneur who learns over time. At the micro-level, a venture is related to other ventures through the entrepreneur.

To make some headway into modeling this situation, we confidently decide that the entrepreneur can be modeled by a single variable: confidence. Confidence acts as a catch-all variable for the entrepreneur's state of mind. Confidence along with the available capital determines the likelihood of her starting a new venture. We also decide that the likelihood of a venture having a positive or negative outcome in a unit time interval is controlled by a single variable. This variable is associated with the venture, not the entrepreneur. A success outcome increases the available capital while a failure outcome reduces it. We should call this venture-related variable something like 'the entrepreneur's competence with respect to the venture,' but, for simplicity's sake, we shall just refer to the venture's 'competence.' The hope is that these three variables – the entrepreneur's confidence, a venture's competence and the available capital – suffice to generate the frequency distribution of all firms within an economy over a specific period of time.

For formal purposes this is still too vague, so a first step is to parameterize these variables into real-valued functions. A firm's likelihood of failure – its death rate – is fixed at the moment of creation t and depends on its competence at time t. Similarly, the likelihood of a new venture at time t – the birth rate – is dependent on the entrepreneur's confidence at time t. Then it is a simple matter to view serial entrepreneurship as a time-dependent birth-and-death process. Simulations of the model show that the frequency distribution of the number of ventures started is bimodal. Entrepreneurs either start very few ventures or a great number of them.

The mathematical analysis of such time-dependent stochastic models is not simple. Fortunately, a deep insight by the statistician Ronald Fisher shows one way to avoid such analysis. Fisher realized that the statistics of contagion or learning processes (Polya distributions) are formally indistinguishable from those produced by heterogeneity (mixing distributions) (Feller, 1943). The simplest example of this equivalence is an ergodic process. As an example, take the tossing of an unbiased coin. If we are shown, say, the record 'TTHTHH,' it is not possible to say whether this represents the record of six consecutive tosses of a single coin or the outcomes of six coins tossed simultaneously just once. This linguistic ambiguity extends to the quantitative level. The statistics of a coin tossed, say, a thousand times are indistinguishable from the statistics of an ensemble of a thousand coins simultaneously tossed, exactly once.

Similarly, the serial entrepreneur can be thought of as someone who tosses one coin (= starts a new venture) after another, perhaps becoming better and better at getting a desired outcome. Fisher's insight shows that this situation is equivalent to the classical portfolio allocation problem where risk is mitigated by optimizing over an ensemble of heterogeneous assets. In other words, the serial entrepreneur who starts one firm ('asset') after another, perhaps gaining in competence and confidence, is formally identical to the portfolio manager who seeks to minimize risk and maximize return. Thus risk/return tradeoffs in serial entrepreneurship with its 'temporal portfolio' may be expected to be similar to the conventional portfolio allocation problem (Sarasvathy, Menon and Kuechle, 2013).

The replacement of temporal averages with ensemble averages was first exploited by the nineteenth-century physicist Josiah Gibbs in his development of statistical mechanics. In the context of formal models of entrepreneurial cognition, it is useful in that class of problems where the entrepreneur makes the same decision repeatedly. Another example comes from the Bose–Einstein distributions that pop up in the modeling of film revenues (De Vany, 2003).

This model shows the power and limitations of the formal modeling approach to entrepreneurial cognition. The idea that serial entrepreneurship is a sort of dual to asset portfolio management is an idea rich in implications. Yet it comes at a cost. The thinking, actively learning, purposeful entrepreneur has been transformed into a mathematical gas molecule, so to speak. The recurrence of the ghost of statistical mechanics past is a hint that this model and others of its kind belong to the tradition of studying l'homme moyen, the 'average man', a tradition that stretches all the way back to Quetelet and 1831 (Gigerenzer, 1989).

As mentioned, formal approaches are satisfying, useful and often valuable. For when they get things wrong, it is usually clear where the models have gone wrong. However, the same is not true when they get things right. Since false premises can generate true consequences, success works against progress. For example, in the serial entrepreneurship model, it is easy to conclude that confidence is necessary, overconfidence is harmful, and optimal confidence is possible. None of this is true. Unfortunately, it is tempting to conclude such non-truths. Suppose the model is well received and accretes an impressive literature. The variables, confidence and competence, introduced to solve a particular problem – the shape of certain frequency distributions – may slowly acquire a human dimension, that is, turn into traits. We can imagine psychological tests to measure these two traits and oppress potential entrepreneurs with these numbers. And so on and so forth.

One solution is not to formalize what cannot be formalized and work with informal models. Mathematics is a wonderful tool, but built into it are myriad assumptions about the nature of containment, contingency and cardinality. A formal model's startling 'insights' may often be just these assumptions brought to light. For example, Arrow's wonderful Impossibility theorem – so counter-intuitive and vexing at first, second and third sight – can be shown to be a distant consequence of the inverse function theorem in real analysis (Saari, 1987).

Informal models may also enter as a complementary approach. Once formal models have brought the minimal scarecrow into relief and offered specific possibilities for embodiment, informal models offer ways around some of the pitfalls of formal models.

Informal Models

As indicated in an earlier section, informal models differ from formal models in that they prefer to characterize their explanatory units in mostly non-mathematical ways. This means translating the problem into a language that is credible according to logical rather than mathematical criteria. Informal theories of cognition fall into two broad groups. The first – constitutive models – considers entrepreneurial cognition as a manifestation of what entrepreneurs *are*; the second – constructive models – considers entrepreneurial cognition to be something humans *do* when they are being entrepreneurial. The first species has a longer history.

Constitutive models

Constitutive models provide a set of explanatory units that are part of a person's essence. For example, it used to be a very popular notion, generated mostly by early travel accounts, that climate determined personality (Boia, 2005). This was an outcome of the fact that most people rarely moved very far from their birth-places in ancient times. The essence of being a Greek was to have been born and raised in Greece. The essence of being a Chinese was to have been born and raised in China. Since widely separated regions often differed in climates, early travelers struck upon climate as an explanatory variable for why the people they encountered on their travels were so different, even freakish. Here is a medieval Arab source on Europeans:

> As the sun, because of its distance, exerts only a feeble power over these regions, cold and humidity dominate and the snow and the ice rarely melt. The humours there have little warmth. The men are tall, of wild character, coarse manners, low intelligence and heavy speech. Their complexion is so extraordinary pale that it is almost blue. Their skin is fine and their flesh heavy; their eyes are blue and harmonize with the nuances of their complexion, and their long, loose hair is red through the effect of the humid vapors. Their religious beliefs are without solidity, owing to the nature of the cold and the lack of warmth. (Boia, 2005: 31)

From a medieval point of view, the explanation is quite persuasive. First, the explanation translates the problem of characterizing Europeans into the best biomedical language available at the time, namely, the theory of humours. Few things were more solid than the theory of humours (Noga, 2007). Second, the explanation is parsimonious. Just four humours suffice to account for the temperament of every human on the planet. Third, the explanation affirms what most of the intended audience – other Arabs – probably believed to be true of Europeans. Fourth, the explanation uses a feature that is unique to Europeans: they live in Europe. They are a product of the European climate. Since humour theory is intimately related to diet and climate, and the evidence meshes with the theory's predictions for people living in Europe's climate, we have the case of a perceived-to-be-true fact predicted by a well-established theory.

The theory of humours was not replaced because there were better models of explaining human temperament. As a cognitive theory of temperaments, the theory of humours was a smash success for centuries. As a medical theory, it seemed to cure just enough people to fend off competing theories (traditional forms of medicine such as Ayurved and Unani continue to be very popular even today). And structurally the germ theory was not all that different at first from the theory of humours. Introduced in an age in which microscopes were very much a novelty, germs were as mysterious as humours (perhaps even more so, since people had had several centuries to get used to the idea of invisible fluid essences glugging around in one's pipes). But the germ theory of disease replaced the theory of humours because the germ theory eventually turned out to be much better at making patients better.

I have dwelled on the theory of humours at some length not because it is representative of the kind of problems inherent in a trait-based theory of temperaments, but because it is representative of the kind of solutions offered by trait-based theories in general. What trait-based theories offer is a taxonomy. In general, the taxonomy is not a logical taxonomy of natural kinds. In other words, there are not many objective criteria by which to decide if an object belongs to this group or that group of the taxonomy. In modern trait-based theories, the classification comes with statistical error estimates.

Yet there is also an element of idealization. For example, classifying a weekday as 'Tuesday' requires one to idealize the actual orbit of the earth around the sun. 'Tuesday' does not objectively exist in nature. Similarly, when an entrepreneur is called a 'risk-taker' it requires a certain idealized notion of what constitutes a risk and what it means to take a risk. Indeed, the classificatory tag 'entrepreneur' is also based on idealized notions of certain kinds of human activity. Lakoff demonstrates how natural-language taxonomies are outcomes of idealized cognitive models, or ICMs (Lakoff, 1987).

The taxonomies generated by ICMs can be useful in at least two ways. Sometimes they are useful because they correspond to physical features of the natural world without too much distortion. For example, 'dyslexia' is a classificatory tag that depends on idealized notions of what constitutes a 'correct' reading but it also has a demonstrable physical correlate (Eckert et al., 2003).

Other times taxonomies are useful because they are linked to sensible bodies of practice that achieve desired goals. The taxonomy survives not because it cannot be disproved – the germ theory did just that with the Western theory of four humours – but because it hitchhikes a ride with practices that are effective at achieving desired goals. For example, humour theory does not have a persuasive explanation of why some diseases, such as smallpox, infect only once. Nevertheless, by the seventeenth century, traditional Chinese medicine had incorporated variolation – the deliberate infection with smallpox – into its prophylactic arsenal.

Taxonomies are most effective when they are built over natural kinds. For example, the periodic table is a logical taxonomy over chemical elements. Chemical elements are natural kinds because there are precise criteria for an element to be classified as, say, 'hydrogen' rather than 'helium.' Because logic entails inference, such logical taxonomies enable predictions, not just descriptions. In contrast, the flexible taxonomies of trait-based models over artificial kinds are mainly descriptive tools. They have to be. Unlike natural kinds, artificial kinds are social constructions. The taxonomies are not independent of the reality they seek to describe. To use a metaphor credited to Sewall Wright, it is akin to walking a sand dune: every step alters the landscape. For example, the ICM of caste led to a taxonomy of people types and their inherited capacities for various tasks. Since the theory was taken seriously on the South Asian subcontinent and every institution worked and reworked over centuries to fit the theory, any empirical observation could only prove the theory's claims.

Periodically, there are attempts to ground trait-based theories in a more empirical – more physical – framework. The current favorites are genetics and neuroanatomy. However, it is doubtful whether these attempts will fare much better than many other failed attempts in history that tried to link behavior to some fundamental human essence (see Baron, Chapter 1, this volume). The lack of a travel literature for sand dunes is not because they are not grounded in physics: sand dunes are transient objects precisely because of physics. Similarly, there might well be gene-complexes corresponding to the taxonomic states of being a 'go-getter,' 'dynamic,' 'control-freak,' 'born leader,' 'visionary,' 'results-oriented' and so on. But genes are flexible enough and the ways organisms use them are sophisticated enough that the only relevant reality is that human behavior is beyond taxonomic grasp. In other words, for every taxonomy we can think up (whether based on natural phenomena such as genes and neuroanatomy, or artificial ones such as risk-taking), variation within category is likely to be as large as variation across categories. In any event, behavioral neuroanatomy is still too nascent a field to merit discussion, and the arguments against genetic determinism have been too well stated to merit any paraphrase (Kitcher, 2000; Moss, 2004).

Instead, I will discuss an alternative approach to entrepreneurial cognition. It is an 'alternative' in the sense that it emerges out of the sciences of the artificial rather than the natural sciences. Instead of striving for descriptive or even predictive models, the aim is to reach for constructive theories of cognition.

Constructive models

Recall that, in a constructive model, entrepreneurial cognition is taken to be an aspect of what entrepreneurs *do*, rather than what they *are*. The first modern work that clearly spelled out the constructive approach to model building was Herbert Simon's *Sciences of the Artificial* (1981).

Drawing on ideas emerging from several research programs, Simon outlined a theory of artifacts. Artificial intelligence contributed the idea of programs as theories. Systems theory contributed the idea of fully decomposable, near-decomposable and non-decomposable systems. Economics contributed the idea of satisficing. Learning theory contributed the idea of learning from examples and empirical evidence for bounded rationality, the basis for satisficing. Organizational theory contributed the idea of modular designs.

It is philosophically harder than one might think to define the notion of an artifact and Simon's definition is not particularly helpful:

> An artifact can be thought of as a meeting point – an 'interface' in today's terms, between an 'inner' environment and an 'outer' environment, the surroundings in which it operates . . . Notice that this way of viewing artifacts applies equally well to many things that are not man-made – to all things in fact that can be regarded as adapted to some situation; and in particular it applies to the living systems that have evolved through the forces of organic evolution. (Simon, 1981: 9)

Contrary to his notice that the concept of artifacts could apply to objects with non-human origins, Simon had made it clear earlier in the book that by 'artifacts' he refers to human-made objects. In fact he identifies four indicia to distinguish artifacts from natural objects. Artificial things are synthetic and of human origin; they may sometimes imitate natural things; they can be characterized in teleological as well as evolutionary terms; and, finally, they are often discussed in modal imperatives such as 'ought,' 'should' and so on. This is hardly a compelling list of criteria; for example, is a test-tube baby an artificial thing? And what exactly is 'synthetic'?

It is odd that a concept so central to the book such as artifact should be treated so casually. True, Simon was not a philosopher and the book was meant for a mass audience, but there also seems to be a genuine reluctance to pin down the notion of the artifact with a definition. Simon has adopted this stance before – in his work on organizations. I believe this is another instance of Simon's gift for working on deep ideas without locking himself into a particular worldview. *Sciences of the Artificial* is indifferent to what constitutes an artifact because Simon is not interested in a constitutive theory. His book is about what can be done using the tools, metaphors and insights offered by the most exemplary artifact of all, the digital computer. Since the theory of artifacts was at an early stage, definitions could wait. This is the exact inverse of a formal or a constitutive approach, where axioms, definitions and notation form the cautious head of the beast that follows.

Is it possible to look at entrepreneurship as a science of the artificial? Or, as far as this chapter is concerned, is it possible to formulate a cognitive theory that is interested in what people can *do* to be entrepreneurial? What would such a theory look like?

First, the theory would not care too much about the exact definitions of its basic explanatory units. For example, what does it mean 'to be entrepreneurial'? After all, before we subscribe to a theory of X, shouldn't we know *exactly* what X is? No. Not always. Not when we are unsure whether we are inventing X's meaning or discovering it. There is a solid framework for such an approach. Prototype theory, originally proposed by Rosch and her collaborators (cf. Rosch, 1983) as an explanation for how color-words are used and perceived, offers an alternate way to look at categories. Briefly, members of a category may be related to each other despite having a set of attributes common to all: some members are better examples of the category than others. Its roots are considerably older, going back to the famous 'family resemblances' passage in Wittgenstein (1953). We can decide whether something is or is not entrepreneurial in the same manner we decide whether a robin is more of a bird than a penguin is. The decision is based on a judgment, not on a logical decision (Lakoff and Johnson, 1980). It should come as no surprise that in practice a great many taxonomies, even those purporting to be constitutive, are used in a prototypical manner. Note that Pluto is no longer a planet.

Another rationale behind postponing the definition of key words is to leverage natural language's power to suggest possibilities. The price is an increased uncertainty, but surely it would be peculiar if *entrepreneurship* researchers could not handle a little extra uncertainty.

Second, a theory that looks at entrepreneurship cognition as a science of the artificial would focus on heuristics rather than as-if explanations. In a paper on risky choices, Milton Friedman and L.J. Savage (1948) made a digression to discuss the realism of their hypotheses. They commented:

> An objection to the hypothesis just presented is likely to be raised by many, if not most, readers, is that it conflicts with the way humans choose . . . While entirely natural and understandable, this objection is not strictly relevant . . . The hypothesis asserts rather that, in making a particular class of decisions, individuals behave *as if* they calculated and compared expected utility and *as if* they knew the odds . . . Stated differently, the test by results is the only possible method of determining whether the *as if* statement is or is not a sufficiently good approximation to reality for the purpose at hand. (Friedman and Savage, 1948: 297–8)

Friedman, in a later, better-known essay on 'positive economics' (1953), reused an analogy from the earlier paper: of an expert billiard ball player playing as if he were computing complex kinematical formulae and then went on to defend as-if models in economics as needing to be judged solely by the results they predicted.

Herbert Simon took the opposite view; and he explicitly stated that view in a response to Friedman (Simon, 1984). His clearest statement on the matter is perhaps the one he offered in his rebuttal to the discussion following his Raffaele Mattioli Lecture:

> We have no reason to accept Friedman's billiard ball argument, or the bullet that knows where to go although it cannot solve differential equations. The differential equations governing the bullet's flight are not supposed to represent the thought processes of a bullet, because a bullet does not have thought processes and its motion is not determined by thought processes. But the motions of human beings are determined by thought processes and we should not take much comfort or satisfaction from the billiard ball or bullet analogies or other analogies of that kind. If we want a theory explaining how people play billiards, we do not want a theory of perfect billiard balls; we want a theory of what heuristics a human billiard ball player uses in order to plan and make a (often not quite accurate) shot. These heuristics and actions do not involve solving the differential equations of the billiard board; they involve rules of thumb, and it is these practical guides to action we are trying to discover in order to explain the behavior. (Simon, 1997: 173)

The idea of a heuristics as 'rules of thumb' echoes Kahneman, Slovic and Tversky's (1982) notion of heuristics as 'natural intuitions,' but for Simon they are also something more, namely, 'practical guides to action.'

A cognitive theory based on heuristics is desirable not only because it has proven to be enormously effective in clarifying the judgments people tend to make under conditions of uncertainty, but also, as Simon foresaw, some heuristics are also practical guides to (entrepreneurial) action.

It should be noted that heuristics have acquired a slightly negative connotation in recent work on human cognition. As Goldstein and Gigerenzer remark in a paper to re-establish the original meaning of the heuristic as useful strategies for action (Goldstein and Gigerenzer, 2002: 75):

> In the past 30 years, however, the definition of heuristics has changed almost to the point of inversion. In research on reasoning, judgment, and decision making, heuristics have come to denote strategies that prevent one from finding out or discovering correct answers to problems that are assumed to be in the domain of probability theory. In this view, heuristics are poor substitutes for computations that are too demanding for ordinary minds to carry out.

An alternative to Friedman's positivist prediction-based approach is to look for models that are even-if rather than as-if. A good example is heuristics identified by Gerd Gigerenzer and his colleagues (Gigerenzer and Todd, 1999). They were inspired partly by Simon's demonstration that bounded rationality led humans to satisfice rather than to optimize. It was not that satisficing was better or worse than optimization. It was simply a solution that fit the innate cognitive constraints. In short: even if humans are boundedly rational . . .

Similarly, Gigerenzer observed that, instead of saying that humans are not good at probability, one could just as well say: classical probability is not a good model for how humans reason about uncertainty. By looking at how humans actually think about uncertain events and then demonstrating the underlying formal logic, often a practical will-do fit for the problem at hand, Gigerenzer and his colleagues were able to argue that classical rationality is just one species in an ecology of rational forms (Todd and Gigerenzer, 2012). In short: even if humans are bad at classical probability . . .

Even-if models are subjunctive models. One way of interpreting them is to say that deviations from the ideal case are not seen as limitations, discrepancies, paradoxes, biases or errors. For example, the famous quote from Adam Smith about 'butchers, bakers and candlestick makers' is usually interpreted to mean that human beings are *de facto* selfish and perhaps ought to be so. In other words, acts of altruism become deviations and, hence, anomalies or errors. In even-if models, the quote would not be interpreted as an ideal. Instead here so-called 'deviations' are seen as sources of possibility. In short: even if human beings are selfish . . .

The constraints permit humans to do what the supposedly ideal cannot.

Just as abstraction is advantageous for the formalist trying to model entrepreneurial cognition, it is to the entrepreneur's advantage to realize that it is their irreducible specificity – what they uniquely bring to the table, their embodiment – that gives them an edge in a situation. Because cognition is embodied, and we have unique bodies, every entrepreneur brings a unique perspective to a situation. Indeed, there is no situation; there is only a perceived situation. A truly embodied theory of cognition would take the perceived situation seriously.

In what follows I will outline one extended example of what I believe to be a constructive model of entrepreneurial cognition, namely, the theory of effectuation.

EFFECTUATION AS AN EXAMPLE OF CONSTRUCTIVE MODELS

Before examining effectuation as a constructive model of entrepreneurial cognition, it may be useful to summarize the growing literature on the subject. The aim of the summary is not to provide a comprehensive review of the literature but to highlight just enough of the theoretical content and empirical support to date to give meaning to the model as an example.

Effectuation consists of five heuristics, each of which emphasizes ways to make decisions without relying on predictive information. The five heuristics that comprise effectuation were induced from a protocol analysis study of expert entrepreneurs, each of whom had encountered the entire gamut of possible experiences that a career in entrepreneurship could offer (Sarasvathy, 2008). The expert entrepreneurs in the study had done everything from starting, running, failing and succeeding at multiple ventures, including at least one company that went public. The original study has since been replicated with novices (Dew, Read, Sarasvathy and Wiltbank, 2009) and expert corporate managers (Read, Dew, Sarasvathy, Song and Wiltbank, 2009). Other studies have extended findings from these into private equity investing (Wiltbank, Read, Dew and Sarasvathy, 2009), R&D management (Brettel, Mauer, Engelen and Küpper, 2012) and social media (Fischer and Reuber, 2011). Additional studies have examined the use of causal and effectual heuristics in internationalization, family business and organic growth, and have incorporated a variety of methods including experiments, simulations, case studies, interviews and conjoint analyses (Read, Song and Smit, 2009; Schweizer, Vahlne and Johanson, 2010); see also Mitchell, Mitchell, Zachary and Ryan, Chapter 13, this volume).

Naturally, such a set of logically consistent *non-predictive* heuristics are particularly useful in the face of Knightian uncertainty, a situation characterized by risks that are literally immeasurable, where event probabilities are not only unknown, but also unknowable (Knight, 2012 (1921)).

The five heuristics, worded below in simple informal terminology, offer mechanisms that leverage things within the entrepreneur's control to shape and influence an unpredictable future:

1. Bird-in-hand heuristic: Start with who you are, what you know and whom you know. Do what you can do with what you have.
2. Affordable-loss heuristic: Invest no more than you can afford to lose. Even slack and waste can be transformed into resources.
3. Crazy-quilt heuristic: Work with self-selected stakeholders who make actual commitments to your enterprise. Who comes on board may not only help shape what you end up doing, but can also reduce costs and increase innovative outcomes.
4. Lemonade heuristic: Embrace and leverage unexpected contingencies, whether positive or negative. Opportunities often come disguised as misfortune.
5. Pilot-in-the-plane heuristic: Co-create the future without worrying about so-called inevitable trends. The future comes from what people do; history does not run on autopilot.

Note that the above heuristic principles are logically consistent. They overlap and may be mixed and matched and work together to reduce reliance on prediction. Each offers ways to shape and influence one's environment, whether defined in spatial, technological, demographic, industrial, social, cultural, institutional or temporal terms.

Effectual entrepreneurs willingly investigate and learn from the past, divide the known and present into things within their control and things outside their control, then seek to construct the future with whatever they have within their control. Things deemed outside their control are accepted as temporary constraints that may later become malleable. The process is iterative, evolutionary and interactive. However, the stance toward the environment is not merely adaptive. It is proactive, co-creative (social) and transformative. Using the terminology of probabilities, effectual entrepreneurs are not merely Bayesian. They seek to reify or falsify conditioning assumptions, viewing event spaces as subject to varying degrees of manipulation and not independent of human actions.

As specified earlier, constructive models do not overspecify definitions and consist of heuristics that embody even-if rather than as-if conceptions of situations and behaviors. And, as summarized above, the literature on effectuation began as an empirically induced model of how expert entrepreneurs make typical decisions that occur in startup situations. Because the model consisted primarily of heuristics, it was possible to develop it as a constructive model characteristic of a science of the artificial. In other words, it became feasible to unpack its behavioral assumptions into embodied even-if conceptualizations.

All four characteristics of good models explicated earlier are attributes of effectuation as well. First, the theory does not unnecessarily multiply explanatory variables. It seeks only to describe and predict one relationship, namely, the positive correlation between expertise and the use of effectual or non-predictive heuristics. Second, it translates entrepreneurial experience into decision making under Knightian uncertainty. Third, all variables, including experience, expertise, effectual heuristics and perceptions of situational uncertainty may be operationalized. Finally, because the heuristics were derived from cognitive studies of real entrepreneurs, that is, from evidence-based model building, application of these heuristics should demonstrate what we intuitively expect entrepreneurial behavior to be, with all the concomitant consequences.

In the effectuation model there are no prior assumptions about who the entrepreneur *is*. However, it is a different matter as regards application domains. Effectual heuristics are about guiding non-predictive control of situations. Accordingly, effectuation is best suited for those problems in which prediction is either impractical or impossible. When applied to situations in which human action can shape, transform and co-create, the effectual model may result in a variety of artifacts several of which formal models take as given and exogenous. The decision to start a new venture may itself be an artifact arising out of the effectual process. Another set of artifacts could consist in the recognition, discovery and creation of opportunities, a theoretically accepted 'definition' of entrepreneurship as a field of research (Shane and Venkataraman, 2000; Venkataraman, 1997). A third set could consist in the formation of constitutive models of entrepreneurship such as an entrepreneurial personality, or an entrepreneurial culture or ecology or institutional context. Let us examine each of these three sets of artifactual consequences in turn, with a view to deepening our understanding of constructive models.

To examine the notion of the decision to become an entrepreneur itself being an artifact of the effectual process, we can delve deeper into the first heuristic of effectuation. Dubbed the bird-in-hand heuristic, this consists of using who you are (history), what you know (know-how) and whom you know (social network) as the primary resources to drive action. Take the case of any new venture, a line of beauty products, for instance. Who are the sort of people who could construct such an enterprise? One can imagine a professional beautician who starts one as a career move (e.g., Florence

Graham, founder of Elizabeth Arden). Or perhaps someone could choose to follow in the footsteps of her uncle from whom she inherits both chemistry and industry know-how that she uses to build a line of cosmetics (e.g., Josephine Esther Mentzer, cofounder of Estee Lauder). Or a newly freed daughter of slaves could suffer hair-loss problems and experiment with solutions that end up in a product for straightening black women's hair (e.g., Sarah Breedlove, founder of Madam CJ Walker Manufacturing). Or a trained skin therapist could start a school that forces her to face the fact that beauty products do not usually overlap with skin health, and if she wanted to practice what she preached/taught, she would have to invent her own line of products (e.g., Jane Wurwand, founder of Dermalogica). Or a female employee could get passed over for a promotion, decide to write a book to help women in business that morphs into a plan for the perfect company that happens to become a line of beauty products (e.g., Mary Kay Ash, founder of Mary Kay Cosmetics). Or a former gang member turned cosmetics company employee could get fired, become homeless, team up with an inventor of hair-care products to sell them door to door and end up building a billion-dollar business (e.g., John Paul Dejoria, cofounder of Paul Mitchell).

For every one of these stories, we can imagine other people in similar circumstances who did not start a successful cosmetics venture, started an unsuccessful one or did not start a venture at all. Similarly, one can also imagine cosmetic ventures started by all kinds of other people in other places and cultures for other reasons – for example, Afghani women secretly building beauty salons as a way to circumvent the Taliban's horrendous punishments for wearing lipstick. In actuality, these 'salons' also served to educate young women who were not allowed to go to school under the Taliban regime. In retrospect, the origins story of an enterprise has the feel of a just-so story, a miracle of happy contingencies. But, seen prospectively, it is an even-so story, a story of making use of the resources that all humans have – namely, a life history, some specific know-how and a social network – to trump the many obstacles that litter all effort.

The point is, the bird-in-hand heuristic neither requires a particular set of characteristics for people to start new ventures nor a new venture as the only possible outcome of using the heuristic. In the case of Mary Kay Ash, the initial outcome was a book, not a company. The founders of both U-Haul and Terracycle used hairdressing as a way to financially support themselves while their ventures struggled to become cash flow positive. Thus what became ultimate opportunities in some cases were mere stopgap inputs in other cases. And in yet other cases, something created as a stopgap arrangement became the ultimate opportunity. For example, Stacy Madison did not want to lose customers waiting in long lines at her sandwich cart. So she and her then husband and cofounder of Stacy's Pita Chips decided to bake leftover pitta bread into snacks to hand out to those waiting in line. The rest, as they say, is history.

In sum, while the use of effectual heuristics is highly correlated with entrepreneurial expertise and the theory itself is derived from an empirical investigation of expert entrepreneurs, the theory is not limited to any specific definition of entrepreneur or entrepreneurship *per se*. Instead it becomes capable of generalization beyond standard definitions of entrepreneurship and becomes a useful pedagogical and practical toolbox – a constructive model of reality, a design logic that brings to light entrepreneurship as a science of the artificial. Several variables previously related to entrepreneurship through formal and constitutive models – be they ventures, markets, opportunities or institutions, or even personality and culture – now become artifacts, in other words, products of design resulting from contingent or intentional human action including entrepreneurial action.

A second aspect of effectual heuristics is that they do not require as-if assumptions about human behavior including assumptions about how decision makers perceive their situations. They differ in that regard from formal and constitutive models.

Formal models tend to make stringent assumptions about human behavior – for example, that the decision maker is risk averse, or opportunistic (i.e., self-interested with guile) or overconfident and the like. The claim is typically not empirically based, but rather a pragmatic Friedmanian one, as laid out earlier when describing constitutive models. In this view, it does not matter whether entrepreneurs are indeed risk averse or opportunistic or overconfident, so long as the models that assume they are so make reasonably accurate predictions that cohere with empirical outcomes down the road.

Similarly, a constitutive model might divide the population of entrepreneurs into (a) risk-loving, (b) risk-averse and (c) risk-neutral groups, and try to operationalize these terms, which then allows predicted relationships between each of these groups and relevant outcome variables such as starting a venture or succeeding with one. And to the extent that risk propensity as a categorical variable explains performance, the risk taxonomy is accepted as both real and useful. So, once again, this approach is tantamount to saying that it is as if entrepreneurs come in three flavors – loving, neutral and averse to risk.

In contrast, the effectuation model does not require any particular constitutive taxonomy of entrepreneurial types. On the other hand, even if entrepreneurs are truly divisible into essentialist types, the heuristics are still available to all three types. In short, the effectuation model does not require people to be classified into constitutive groups, but may work with such categorization. A risk-averse person may have a different level of affordable loss than a risk-loving one, but both can use the affordable-loss heuristic to make investments. For more on the affordable-loss heuristic, see Dew, Sarasvathy, Read and Wiltbank (2009). Similarly, while different types of behavior toward others – selfish, altruistic, opportunistic or docile – may induce or repel different types of stakeholders from self-selecting into the effectual process with the entrepreneur, anyone can use the crazy-quilt heuristic to create corridors for self-selection. For more on how the stakeholder self-selection heuristic works, see Sarasvathy and Dew (2005).

Another way to understand this sidestepping of as-if assumptions in effectuation is to see the even-if logic embedded in effectual heuristics. Thus the statement 'effectuation does not require people to be classified into constitutive groups, but may work with such categorization' may be restated as: even if someone is risk averse, he can have a high likelihood of starting a venture; even if someone is overconfident, she may succeed in opening up a new market; even if you are altruistic, you can build a highly profitable business.

Notice how a heuristic that may be used by all allows any particular entrepreneur – with every peculiar vagary of his or her body, gender, intellect, emotion and cognition – to become crucial to constructive models. Mary Kay Ash can no longer be explained away in terms of statistics correlated with successful ventures; nor can Mary Kay Cosmetics be seen simply as a surviving firm in the cosmetics industry. In fact, survival and success themselves have to be considered in terms of the founder's intents and how these were formed and evolved over time and through interactions with particular other stakeholders, each of whom has also influenced yet also been influenced and changed through the process. Constitutive models such as effectuation, therefore, do not require the Disembodiment Assumption. Yet they are more generalizable than formal and constitutive models precisely because they thwart this assumption.

An even-if approach to behavioral and situational assumptions ruptures the fabric of inevitability that links taken-for-granted exogenous givens such as personality, culture, resources and so on, on the one hand, and performance, on the other. Instead, these links are no longer causal; they are contingent on what the effectual entrepreneur does. In other words, constructive models compel us to see these taken-for-granted givens from formal and constitutive models as mere ingredients in a transformative process where design and co-creation are not only feasible but become the primary phenomena of interest. Entrepreneurial ventures are acts of design, and constructive models of venturing, therefore, make entrepreneurship a science of the artificial.

Does this mean that constitutive models are false or useless or that constructive models are not formalizable? The answer to both of these concerns is a resounding no. As mentioned in the section on models in general, models are useful to the extent that we are careful and explicit in our choices of what to include, what to exclude, and how to translate back and forth from reality to model. The approaches are complementary. Once formal models disown embodiment, they acquire a scarecrow body of sorts. Constitutive models seek to provide these scarecrows with a reasonably accurate skeleton based on observed categories of human behavior and circumstances. Constructive models take bodies and situations as they come, fully enfleshed and with idiosyncratic and changing personalities difficult to fit into strict taxonomies. Yet all three kinds of models can work with each other, through both coherence and contradiction to deepen our understanding of phenomena we care about.

Take the oft-studied formal model of Occupational Choice (OCM) in economics. In OCM, the individual decision maker is modeled as choosing between wage employment and self-employment. Over the years, the model has been modified to incorporate constitutive elements such as risk propensity and labor market imperfections (Åstebro, Chen and Thompson, 2011). Also, recently, in an effort to enflesh the model through data from actual entrepreneurs, Folta, Delmar and Wennberg (2010) introduced a third category of 'hybrid' entrepreneurs – people who stay in their jobs while working on their ventures nights and weekends or those who take up part-time jobs to pay the bills while they work full time on the venture they hope to grow into financial success. We can envisage a similar movement of the three types of models in the opposite direction as well. Some scholars seek to define 'effectual' itself as a constitutive category (Chandler, DeTienne, McKelvie and Mumford, 2011), presumably correlated to other constitutive categories such as 'optimal trust' (Goel and Karri, 2006). Others are beginning to build formal models in terms of a dual of loss aversion, or simulated comparisons between predictive and non-predictive games (Kuechle, 2008).

Formal models are useful in identifying stable theoretical relationships between key variables relevant to the phenomena being studied. Constitutive models seek to incorporate empirically useful variables such as traits and success factors. And constructive models are useful for isolating particular mechanisms that can guide design leading to the fabrication of artifacts including the future itself. It should also be kept in mind that the classification of models into formal, constitutive and constructive is itself a constitutive exercise. In real life, models will rarely fall neatly into such pre-assigned slots. As I've tried to argue, it is wise not to get overly attached to such classifications. From a research perspective, it is good effectual practice to develop and use all three types of models to understand entrepreneurial cognition.

REFERENCES

Åstebro, T., Chen, J. and Thompson, P. 2011. Stars and misfits: Self-employment and labor market frictions. *Management Science*, **57**(11): 1999–2017.

Bisson, T. 1991. They're Made out of Meat. *OMNI*.

Boia, L. 2005. *The Weather in the Imagination*. London: Reaktion Books Ltd.

Boring, E. 1953. A history of introspection. *Psychological Bulletin*, **50**: 169–89.

Brettel, M., Mauer, R., Engelen, A. and Küpper, D. 2012. Corporate effectuation: Entrepreneurial action and its impact on R&D project performance. *Journal of Business Venturing*, **27**(2): 167–84.

Chandler, G.N., DeTienne, D.R., McKelvie, A. and Mumford, T.V. 2011. Causation and effectuation processes: A validation study. *Journal of Business Venturing*, **26**(3): 375–90.

De Vany, A. 2003. *Hollywood Economics: How extreme uncertainty shapes the film industry*. New York: Routledge.

Dew, N., Read, S., Sarasvathy, S.D. and Wiltbank, R. 2009. Effectual versus predictive logics in entrepreneurial decision-making: Differences between experts and novices. *Journal of Business Venturing*, **24**(4): 287–309.

Dew, N., Sarasvathy, S.D., Read, S. and Wiltbank, R. 2009. Affordable loss: Behavioral economic aspects of the plunge decision. *Strategic Entrepreneurship Journal*, **3**(2): 105–26.

Eckert, M.A., Leonard, C.M., Richards, T.L., Aylward, E.H., Thomson, J. and Berninger, V.W. 2003. Anatomical correlates of dyslexia: frontal and cerebellar findings. *Brain*, **126**(2): 482–94.

Feller, W. 1943. On a general class of contagious distributions. *Annals of Mathematical Statistics*, **14**: 389–400.

Fischer, E. and Reuber, A.R. 2011. Social interaction via new social media: (How) can interactions on Twitter affect effectual thinking and behavior? *Journal of Business Venturing*, **26**(1): 1–18.

Folta, T.B., Delmar, F. and Wennberg, K. 2010. Hybrid entrepreneurship. *Management Science*, **56**(2): 253–69.

Friedman, M. 1953. The methodology of positive economics. In *Essays in Positive Economics*: 3–43. Chicago, IL: University of Chicago Press.

Friedman, M. and Savage, L.J. 1948. The utility analysis of choices involving risk. *The Journal of Political Economy*, **56**(4): 279–304.

Gigerenzer, G. 1989. *The Empire of Chance*. New York: Cambridge University Press.

Gigerenzer, G. and Todd, P.M. 1999. *Simple Heuristics that make us Smart*. Oxford: Oxford University Press.

Goel, S. and Karri, R. 2006. Entrepreneurs, effectual logic, and over-trust. *Entrepreneurship Theory and Practice*, **30**(4): 477–93.

Goldstein, D.G. and Gigerenzer, G. 2002. Models of ecological rationality: The recognition heuristic. *Psychological Review*, **109**(1): 75–90.

James, W. 1950 (1890). *The Principles of Psychology (Vol. 1)*. New York: Dover.

Kahneman, D., Slovic, P. and Tversky, A. 1982. *Judgment under Uncertainty: Heuristics and biases*. Cambridge, UK: Cambridge University Press.

Kitcher, P. 2000. Battling the undead: How and how not to resist genetic determinism. In R.S. Singh, C.B. Krimbas, D.B. Paul and J. Beatty (eds), *Thinking about evolution: Historical, philosophical, and political perspectives*: 396–414. New York: Cambridge University Press.

Knight, F. 2012 (1921). *Risk, Uncertainty and Profit*. New York: Courier Dover Publications.

Krugman, P. 1994. The rise and fall of developmental economics. In L. Rodwin and

D.A. Schön (eds), *Rethinking the Development Experience: Essays provoked by the work of Albert O. Hirschman*: 39–58. Washington, DC: Brookings Institution Press.
Kuechle, G. 2008. Evolution, games and entrepreneurship theory. *Frontiers of Entrepreneurship Research*, **28**(5): Article 2.
Lakoff, G. 1987. *Women, Fire, and Dangerous Things: What categories reveal about the mind.* Chicago, IL: University of Chicago Press.
Lakoff, G., and Johnson, M. 1980. *Metaphors we Live by*. Chicago, IL: University of Chicago Press.
MacKay, A.F. 1980. *Arrow's Theorem: The paradox of social choice*. New Haven, CT and London: Yale University Press.
Moss, L. 2004. *What Genes can't Do*: Cambridge, MA: The MIT Press.
Newell, A. and Simon, H.A. 1972. *Human Problem Solving*. Englewood Cliffs, NJ: Prentice-Hall.
Noga, A. 2007. *Passions and Tempers: A history of the humours*. New York: HarperCollins.
Read, S., Dew, N., Sarasvathy, S.D., Song, M. and Wiltbank, R. 2009. Marketing under uncertainty: The logic of an effectual approach. *Journal of Marketing*, **73**(3): 1–18.
Read, S., Song, M. and Smit, W. 2009. A meta-analytic review of effectuation and venture performance. *Journal of Business Venturing*, **24**(6): 573–87.
Rosch, E. 1983. Prototype classification and logical classification: The two systems. In E. Scholnick (ed.), *New Trends in Cognitive Representation: Challenges to Piaget's theory*: 73–86. Hillsdale, NJ: Lawrence Erlbaum Associates.
Saari, D.G. 1987. The source of some paradoxes from social choice and probability. *Journal of Economic Theory*, **41**: 1–22.
Sarasvathy, S.D. 2008. *Effectuation: Elements of entrepreneurial expertise*. Cheltenham, UK and Northampton, MA, USA: Edward Elgar Publishing.
Sarasvathy, S.D. and Dew, N. 2005. New market creation as transformation. *Journal of Evolutionary Economics*, **15**(5): 533–65.
Sarasvathy, S.D., Menon, A. and Kuechle, G. 2013. Failing firms and successful entrepreneurs: Serial entrepreneurship as a temporal portfolio. *Small Business Economics*, **40**(2): 417–34.
Schweizer, R., Vahlne, J.E. and Johanson, J. 2010. Internationalization as an entrepreneurial process. *Journal of International Entrepreneurship*, **8**(4): 343–70.
Shane, S. and Venkataraman, S. 2000. The promise of entrepreneurship as a field of research. *Academy of Management Review*, **25**(1): 217–26.
Simon, H.A. 1981. *The Sciences of the Artificial*. Cambridge, MA: The MIT Press. Further editions in 1982 and 1996.
Simon, H.A. 1984. Testability and approximation. In D.M. Hausman (ed.), *The Philosophy of Economics*: 245–8. London: Cambridge University Press.
Simon, H.A. 1993. Altruism and economics. *The American Economic Review*, **83**(2): 151–61.
Simon, H.A. 1997. *An Empirically Based Microeconomics*. Cambridge, New York and Melbourne: Cambridge University Press.
Todd, P.A. and Gigerenzer, G. 2012. *Ecological Rationality: Intelligence in the world*. Oxford: Oxford University Press.
Varela, F., Thompson, E. and Rosch, E. 1991. *The Embodied Mind: Cognitive science and human experience*. Cambridge, MA: MIT Press.
Venkataraman, S. 1997. The distinctive domain of entrepreneurship research. In J.A. Katz (ed.), *Entrepreneurial Action: Advances in entrepreneurship, firm emergence and growth*, vol. 3: 119–38. Greenwich, CT: JAI Press.
Wiltbank, R., Read, S., Dew, N. and Sarasvathy, S.D. 2009. Prediction and control under uncertainty: Outcomes in angel investing. *Journal of Business Venturing*, **24**(2): 116–33.
Wittgenstein, L. 1953. *Philosophical Investigations*. New York: Macmillan.

Afterword

Ron Mitchell

Such writers as William James and George Bernard Shaw have made noteworthy the contribution that thinking makes to doing. In entrepreneurship research, this thinking–doing link has developed into the entrepreneurial cognition research stream; and as theories of social cognition have developed in general, the entrepreneurial cognition research stream has specifically, for example, been moving from static to dynamic conceptualizations. Thus, of necessity, both dynamism in conceptualizations of entrepreneurial cognition and dynamism in the research stream itself position this volume at an intermediate point. What can a reader glean from a work so positioned?

It has been our hope as editors of this volume to enable several uses. Included among these are: multiple perspectives on the phenomenon of entrepreneurial cognition; a snapshot of some of the current 'thinking about entrepreneurial thinking'; the provision of a common reference core for those seeking to 'read into' this field to enable their own research; a point of reflection about where this stream has been and may be going; and possibly also a point of inflection when, due to this collection, some spark of a new idea can trigger some hitherto unforeseen but important change in direction, emphasis, substance or form of the research. Our hope, as we expressed it to each of the contributors to this volume when we discussed the call that brought these chapters together, is that this volume can provide for all concerned a platform from which to proceed regardless of a reader's initial purpose in picking it up.

What are some of the open questions in entrepreneurial cognition research? Within each chapter can be found imaginative ideas from which these have been and may continue to be formed. What is the future of this research stream? I believe that it is open to all interested scholars, and I support this belief with reference to its short history as a way to chronicle a few of the key events and to recognize a few of the key contributors (some of whom have been able to contribute to this volume; and others who have not).

One series of events that led to the coalescence of the entrepreneurial cognition research stream resulted from a suggestion by Tricia MacDougall (while she was a Visiting Winspear Scholar at the University of Victoria, BC, Canada) to Ray Bagby, Editor of *Entrepreneurship*

Theory and Practice, that Lowell Busenitz and I be asked to hold a conference and edit a special issue on entrepreneurial cognition. This led to the Victoria 2002, and the Ivey 2005 conferences, and the 2002, 2004 and 2007 special issues of *Entrepreneurship Theory and Practice*. Following this initiative, Lowell and I made a conscious decision to invite several highly talented rising entrepreneurial cognition scholars to take a special interest in the progress of the field; and this is the place to acknowledge that they have done so admirably. So thanks to all who participated in and enabled the conferences and special issues that helped to establish the early foundations; and thanks also to the many additional researchers, journal editors and reviewers who have helped this research stream to proceed. And, of course, thanks to the authors and reviewers who have contributed to this volume, and who through their efforts have further enabled a bright and constructive future for the entrepreneurial cognition research stream.

Citation index

Index